AF782023

Electromagnetic Fields and Biological Systems

Electromagnetic Fields and Biological Systems

Edited by **Edgar Wilson**

Willford Press
New York

Published by Willford Press,
118-35 Queens Blvd., Suite 400,
Forest Hills, NY 11375, USA
www.willfordpress.com

Electromagnetic Fields and Biological Systems
Edited by Edgar Wilson

© 2016 Willford Press

International Standard Book Number: 978-1-68285-310-8 (Hardback)

This book contains information obtained from authentic and highly regarded sources. Copyright for all individual chapters remain with the respective authors as indicated. All chapters are published with permission under the Creative Commons Attribution License or equivalent. A wide variety of references are listed. Permission and sources are indicated; for detailed attributions, please refer to the permissions page and list of contributors. Reasonable efforts have been made to publish reliable data and information, but the authors, editors and publisher cannot assume any responsibility for the validity of all materials or the consequences of their use.

The publisher's policy is to use permanent paper from mills that operate a sustainable forestry policy. Furthermore, the publisher ensures that the text paper and cover boards used have met acceptable environmental accreditation standards.

Trademark Notice: Registered trademark of products or corporate names are used only for explanation and identification without intent to infringe.

Printed in China.

Contents

Preface

This book aims to highlight the current researches and provides a platform to further the scope of innovations in this area. This book is a product of the combined efforts of many researchers and scientists from different parts of the world. The objective of this book is to provide the readers with the latest information in the field.

It has been observed in many researches conducted in the past that electromagnetic fields can have certain effects on biological systems. This innovative and comprehensive book integrates the well-developed theory and practical applications of electromagnetic field and associated theories. It consists of contributions made by international experts and lucidly elaborates various concepts like radiation, multi-frequency sensors, microwave sensors, numerical modeling and wavefield inversion in nondestructive testing, etc. It will provide comprehensive knowledge to the readers. This book will serve as a valuable source of reference for graduate and post graduate students pursuing physics, biophysics, engineering, and associated disciplines.

I would like to express my sincere thanks to the authors for their dedicated efforts in the completion of this book. I acknowledge the efforts of the publisher for providing constant support. Lastly, I would like to thank my family for their support in all academic endeavors.

Editor

Radiation Efficient Unidirectional Low-Profile Slot Antenna Elements for X-Band Application

C. Löcker, T. Vaupel, and T. F. Eibert

FGAN-Research Institute for High Frequency Physics and Radar Techniques (FHR), Neuenahrer Str. 20, 53343 Wachtberg-Werthhoven, Germany

Abstract. Slots in metallic ground planes are very promising candidates for conformal antenna applications. However, a low-profile unidirectional antenna requires a back reflector close to the slot and the resulting stripline feed causes strong excitation of parallel-plate modes. In this contribution, we consider unidirectional reflector-backed slot configurations with parallel-plate mode suppression by shorting pins. Starting from a parametric study with respect to shorting pin location and back reflector distance, we present a stripline-fed rectangular slot element with radiation efficiency of more than 80% and a bandwidth of about 5% at centre frequency 10 GHz. A careful optimisation of shorting pin locations guarantees reliable parallel-plate mode suppression without deteriorating the slot radiation behaviour. Coupling coefficients between parallel and aligned rectangular slot elements are presented. For increased bandwidth applications, a bow-tie slot element with about 8% bandwidth and radiation efficiency of close to 80% is proposed.

1 Introduction

Due to their low profile, light weight, robustness and low cost, antennas in printed circuit technology are widely used in electronic warfare, communication, automotive, etc. Traditionally, many design instructions can be found in literature for patch antennas (James et al., 1981). However, due to its resonant behaviour the radiating patch element has only narrow bandwidth, unless thick substrates are used, which is disadvantageous with respect to integration into non-planar surfaces. The surrounding of the patch is inherently dielectric, which benefits propagation of surface waves resulting in low radiation efficiency.

The radiating slot element is an attractive alternative. A slot in a metallic ground plane is the complementary structure of the corresponding electric dipole. Since the radiation pattern of the slot is bidirectional, a back reflector must be used to avoid back radiation. With respect to antenna integration in curved surfaces, the reflector should be placed closely to the slot, resulting in a typical stripline or triplate configuration. Unfortunately, this structure admits parallel-plate propagation and the discontinuity slot/stripline excites the corresponding modes. Higher order modes can be avoided by choosing a thin substrate, because their cut-off frequencies increase by decreasing the distance between the ground planes. The suppression of the fundamental mode, however, is a challenging task, since it is a TEM type mode starting at 0 Hz. One method to suppress the TEM mode is the use of twin slot configurations for destructive interference of the excited waves. An alternative method is the cavity backed slot (Galejs, 1963; Biebl and Friedsam, 1995; Omiya et al., 1998). Metallic walls force the electric field of the TEM wave to zero. A similar procedure is the placement of shorting pins around the slot aperture, connecting the slot's ground plane with the back reflector (Sommers, 1955). Compared to cavity backed slots, only little information about guidelines on positioning of vias is available in literature. Some papers treat stipline-slot-fed patch antennas (Brachat and Baracco, 1995; Bhattacharyya et al., 1998), where a theoretical analysis for via-slot interaction is given in the latter.

In this contribution, a parametric study is concerned with the influence of via and reflector positions on the input impedance and radiation efficiency of a single rectangular slot element. Then, an optimised unidirectional slot element printed on a very thin substrate with parallel-plate mode suppression by shorting pins is presented by simulated and measured data, based on the approach proposed in (Sommers, 1955). Towards array applications, we present mutual coupling investigations for different slot arrangements. A second radiating bow-tie like aperture for bandwidth enhancement is presented by numerical and experimental results.

2 Rectangular Slot Element

The geometry of the investigated rectangular slot element is depicted in Fig. 1. The slot is located in the upper ground plane and is excited by a stripline located in the centre of the substrate with ϵ_r=2.2. In order to suppress the parasitic parallel plate mode, the upper and lower ground planes are

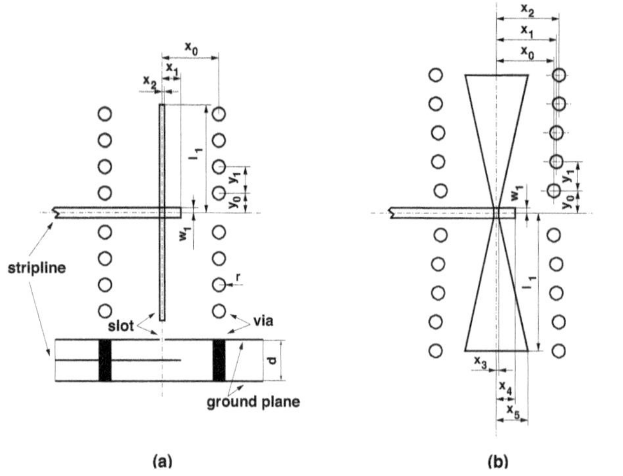

Fig. 1. Geometries of investigated apertures: **(a)** stripline-fed rectangular slot element, **(b)** bow-tie slot element.

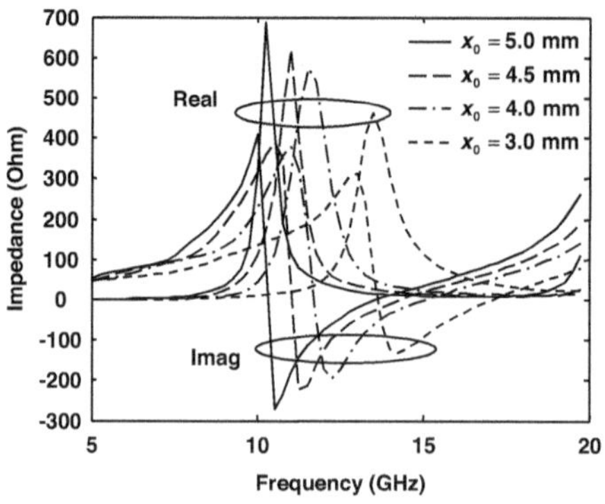

Fig. 2. Parametric study: Input impedance with respect to via-position x_0.

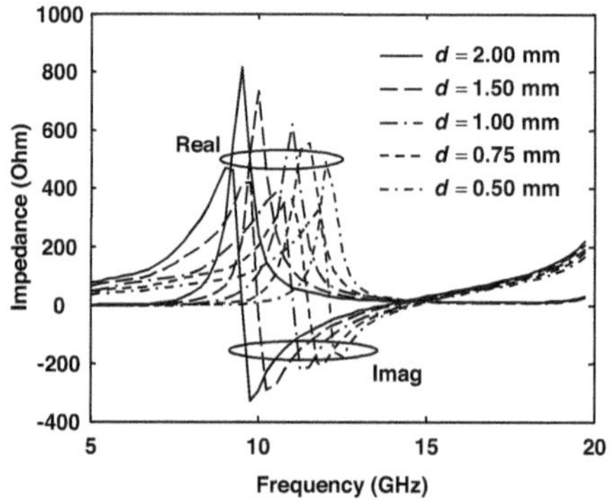

Fig. 3. Parametric study: Input impedance with respect to reflector distance d.

connected by shorting pins or vias. Antenna dimensions are (unit: millimetre): x_0=4.5, x_1=1.5, x_2 =0.25, y_0=1.5, y_1=2.2 (constant distance in y-direction), r=0.5, d=1, w_1=0.435 and l_1=8.25.

To gain a broad understanding of the influence of via positions and reflector distances on the resonance behaviour, the slot was investigated by a parametric study. To avoid transformation and compensation effects by coupling from the slot to the stripline and the open stripline stub, the stripline was removed and the slot was excited by a magnetic δ-gap voltage source. The terminal impedance was computed by a surface/volume integral equation method for the analysis of planar circuits with additional 3-D-components embedded in a multilayered medium of infinite extent (Vaupel and Hansen, 1999; Vaupel et al., 2003). Thus, the layered medium environment is analytically considered by its spectral domain Green's function and only the magnetic surface currents in the slot aperture as well as the electric via currents (modelled as volume currents) must be discretised.
Numerical results on the input impedance and radiation efficiency show Figs. 2, 3 and Figs. 4, 5, respectively, with respect to via-slot distance x_0 and back reflector-slot distance d. The curves for the real and imaginary part of the input impedance show a typical resonator behaviour. The reflector and the vias cause a shift of the input impedance to the inductive regime. The frequency of the first resonance (corresponding to the peak input resistance) increases and the peak resistance decreases and broadens with decreasing via and reflector distance, resulting in a reduced effective length of the slot. The efficiency curves show that the parallel plate mode can be effectively suppressed by the vias. At the frequency band edges, Fig. 4 shows a decreasing efficiency for vias placed too far as well as too close to the slot. A distance x_0 smaller than approx. 0.25λ should ensure efficiencies over 90%. If the back reflector is too close to the slot, the parallel plate mode is excited stronger, yielding a lower efficiency.

The fabricated single slot element with dimensions as described in Fig. 1 is excited by a stripline, which is open at its end. Thus, the stub can be used to optimise impedance matching. The slot was printed on a 90 mm×90 mm Rogers 5880 Duroid substrate and should have a theoretical directivity of about 5.15 dB (gain of a lossless $\lambda/2$-dipole increased by 3 dB due to missing back radiation). In order to decrease the number of degrees of freedom, the width of the feeding stripline was chosen for 50Ω. The slot's dimensions were carefully optimised by Agilent's ADS2003A Momentum (ADS), including metallisation (copper) and dielectric losses. The directivity was estimated by integrating the radiation fields over all radiation directions and the gain is determined by the gain transfer method. A slightly higher resonance frequency was chosen in ADS to obtain a resonance frequency of 10 GHz. Figure 6 depicts simulated (ADS) and measured return loss data. The measured bandwidth (return loss <-10 dB) is about 5%. Simulated and measured data are in good agreement. The efficiency curve in Fig. 7 indicates, that the efficiency does not decrease below about 80% in the band of operation.

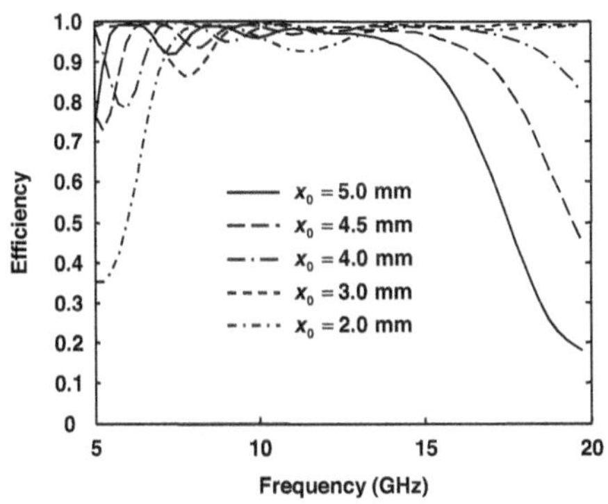

Fig. 4. Parametric study: Efficiency with respect to via-position x_0.

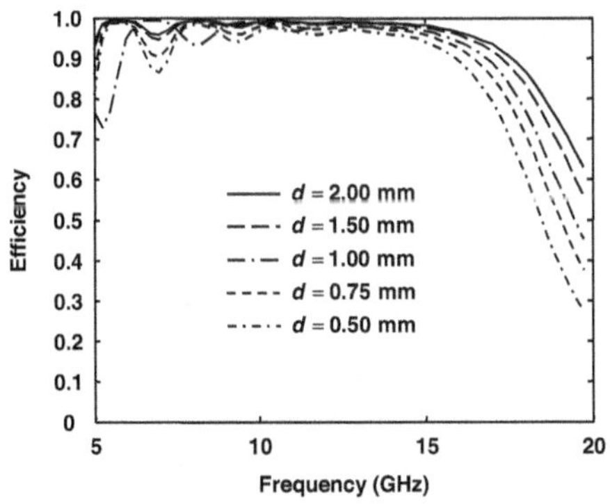

Fig. 5. Parametric study: Efficiency with respect to reflector distance d.

3 Mutual coupling between rectangular slot elements

Mutual coupling between antenna elements is an important criterion in conformal array antenna applications. We investigated collinear and broadside slot arrangements. Three slots were considered in each configuration. The single slot dimensions are taken from Fig. 1a. In the collinear configuration, centre-centre distances with respect to the first slot are 22.5 mm and 52.5 mm. In the broadside slot configuration, the slots could be arranged closer to each other, resulting in separations of 15 mm and 45 mm. The mutual coupling was determined by transmission measurements between neighbouring slots with a HP8510 network analyser. The remaining third port was terminated by a 50 Ω match. Figures 8 and 9 show simulated and measured scattering coefficients S_{ij}, which describe the coupling behaviour. For collinear alignment the radiation null in the corresponding direction causes a smaller coupling compared to the coupling for broadside configuration. In this case, the radiated space wave travelling along the slot ground plane creates the relative large coupling. By placing a metallic plate between the slot aperture, it can be demonstrated that the mutual coupling due to parallel-plate mode propagation is effectively suppressed by the vias.

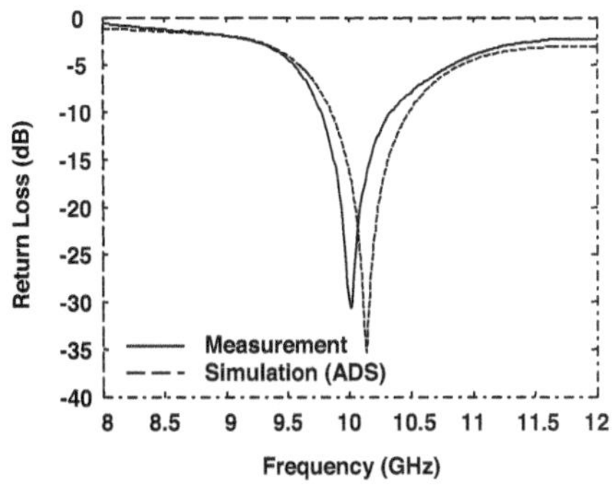

Fig. 6. Rectangular slot element, simulated and measured return loss.

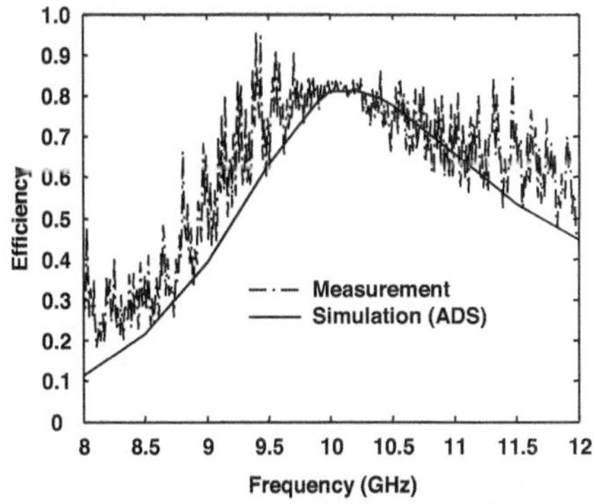

Fig. 7. Rectangular slot element, simulated and measured efficiency.

4 Bow-Tie Slot Element

Since the bandwidth of the rectangular slot element is relatively small, we also investigated bow-tie like slot elements (Kraus, 1988). The slot was widened at its ends and the length was increased to about $0.7\lambda_0$ for a centre frequency of 10 GHz, see Fig. 1. In order to reduce edge diffraction effects, the side length of the substrate was increased to 100 mm. In this structure the vias positions are not parallel to the aperture axis. The vias positions were carefully optimised to achieve the design requirements. The bow-tie dimensions are (units: millimetre): x_0=4.25, x_1=4.45, x_2=4.65, x_3=0.25, x_4=1.6, x_5=2.5, y_0=1.7, y_1=2.2, w_1=0.425 and l_1=10.5. As for the rectangular slot element, the substrate has an permittivity of 2.2. Figure 10 depicts the measured and simulated return loss. The 10 dB bandwidth amounts to 0.8 GHz and is higher than for the rectangular slot element. The measured antenna efficiency is about 75–80%, see Fig. 11, in the band of operation.

5 Conclusion

Radiating slots in a metallic ground plane were investigated. The upper ground plane was connected with a back reflector

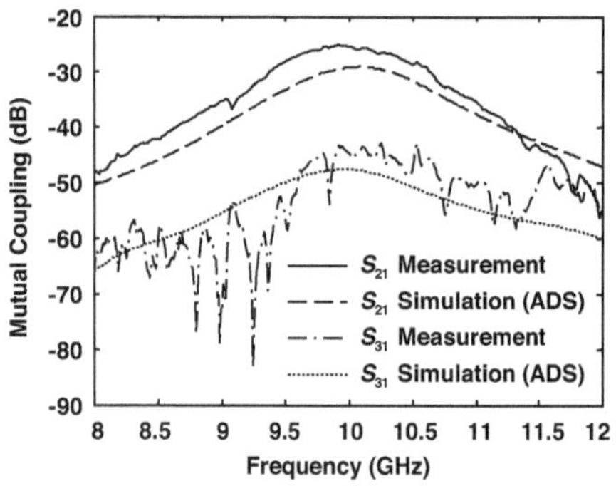

Fig. 8. Mutual coupling, collinear slot arrangement.

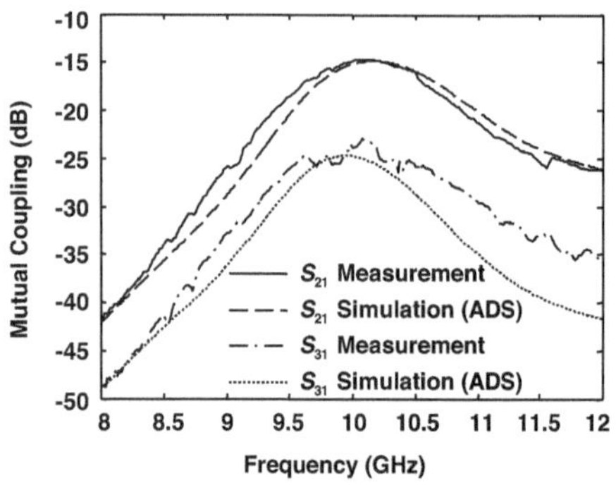

Fig. 9. Mutual coupling, broadside slot arrangement.

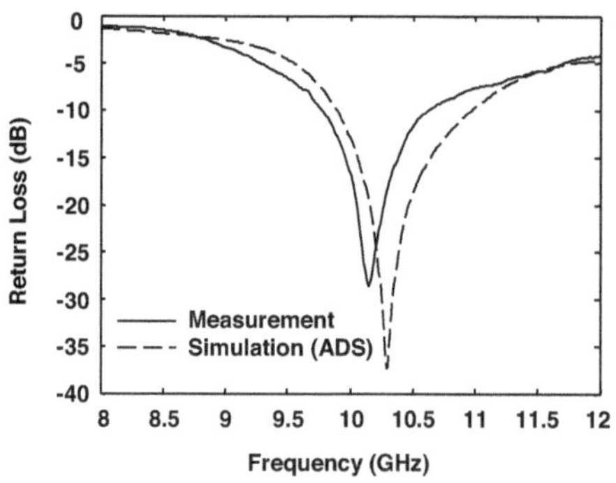

Fig. 10. Bow-Tie element, measured and simulated return loss.

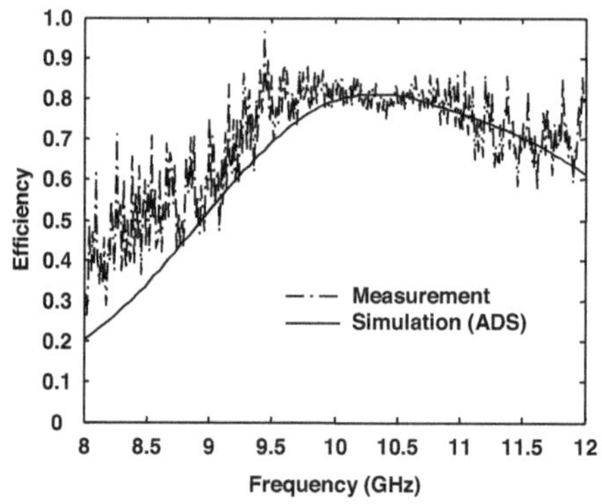

Fig. 11. Bow-Tie element, measured and simulated efficiency.

by shorting pins. It is shown by simulation and measurement, that the parasitic TEM parallel plate mode can be effectively suppressed by shorting pins. However, the slot design must pass a careful optimisation process, because of the strong influence of the vias on the resonance and broadband behaviour. A stripline-fed rectangular slot element with a 10 dB-bandwidth of 5% at centre frequency of 10 GHz and an efficiency of more than 80% was fabricated. Studies on coupling effects between aligned slots show that collinear slots are non-critical. For bandwidth enhancement, a bow-tie like slot element with 10 dB bandwidth of 8% and radiation efficiency of close to 80% was presented.

References

Bhattacharyya, A., Fordham, O., and Liu, Y.: Analysis of Stripline-Fed Slot-Coupled Patch Antennas with Vias for Parallel-Plate Mode Suppression, IEEE Transactions on Antennas and Propagation, 46, 538–545, 1998.

Biebl, E. M. and Friedsam, G. L.: Cavity-Backed Aperture Antennas with Dielectric and Magnetic Overlay, IEEE Transactions on Antennas and Propagation, 43, 1226–1232, 1995.

Brachat, P. and Baracco, J. M.: Dual-Polarization Slot-Coupled Printed Antennas Fed by Stripline, IEEE Transactions on Antennas and Propagation, 43, 738–742, 1995.

Galejs, J.: Admittance of a Rectangular Slot Which is Backed by a Rectangular Cavity, IEEE Transactions on Antennas and Propagation, 119–126, 1963.

James, J. R., Hall, P. S., and Wood, C.: Microstrip Antenna Theory and Design, Peter Peregrinus Ltd., 1981.

Kraus, J. D.: Antennas, McGraw Hill, 1988.

Omiya, M., Hikage, T., Ohno, N., Horiguchi, K., and Itho, K.: Design of Cavity-Backed Slot Antennas Using the Finite-Difference Time-Domain Technique, IEEE Transactions on Antennas and Propagation, 46, 1853–1858, 1998.

Sommers, D. J.: Slot Array Employing Photoetched Tri-Plate Transmission Lines, IEEE Transactions on Microwave Theory and Techniques, 3, 157–162, 1955.

Vaupel, T. and Hansen, V.: Electrodynamic Analysis of Combined Microstrip and Coplanar/Slotline Structures with 3-D Components Based on a Surface/Volume Integral Equation Approach, IEEE Transactions on Microwave Theory and Techniques, 47, 1788–1800, 1999.

Vaupel, T., Hansen, V., and Schäfer, F.: Radiation Efficiency Analysis of Submm-Wave Receivers Based on a Modified Spectral Domain Technique, Radio Sience, 4, 13-1-13-2, 2003.

2

Radiation Pattern Analysis of Antenna Systems for MIMO and Diversity Configurations

O. Klemp and H. Eul

University of Hannover, Department of High Frequency Technology and Radio Systems, Hannover, Germany

Abstract. Multiple-input multiple-output (MIMO) antenna systems and antenna configurations for wideband multimode diversity rank among the emerging key technologies in next generation wireless communication systems. The analysis of such transmission systems usually neglects the influences of real antenna radiation characteristics as well as the influences of mutual coupling in a multielement antenna arrangement. Nevertheless, to achieve a detailed description of diversity gain and channel capacity by using several transmit- and receive antennas in a wireless link, it is essential to take all those effects into account. The expansion of the radiation fields in terms of spherical eigenmodes allows an analytical description of the antenna radiation characteristics and accounts for all the coupling effects in multielement antenna configurations. Therefore the radiation pattern analysis by spherical eigenmode expansion provides an efficient alternative to establish an analytical approach in the calculation of envelope correlation or channel capacity.

1 Introduction

The spherical mode expansion (SME) is a well known technique in terms of the characterization of a various number of applications of classical electromagnetic scattering and radiation problems. Numerous authors such as Stratton (1941) and Werner and Mittra (2000) have presented a generalized theory employing spherical mode analysis and adopted this method to study a multiplicity of wave propagation problems. Applications of the spherical mode expansion have been applied e.g. to the numerical analysis of Cassegrainian-fed paraboloids in Potter (1967), the investigation of multiple scattering of electromagnetic waves in Bruning and Lo (1971), the computation of antenna radiation patterns of arbitrary wire antennas in Chen and Simpson (1991b,a) and the modeling of base station antenna equipment in Adane et al. (2002). Furthermore the application of spherical mode expansion to perform a transformation between near- and far-field related field quantities as described in Ludwig (1971) is a fundamental basis for todays techniques of spherical antenna near field measurement.

In this article the analysis of the radiation behavior using spherical eigenmodes will be applied to the characterization of the diversity performance of multielement antenna configurations in terms of interelement correlation. Therefore we present a modified technique of spherical mode expansion as in Chen et al. (1992) for the analysis of arbitrary antennas and multielement antennas. Based upon a suitable statistical model, e.g. provided in Waldschmidt and Wiesbeck (2004) to include the spatial properties of a multipath propagation scenario, a closed form representation for antenna correlation will be derived.

The article is organized as follows, Sect. 2 reviews the fundamental properties of the spherical mode expansion technique with respect to the electric- and magnetic vector potential for a given elementary source distribution. Section 3 derives a generalized representation of antenna radiation pattern due to the Fraunhofer approximation of the Green's radiation function. Subsequently the modal representation of antenna radiation patterns is used in Sect. 4 to specify the power correlation between interactive antenna elements in a diversity configuration to establish a measure for their diversity performance. The presented analysis is adopted to an array of resonant $\lambda/2$ dipoles given in Sects. 5.1 and 5.2 and a configuration of multi-port wideband logarithmically periodic antennas in Sect. 6. The latter analysis will be completed by results derived from spherical near-field measurement. Section 7 concludes this article.

2 Modal Solutions of Vector Potentials

The spherical mode expansion as given in this section will be used to accomplish the transformation between a known distribution of electric and magnetic currents (e.g. on the metalization surfaces of an antenna structure) and the resulting electromagnetic field distribution in the field region, which is considered to be source-free. In Ludwig (1971) this transformation technique was applied to derive a generalized mapping rule between field quantities in different coordinate systems. Therefore in SME analysis a separation of coordinate systems for the given source-region and the requested field-region is required. For a given volume V containing electric- and magnetic current densities $\boldsymbol{J}$ and $\boldsymbol{J}_{\mathrm{m}}$ as shown in Fig. 1 the magnetic- and electric vector potentials $\boldsymbol{A}$ and $\boldsymbol{F}$ can be

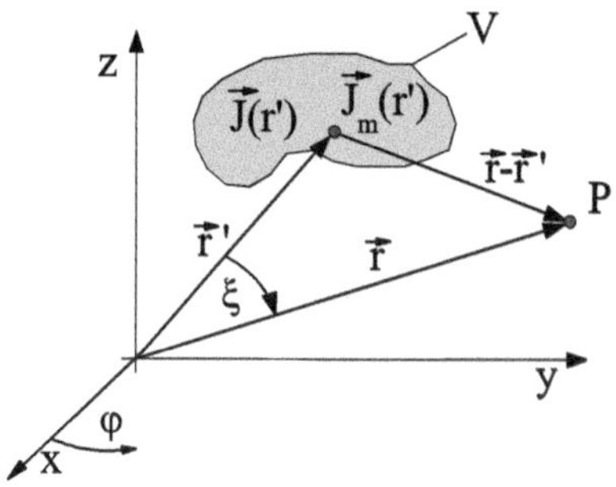

Fig. 1. Distribution of electric and magnetic current densities $\boldsymbol{J}$ and $\boldsymbol{J}_{\mathrm{m}}$ in volume V.

represented using Chen et al. (1992) as follows:

$$\boldsymbol{A}(\boldsymbol{r}) = \frac{1}{4\pi}\iiint_{(V')}\frac{\boldsymbol{J}(\boldsymbol{r}')}{|\boldsymbol{r}-\boldsymbol{r}'|}e^{-jk|\boldsymbol{r}-\boldsymbol{r}'|}\,dV' \tag{1}$$

$$\boldsymbol{F}(\boldsymbol{r}) = \frac{1}{4\pi}\iiint_{(V')}\frac{\boldsymbol{J}_m(\boldsymbol{r}')}{|\boldsymbol{r}-\boldsymbol{r}'|}e^{-jk|\boldsymbol{r}-\boldsymbol{r}'|}\,dV' .$$

In Eq. (1) $k=2\pi\sqrt{\epsilon\mu}f$ denotes the wave number of the electromagnetic field, where ϵ represents the permittivity of free space and μ the respective permeability. Harmonic time dependence will be assumed, so that f is equivalent to the operational frequency of any elementary source considered in the volume V. The vector $\boldsymbol{r}$ represents the distance between the origin and the point of observation and $\boldsymbol{r}'$ gives the displacement of the source point in relation to the origin. Therefore $\boldsymbol{A}$ and $\boldsymbol{F}$ represent a solution to the vector Helmholtz equation of the total electric field $\boldsymbol{E}$ and magnetic field $\boldsymbol{H}$. Subsequently, the analysis will be restricted to the distribution of electric current densities, where the total radiation field is exclusively given in terms of the magnetic vector potential. Using the theorem of dual fields in electrodynamics an arbitrary distribution of the magnetic current density $\boldsymbol{J}_{\mathrm{m}}$ can be related to an electric current density $\boldsymbol{J}$, for this purpose (1) represents a complete solution of the radiation problem. Provided the field solution of the source distribution is considered in terms of the spherical coordinate system (r, ϑ, φ), the Green's function may be represented as in Chen and Simpson (1991a) using the zero-order spherical Hankel function $h_0^{(2)}$ of the second kind as given in Eq. (2) with $R=|\boldsymbol{r}-\boldsymbol{r}'|$:

$$\frac{e^{-jkR}}{R} = -jk\,h_0^{(2)}(kR) . \tag{2}$$

For the validity of Eq. (2), the case where $\boldsymbol{r}>\boldsymbol{r}'$ has to be considered, where the antennas will be analyzed in the transmit case. Applying the principle of reciprocity in antenna analysis, the receive case is implicitly given in Eq. (2). To derive a modal representation of the electromagnetic field, first of all Eq. (2) can be re-formulated in terms of an infinite series as given in Eq. (3). Using n.-order spherical Bessel functions $j_{\mathrm{n}}(kr')$ in terms of the source-point dependence, n.-order spherical Hankel functions $h_{\mathrm{n}}^{(2)}(kr)$ of the second kind for the dependence of the observation-point and Legendre polynomials $P_{\mathrm{n}}(\cos\xi)$ of degree n, Eq. (2) is the first step to derive a separation of source- and field region.

$$h_0^{(2)}(kR) = \sum_{n=0}^{\infty}(2n+1)\,j_{\mathrm{n}}(kr')\,h_{\mathrm{n}}^{(2)}(kr)\,P_{\mathrm{n}}(\cos\xi) . \tag{3}$$

Using the addition theorem for Legendre polynomials as in Abramovitz and Stegun (1972),

$$P_{\mathrm{n}}(\cos\xi) = \sum_{m=0}^{n}\epsilon_{\mathrm{m}}\frac{(n-m)!}{(n+m)!} \cdot P_{\mathrm{n}}^{\mathrm{m}}(\cos\vartheta)\,P_{\mathrm{n}}^{\mathrm{m}}(\cos\vartheta')\cos\left(m(\varphi-\varphi')\right) \tag{4}$$

Eq. (5) gives a generalized solution to the magnetic vector potential. $P_{\mathrm{n}}^{\mathrm{m}}(\cos\vartheta)$ denotes the associated Legendre function of degree n and order m, ϵ_{m} specifies Neumann's number (1 for m=0, 2 for m>0).

$$\begin{aligned}\boldsymbol{A}(\boldsymbol{r}) = &\\ -jk&\sum_{n=0}^{\infty}\sum_{m=0}^{n}\boldsymbol{a}_{\mathrm{n}}^{\mathrm{m}'}N_{\mathrm{n}}^{\mathrm{m}}h_{\mathrm{n}}^{(2)}(kr)\,P_{\mathrm{n}}^{\mathrm{m}}(\cos\vartheta)\cos(m\varphi)\\ -jk&\sum_{n=0}^{\infty}\sum_{m=0}^{n}\boldsymbol{b}_{\mathrm{n}}^{\mathrm{m}'}N_{\mathrm{n}}^{\mathrm{m}}h_{\mathrm{n}}^{(2)}(kr)\,P_{\mathrm{n}}^{\mathrm{m}}(\cos\vartheta)\sin(m\varphi) .\end{aligned} \tag{5}$$

In Eq. (5) the response to the given electric current distribution is separated with respect to source point- and field point dependent quantities using the normalization $N_{\mathrm{n}}^{\mathrm{m}}$ given in Eq. (6).

$$N_{\mathrm{n}}^{\mathrm{m}} = \sqrt{\epsilon_{\mathrm{m}}\frac{(2n+1)(n-m)!}{4\pi(n+m)!}} \tag{6}$$

Orthogonality between the axes of the spherical coordinate systems approves a separation of functional quantities that are solely related to a single coordinate direction. Therefore the sources of the radiation fields (given in primed coordinates) may be analyzed independently from the respective electromagnetic fields at the field-points. Any source dependence is given in terms of the vector expansion coefficients $\boldsymbol{a}_{\mathrm{n}}^{\mathrm{m}'}$ and $\boldsymbol{b}_{\mathrm{n}}^{\mathrm{m}'}$ of the field expansion as given in Eqs. (7) and (8).

$$\boldsymbol{a}_{\mathrm{n}}^{\mathrm{m}'} = \iiint_{(V')}N_{\mathrm{n}}^{\mathrm{m}}\boldsymbol{J}(\boldsymbol{r}')\,j_{\mathrm{n}}(kr') \cdot P_{\mathrm{n}}^{\mathrm{m}}(\cos\vartheta')\cos(m\varphi')\,dV' \tag{7}$$

$$\boldsymbol{b}_{\mathrm{n}}^{\mathrm{m}'} = \iiint_{(V')}N_{\mathrm{n}}^{\mathrm{m}}\boldsymbol{J}(\boldsymbol{r}')\,j_{\mathrm{n}}(kr') \cdot P_{\mathrm{n}}^{\mathrm{m}}(\cos\vartheta')\sin(m\varphi')\,dV' \tag{8}$$

Therefore, Eqs. (7) and (8) denote a generalized description of all considered field sources in volume V in terms of elementary spherical wave contributions.

3 Antenna Radiation pattern analysis

Considering the radiation behavior of antenna elements usually the far-field approximation by Fraunhofer is applied. This approximation given in Eq. (9) neglects differences in amplitude of all incoming elementary waves at a certain far field distance r so that the response to all field sources leads to a wave propagation with planar amplitude distribution. Thus effects of interference as a consequence of spatial separation between the field sources lead to a maximum phase deviation of $\pi/8$ for all field points that adhere to the far field conditions.

$$|r| > 8\frac{a^2}{\lambda}. \tag{9}$$

In Eq. (9) a denotes the radius of a sphere that completely encloses the region of distributed source points. As a consequence of the Fraunhofer approximation, Eq. (3) can be simplified with respect to the spherical Hankel function:

$$\lim_{|r|>8\frac{a^2}{\lambda}} h_{\mathrm{n}}^{(2)}(kr) = \frac{j^{n+1}}{kr} e^{-jkr}. \tag{10}$$

In spherical coordinates (ϑ, φ), the vector potentials $\boldsymbol{A}$ or $\boldsymbol{F}$ may be directly mapped to the respective field quantities $\boldsymbol{E}$ or $\boldsymbol{H}$. Under the terms of an exclusive distribution of electric currents $\boldsymbol{J}$ in the considered source region and a field point adhering to the far field condition by Fraunhofer, Eq. (11) represents the complete solution of the electric field in terms of the spherical components E_ϑ and E_φ.

$$\boldsymbol{E}(\vartheta, \varphi) = \begin{bmatrix} E_\vartheta \\ E_\varphi \end{bmatrix} = -jkZ \begin{bmatrix} A_\vartheta \\ A_\varphi \end{bmatrix}. \tag{11}$$

In Eq. (11) Z denotes the value of characteristic free-space impedance given by $Z=120\pi\ \Omega$.

$$\boldsymbol{E}(\vartheta, \varphi) = \frac{e^{-jkr}}{r} \sum_{n=0}^{\infty} \sum_{m=0}^{n} N_{\mathrm{n}}^{\mathrm{m}} \left\{ \boldsymbol{a}_{\mathrm{n}}^{\mathrm{m}} \cos(m\varphi) + \boldsymbol{b}_{\mathrm{n}}^{\mathrm{m}} \sin(m\varphi) \right\} j^{\mathrm{n}} P_{\mathrm{n}}^{\mathrm{m}}(\cos\vartheta). \tag{12}$$

Therefore, Eq. (12) represents the far-field solution of the electromagnetic field for a transmit antenna in terms of a series representation of spherical harmonics. The respective source distribution is formulated in terms of an electric current distribution in accordance to Fig. 1 and Eq. (1). Due to the limitation of far field conditions, the spatial orientation of transverse electromagnetic field components becomes orthogonal in the far field plane, that again exhibits a normal orientation to the radial direction of wave propagation at any point of observation. Therefore the respective dependence of electrical field $\boldsymbol{E}$ and magnetic vector potential $\boldsymbol{A}$ is specified by Eq. (11) for the special case of spherical coordinates. Applying Eq. (12) a generalized description of antenna radiation pattern is derived by means of an infinite series expansion implying spherical harmonics. For this reason Eq. (13) includes the spatial properties of antenna radiation in terms of the vector antenna radiation pattern $\boldsymbol{C}(\vartheta, \varphi)$.

$$\boldsymbol{E}(\vartheta, \varphi) = \frac{e^{-jkr}}{r} \boldsymbol{C}(\vartheta, \varphi). \tag{13}$$

The vector expansion coefficients of the field approach, $\boldsymbol{a}_{\mathrm{n}}^{\mathrm{m}}$ and $\boldsymbol{b}_{\mathrm{n}}^{\mathrm{m}}$ in terms of the corresponding antenna radiation pattern, are related to the expansion coefficients of the magnetic vector potential $\boldsymbol{a}_{\mathrm{n}}^{\mathrm{m}'}$ and $\boldsymbol{b}_{\mathrm{n}}^{\mathrm{m}'}$ employing Eqs. (9) and (11) as follows:

$$\begin{Bmatrix} a_{\mathrm{n},\vartheta}^{\mathrm{m}} \\ a_{\mathrm{n},\varphi}^{\mathrm{m}} \end{Bmatrix} = -jkZ \begin{Bmatrix} a_{\mathrm{n},\vartheta}^{\mathrm{m}'} \\ a_{\mathrm{n},\varphi}^{\mathrm{m}'} \end{Bmatrix}, \tag{14}$$

and

$$\begin{Bmatrix} b_{n,\vartheta}^{m} \\ b_{\mathrm{n},\varphi}^{\mathrm{m}} \end{Bmatrix} = -jkZ \begin{Bmatrix} b_{\mathrm{n},\vartheta}^{\mathrm{m}'} \\ b_{\mathrm{n},\varphi}^{\mathrm{m}'} \end{Bmatrix}. \tag{15}$$

The normalized eigensolutions to the vector Helmholtz equation are given in terms of the normalized spherical harmonics $Y_{\mathrm{nm}}^{\mathrm{e}}(\vartheta, \varphi)$ and $Y_{\mathrm{nm}}^{\mathrm{o}}(\vartheta, \varphi)$ as given in Stratton (1941) due to:

$$\begin{Bmatrix} Y_{\mathrm{nm}}^{\mathrm{e}}(\vartheta, \varphi) \\ Y_{\mathrm{nm}}^{\mathrm{o}}(\vartheta, \varphi) \end{Bmatrix} = N_{\mathrm{n}}^{m} P_{\mathrm{n}}^{\mathrm{m}}(\cos\vartheta) \begin{Bmatrix} \cos(m\varphi) \\ \sin(m\varphi) \end{Bmatrix}. \tag{16}$$

With respect to the orthogonality property between spherical harmonics of different degree n or order m as shown in Werner and Mittra (2000), the field expansion as given in Eq. (12) represents a complete solution to the radiation behavior of an arbitrary source configuration represented in terms of vector expansion coefficients $\boldsymbol{a}_{\mathrm{n}}^{\mathrm{m}}$ and $\boldsymbol{b}_{\mathrm{n}}^{\mathrm{m}}$.

For practical considerations the infinite upper summation bound of Eq. (12) has to be limited to a finite value N. Impairments resulting from the limitation of the infinite summation can be isolated with respect to the requirement of approximative field convergence e.g. given by Narasimhan et al. (1985). Therefore modal indices greater $N_{\mathrm{min}}=ka$ are regarded to have little significance to the field composition, where a denotes the minimum radius of a sphere that completely encloses the source antenna. A numerical study of the convergence behavior of the field expansion is given e.g. in Chen and Simpson (1991a) and Chen et al. (1992).

The vector coefficients of the field expansion can be determined according to Witte et al. (2003) by methods of numerical point matching as given in Hafner (1990). As described in Ludwig (1971) the matching procedure will be applied on the surface on which the fields are known. Subsequently the far field sphere is regarded to be the surface where the field quantities are given in terms of the far field components of antenna radiation pattern. In order to determine the requested modal weights, an equiangular grid is placed upon the far field sphere providing constant angular steps of $\Delta\varphi=2\pi/\mathrm{P}$ and $\Delta\vartheta=\pi/\mathrm{Q}$. P and Q correspond to the number of field points in the directions of azimuth and elevation, respectively. Antenna radiation patterns are derived from numerical analysis employing the FEM.

4 Computation of Multielement Antenna Correlation

Antenna diversity reception is an effective technique for suppressing multipath fading in a mobile antenna system. Diversity reception attains effective fading mitigation by the synthesis, selection, and switching of several received signals

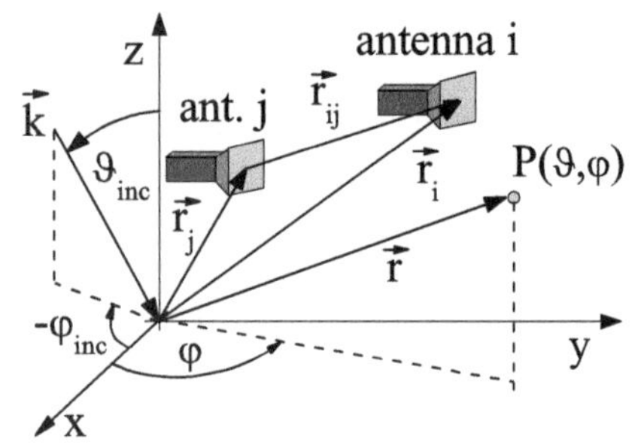

Fig. 2. Two-element antenna configuration in free space with antennas i and j.

that are correlated to the smallest possible extent. In the case of antenna diversity, reduction of the correlation coefficient between the receiving antenna branches leads to an enhanced diversity effect. Correlation analysis may be performed using a representation of transmission line parameters like input impedances and scattering parameters at the individual diversity branches as demonstrated in Kildal and Rosengren (2003) and Derneryd and Kristensson (2004). The derived correlation coefficients do not account for the spatial properties of the transmit channel, where the receive antenna array is applied, but inherently assume a uniform distribution of the angle of arrival (AOA). Essentially, interelement correlation coefficients in a multielement antenna configuration are strongly affected by the properties of the considered multipath propagation environment as will be shown in the following. In Jakes (1974) the spatial properties of a multielement diversity antenna configuration may be acquired by applying two-dimensional probability distributions $p(\vartheta, \varphi)$ in terms of the AOA distribution of the incoming electromagnetic field in elevation- and azimuth direction. Examples for correlation analysis in multielement antennas may be found in Tsunekawa and Kagoshima (1990), Svantesson (2002) and Waldschmidt and Wiesbeck (2004).

$$\rho_{P_{i,j}} \simeq \left|\rho_{i,j}\right|^2 = \left|\frac{R_{i,j}}{\sqrt{\sigma_i^2\sigma_j^2}}\right|^2 . \tag{17}$$

Subsequently, the relation between power correlation coefficient $\rho_{P_{i,j}}$ and complex correlation coefficient $\rho_{i,j}$ as given in Eq. (17) between antenna elements i and j will be considered. The power correlation therefore depends on the covariance $R_{i,j}$ between the two antenna elements and the variances σ_i^2 and σ_j^2 that are detected at the respective diversity branches. Figure 2 depicts the corresponding configuration of a two-element antenna in receive mode. The distance vector $\boldsymbol{r}_{ij}$ accounts for the spatial separation of antennas i and j. As shown in Leifer (2002) the spatial separation between the antenna elements results in a receive gain due to spatial diversity. The direction of plane waves impinging at a direction ϑ_{inc} and φ_{inc} are represented in terms of $\boldsymbol{k}$.
Provided an antenna far field representation that may be decomposed into a ϑ- and φ dependent fraction which are given by $C_{\vartheta,i}(\vartheta,\varphi)$ and $C_{\varphi,i}(\vartheta,\varphi)$ in terms of antenna element i and a 2-D polarization invariant probability distribution $p_{\vartheta,\varphi}(\vartheta,\varphi)$ of the AOA with the appropriate normalization given in Waldschmidt and Wiesbeck (2004), Eq. (18) denotes antenna covariance.

$$R_{i,j} = \mathrm{K}\int_0^{2\pi}\int_0^{\pi}\Big[\mathrm{XPR}\, C_{\vartheta,i}(\vartheta,\varphi)\, C^*_{\vartheta,j}(\vartheta,\varphi) + C_{\varphi,i}(\vartheta,\varphi)\, C^*_{\varphi,j}(\vartheta,\varphi)\Big] e^{j\boldsymbol{k}\boldsymbol{r}_{ij}} p_{\vartheta,\varphi}(\vartheta,\varphi)\sin\vartheta d\vartheta d\varphi \tag{18}$$

The ratio of the mean incident power in terms of a vertical polarized receive signal to the mean incident power of a horizontal polarized receive signal is denoted by the factor of cross-polarization ratio XPR. In Eq. (18), "*" means conjugate operation, K is a proportionality constant.

$$\sigma_i^2 = \mathrm{K}\int_0^{2\pi}\int_0^{\pi}\Big[\mathrm{XPR}\left|C_{\vartheta,i}(\vartheta,\varphi)\right|^2 + \left|C_{\varphi,i}(\vartheta,\varphi)\right|^2\Big] p_{\vartheta,\varphi}(\vartheta,\varphi)\sin\vartheta d\vartheta d\varphi . \tag{19}$$

As the spatial properties of the considered transmit channel strongly influence to the properties of antenna correlation, they have to be inevitably taken into account. An analytical formulation of antenna radiation patterns therefore simplifies the computation of correlation coefficients for diversity analysis. In Leifer (2002) a simplified multipole computational model was published in terms of parallel oriented dipole antennas. It is therefore straightforward to express the vector antenna radiation patterns $\boldsymbol{C}_{i,j}(\vartheta,\varphi)$ of antennas i and j in terms of a spherical mode expansion as given in Sect. 3. Provided the series expansion for the radiation fields as in Eq. (12), the expressions for antenna covariance and variance become a function of modal expansion coefficients $\boldsymbol{a}_n^m$ and $\boldsymbol{b}_n^m$. Equation (20) accounts for the series representation of antenna radiation patterns of antennas i and j by means of spherical eigenmodes and represents a generalized description of antenna covariance in terms of a spherical mode expansion.

$$\begin{aligned} R_{i,j} = \mathrm{K}\sum_{n_i=0}^{\infty}\sum_{m_i=0}^{n_i}\sum_{n_j=0}^{\infty}\sum_{m_j=0}^{n_j}\Big\{ &\left(\mathrm{XPR}a_{n_im_i,\vartheta}a^*_{n_jm_j,\vartheta} + a_{n_im_i,\varphi}a^*_{n_jm_j,\varphi}\right) I^{ee}_{n_im_in_jm_j} \\ &+\left(\mathrm{XPR}b_{n_im_i,\vartheta}b^*_{n_jm_j,\vartheta} + b_{n_im_i,\varphi}b^*_{n_jm_j,\varphi}\right) I^{oo}_{n_im_in_jm_j} \\ &+\left(\mathrm{XPR}a_{n_im_i,\vartheta}b^*_{n_jm_j,\vartheta} + a_{n_im_i,\varphi}b^*_{n_jm_j,\varphi}\right) I^{eo}_{n_im_in_jm_j} \\ &+\left(\mathrm{XPR}b_{n_im_i,\vartheta}a^*_{n_jm_j,\vartheta} + b_{n_im_i,\varphi}a^*_{n_jm_j,\varphi}\right) I^{oe}_{n_im_in_jm_j}\Big\} . \end{aligned} \tag{20}$$

As can be seen in (20), covariance computation requires a fourfold summation of 2-D coupling integrals $I^{e,o\,e,o}_{n_im_in_jm_j}$ that are weighed by the respective expansion coefficients. Therefore an evaluation of Eq. (20) might become quite time consuming with the advantage to derive an analytical formulation of the correlation properties in multielement antenna configurations. The coupling integrals are given in terms of a product of two normalized spherical harmonics $Y^{e,o}_{nm}(\vartheta,\varphi)$

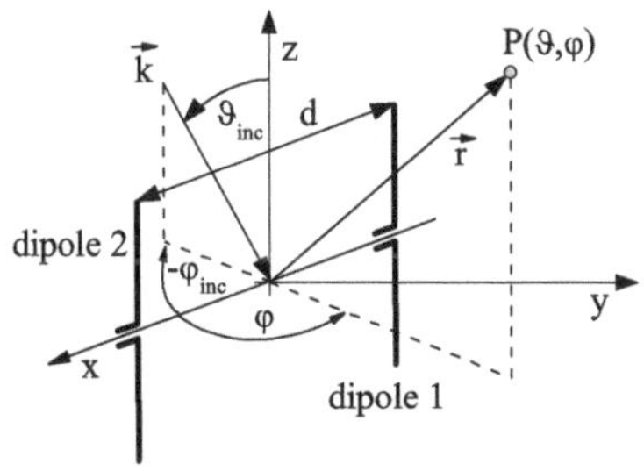

Fig. 3. Configuration of two parallel oriented $\lambda/2$ dipoles with a relative displacement of d along the x-axis.

with the angular distribution of incoming waves, $p_{\vartheta,\varphi}(\vartheta,\varphi)$ as shown in Eq. (21).

$$I^{e,o\,e,o}_{n_i m_i n_j m_j} = \int_0^{2\pi}\int_0^{\pi} Y^{e,o}_{n_i m_i}(\vartheta,\varphi)\, Y^{e,o}_{n_j m_j}(\vartheta,\varphi) \cdot e^{jkr_{ij}}\, p_{\vartheta,\varphi}(\vartheta,\varphi) \sin\vartheta\, d\vartheta\, d\varphi \,. \qquad (21)$$

Using the orthogonality relation as in Werner and Mittra (2000) for spherical harmonics, Eq. (20) may be simplified with respect to a certain symmetry of the AOA distribution $p_{\vartheta,\varphi}(\vartheta,\varphi)$ and the antenna array topology.

Applying the spherical wave expansion to the general formulation of variances as in Eq. (19) for antenna element ν, ν=1,2 yields:

$$\sigma_\nu^2 = \mathrm{K}\int_0^{2\pi}\int_0^{\pi}\Bigg\{\mathrm{XPR}\left|\sum_{n_\nu=0}^{\infty}\sum_{m_\nu=0}^{n_\nu} a_{n_\nu m_\nu,\vartheta} Y^{e}_{n_\nu m_\nu} + b_{n_\nu m_\nu,\vartheta} Y^{o}_{n_\nu m_\nu}\right|^2 + \left|\sum_{n_\nu=0}^{\infty}\sum_{m_\nu=0}^{n_\nu} a_{n_\nu m_\nu,\varphi} Y^{e}_{n_\nu m_\nu} + b_{n_\nu m_\nu,\varphi} Y^{o}_{n_\nu m_\nu}\right|^2\Bigg\}\, p_{\vartheta,\varphi}(\vartheta,\varphi)\sin\vartheta\, d\vartheta\, d\varphi \,. \qquad (22)$$

5 Analysis of Two-Element Dipole Array

The correlation analysis of an arrangement of two parallel oriented, resonant $\lambda/2$ dipoles in free space depicted in Fig. 3 will be accomplished in this section. In order to compare the results from the customary computation of antenna correlation by an evaluation of the twofold Riemann sum in Eqs. (18) and (19) with the method based upon a spherical mode expansion of antenna radiation patterns as given in Sect. 4 the patterns are evaluated in terms of a spherical mode expansion delimited by N. In terms of the correlation analysis of the two-element dipole configuration two different cases have to be considered. The first case assumes both dipole antennas isolated from each other, so that the effect of electromagnetic interaction is neglected by definition. The assumption justifies a modification of the field expansion applying the array factor of the antenna configuration into the correlation analysis. The results of this study are summarized in Sect. 5.1. In the second approach electromagnetic coupling between the dipole antennas is taken into account.

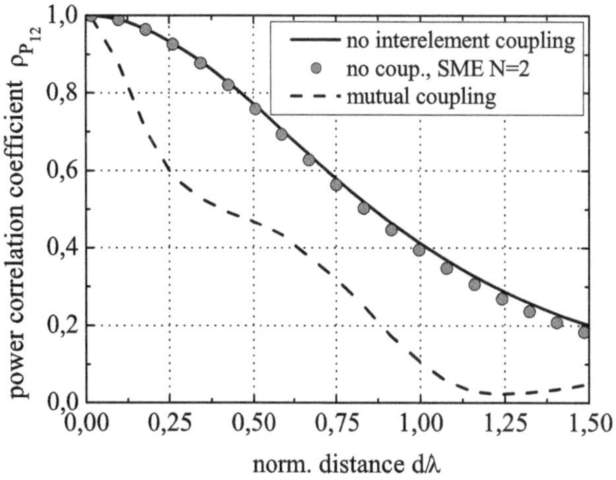

Fig. 4. Power correlation of isolated- and coupled dipole antennas for an angular spread of σ_φ=10°.

Section 5.2 provides the results for the coupling-case of the dipole array.

5.1 Analysis of isolated Dipoles

In terms of a configuration of identical isolated antenna elements, the representation for the total radiation pattern can be simplified with respect to the appropriate array factor $G(\vartheta,\varphi)$ of the antenna configuration. Therefore the computation of antenna correlation simplifying requires the spherical mode expansion of the radiation fields of the isolated antenna element that is identical for all elements within the arrangement. The consideration of an array of identical isolated antenna elements therefore leads to an effective reduction of the computational resources for correlation analysis. For convenience, the AOA distribution will be restricted to the xy-plane and hence be given in terms of a Laplacian distribution as in Svantesson (2002) that is properly described by a mean value m_φ and the respective spread σ_φ. All simulations are performed using a mean value of m_φ=90° using the coordinate system depicted in Fig. 3. Three different values of angular spread σ_φ are considered. Figure 4 depicts antenna power correlation coefficient with respect to the normalized interelement distance d/λ and a narrow angular spread σ_φ=10°. In order to address to the influences of electromagnetic coupling on antenna correlation, the dashed lines in Fig. 4 represent the correlation coefficient in terms of coupled dipole antennas derived from the customary evaluation of Eqs. (18) and (19). The results derived from a spherical wave expansion of the radiation pattern of the isolated dipole antennas are given as circles. The series representation in (Eq. 12) is restricted with respect to N=2 leading to a maximum of 12 complex vector coefficients in the field expansion. Comparing the processes of antenna correlation for the uncoupled arrangement and the case of an electromagnetically coupled antenna configuration, interelement coupling leads to a decrease of antenna correlation for

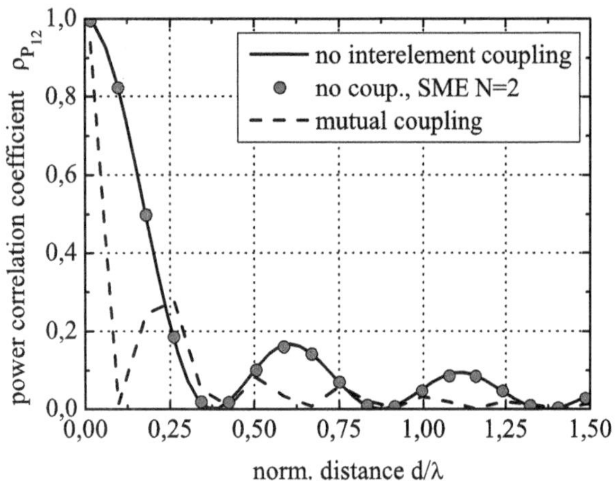

Fig. 5. Power correlation of isolated- and coupled dipole antennas in terms of equally distributed AOAs in the horizontal plane.

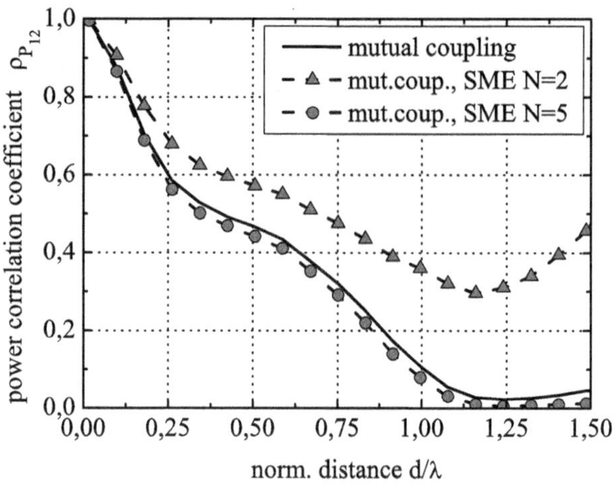

Fig. 6. Power correlation of coupled dipole antennas for an angular spread of σ_φ=10° and different degrees of SME.

all interelement coupling distances. This observation is in accordance to Tsunekawa and Kagoshima (1990) as well as Leifer (2002). The degradation of antenna correlation is due to the effect of pattern diversity that can only be contributed of an arrangement of electromagnetically coupled antenna elements. Antenna correlation of the electromagnetically decoupled dipole antennas is restricted to the occurrence of spatial diversity due to antenna separation $\boldsymbol{r}_{\mathrm{ij}}$. The results for the correlation analysis based upon the spherical mode expansion of the radiation field matches very well with the results derived from customary correlation analysis, unless only 12 vector coefficients are involved in the field expansion. This may be explained in terms of the convergence criterion given by Narasimhan et al. (1985) that claims modal indices $N_{\mathrm{min}}=ka$, with $a=\lambda/2$ with respect to the configuration of $\lambda/2$-resonant, uncoupled dipole antennas. Therefore the convergence criterion requires integer indices of N_{min}=2. It has to be clarified, that the processes of antenna correlation derived from the customary numerical analysis of Eq. (17) using Eqs. (18) and (19) may only be considered to be approximative solutions. As a result of numerical simulation applying the FEM, antenna far field representation may not be given in terms of an analytical formulation. The results were obtained by applying an equiangular grid on the far field sphere, where the radiation fields were calculated. Throughout this paper equal grid distances in the directions of azimuth and elevation, $\Delta\varphi$=3° and $\Delta\vartheta$=3° were applied.
Figure 5 considers the case of an equal angle of arrival distribution in the xy-plane. Neglecting the effects of interelement coupling between the dipole antennas, the observed antenna correlation corresponds to the well known result for the power correlation of Jakes (1974) between antennas i and j using the Bessel function of 0. order, with $\rho_{P_{\mathrm{i,j}}}=J_0^2(k\,d)$. As observed from Fig. 4 mutual coupling between adjacent antenna elements also leads to a decrease of power correlation (e.g. Fig. 5) due to pattern diversity gain, especially for small antenna displacements. The results of correlation analysis with respect to the SME of the radiation fields for N=2 again reveals a good agreement to the customary computation technique.

5.2 Analysis of coupled Dipoles

From Sect. 5.1 it can be concluded, that mutual coupling between adjacent elements within a multielement antenna arrangement may result in an improvement of diversity performance. This increase of diversity gain is a result of a degradation of antenna radiation patterns losing their axial symmetry. Therefore the convergence of antenna radiation pattern analysis is influenced by the appearance of mutual coupling causing higher order modes in the field expansion. As a result, a larger number of vector expansion coefficients has to be taken into account to guarantee an accurate approximation of antenna radiation behavior.
This effect can be derived from Fig. 6 that shows the results of antenna correlation for an angular spread of σ_φ=10°. The computation of interelement antenna correlation is based upon field expansions dealing with summation limits of N=2 and N=5 in accordance with Eq. (12). The differences in simulation results between the customary computation technique by sampling the far field sphere and the series expansion in terms of spherical harmonics are minimized with respect to an increase of N. This is due to the violation of the convergence criterion $N_{\mathrm{min}}=ka$ for low degrees of the field expansion. With an increase of the normalized interelement distance d/λ of the coupled dipole configuration, the effective antenna aperture increases. Therefore an enhanced number of spherical modes has to be taken into account, to ensure sufficient field accuracy. E.g. for a given interelement spacing of d/λ=1.5, modal indices of degree N_{min}=4 should be considered in the field approach.
Figure 7 represents the results of antenna correlation for an equally distributed AOA in the horizontal plane. As expected the results show the same bias. Due to the increase

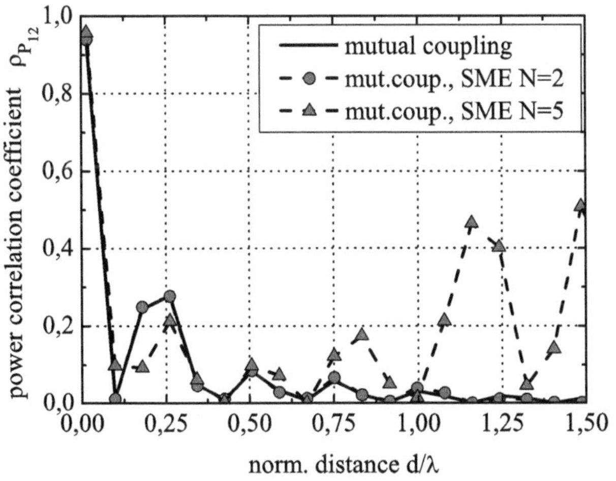

Fig. 7. Power correlation of coupled dipole antennas in terms of equally distributed AOAs in the horizontal plane and different degrees of SME.

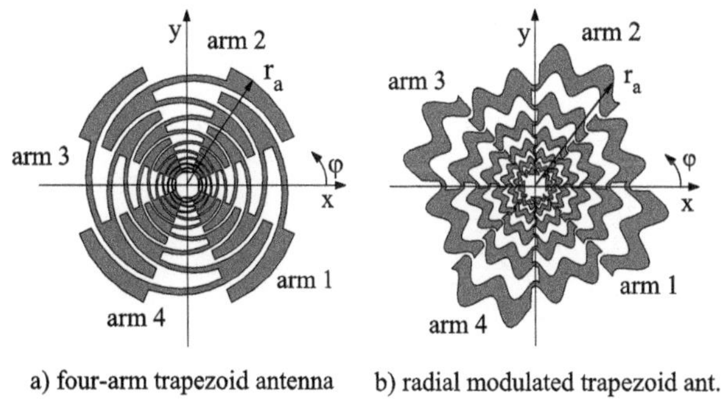

Fig. 8. Trapezoidal antenna geometries for the realization of frequency-independent diversity antennas.

of effective antenna aperture and the degradation of antenna radiation behavior from the axial symmetry of the isolated dipole antenna, a larger series limit N is required to account for a sufficient approximation of antenna radiation fields. With an increase of the number of vector expansion coefficients in Eq. (12) the curves of antenna correlation tend to converge.

6 Analysis of Frequency-Independent Planar Antennas

Compact multimode antennas as the biconical antenna that has been reported by Svantesson (2002) or the planar multiterminal spiral- and sinuous antennas reported by Waldschmidt and Wiesbeck (2004) seem to be promising for the integration in MIMO transmission systems with a combination of polarization-, pattern- and modal diversity. In this section the four-arm log.-per. trapezoid antenna, that was reported by Klemp et al. (2004a) and a miniaturized derivative of this antenna type, the radial modulated four-arm log.-per. trapezoid antenna reported by Klemp et al. (2004b) will be analyzed with respect to their property of polarization diversity. The antennas adhere to the principle of self-complement and therefore provide large operational bandwidths. Both antenna types are depicted in Fig. 8. The ability of a good polarization diversity behavior is based on the degeneration of two spatially orthogonal eigenfield distributions, that are supplied by the symmetrical four-arm antennas in Fig. 8. In the ideal case the four arms of the log.-per. antenna structures can be decomposed into two orthogonal arm pairs that represent two decoupled transmit channels with orthogonal linear polarization leading to uncorrelated output voltages at the feeding bridge of the antenna structure.

For the evaluation of polarization diversity performance a 2-D probability distribution of the angle of arrival is presumed. In analogy to the receive diversity case of the dipole antenna array, the azimuthal dependence of AOA is modeled using a Laplacian distribution. In terms of the elevational dependence, a Gauss distribution is assumed in accordance with Waldschmidt and Wiesbeck (2004). Mean value and angular spread in the elevation are given as m_ϑ=45°, σ_ϑ=15°. The Laplace distribution is determined by a mean value of m_φ=10° and a spread of σ_φ=20°.

Figure 9 shows the result of power correlation coefficient $\rho_{P_{xy}}$ in terms of polarization diversity performance with equal mean branch power in each polarization component (x-polarization or y-pol.) in the specified frequency range from 1 GHz to 6 GHz. Due to the high degree of cross polarization discrimination of each pair of opposite antenna arms, a low degree of power correlation can be observed in the entire frequency range of operation. ρ_P always remains below 0.12. The results of an ordinary computation of power correlation are compared to the results of a spherical mode expansion (N=5) for the trapezoid antenna leading to a good agreement. The low degree of correlation can be justified by using the results derived from spherical antenna near field measurement at the distinct frequency positions f_1=1.8 GHz, f_2=2.45 GHz and f_3=5.85 GHz. Based upon the measurement results with an angular resolution of 1° in ϑ- and φ-direction the power correlation coefficient was computed using the twofold Riemann approximation of Eqs. (17) to (19). As for the miniaturized derivative of the four-arm trapezoid antenna depicted in Figs. 8b, 10 represents the results of power correlation against operating frequency. Results derived from an approach of Eq. (17) by taking the equivalent Riemann sums of Eqs. (18) and (19) as well as the evaluation of power correlation by spherical mode expansion of the individual radiation patterns yield a good agreement. Numerical data of power correlation can be verified very well by results from spherical near field measurement of the miniaturized trapezoid antennas at the respective frequency positions f_1 to f_3 as shown in Fig. 10.

7 Conclusions

In this paper, a generalized approach for the computation of power correlation in multielement antenna arrangements was

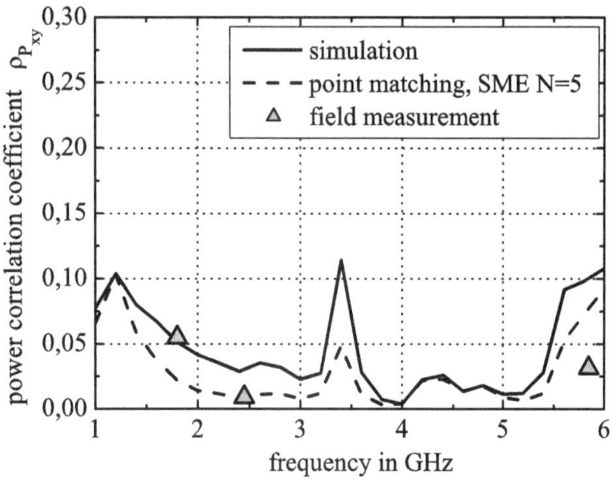

Fig. 9. Power correlation for log.-per. trapezoid antenna. Simulation and measurement results.

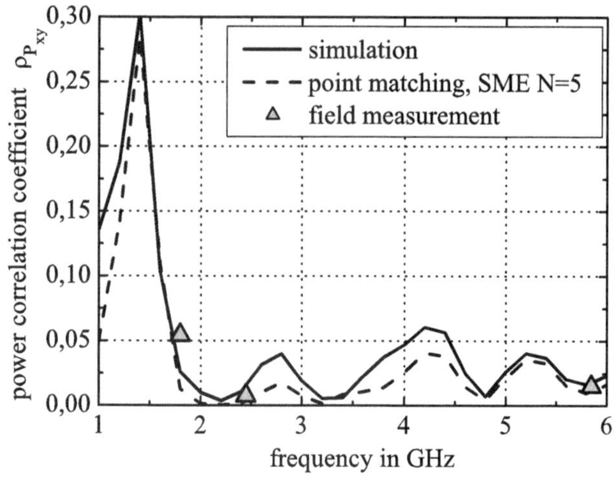

Fig. 10. Power correlation for miniaturized log.-per. trapezoid antenna. Simulation and measurement results.

presented, which is based upon the expansion of antenna radiation fields in terms of spherical harmonics. Therefore antenna radiation behavior was considered in receive-mode. The spatial properties of a multipath propagation scenario were modeled using 2-D probability density distributions. In terms of a finite series representation of antenna radiation fields, an analytical investigation of antenna correlation properties can be accomplished. An analytical formulation of antenna power correlation was presented involving elementary 2-D coupling integrals over a product of spherical harmonics, that are weighed by the vector series expansion coefficients of the corresponding antenna radiation patterns. Due to the fact that the contributions of elementary coupling integrals remain fixed for a given array topology and propagation environment, the amount of antenna correlation only depends on the expansion coefficients of their radiation pattern and therefore from the current distribution on the corresponding metalization surfaces. Hence the presented analysis provides a major simplification in order to derive the respective correlation properties of individual antenna elements in a multipath propagation scenario. Furthermore the analysis may serve as an initial point for the optimization of antenna radiation properties with respect to low interelement correlations by adjusting the modal content of the respective radiation fields.

The presented correlation analysis was applied on a configuration involving two parallel resonant $\lambda/2$ dipole antennas in receive mode differentiating between isolated and electromagnetically coupled antennas. The results of customary correlation analysis and SME based computation of antenna correlation agree very well. Likewise the two different approaches in correlation analysis were adopted to the characterization of two wideband log.-per. planar antennas in polarization diverse receive mode. The results also reveal a very good agreement and could be verified very well by results derived from spherical antenna near field measurement. The self-complementary log.-per. four-arm trapezoid antennas may be considered for an implementation in a transmission system with wideband polarization diversity reception and seem promising for the integration in MIMO transmission systems with multimode operation due to the multi-port antenna concept and low correlations between the different operational modes.

References

Abramovitz, M. and Stegun, I. A.: Handbook of Mathematical Functions, Dover Publications, New York, 1972.

Adane, Y., Gati, A., Wong, M.-F., Dale, C., Wiart, J., and Hanna, V. F.: Optimal Modeling of Real Radio Base Station Antennas for Human Exposure Assessment Using Spherical-Mode Decomposition, IEEE Antennas and Wireless Propagation Letters, 1, 215–218, 2002.

Bruning, J. H. and Lo, Y. T.: Multiple Scattering of EM Waves by Spheres Part I-Multipole Expansion and Ray-Optical Solutions, IEEE Transactions on Antennas and Propagation, 19, 378–390, 1971.

Chen, Y. and Simpson, T.: Radiation Pattern Analysis of Arbitrary Wire Antennas Using Spherical Mode Expansions with Vector Coefficients, IEEE Transactions on Antennas and Propagation, 39, 1716–1721, 1991a.

Chen, Y. and Simpson, T. L.: Compact Representation of Radiation Patterns using Vector Spherical Mode Expansions, Antennas and Propagation Society International Symposium, 24–28 June 1991, 3, 1632–1634, 1991b.

Chen, Y., Simpson, T. L., and Ho, T. Q.: Highly efficient technique for solving radiation and scattering problems, IEE Proceedings-H, 139, 7–10, 1992.

Derneryd, A. and Kristensson, G.: Antenna signal correlation and its relation to the impedance matrix, Electronics Letters, 40, 401–402, 2004.

Hafner, C.: The Generalized Multipole Technique for Computational Electromagnetics, Artech House, Inc., Boston, London, 1990.

Jakes, W. C.: Microwave Mobile Communications, IEEE Press, Inc., New York, 1974.

Kildal, P.-S. and Rosengren, K.: Electromagnetic Analysis of Effective and Apparent Diversity Gain of Two Parallel Dipoles, IEEE Antennas and Wireless Propagation Letters, 2, 9–13, 2003.

Klemp, O., Schultz, M., and Eul, H.: A Planar Broadband Antenna for Applications in WLAN and 3G, International Symposium on Signals, Systems and Electronics ISSSE, Linz, Austria, August 10–13, 2004a.

Klemp, O., Schultz, M., and Eul, H.: Miniaturization Techniques For Logarithmically-Periodic Planar Antennas, IEEE International Symposium on Personal Indoor and Mobile Communications PIMRC, Barcelona, Spain, September 5–9, 2004b.

Leifer, M. C.: Signal Correlations in Coupled Cell and MIMO antennas, Antennas and Propagation Society International Symposium, 16–22 June 2002, 3, 194–197, 2002.

Ludwig, A. C.: Near-Field Far-Field Transformations Using Spherical-Wave Expansions, IEEE Transactions on Antennas and Propagation, 19, 214–220, 1971.

Narasimhan, M. S., Christopher, S., and Varadarangan, K.: Modal Behavior of Spherical Waves from a Source of EM Radiation with Application to Spherical Scanning, IEEE Transactions on Antennas and Propagation, 33, 350–354, 1985.

Potter, P. D.: Application of Spherical Wave Therory to Cassegrainian-Fed Paraboloids, IEEE Transactions on Antennas and Propagation, 15, 727–736, 1967.

Stratton, J. A.: Electromagnetic Theory, McGraw Hill, New York, 1941.

Svantesson, T.: Correlation and Channel Capacity of MIMO Systems Employing Multimode Antennas, IEEE Transactions on Vehicular Technology, 51, 1304–1312, 2002.

Tsunekawa, K. and Kagoshima, K.: Analysis of a Correlation Coefficient of Built-In Diversity Antennas for a Portable Telephone, Antennas and Propagation Society International Symposium, 7–11 May 1990, 1, 543–546, 1990.

Waldschmidt, C. and Wiesbeck, W.: Compact Wide-Band Multimode Antennas for MIMO and Diversity, IEEE Transactions on Antennas and Propagation, 52, 1963–1969, 2004.

Werner, D. H. and Mittra, R.: Frontiers in Electromagnetics, IEEE Press Series on Microwave Technology and RF, New York, 2000.

Witte, E. D., Griffiths, H. D., and Brennan, P. V.: Phase mode processing for spherical antenna arrays, Electronics Letters, 39, 1430–1431, 2003.

3

Time-Domain Techniques for Computation and Reconstruction of One-Dimensional Profiles

M. Rahman and R. Marklein

Fachgebiet Theoretische Elektrotechnik, Fachbereich Elektrotechnik/Informatik, Universität Kassel (UNIK), 34109 Kassel, Germany

Abstract. This paper presents a time-domain technique to compute the electromagnetic fields and to reconstruct the permittivity profile within a one-dimensional medium of finite length. The medium is characterized by a permittivity as well as conductivity profile which vary only with depth. The discussed scattering problem is thus one-dimensional. The modeling tool is divided into two different schemes which are named as the forward solver and the inverse solver. The task of the forward solver is to compute the internal fields of the specimen which is performed by Greens function approach. When a known electromagnetic wave is incident normally on the media, the resulting electromagnetic field within the media can be calculated by constructing a Greens operator. This operator maps the incident field on either side of the medium to the field at an arbitrary observation point. It is nothing but a matrix of integral operators with kernels satisfying known partial differential equations. The reflection and transmission behavior of the medium is also determined from the boundary values of the Green's operator. The inverse solver is responsible for solving an inverse scattering problem by reconstructing the permittivity profile of the medium. Though it is possible to use several algorithms to solve this problem, the invariant embedding method, also known as the layer-stripping method, has been implemented here due to the advantage that it requires a finite time trace of reflection data. Here only one round trip of reflection data is used, where one round trip is defined by the time required by the pulse to propagate through the medium and back again. The inversion process begins by retrieving the reflection kernel from the reflected wave data by simply using a deconvolution technique. The rest of the task can easily be performed by applying a numerical approach to determine different profile parameters. Both the solvers have been found to have the ability to deal with different types of slabs and incident electromagnetic pulses. Slabs having continuous and discontinuous relative permittivity have already been tested successfully. The tested electromagnetic pulses are a Dirac, Gaussian and sinusoidal pulse. Due to sampling, the resolution of the system also plays a significant role in obtaining better outputs from this scheme.

1 Introduction

Recently, 1-D profiling has become a point of high interest, for instance in geophysics and Non-Destructive Testing (NDT), not only due to the ease to solve the problems analytically but also due to high accuracy level. They have large importance in Time Domain Reflectometry (TDR) such as moisture meter and material analysis (Connor and Dowding, 1999; Schlaeger, 2002). The forward solver of such schemes works on a known slab and determines the electromagnetic field. On the other hand, the inverse solver takes the reflection data from the slab as an input and determines the properties of the unknown slab. The known methods to solve the forward problem are

- Finite-Difference Time-Domain (FDTD) method (Taflove and Hagness, 2000)
- Finite Integration Technique (FIT) (Marklein, 2002)
- Green's function approach (Krueger and Ochs Jr., 1989).

The Green's function approach has some advantages over the other methods. Firstly, in this method the wave equation needs not to be solved for each incident wave form. Secondly, wave field throughout the entire medium does not have to be computed. For these advantages, this method has been implemented in the forward solver. The inverse problem can be solved by

- invariant embedding (layer stripping) (Corones et al., 1983)
- downward continuation (Green's function approach) (Kristensson and Krueger, 1986, 1987).

The less amount of input data could be mentioned as main advantage of layer stripping method. Obviously the reflection data should be processed before feeding them into the inverse scheme.

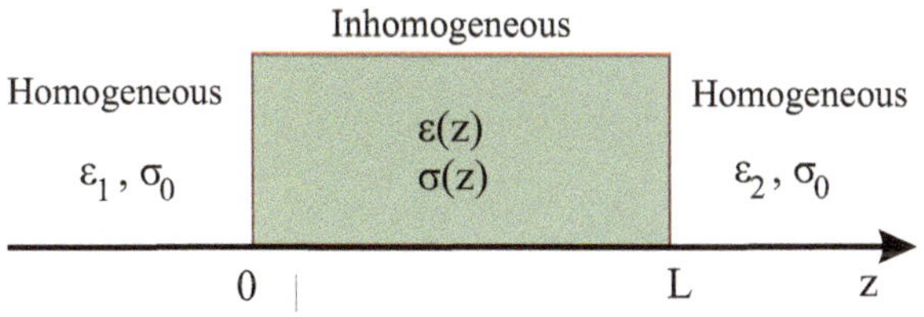

Fig. 1. Geometry of the inhomogeneous slab.

2 Statement of the problem

The geometry of the problem is shown in Fig. 1. An inhomogeneous slab occupies the region $0 \leq z \leq L$. The permittivity and conductivity profiles are functions of depth z only.

A homogeneous, lossless medium is situated on either side of the slab. The electric field strength $E(z,t)$ within the slab can be expressed by

$$\frac{\partial^2}{\partial z^2}E(z,t) - \frac{1}{c^2(z)}\frac{\partial^2}{\partial t^2}E(z,t) - b(z)\frac{\partial}{\partial t}E(z,t) = 0\,, \quad (1)$$

where

$$c^{-2}(z) = \varepsilon(z)\,\mu_0 \quad , \quad b(z) = \sigma(z)\,\mu_0\,. \quad (2)$$

Here μ_0, $\sigma(z)$ and $\varepsilon(z)$ represent the permeability of vacuum, conductivity and permittivity respectively. The phase velocity $c(z)$ is assumed to be continuous at the boundary, which means

$$c(z) = c(0^+),\ z \leq 0 \quad ; \quad c(z) = c(L^-),\ z \geq L\,. \quad (3)$$

2.1 Normalization

In order to facilitate the numerical computations a conversion to travel time coordinates is made. These coordinates are defined as

$$l = \int_0^L \frac{1}{c(z)}\mathrm{d}z \quad , \quad s(t) = t/l \quad (4)$$

$$x = x(z) = \int_{z'=0}^{z} \frac{1}{lc(z')}\mathrm{d}z' \quad , \quad u(x,s) = E(z,t)\,, \quad (5)$$

where x is normalized distance and s is normalized time. The slab occupies the region $0 \leq x \leq 1$ and a round trip time is described by $0<s<2$ which is equivalent to $2l$. In fact, l represents the time taken by the wave front to travel through the slab once. So the wave equation will be transformed into

$$\frac{\partial^2}{\partial x^2}u(x,s) - \frac{\partial^2}{\partial s^2}u(x,s) + A(x)\frac{\partial}{\partial x}u(x,s) + B(x)\frac{\partial}{\partial s}u(x,s) = 0\,, \quad (6)$$

where

$$A(x) = -\frac{\mathrm{d}}{\mathrm{d}x}\ln c[z(x)] \quad (7)$$

$$B(x) = -l\,b[z(x)]\,c^2[z(x)]\,. \quad (8)$$

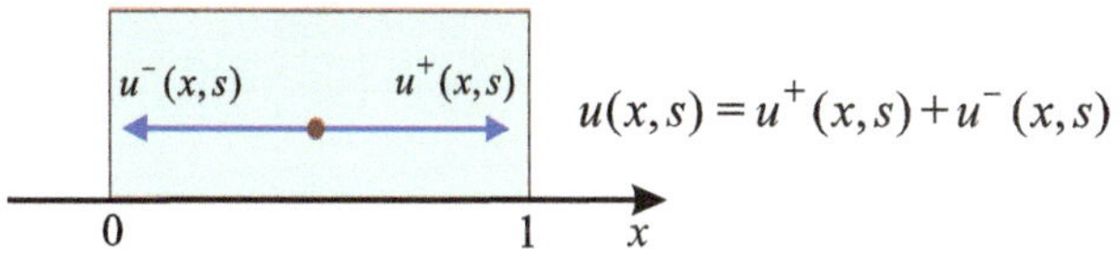

Fig. 2. Wave splitting.

2.2 Wave splitting

The wave travelling in the non-homogeneous slab can be splitted into left- and right-going parts. In Fig. 2, $u^+(x,s)$ is the right going wave and $u^-(x,s)$ is the left going wave. For all time s, the initial and boundary conditions are

$$u^+(x,0) = u^-(x,0) = 0 \quad ; \quad 0 < x < 1 \quad (9)$$

$$u^+(0,s) = f(s) \quad , \quad u^-(1,s) = 0 \quad ; \quad s > 0\,. \quad (10)$$

An incident wave is impinged only from the left side which is denoted by $f(s)$.

3 Forward solver: introduction to the Green's kernel

The known parameters are incident pulse, slab length L, ε_r and σ profile. The goal is to determine the internal fields. A matrix operator $[\mathbf{G}](x)$ can be defined to map the incident waves $u^+(0,s), u^-(1,s)$ to internal fields $u^+(x,s), u^-(x,s)$

$$\begin{Bmatrix} u^+(x,s) \\ u^-(x,s) \end{Bmatrix} = \underbrace{\begin{bmatrix} G_{11}(x) & G_{12}(x) \\ G_{21}(x) & G_{22}(x) \end{bmatrix}}_{=[\mathbf{G}](x)} \begin{Bmatrix} u^+(0,s) \\ u^-(1,s) \end{Bmatrix}. \quad (11)$$

Considering an incident wave only from left side, we find

$$\left.\begin{array}{l} u^+(x,s) = G_{11}(x) * f(s) \\ u^-(x,s) = G_{21}(x) * f(s) \end{array}\right\}; \text{ as } u^+(0,s) = f(s)\,, \quad (12)$$

where "$*$" denotes convolution. G_{11}, G_{21} are related to the forward and backward travelling wave, respectively, when the incident wave is applied from the left side of the slab. G_{12}, G_{22} are not considered here as they are only applicable for an incident wave coming from the right side of the slab. Now by applying Duhamel's integral,

$$u^+(x,s) = t^+(0,x)\left\{f(s-x) + \int_0^{s-x} f(s')\,G_{11}(x,s-s')\mathrm{d}s'\right\} \quad (13)$$

$$u^-(x,s) = \frac{1}{t^-(0,x)}\int_0^{s-x} f(s')\,G_{21}(x,s-s')\mathrm{d}s'\,, \quad (14)$$

where

$$t^{\pm}(x_1,x_2) = \exp\left\{\pm\frac{1}{2}\int_{x_1}^{x_2}[A(x')\mp B(x')]\mathrm{d}x'\right\}. \quad (15)$$

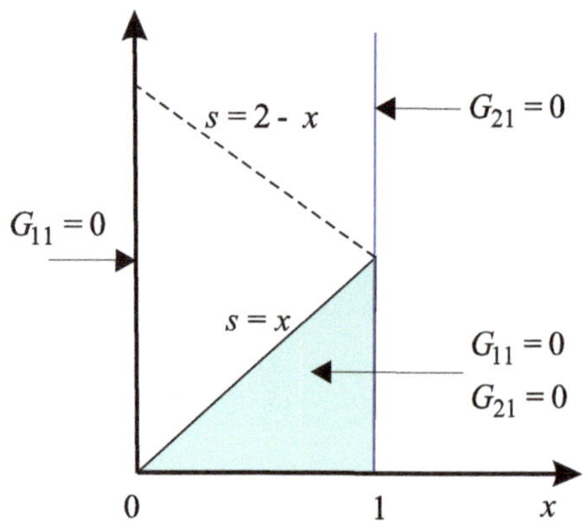

Fig. 3. Conditions of the Green's kernel.

3.1 Equations of the Green's kernel

The fields $u^+(x,s)$ and $u^-(x,s)$ satisfy the following equation

$$\frac{\partial}{\partial x}\begin{Bmatrix} u^+(x,s) \\ u^-(x,s) \end{Bmatrix} = \begin{bmatrix} \alpha(x) & \beta(x) \\ \gamma(x) & \delta(x) \end{bmatrix} \begin{Bmatrix} u^+(x,s) \\ u^-(x,s) \end{Bmatrix}, \tag{16}$$

where $\alpha(x)$, $\beta(x)$, $\gamma(x)$, and $\delta(x)$ are the functions of $A(x)$, $B(x)$ and incorporate time derivative $\frac{\partial}{\partial s}$. From Eqs. (11) and (10), it can be summarized that

$$\frac{\partial}{\partial x}[\mathbf{G}](x) = \begin{bmatrix} \alpha(x) & \beta(x) \\ \gamma(x) & \delta(x) \end{bmatrix} \begin{bmatrix} G_{11}(x) & G_{12}(x) \\ G_{21}(x) & G_{22}(x) \end{bmatrix}. \tag{17}$$

Again using Eq. (7), it can be shown that

$$\frac{\partial}{\partial x}[G_{11}(x)\,f](s) = \alpha(x)\,[G_{11}(x)\,f](s) + \beta(x)\,[G_{21}(x)\,f](s). \tag{18}$$

Substituting $[G_{11}(x)\,f](s)$ by the right side of the Eq. (13), a simple expression of G_{11} can be obtained, as follows

$$\frac{\partial}{\partial x}G_{11}(x,s) + \frac{\partial}{\partial s}G_{11}(x,s) = \frac{1}{2}[A(x)+B(x)]\exp\left\{-\int_0^x B(x')\mathrm{d}x'\right\}G_{21}(x,s). \tag{19}$$

Similarly using Eq. (14), G_{21} can be defined as

$$\frac{\partial}{\partial x}G_{21}(x,s) + \frac{\partial}{\partial s}G_{21}(x,s) = \frac{1}{2}[A(x)-B(x)]\exp\left\{\int_0^x B(x')\mathrm{d}x'\right\}G_{11}(x,s). \tag{20}$$

3.2 Initial, boundary, and jump conditions

G_{11} is continuous everywhere except on $s{=}x$ (see Fig. 3). On this line the kernels can be expressed as

$$G_{11}(x,x^+) = -\frac{1}{8}\int_0^x [A^2(x') - B^2(x')]\mathrm{d}x' \tag{21}$$

$$G_{21}(x,x^+) = -\frac{1}{4}[A(x)-B(x)]\exp\left\{\int_0^x B(x')\mathrm{d}x'\right\}. \tag{22}$$

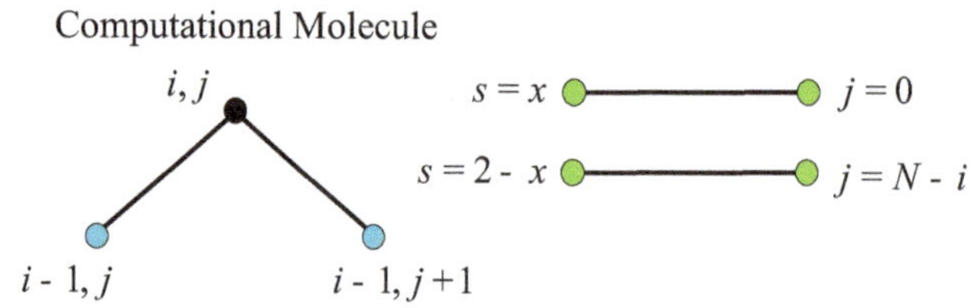

Fig. 4. Computational molecule and variable transformation.

Again on the line $s{=}2{-}x$,

$$G_{21}(x,(2-x)^+) - G_{21}(x,(2-x)^-) = \frac{1}{4}[A(1)-B(1)]\exp\left\{\int_0^1 B(x')\mathrm{d}x'\right\}. \tag{23}$$

The initial and boundary conditions can be defined as

$$G_{11}(x,s) = G_{21}(x,s) = 0 \quad ; \quad s < x \tag{24}$$

$$G_{11}(0,s) = G_{21}(1,s) = 0 \quad ; \quad s > 0. \tag{25}$$

3.3 Numerical implementation of the forward solver

The Green's kernels explained by Eqs. (19) and (20) can also be implemented numerically. By fixing a constant Δ, the first equation can be integrated from $(x{-}\Delta, s{-}\Delta)$ to (x,s) along the characteristics $s{-}x{=}$const. and the second one from (x,s) to $(x{+}\Delta, s{-}\Delta)$ along the characteristics $s{+}x{=}$const. with approximating the right-hand side by trapezoidal rule. So the equations will become

$$G_{11}(x,s) - G_{11}(x-\Delta, s-\Delta) = a(x)G_{21}(x,s) + a(x-\Delta)G_{21}(x-\Delta,s-\Delta) + O(\Delta^3) \tag{26}$$

$$G_{21}(x+\Delta, s-\Delta) - G_{21}(x,s) = b(x)G_{11}(x,s) + b(x+\Delta)G_{11}(x+\Delta,s-\Delta) + O(\Delta^3), \tag{27}$$

where

$$a(x) = \frac{1}{4}\Delta[A(x)+B(x)]\exp\left(-\int_{x'=0}^x B(x')\mathrm{d}x'\right) \tag{28}$$

$$b(x) = \frac{1}{4}\Delta[A(x)-B(x)]\exp\left(\int_{x'=0}^x B(x')\mathrm{d}x'\right). \tag{29}$$

Let us now introduce a grid of points as shown in Fig. 4: $(x_i, s_{i+2j}) {=} (i\,\Delta, (i{+}2j)\Delta)$, $i{=}0,\ldots,N$; $j{=}0,\ldots$ Here N represents the total number of grid points. Considering $\Delta{=}1/N$, it can be stated that

$$G_{11}^{i,j} = G_{11}(i\,\Delta, (i+2j)\Delta)\ , \quad a^i = a(i\,\Delta) \tag{30}$$

$$G_{21}^{i,j} = G_{21}(i\,\Delta, (i+2j)\Delta)\ , \quad b^i = b(i\,\Delta). \tag{31}$$

Using these approximations, Eqs. (26) and (27) can be rewritten as

$$G_{11}^{i,j} = d^i\left[G_{11}^{i-1,j} + a^{i-1}G_{21}^{i-1,j} - a^i b^{i+1}G_{11}^{i+1,j-1} + a^i G_{21}^{i+1,j-1}\right] \tag{32}$$

$$G_{21}^{i,j} = d^i\left[-b^i G_{11}^{i-1,j} - a^{i-1}b^i G_{21}^{i-1,j} - b^{i+1}G_{11}^{i+1,j-1} + G_{21}^{i+1,j-1}\right], \tag{33}$$

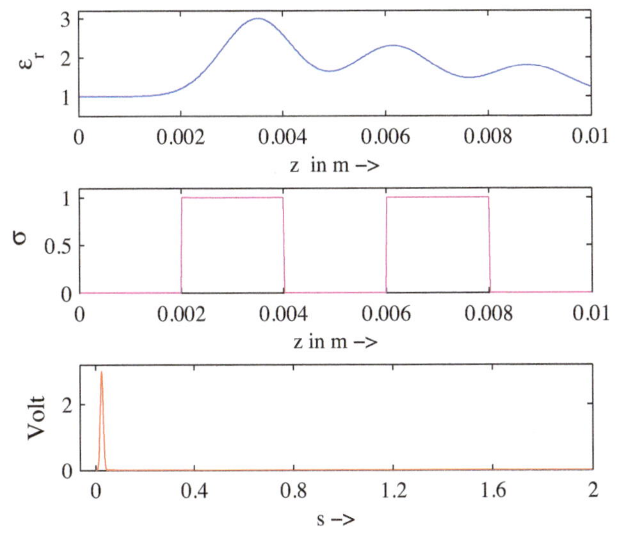

Fig. 5. Top: ε_r profile; Middle: σ profile; Bottom: incident pulse. Slab length L=1 cm, number of cells in both z and x direction is 512, grid size Δ=1.95×10^{-5} m and Δt=85 ps.

where

$$d^i = [1 + a^i b^i]^{-1} = \left[1 + \frac{1}{16}\Delta^2(A^2(i\,\Delta) - B^2(i\,\Delta))\right]^{-1} \quad (34)$$

and i=1, 2, ..., $N-1$; j=1, 2, ... The initial condition represented by s=x, i.e., j=0, can be determined by Eqs. (21) and (22). The boundary conditions are

$$G_{11}^{0,j} = 0, \quad G_{21}^{N,j} = 0; \qquad j = 0, 1, 2, ... \quad (35)$$

To incorporate the discontinuity of G_{21} on the line s=$2-x$, i.e. j=$N-i$, $G_{21}^{i,N-i}$ below the line can be computed by Eq. (33). Above this line, $(G_{21}^{i,N-i})^+$ can be determined by

$$\left(G_{21}^{i,N-i}\right)^+ = \left(G_{21}^{i,N-i}\right)^- + \frac{1}{4}[A(1) - B(1)] \exp\left\{\int_0^1 B(x)\mathrm{d}x\right\}. \quad (36)$$

This value is used in place of $G_{21}^{i,N-i}$ determined by Eq. (33) while computing $G_{11}^{i,N-i+1}$ and $G_{21}^{i,N-i+1}$.

3.4 Software implementation of the forward solver

This scheme has already been tested to simulate sinusoidal, square and multiple Gaussian shaped ε_r profiles. Considered electromagnetic pulses are: Dirac pulse, Gaussian pulse, five pulses of sinusoidal wave, raised cosine and step function. The results have been tested in both lossless and lossy condition. A multiple Gaussian shaped slab with incident Gaussian pulse has been simulated here (see Figs. 5 and 6).

4 Inverse solver: layer stripping

The necessary data to start inversion are the length of the slab L, incident pulse (as reference), relative permittivity ε_1 of the first medium (surrounding medium) and measured reflection data. The desired goal is to determine the unknown relative permittivity ε_r profile of the slab. In the inversion scheme, the slab is assumed to be lossless, so the conductivity σ becomes zero. From the wave splitting phenomenon, it can be shown that

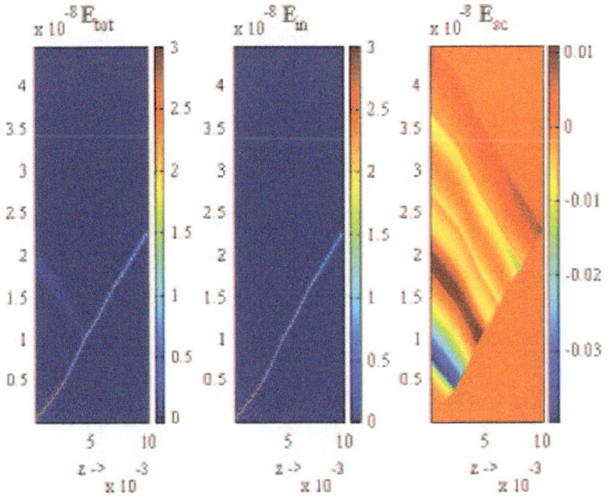

Fig. 6. Left: Total field E_{total}; Middle: Incident field E^+; Right: Reflected field E^-.

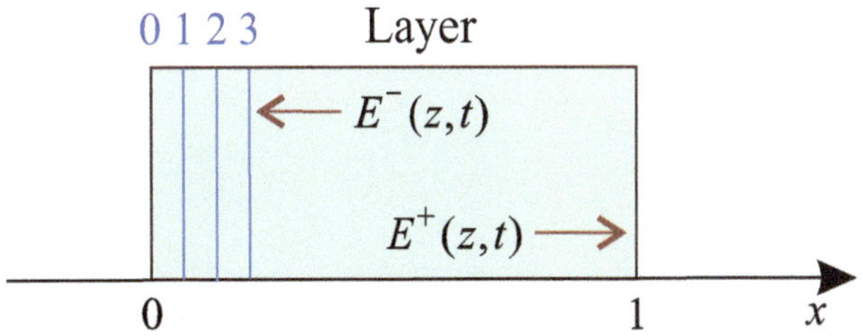

Fig. 7. Wave splitting and layer approach.

$$E(z,t) = E^+(z,t) + E^-(z,t)\,. \quad (37)$$

The backward travelling wave $E^-(z,t)$ can be associated with reflection kernel $R^+(z,t)$ in the following way

$$E^-(z,t) = E^+(z,t) * R^+(z,t)\,. \quad (38)$$

The reflection data obtained from measurement can be mentioned as $E^-(0,t)$ at the first layer (see Fig. 7). To find the reflection kernel $R^+(0,t)$ at this layer, we have to perform deconvolution. As a result

$$R^+(0,t) = \text{decon}\{E^-(0,t), f(t)\} \text{ as } f(t) = E^+(0,t)\,. \quad (39)$$

4.1 Input data processing

As deconvolution is much easier to perform in frequency domain, the time domain parameters $f(t)$ and $E^-(0,t)$ will be transformed into frequency domain parameters $F(\omega)$ and $E^-(\omega)$ at first by using Fast Fourier Transform (FFT). Then in frequency domain it can be written as

$$E^-(\omega) = R^+(\omega)\,F(\omega) \;\Rightarrow\; R^+(\omega) = E^-(\omega)/F(\omega)\,. \quad (40)$$

The obtained data should be filtered in order to avoid zeros which might occur at the deconvolution step. A Hanning window is used in this scheme as a filter. At the end, $R^+(\omega)$ is converted into $R^+(0,t)$ by Inverse Fast Fourier Transform (IFFT). Then the reflection data from the first layer will be used to initiate the reconstruction process.

4.2 Reflection kernel

Equation (38) can be rewritten as

$$E^-(z,t) = \int_{-\infty}^{t} R^+(z,t-t')\,E^+(z,t')\,\mathrm{d}t'\,. \quad (41)$$

Solving this equation for $R^+(z,t)$,

$$\frac{\partial}{\partial z}R^+(z,t) - \frac{2}{c(z)}\frac{\partial}{\partial t}R^+(z,t) = \frac{c'(z)}{2c(z)}\int_0^t R^+(z,t-t')\,R^+(z,t')\mathrm{d}t' \quad (42)$$

$$R^+(z,0) = \frac{1}{4}c'(z) \quad (43)$$

$$R^+(L,t) = 0\,. \quad (44)$$

Before entering the reconstruction step, normalization will be performed according to the principle explained in Sect. 2.1. So the reflection kernel $R^+(z,t)$ will be transformed into $R^+(x,s)$ according to the relationship $R^+(x,s){=}l R^+(z,t)$. Equations (42)-(44) will become

$$\frac{\partial}{\partial x}R^+(x,s) - 2\frac{\partial}{\partial s}R^+(x,s) = -\frac{A(x)}{2}\int_0^s R^+(x,s-s')R^+(x,s')\mathrm{d}s' \quad (45)$$

$$R^+(x,0) = -\frac{1}{4}A(x) \quad (46)$$

$$R^+(1,s) = 0\,. \quad (47)$$

4.3 Reconstruction basics

The goal of this part is to determine the mapping parameter $z(x)$ and the permittivity profile $\varepsilon[z(x)]$. From Eq. (7)

$$A(x) = -\frac{\mathrm{d}}{\mathrm{d}x}\ln c[z(x)] \quad (48)$$

$$\Rightarrow \ln c[z(x)] = -\int_0^x A(x')\mathrm{d}x' + \text{const.} \quad (49)$$

$$\Rightarrow c[z(x)] = k\,\exp\left\{-\int_0^x A(x')\mathrm{d}x'\right\}. \quad (50)$$

The constant k represents $c(0)$, the phase velocity of the first layer, i.e at $x=0$. Now from normalization principle,

$$x(z) = \int_0^z \frac{1}{l\,c(z')}\mathrm{d}z' \quad \Rightarrow \frac{\mathrm{d}}{\mathrm{d}z}x(z) = -\frac{1}{l\,c(z)} \quad (51)$$

$$\Rightarrow c(z(x)) = -\frac{1}{l}\frac{\mathrm{d}}{\mathrm{d}x}z(x)\,. \quad (52)$$

From Eqs. (50) and (52) it can be summed up that

$$z(x) = c(0)l\int_0^x\left[\exp\left\{-\int_0^{x'} A(x'')\mathrm{d}x''\right\}\mathrm{d}x'\right]. \quad (53)$$

Replacing $c(z)$ by $1/\sqrt{\mu\varepsilon(z)}$, Eq. (50) can be rewritten as

$$\frac{1}{\mu\varepsilon(z)} = \frac{1}{\mu\varepsilon(0)}\exp\left\{-2\int_0^x A(x')\mathrm{d}x'\right\} \quad (54)$$

$$\Rightarrow \varepsilon[z(x)] = \varepsilon_1\exp\left\{2\int_0^x A(x')\mathrm{d}x'\right\};\quad 0<x<1\,, \quad (55)$$

where $\varepsilon_1{=}\varepsilon(0)$ represents the permittivity of the first layer or the first medium. Eqs. (53) and (55) are the basic equations for reconstruction.

4.4 Numerical implementation of the inverse solver

The embedding equation (Eq. 45) can be rewritten as

$$\frac{\partial}{\partial x}R^+(x,s-2x) = -\frac{1}{2}A(x)\int_0^{s-2x} R^+(x,s-2x-s')\,R^+(x,s')\,\mathrm{d}s'\,. \quad (56)$$

Let us fix a constant $\Delta(=\Delta x{=}\Delta s)$. Integrating this equation from $x-\Delta$ to x and keeping time at $s{+}2x$,

$$R^+(x,s) - R^+(x-\Delta,s+2\Delta) = -\frac{1}{2}\int_{x-\Delta}^{x} A(x')(R^+ * R^+)\left(x',s+2(x-x')\right)\mathrm{d}x'\,, \quad (57)$$

where

$$(R^+ * R^+)(x,s) = \int_0^s R^+(x,s-s')\,R^+(x,s')\mathrm{d}s'. \quad (58)$$

Let us introduce a uniform grid of points (x_i, s_j):

$$x_i = i\Delta\,,\quad i=0,1,...,N \quad (59)$$

$$s_j = 2j\Delta\,,\quad j=0,1,...,N-i\,. \quad (60)$$

Considering $\Delta{=}1/N$, where N is the number of grid points,

$$R_{i,j} = R(x_i,s_j)\;;\quad A_i = A(x_i)\,. \quad (61)$$

Using this grid point, Eq. (56) will be transformed into

$$R^+_{i,j} = \left[R^+_{i-1,j+1} - \frac{\Delta^2}{2}\left\{A_i\sum_{k=1}^{j-1}R^+_{i,j-k}R^+_{i,k} + A_{i-1}\sum_{k=1}^{j+1}R^+_{i-1,j+1-k}R^+_{i-1,k}\right\}\right]\left(1-\frac{\Delta^2}{8}A_i^2\right)^{-1}. \quad (62)$$

Again from Eq. (46),

$$A_i = -4R^+_{i-1,1}\left\{1+\frac{\Delta^2}{8}A_{i-1}^2\right\}. \quad (63)$$

The error made in Eqs. (62) and (63) is of order $O(\Delta^3)$. The initialization of the algorithm is made by assigning,

$$R^+_{0,j} = R^+(0,2j\Delta) = lR^+(2jl\Delta);\quad j=0,1,...,N\,. \quad (64)$$

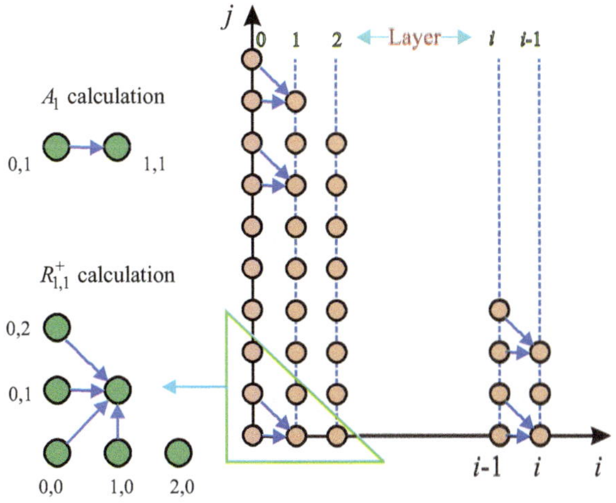

Fig. 8. Numerical algorithm of inversion scheme.

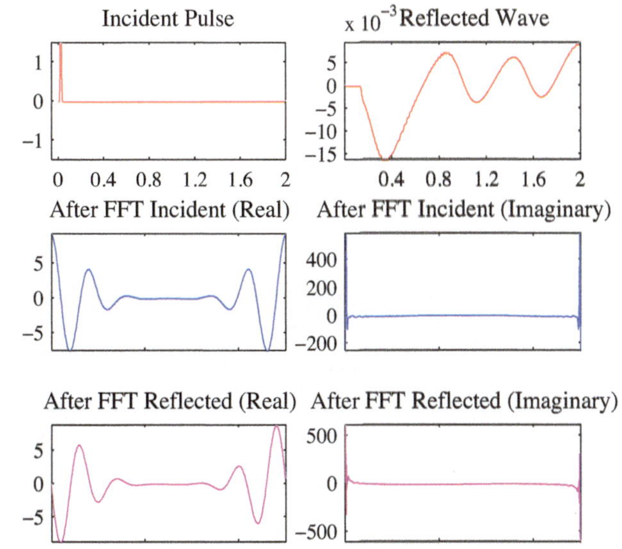

Fig. 9. Incident wave, reflected wave and FFT results. L=1 cm, N_z=N_x = 512, Δ=1.95×10^{-5} m, Δt=85 ps.

The reconstruction is done in two steps as shown in in Fig. 8. At first, $A(x_i)$ is calculated from $(i-1)$-th grid. Secondly $R^+_{i,j}$ is calculated from current time step data of $(i-1)$-th grid, old time step data of i-th grid and next time step data of $i-1$-th grid.

4.5 Software implementation of the inverse solver

The inversion scheme has already been tested to reconstruct sinusoidal shaped, square shaped and multiple Gaussian shaped relative permittivity profiles using the reflection data from Dirac pulse, Gaussian pulse, five pulses of sinusoidal wave, raised cosine and step function in both noisy and noise free environment. A multiple Gaussian shaped ε_r profile has been reconstructed here as an example simulation (see Figs. 9 and 10).

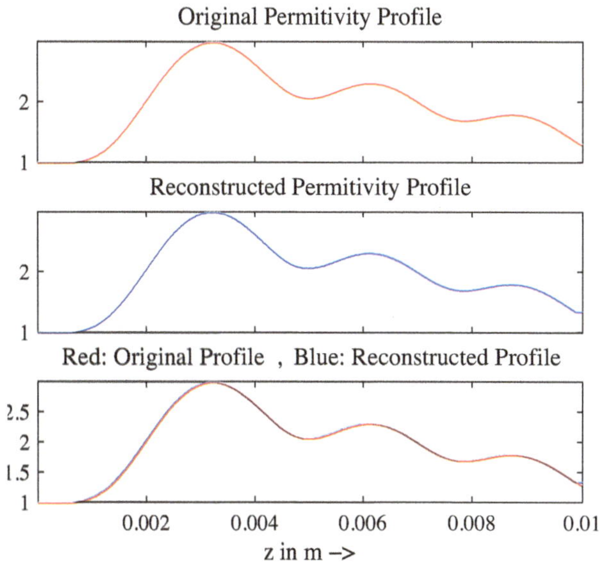

Fig. 10. Top: Original profile; Middle: Reconstructed profile; Bottom: Superposition of original and reconstructed profile.

5 Conclusions

The inverse solver presented in this paper shows almost accurate reconstruction. Better output can be generated by using higher resolutions. The tested resolutions are 128, 256, 512, 1024 and 2048. Besides all the combinations of relative permittivity and incident pulses have already been checked with synthetic and measured data. From the recent observations it has been proved that the inverse solver shows satisfactory result under noisy environment, too. Further modifications of the inversion algorithm are still going on. The next step is to reconstruct the permittivity and conductivity profiles from reflection and transmission data in lossy case.

References

Connor, K. M. and Dowding, C. H.: GeoMeasurements by Pulsing TDR Cables and Probes, CRC Press, Boca Raton, USA, 1999.

Corones, J. P., Davison, M. E. and Krueger, R. J.: Wave splittings, invariant embedding and inverse scattering, Inverse Optics, Proc. SPIE, 413, 102–106, 1983.

Kristensson, G. and Krueger, R. J.: Direct and inverse scattering in the time domain for a dissipative wave equation. Part 1 and Part 2, Journal of Mathematical Physics, 27, 1667–1693, 1986.

Kristensson, G. and Krueger, R. J.: Direct and inverse scattering in the time domain for a dissipative wave equation. Part 3: Scattering operators in presence of phase velocity mismatch, Journal of Mathematical Physics, 28, 360–370, 1987.

Krueger, R. J. and Ochs Jr., R. L.: A Green's function approach to the determination of internal fields, Applied Mathematical Sciences, 11, 525–543, 1989.

Marklein, R.: The Finite Integration Technique as a General Tool to Compute Acoustic, Electromagnetic, Elastodynamic, and Coupled Wave Fields, in: Review of Radio Science: 1999–2002

URSI, edited by: Stone, W. R., IEEE Press, Piscataway, 201–244, 2002.

Schlaeger, S.: Inveriosn of TDR Measurements to Reconstruct Spatially Distributed Geophysical Ground Parameter, Ph.D. Thesis (in German), Karlsruhe, Germany, 2002.

Taflove, A. and Hagness, S. C.: Computational Electrodynamics: The Finite-Difference Time-Domain Method, 2nd ed., Artech House, Boston, 2000.

4

Employing beam-forming for estimating the direction of arrival in a multi-path propagation environment

M. Bresch[1,2], J. Shi[1], and R. Kokozinski[1]

[1]Fraunhofer-Institut für Mikroelektronische Schaltungen & Systeme, Finkenstraße 61, 47057 Duisburg, Germany
[2]now at: Infineon Technologies Austria AG, Siemensstrasse 2, A-9500 Villach, Austria

Abstract. Due to recent researches on traffic accidents with vulnerable road users (VRUs), several measures revealed a great opportunity of reduction. However, all measures applied so far failed to reduce the number of traffic accidents if there is no line-of-sight. Therefore, a transponder signal is utilized to make the VRU visible. The motor vehicle carries a mobile receiver for VRU detection and location. The receiver employs digital beam-forming for estimating the direction of arrival (DOA) with an antenna array for RF ISM band. A sequence of DOA estimations is used for location and motion estimation purposes.

1 Introduction

In the recent couple of years, the number of traffic casualties has been vastly reduced by several methods. However, none of all methods applied so far has an impact on the traffic accidents if there is no line-of-sight between a driver of a motor vehicle and a vulnerable road user. Therefore, a transponder is utilized to make the road user visible. Thus, traffic accidents may be avoided or at least mitigated in case of no (or poor) visibility of a VRU, i.e. at night, in foggy environment, rain, mist, or if the VRU is occluded by parking cars, trees, houses, or if the driver just does not look into the direction of the VRU. The VRU carries a transponder, which transmits a unique signature, while a mobile receiver is installed on the motor vehicle. The receiver detects and locates the VRU by receiving his transponder signature in order to warn the driver. In Sect. 2 a system for transponder-based DOA estimation is introduced for which the carrier frequency is determined in Sect. 3. Section 4 deals with the range requirements while Sect. 5 determines the region of interest. Section 6 introduces the warning system, Sect. 7 presents the location and motion estimation procedure, Sect. 8 deals with multi-path propagation, Sect. 9 gives a closer look on the beam-forming system, Sect. 10 introduces the signature and Sect. 11 presents location precision results.

2 Estimating the direction of arrival

The mobile receiver locates the VRU of interest by estimating the direction of arrival (DOA) of the received signal. VRUs – mostly pedestrians and bicyclists – can be detected and located even if they are not visible, e.g. when they are occluded by parking cars, or by trees or houses. For DOA estimation, antenna arrays are employed. Such an antenna array achieves a much higher gain than only one single antenna. This leads to a very high angular resolution and precision of DOA estimation. The linear antenna array can be mounted on the car front and consists of vertically oriented dipoles. Mounting on the front can be combined with shading the rear since the front region is much more important. This avoids an angular ambiguity in the direction of arrival. The typical distance between adjacent antenna elements is half the wavelength. However, according to simulations, antenna element insertion in between to a distance of one eighth of the wavelength while maintaining the overall array size and increasing the number of elements by a factor of four significantly enlarges the gain and the angular precision. Since the distance between adjacent antenna elements is important for the angular precision, mechanical oscillations of the antenna array would be disturbing. However, typical amplitudes of mechanical oscillations of the car body are small enough so that these do not interfere the quality of DOA estimation. A circular antenna array would lead to a constant angular resolution and precision at any angle, but first of all, this cannot be suitably mounted on the car and, moreover, achieves less gain and angular resolution. A 2-D positioning is sufficient, since the VRU can be assumed to move on the ground. Therefore, azimuthal determination suffices and elevation determination is dispensable. For positioning purposes, a TOA (time-of-arrival) estimation as used e.g. in the global positioning system is too imprecise by far, since it would require analog-to-digital converters with a sampling rate of more than 100 MHz to achieve three meters precision. The determination of the RSS (received signal strength) is not feasible as well, since the VRU shall be located behind attenuating obstacles while recent developments of positioning systems based on RSS estimation (which feature high precision) can only be employed for VRU location purpose, if there is no occluding obstacle. Utilizing a radar detector would inconveniently require the VRU to wear some metallic burden. Moreover, a VRU cannot be located behind a metallic obstacle.

3 Carrier frequency determination

For the determination of the carrier frequency, only ISM bands are eligible. The carrier frequency must not be too high, since then the attenuation by occluding objects is too high and an occluded VRU is not visible. On the other hand, the carrier frequency shall not to be too low either, since in this case, the antenna dimensions are too large to fit on the car front. Therefore, 868/915 MHz or 2.45 GHz are appropriate carrier frequencies for the data transmission. Measurements employing a carrier frequency of 2.45 GHz have been carried out. The results show that the transponder signal penetrates obstacles at reasonable intensity. For example, the attenuation by parking cars or trees is less than 15 dB, which is sufficiently low to enable detecting occluded VRUs, since a sufficiently strong signal descending from a line-of-sight propagation, which is necessary for the DOA estimation in a multi-path environment, can be detected by the receiver. At a carrier frequency of 868/915 MHz, the attenuation is even lower.

4 Range requirements

Since the typical speed limit in European cities is 50 kph (about 30 mph), the maximum vehicle velocity is assumed to be roughly 70 kph since the speed limits are often violated. As the overall stopping distance at this vehicle velocity is about 70 m, the transponder data transmission range has to be about 70 m as well. Hence, only an active transponder is eligible. A major problem is the implementation of a sufficiently high market penetration of active transponders for the reduction of traffic accidents, since these are heavier than passive transponders. Considerations concerning these issues are still in their early stage. One of the suggestions is utilizing radio controlled clocks. Another proposal is the use of mobile phones as a host for the active transponder. Imaginations also include certain labelled identity cards which, however, will require special batteries, the technology of which is just emerging. Bicyclists' transponders could be implemented in the reflectors.

5 Region of interest

The VRU detection systems employing a mobile receiver focuses on the front (the antenna array is mounted on the car front and is blind in the rear). The main region of interest is from 15 m (this is the reaction distance at 70 kph) to the stopping distance mentioned above. Warning the driver is senseless if the VRU is inside the reaction distance. As pedestrians and bicyclists are most likely to appear about 8 m or less aside the car, the angle of interest is within 30° on both sides of the direction of vehicle velocity. Inside the region of interest, the scene to the right of the car is of more importance since pedestrians crossing from the right are mostly invisible whereas pedestrians crossing the street from the left can often be observed accurately by the driver while they are passing the lane for the oncoming traffic. In countries where the traffic must keep to the left it is vice versa.

6 Three-stage driver warning

As in an urban environment, the presence of vulnerable road users has to be assumed all the time, a system that reacts on each detected VRU would lead to a constant warning weakening the driver's attention. Therefore, a three-stage warning system shall avoid the constant warning. In the first stage, only a notice "there is a VRU about" giving the VRU location shall be placed on the screen. In the second stage, the driver receives a warning, that an accident is likely to happen. In the third stage, the driver is strongly encouraged to react, since otherwise an accident cannot be avoided. The third stage could be combined with a pre-crash system.

7 Positioning and motion detection

As the vehicle moves on, a sequence of multiple DOA estimations is utilized to spot the VRU location. This kind of sequential triangulation can be used to locate the VRU. The VRU is assumed to move much slower than the vehicle. A precise location can be used for motion estimation, which can be help to reduce traffic accidents, since a VRU crossing the street causes more danger of accident than a VRU moving along the street.

8 Multi-path propagation

A very important problem concerning the location by DOA estimation is the multi-path propagation. A discrimination between a line-of-sight propagation and reflection is inevitable in typical urban environment, since a lot of reflections are highly probable. For this discrimination, the sequence of angles-of-arrival of all received signals is analysed. For example, if a certain signal is received from the same direction all the time, it is very likely to be a reflection from the vehicle carrying the receiver or maybe from the car in front, so it is not a signal of interest. Such reflection can be suppressed by means of signal processing. A directly propagated wave moves in a characteristic way as the car moves on. This information can be exploited for echo discrimination. The typical motion of the line-of-sight propagation is stored in the receiver's data base and is compared with all received signals. A time sequence of DOA estimations is used to generate one angle-in-time function for each received signal. The generated angle-in-time functions are analysed by means of supervised pattern recognition. During a receiver's training phase, a car moves passing a transmitter and estimates

Fig. 1. Standard simulation scenario.

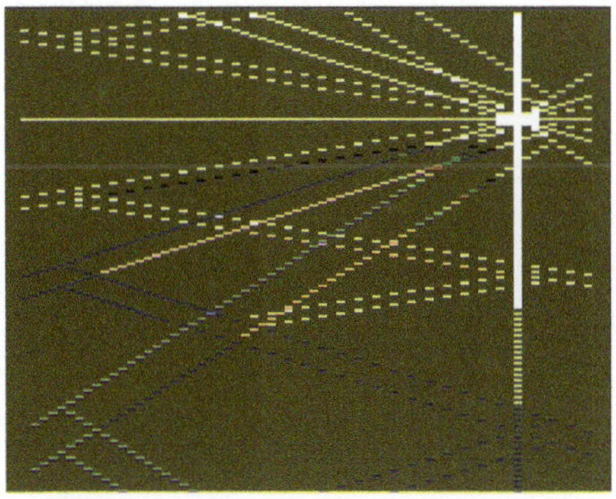

Fig. 2. Multi-path propagation.

the DOA sequence. A feature vector is extracted from the generated angle-in-time function and stored in the receiver's data base. From each angle-in-time function of the received signals, feature vectors are extracted in the same manner as the stored vector and compared with the vector stored in the data base. As classifier, a maximum-likelihood classification can be used. Another approach is a (symmetric) phase-only matched filtering, which performs a correlation of the phases of two signals to be compared. For the feature extraction, a multi-channel filtering can be used, where the power of each channel output results in one feature. In case of feature extraction using multi-channel filtering, Gabor filters are well suited, since these exhibit optimal time-frequency resolution product. In this case, (symmetric) phase-only matched filtering is likely to be the best classification algorithm. This type of phase-correlating filter is very noise robust, which is important, if the feature is not optimally extracted. Furthermore, phase filters exploit the high information content of phases. The feature vector matching best is assumed to be representing the line-of-sight propagation. This sequential DOA analysis fails in case of a stand-still receiver. However, in this case, no VRU detection is necessary, since only a moving vehicle causes the danger of an accident. Furthermore, this analysis fails in case of a VRU which is directly in front of the car, since the DOA is constantly zero degree and cannot be discriminated against a reflection on the car. But a VRU directly in front of the car is very likely to be seen by the driver. During night and in foggy or raining weather conditions, the driver at least receives the notice that a pedestrian is present in the front of the car (although the distance cannot be determined).

9 Digital beam-forming

For DOA estimation using antenna arrays, digital beam-forming is employed. The received signal is simultaneously analysed by several signal processing channels in parallel, each maximizing the sensitivity and output power from a certain direction. For an angular precision of one degree in a range of thirty degrees on both sides, 61 channels are necessary. Therefore, the receiver structure is an antenna array connected to an RF front-end, which performs a modulation to an intermediate frequency using a local oscillator. The local oscillator converts the received signal down to a frequency below 10 MHz in order to enable digitisation. This procedure is feasible, since the down-conversion does not influence the phase-differences of the received signal of different antenna elements. The phase-differences of the antenna elements are exploited for DOA estimation. After digitisation, the beam-forming is performed in the digital domain. This is much more robust and less complex than determining the phase relation at RF frequencies. The digital beam-forming is just a mathematical procedure and can be carried out for example by a suitable computer program. The received signal of each antenna element is multiplied by a complex factor, that belongs to the channel and to the antenna element. Afterwards all complex products are summed up to achieve the high gain of the antenna array. This results in Eq. (1).

$$s_E(x, y, t) = \sum_{i=0}^{I} \widehat{s}(i) e^{j(k_{xi}x + k_{yi}y - \omega t)}, \quad (1)$$

where i represents the i-th channel. I is the number of channels, while k_{xi} and k_{yi} are the wave propagation constants along and across the street, respectively, and ω is the angular frequency of the incident wave. The incident waves of channel i are marked with $\widehat{s}(i)$ and the summed up signal is s_E. This procedure is carried out in each channel in parallel, and the received signal is interpreted as arriving from the direction of the referring channel, which produces the largest output. Simulations show, that arrival of several signals from various directions does not result in a broad smeared peak, but in several sharp peaks, so that a multi-path discrimination as echo cancellation to determine the line-of-sight arrival is feasible. Figure 1 shows a standard simulation scenario with a road and reflecting sidings, a VRU as transponder signal source and an attenuating car. Figure 2 shows a simulation of multi-path propagation.

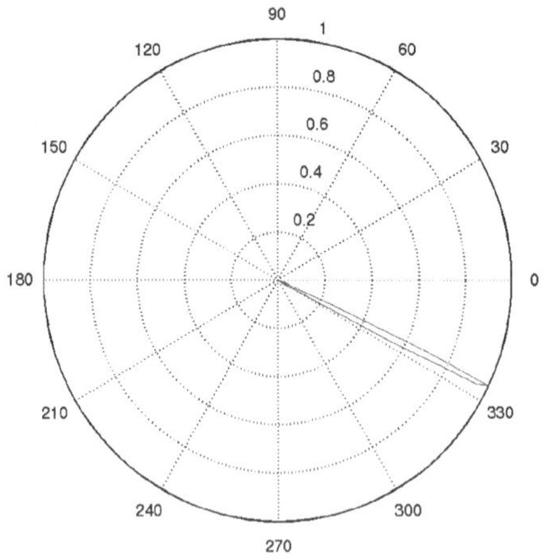

Fig. 3. Directional pattern using 33 antennas at 2.45 GHz (linear).

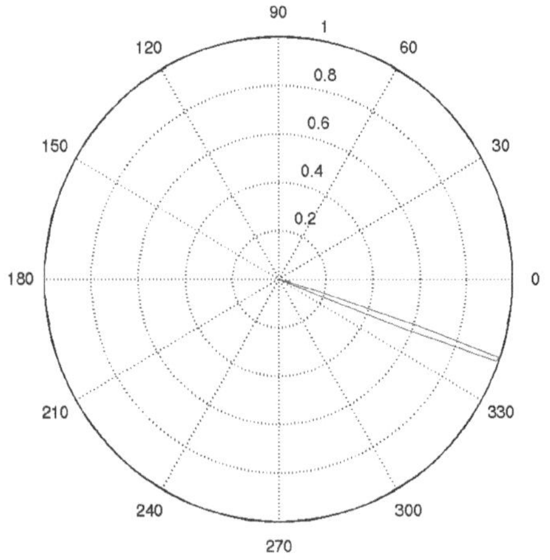

Fig. 4. Directional pattern using 33 antennas at 2.45 GHz (linear).

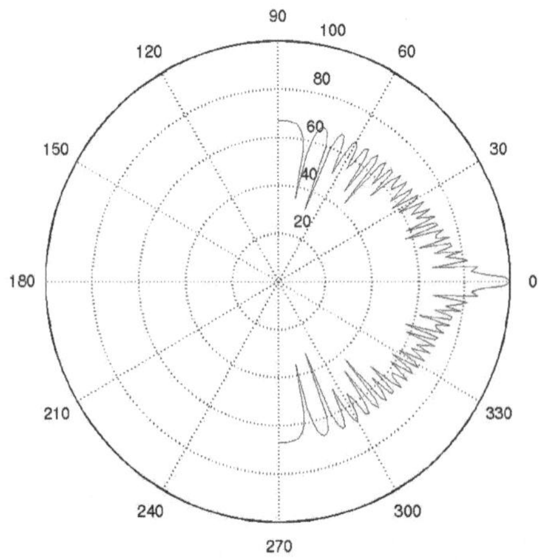

Fig. 5. Directional pattern using 45 antennas at 868/915 MHz (in dB).

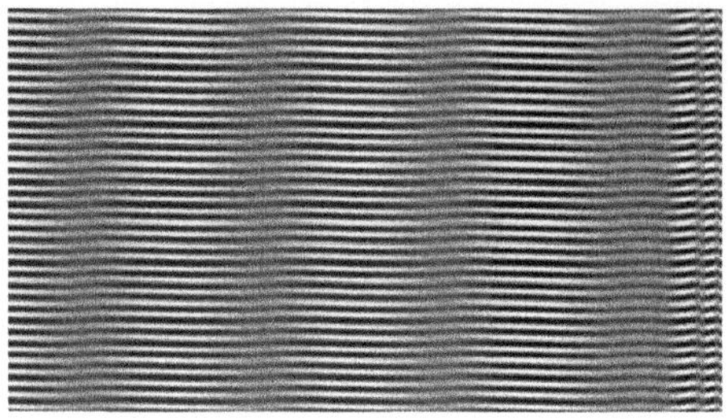

Fig. 6. Simulating three incident waves.

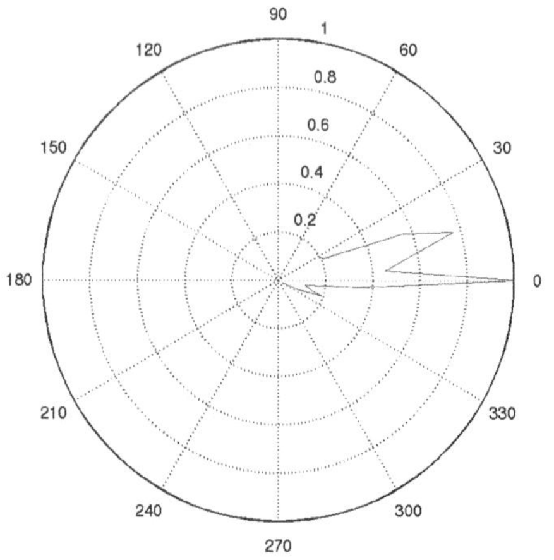

Fig. 7. Simulating received signal strength depending on DOA. Three distinct peaks can be observed.

10 Unique signature

In case of the presence of several vulnerable road-users which is always probable in urban environments, a discrimination between the signals of different transponders is necessary. A world-wide unique signature of at least 32 bit serves for this purpose. This signature can moreover be used to distinguish between a VRU and an interferer (e.g. descending from a bluetooth-signal, another RFID-system, or from a microwave stove). This avoids the near-far problem. The signature can provide important information about the VRU for accident avoidance, e.g. a blind person, a child, a disabled person, bicyclist, bicycle properties and so on.

11 Location precision

If the distance between adjacent antenna elements is half the wavelength, this is about 17 cm at a carrier frequency of 868/915 MHz and about 6 cm at a carrier frequency of 2.45 GHz. Therefore 11 elements at 868/915 MHz and 29 elements at 2.45 GHz fit to a 1.7 m wide car front. As to be seen from Figs. 3 and 4, simulation results show an angular resolution of 3 deg employing 33 antennas at 868/915 MHz. The results are shown for two parallel channels maximizing the received signal strength from two different angles. The resolution at 2.45 GHz is higher. Simulation results based on further antenna element insertion to 45 antennas equally spaced at one eighth of the wavelength (about 4.2 cm) result in an angular resolution of 1 deg. Figure 5 shows the simulation results employing 45 antennas at spacing of one eighth wavelength. At 2.45 GHz, the resolution is less (i.e. better) than one degree. Employing sequential triangulation for VRU location, this leads to a precision in the cross direction of 1.2 m in the worst case of 70 m distance and best case of 0 deg DOA. The precision in the direction of vehicle velocity, the precision depends on the time interval between two successive DOA estimations. In case of a car velocity of 70 kph and a time interval of 200 ms between two consecutive measurements, the longitudinal precision is about 2.5 m, if the VRU is 70 m in front of and 15 m aside the receiver. The precision increases with decreasing distance and decreases with decreasing angle with respect to the vehicle velocity. This is a much higher precision than can be obtained from a TOA based positioning system. Figures 6 and 7 give evidence, that three incident waves lead to three sharp peaks and not to a smeared one.

12 Conclusion

We have presented a novel location system based on DOA estimation employing digital beam-forming. The location algorithm is suitable for detecting occluded objects at a precision of about 2m and utilizes transponder signals in the 868/915 MHz and the 2.45 GHz ISM band.

5

Radiation characteristics of a coaxial waveguide with eccentric inner conductor for application in hyperthermia and microwave reflex therapy

R. Herschmann[1,2] **and O. Büchel**[3]

[1]Leibniz Universität Hannover, Institute of Radiofrequency and Microwave Engineering, Appelstraße 9a, 30167 Hannover, Germany
[2]Smart Devices GmbH & Co. KG, Schönebecker Allee 2, 30823 Garbsen, Germany
[3]Fuba Automotive GmbH & Co. KG, Tec Center, 31162 Bad Salzdetfurth, Germany

Abstract. This paper examines the radiation characteristics of a contact emitter conceived for application in hyperthermia and microwave reflex therapy. It is important to analyse the distribution of power density in the near field area, as the radiator's therapeutic sphere of activity is localized here. The contact emitter is a coaxial radiator with an eccentric course of the inner conductor. According to Huygens principle, a theoretical view of the near field radiation characteristics is made by determining the equivalent current densities in the emitter aperture. It is shown that by an eccentric shift of the inner conductor, an almost isotropic near field radiation pattern and power density can be achieved. For this, the electromagnetic field in the emitter aperture is determined by using a Bipolar coordinate system. This calculation considers only the fundamental TEM mode of the contact emitter. Besides the theoretical results near and far fields are simulated using the programme system Ansoft HFSS.

1 Introduction

Hyperthermia is a technique used in the medical treatments of cancer and other medical therapy. Tumors are heated to therapeutic temperatures (invasive $T > 42.5°C$, non invasive $T = 40°C-42°C$ without overheating the surrounding normal tissue. Non invasive hyperthermia is used often in combination with chemotherapy and radiation therapy for higher complete response compared to conventional therapies alone. Microwave reflex therapy is a relatively new working method in medicine. High frequency, electromagnetic energy is concentrated on a closely restricted, biological active point by use of a contact emitter. The emitter is put directly on the skin. It provides a type of non invasive acupuncture. The contact emitter introduced in this paper was developed for use at the skin surface area.

For application within the medical area the emitter aperture must be as small as possible in order to ensure a defined irradiation of the cancerous tissue or biological active point. Furthermore, the electromagnetic field must be isotropic in direct proximity to the effect. A broad emitter bandwidth is necessary in order to adapt the frequency of radiotherapy individually in the examined range from $f = (8-20)$ GHz. The emitter must rest upon the skin well and the impedance of the contact emitter has to be adapted to the human tissue over a broad frequency range. The described requirement profile is fulfilled, to a large extent, by a coaxial radiator. These radiators have long found various applications within the area of non destructive material testing (see Mosig et al., 1981; Bakhtiari et al., 1994), as well as the determination of the dielectric properties of biological substances (see Stuchly and Stuchly, 1980; Stuchly et al., 1982). However, the directional characteristic of a coaxial emitter has a dip in axial direction. Therefore, the desired concentration of radiation intensity cannot be achieved in the near field. A simple possibility for a remedy offers an eccentric shift of the inner conductor in the emitter aperture. The influence of the inner conductor's eccentricity on the radiation pattern can be illustrated with the following thought experiment. One can understand the inner conductor as a dipole from which a radiation dip in axial direction follows. The consequence of an eccentric shift of the inner conductor is a misalignment of the radiation minimum from the symmetry axis, whereby the radiation pattern of the complete arrangement is influenced.

In this publication, the theory for analysing the near field of the radiator described above is based upon the determination of equivalent current densities in the emitter aperture, according to Huygen's principle. The theoretical investigations serve two purposes: to acknowledge the thought experiment; and to support the later presented results simulated by the programme system Ansoft HFSS within the near field

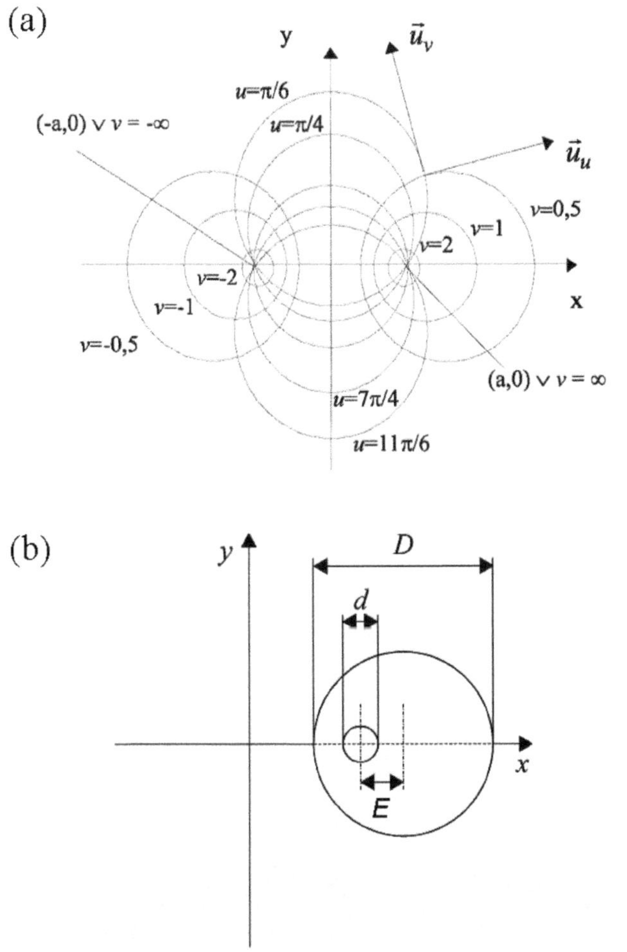

Fig. 1. Interrelation between the geometry of the Bipolar coordinate system (**a**) and the geometry of the contact emitter (**b**).

area. To determinate the equivalent sources only the fundamental TEM mode is considered for propagation inside the line. In the mathematical analysis, higher order modes are not considered. Besides the radiator (i.e. geometrical dimensions, dielectric characteristics), the mathematical model describing the radiation behaviour contains the infinitely expanded, semi-infinite space in front of the aperture filled with an isotropic homogenous material. Additionally, the far field radiation characteristics are represented as a result of the analytical approach and simulation.

2 Theory

With the analysis of conventional antennas, one is usually interested in the far field characteristics. However, the therapeutic sphere of activity of the contact emitter for medical applications is situated in the near field area. The therapeutic area for local hyperthermia and the biological active points stimulated by microwave reflex therapy are located at the skin surface, which can be 2–12 mm thick (see Kalantaewskaja, 1972). Therefore, an analysis of near field characteristics is necessary.

The human body consists of a layered tissue with different electrical characteristics. During the irradiation of electromagnetic energy, reflections occur at the transitions between the individual layers. In the following, the layering of tissue is neglected and a homogenous skin tissue with an infinite expansion in the illuminated semi-infinite space is assumed.

In order to predict the radiation behaviour of the radiator exactly, the field equations of the sub ranges feeder, transition zone and infinite space have to be solved by using boundary conditions. In some publications, approaches and methods for the solution of similar problems were already represented. Sumbar et al. presented solutions for field equations relying on the finite element method (see Sumbar, 1991). However, due to the approximated boundary conditions at the sub zone transitions, the solution can only be approximated. In Mosig et al. (1981) and Nevels and Wheeler (1989), the field equations for special coaxial radiators were solved by applying the method of moments. In Maloney et al. (1990), the FD-TD method (finite-difference time-domain method) was used as a numerical solution for radiation problems in the time domain. The procedures mentioned were used exclusively for far field calculation and contain a multitude of neglects. In the theoretical part of this work, the radiation behaviour is determined with the help of the equivalent current densities in the emitter aperture, according to Huygen's principle. The calculation is executed under the following prerequisites:

1. Inside the line, the TEM fundamental mode propagates.
2. The analysis does not consider higher order modes stimulated in the emitter aperture.
3. Energy is radiated only into the semi-infinite space in front of the aperture.
4. The radiator is lossless.

A Bipolar coordinate system is used in order to determine the field distribution in the aperture of the radiator. In this coordinate system, the conductor surface coincides with a coordinate surface in each case.

In Fig. 1, the Bipolar coordinate system (u, v, z) is compared to the geometry of the emitter aperture to explain the interrelations. For a better orientation, the Cartesian coordinate system is projected onto the Bipolar coordinate system in Fig. 1. The geometry of the emitter aperture is characterised by the inside diameter D of the outer conductor, the outside diameter d of the inner conductor and by the eccentricity $E=e\cdot D/2$ of the inner conductor in the emitter aperture. For instance, if the eccentricity factor $e=0.5$ the centre of the inner conductor is positioned between the centre and the outside margin of the outer conductor.

2.1 Calculation of the field distribution in the emitter aperture

Because of the assumptions given above, only transversal field components exist in the emitter aperture. These components can be written in Bipolar coordinates as follows:

$$\begin{aligned} \underline{\boldsymbol{E}} &= \underline{\boldsymbol{E}}_v = \underline{\boldsymbol{e}}_v\,(u, v) \cdot e^{-\underline{\gamma}_z \cdot z} \\ \underline{\boldsymbol{H}} &= \underline{\boldsymbol{H}}_u = \underline{\boldsymbol{h}}_u\,(u, v) \cdot e^{-\underline{\gamma}_z \cdot z} \end{aligned} \tag{1}$$

If this approach is inserted in Maxwell's second equation, $rot\underline{\boldsymbol{E}}{=}-\mu\partial\underline{\boldsymbol{H}}/\partial t$, then the following conditions result on the assumption of a time harmonic dependence:

$$\mathrm{rot_t}\underline{\boldsymbol{e}}_v\,(u, v) = 0 \text{ and } \underline{\gamma}_z \cdot \boldsymbol{u}_z \times \underline{\boldsymbol{e}}_v\,(u, v) = j\omega\mu \cdot \underline{\boldsymbol{h}}_u\,(u, v) \tag{2}$$

The nonvortical electric field can be expressed as the gradient of a scalar potential:

$$\underline{\boldsymbol{e}}_v\,(u, v) = -\nabla\Phi\,(v) \tag{3}$$

The equipotential lines in the emitter aperture are identical to the parameter lines for constant v as shown in Fig. 1. On the assumption of a charge-free dielectric inside the radiator, as well as a constant permittivity ε_r, we find Laplace's equation in Bipolar coordinates (see Murray, 1979) as follows:

$$\nabla^2\Phi\,(v) = \frac{(\cosh(v) - \cos(u))^2}{a^2} \cdot \frac{\partial^2\Phi\,(v)}{\partial v^2} = 0 \tag{4}$$

The variable a characterises the pole position of the coordinate system (see Fig. 1a). If Eq. (4) is integrated twice, one gets the scalar potential, which with the boundary conditions $\Phi\,(v{=}v_1) = 0$ and $\Phi\,(v{=}v_2) = U_0$ can be expressed as:

$$\Phi\,(v) = \frac{U_0}{v_2 - v_1} \cdot (v - v_1) \tag{5}$$

Substituting Eqs. (5) into (3), the electric field in the emitter aperture can be given by:

$$\underline{\boldsymbol{E}} = A \cdot \frac{1}{a} \cdot (\cosh(v) - \cos(u)) \cdot e^{-j \cdot k \cdot z} \cdot \boldsymbol{u}_v \tag{6}$$

From this the magnetic field can be calculated with application of Maxwell's second equation as follows:

$$\underline{\boldsymbol{H}} = A \cdot \frac{1}{a} \cdot \sqrt{\varepsilon_{emitter}/\mu_{emitter}} \cdot (\cosh(v) - \cos(u)) \cdot e^{-j \cdot k \cdot z} \cdot \boldsymbol{u}_u \tag{7}$$

In Eqs. (6) and (7), A represents an amplitude factor. In addition the propagation constant $\underline{\gamma}_z$ was replaced by:

$$\underline{\gamma}_z = j \cdot k = j \cdot \omega \cdot \sqrt{\mu_{emitter} \cdot \varepsilon_{emitter}} \tag{8}$$

In order to calculate the near field characteristics, the field distribution of the emitter aperture is used to obtain the

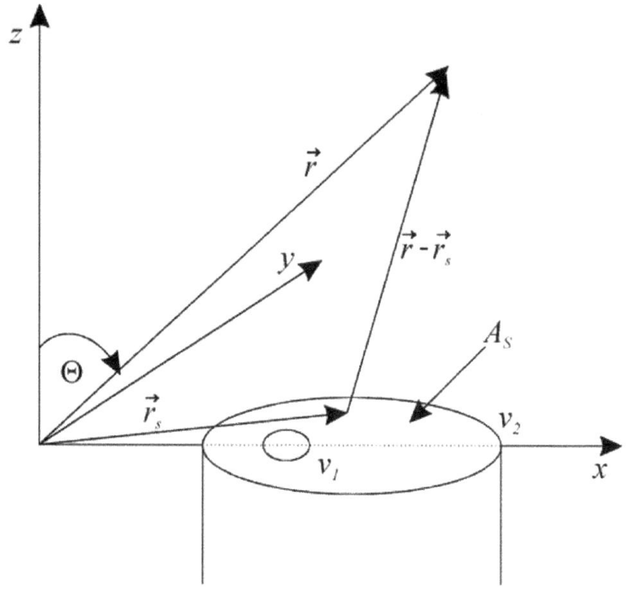

Fig. 2. Emitter arrangement in Cartesian coordinates.

equivalent electric and magnetic current densities, according to Huygen's principle:

$$\begin{aligned} \underline{\boldsymbol{J}}_{\ddot{a}q} &= \boldsymbol{u}_z \times \underline{\boldsymbol{H}}_t \\ \underline{\boldsymbol{J}}_{m\ddot{a}q} &= -\boldsymbol{u}_z \times \underline{\boldsymbol{E}}_t \end{aligned} \tag{9}$$

The transversal field strength results from the interference of an incident and reflected wave component. The reflection factor at the transition "radiator ↔ homogenous tissue" is defined as Γ. Now the transversal electromagnetic field components in the aperture of the contact emitter can be written as:

$$\begin{aligned} \underline{\boldsymbol{E}}_t &= \underline{E}_v(u, v) \cdot (1 + \Gamma) \cdot \boldsymbol{u}_v \\ \underline{\boldsymbol{H}}_t &= \underline{H}_u(u, v) \cdot (1 - \Gamma) \cdot \boldsymbol{u}_u \end{aligned} \tag{10}$$

Considering Eqs. (6) and (7), the equivalent current densities can be determined from the following equations:

$$\begin{aligned} \underline{\boldsymbol{J}}_{\ddot{a}q} &= -\frac{A \cdot (1-\Gamma)}{a \cdot \sqrt{\mu_{Strahler}/\varepsilon_{Strahler}}} \cdot (\cosh(v) - \cos(u)) \cdot \boldsymbol{u}_v \\ \underline{\boldsymbol{J}}_{m\ddot{a}q} &= \frac{A \cdot (1+\Gamma)}{a} \cdot (\cosh(v) - \cos(u)) \cdot \boldsymbol{u}_u \end{aligned} \tag{11}$$

2.2 Calculation of the radiation fields of the contact emitter

The magnetic vector potential

$$\underline{\boldsymbol{A}}(r) = \frac{1}{4 \cdot \pi} \cdot \iint_{A_s} \frac{\underline{\boldsymbol{J}}_{\ddot{a}q}(\boldsymbol{r}_s)}{|\boldsymbol{r} - \boldsymbol{r}_s|} \cdot e^{-j \cdot k \cdot |\boldsymbol{r} - \boldsymbol{r}_s|} \cdot dA_s \tag{12}$$

and the electric vector potential

$$\underline{\boldsymbol{F}}(r) = \frac{1}{4 \cdot \pi} \cdot \iint_{A_s} \frac{\underline{\boldsymbol{J}}_{m\ddot{a}q}(\boldsymbol{r}_s)}{|\boldsymbol{r} - \boldsymbol{r}_s|} \cdot e^{-j \cdot k \cdot |\boldsymbol{r} - \boldsymbol{r}_s|} \cdot dA_s \tag{13}$$

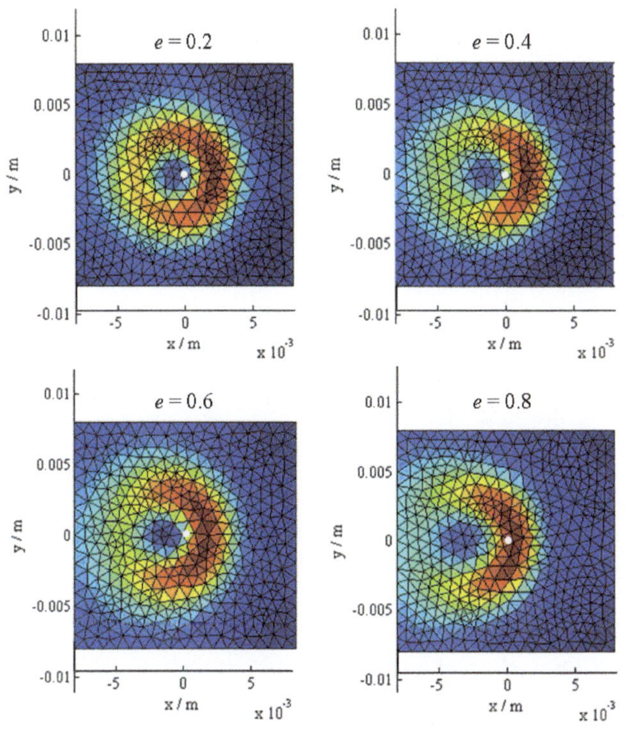

Fig. 3. Distribution of the real part of Poynting's vector in the near field as a function of the inner conductor's eccentricity factor e (theory).

can be evaluated by integration over the emitter cross section A_S. The radiation field follows from:

$$\underline{\boldsymbol{E}} = -\mathrm{rot}\underline{\boldsymbol{F}} + \frac{1}{j \cdot \omega \cdot \varepsilon_{Gewebe}} \cdot \mathrm{rot\,rot}\underline{\boldsymbol{A}} \tag{14}$$

$$\underline{\boldsymbol{H}} = \mathrm{rot}\underline{\boldsymbol{A}} + \frac{1}{j \cdot \omega \cdot \mu_{Gewebe}} \cdot \mathrm{rot\,rot}\underline{\boldsymbol{F}} \tag{15}$$

It is not appropriate to put the origin of the coordinate system into the emitter centre for a simple organization of the integrals (see Fig. 2). The components of the vector potentials are represented in Cartesian coordinates. This facilitates the curl formation regarding the field calculation which is executed later and serves a clear near field representation. The transformation of the Bipolar unit vectors into Cartesian unit vectors gives the magnetic vector potential and the electric potential in Eqs. (16) and (17) using the substitutions $S(v,u) = \frac{\sinh(v)\cdot\sin(u)}{(\cosh(v)-\cos(u))^2}$ and $C(v,u) = \frac{\cosh(v)\cdot\cos(u)-1}{(\cosh(v)-\cos(u))^2}$.

$$\underline{\boldsymbol{A}} = -\frac{A\cdot a\cdot(1-\Gamma)}{4\cdot\pi\cdot\sqrt{\mu_{Strahler}/\varepsilon_{Strahler}}} \cdot \int_0^{2\cdot\pi}\int_{v_1}^{v_2} \left\{ C(v,u)\cdot\frac{e^{-j\cdot k\cdot|\boldsymbol{r}-\boldsymbol{r}_s|}}{|\boldsymbol{r}-\boldsymbol{r}_s|}\cdot\boldsymbol{u}_x + \ldots \right.$$
$$\left. S(v,u)\cdot\frac{e^{-j\cdot k\cdot|\boldsymbol{r}-\boldsymbol{r}_s|}}{|\boldsymbol{r}-\boldsymbol{r}_s|}\cdot\boldsymbol{u}_y \right\}\cdot dv\cdot du \tag{16}$$

$$\underline{\boldsymbol{F}} = \frac{A\cdot a\cdot(1+\Gamma)}{4\cdot\pi} \cdot \int_0^{2\cdot\pi}\int_{v_1}^{v_2} \left\{ S(v,u)\cdot\frac{e^{-j\cdot k\cdot|\boldsymbol{r}-\boldsymbol{r}_s|}}{|\boldsymbol{r}-\boldsymbol{r}_s|}\cdot\boldsymbol{u}_x - \ldots \right.$$
$$\left. C(v,u)\cdot\frac{e^{-j\cdot k\cdot|\boldsymbol{r}-\boldsymbol{r}_s|}}{|\boldsymbol{r}-\boldsymbol{r}_s|}\cdot\boldsymbol{u}_y \right\}\cdot dv\cdot du \tag{17}$$

The integral Eqs. (16) and (17) are not analytically solvable. Therefore the curl of the integrals is determined beforehand. Afterwards, a pure numeric calculation of the electric and magnetic fields is executed. For calculating the curl of a vector, it is necessary to determine the partial derivatives. Analysing the electric and magnetic vector potential, these partial derivatives are limited to the following expression:

$$\underline{P}(x,y,z) = \frac{e^{-j\cdot k\cdot|\boldsymbol{r}-\boldsymbol{r}_s|}}{|\boldsymbol{r}-\boldsymbol{r}_s|} = \ldots$$
$$\frac{e^{-j\cdot k\cdot\sqrt{\left(x-\frac{a\cdot\sinh(v)}{\cosh(v)-\cos(u)}\right)^2+\left(y-\frac{a\cdot\sin(u)}{\cosh(v)-\cos(u)}\right)^2+z^2}}}{\sqrt{\left(x-\frac{a\cdot\sinh(v)}{\cosh(v)-\cos(u)}\right)^2+\left(y-\frac{a\cdot\sin(u)}{\cosh(v)-\cos(u)}\right)^2+z^2}} \tag{18}$$

According to Eq. (9), the vector of source point $\boldsymbol{r}_s$ is represented in Bipolar coordinates. The partial derivatives can be abbreviated by

$$\underline{P}_\eta = \frac{\partial\underline{P}(x,y,z)}{\partial\eta}, \quad \underline{P}_{\eta,\tau} = \frac{\partial\underline{P}(x,y,z)}{\partial\eta\,\partial\tau}$$

where $\eta, \tau \in \{x, y, z\}$.

Using the mentioned expressions, the following terms result from simple and double curl formation:

$$rot\underline{\boldsymbol{A}} = -\frac{A\cdot a\cdot(1-\Gamma)}{4\cdot\pi\cdot\sqrt{\mu_{Strahler}/\varepsilon_{Strahler}}}\cdot\ldots$$
$$\int_0^{2\cdot\pi}\int_{v_1}^{v_2}\left\{\left[-S(v,u)\cdot\underline{P}_z\right]\cdot\boldsymbol{u}_x + \left[C(v,u)\cdot\underline{P}_z\right]\cdot\boldsymbol{u}_y + \ldots\right. \tag{19}$$
$$\left.\left[S(v,u)\cdot\underline{P}_x - C(v,u)\cdot\underline{P}_y\right]\cdot\boldsymbol{u}_z\right\}\cdot dv\cdot du$$

$$\mathrm{rot}\,\underline{\boldsymbol{F}} = \frac{A\cdot a\cdot(1+\Gamma)}{4\cdot\pi}\cdot\ldots$$
$$\int_0^{2\cdot\pi}\int_{v_1}^{v_2}\left\{\left[C(v,u)\cdot\underline{P}_z\right]\cdot\boldsymbol{u}_x + \left[S(v,u)\cdot\underline{P}_z\right]\cdot\boldsymbol{u}_y - \ldots\right. \tag{20}$$
$$\left.\left[C(v,u)\cdot\underline{P}_x + S(v,u)\cdot\underline{P}_y\right]\cdot\boldsymbol{u}_z\right\}\cdot dv\cdot du$$

$$\mathrm{rot\,rot}\,\underline{\boldsymbol{A}} =$$
$$-\frac{A\cdot a\cdot(1-\Gamma)}{4\cdot\pi\cdot\sqrt{\mu_{Strahler}/\varepsilon_{Strahler}}}\cdot\ldots$$
$$\int_0^{2\cdot\pi}\int_{v_1}^{v_2}\left\{\left[S(v,u)\cdot\underline{P}_{xy} - C(v,u)\cdot\underline{P}_{yy} - C(v,u)\cdot\underline{P}_{zz}\right]\right. \tag{21}$$
$$\cdot\boldsymbol{u}_x + \ldots\left[-S(v,u)\cdot\underline{P}_{zz} - S(v,u)\cdot\underline{P}_{xx} + C(v,u)\cdot\underline{P}_{yx}\right]\cdot\boldsymbol{u}_y$$

$$\mathrm{rot\,rot}\,\underline{\boldsymbol{F}} = \frac{A\cdot a\cdot(1+\Gamma)}{4\cdot\pi}\cdot\ldots$$
$$\int_0^{2\cdot\pi}\int_{v_1}^{v_2}\left\{\left[-C(v,u)\cdot\underline{P}_{xy} - S(v,u)\cdot\underline{P}_{yy} - S(v,u)\cdot\underline{P}_{zz}\right]\cdot\boldsymbol{u}_x + \ldots\right.$$
$$\left[-C(v,u)\cdot\underline{P}_{zz} + C(v,u)\cdot\underline{P}_{xx} + S(v,u)\cdot\underline{P}_{yx}\right]\cdot\boldsymbol{u}_y + \ldots \tag{22}$$
$$\left.\left[S(v,u)\cdot\underline{P}_{zx} - C(v,u)\cdot\underline{P}_{zy}\right]\cdot\boldsymbol{u}_z\right\}\cdot dv\cdot du$$

Substituting Eqs. (20) and (21) into Eq. (14), and Eqs. (19) and (22) into Eq. (15) respectively, a pure numeric field calculation of the electric and magnetic field for defined space points (x, y, z) is possible. As the solution is based on the use of the Bipolar coordinate system, the shift of the radiator's aperture from the origin has to be considered prior to selection of the xcoordinate.

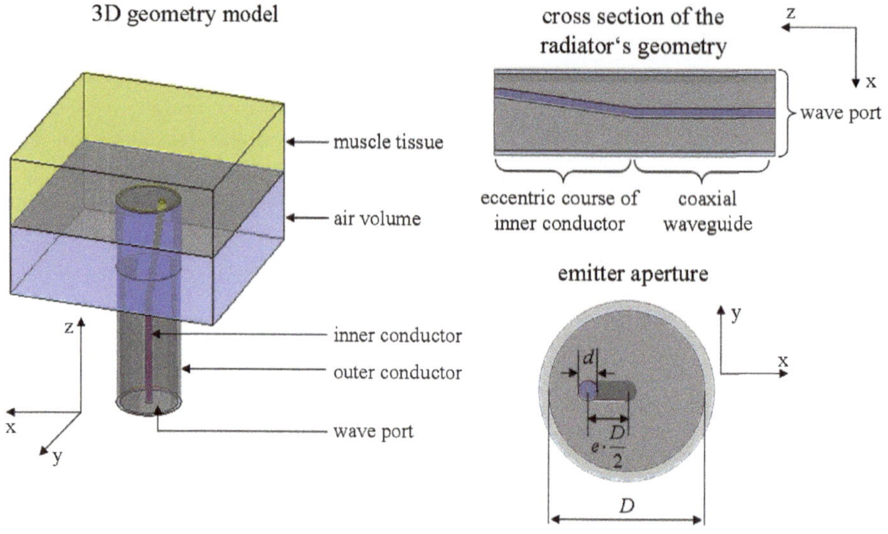

Fig. 4. Ansoft HFSS simulation model of the contact emitter.

2.3 Graphical representation of the near field

The programme system MATLAB is applied for a mathematical evaluation of the formalism presented above. The numerical results constitute the basis for a graphical near field representation of the radiator as a function of the eccentricity factor e of the inner conductor. In Fig. 3, the real part of the complex Poynting vector is presented for defined space points in a plane parallel to the emitter aperture at a frequency of f=10 GHz. This time-independent quantity ensures a direct comparison with the simulation results. The calculated field strength values are normalized to the maximum value of the field strength at each examined eccentricity factor e. The amplitudes are subdivided linearly in colour tones with increasing power density from blue to red. The square plane has an edge length of l=16 mm and is placed Δz=5 mm in front of the emitter aperture. The inner diameter of the outer conductor and the outer diameter of the inner conductor are set to D=10 mm and d=1 mm. The relative dielectric constant inside the radiator is selected to $\varepsilon_{r,waveguide}=1$. The relative dielectric constant of the homogenous tissue in front of the emitter aperture is assigned a value of $\varepsilon_{r,tissue}=50$. This is a typical value found in literature as a crude approximation for muscle tissue (see Stauffer, et al., 1998; Rossetto, et al., 2000). It can be seen clearly by Fig. 3, that the absolute maximum of the power density shifts as a function of the eccentricity factor e of the inner conductor. With increasing misalignment of the inner conductor to the left-hand side of the emitter centre, the maximum moves towards the centre of the emitter aperture, which is marked by a white circle in the graphical representations.

3 Comparison of analytical results and simulation data

To simulate the radiation fields of the contact emitter, the commercially available simulation programme HFSS (High Frequency Stucture Simulator) from Ansoft Inc. is used. Ansoft HFSS applies the finite element method (FEM) in order to analyse the electromagnetic behaviour of complex three dimensional model geometries. The implemented FEM allows accounting for the influence of arbitrary material surrounding the antenna aperture. The FEM directly calculates the field distribution inside a defined simulation volume which is confined by a radiation boundary for calculating far fields. The construction of the radiator's simulation model is summarised in Fig. 4. In contrast to the theoretical analysis, the simulation with Ansoft HFSS considers higher order modes inside the coaxial waveguide with an eccentric course of the inner conductor and in the emitter aperture. The simulation models used in the following are based on the geometry and material dates given under Sect. 2.

3.1 Distribution of power density in the near field area

The results of the near field simulation are shown in Fig. 5. The real part of the complex Poynting vector is calculated for defined space points of the above mentioned plane, as a function of the eccentricity factor e of the inner conductor. Comparing the graphical representation with those in Fig. 3 it is obvious that the analytical and the simulation results correspond quite well. But the influence of the higher order modes neglected in the analytic approach is also obvious. With increasing eccentricity factor e a second maximum appears. This area of high power density is located at the left boundary of the outer conductor. So there is no relevance for the medical application considered in this paper because only very small superficial skin areas of cancerous tissue or

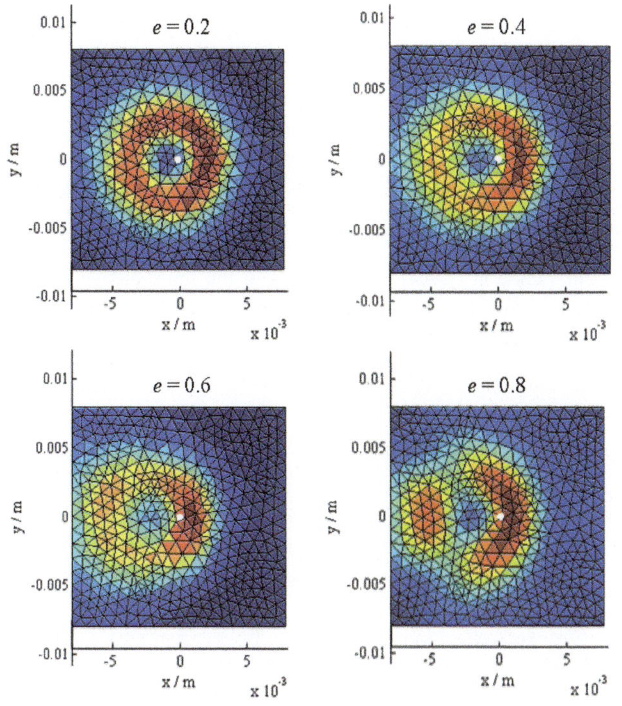

Fig. 5. Simulated distribution of the real part of Poynting's vector in the near field as a function of the inner conductor's eccentricity factor e.

biological active points interacting with the main maximum are considered. In comparison with the analytic results the deviating field distribution is due to the higher order modes. But the overall effect of positioning the absolute maximum of power density at the centre of the contact emitter by an eccentric shift of the inner conductor is confirmed by the simulation results as well.

3.2 Far field characteristics

In this section the differences in far field characteristics caused by ignoring the effects of higher order modes in the analytical approach in comparison to the simulation are investigated. Therefore Fig. 6 contains the far field plots and spherical far field cuts to clarify these differences. Analysis and simulation are performed with an eccentricity factor e=0.5 at f=10 GHz. In contrast to the analytical dates the simulation results describe an antenna with a maximum at Φ=0° and Θ=335°. However the theoretical approach predicts a nearly constant power density surrounding a global minimum in the direction of the contact emitter's centre. The theory based radiation pattern predict maxima at Φ=90°, Θ=30° and Φ=90°, Θ=330°, which are situated point-symmetrically to the focal point of the respective directivity pattern. The distinct minimum in front of the aperture, which is typical for the rotationally symmetric radiation pattern of a coaxial radiator (e=0), also appears in the simulation results at the same position.

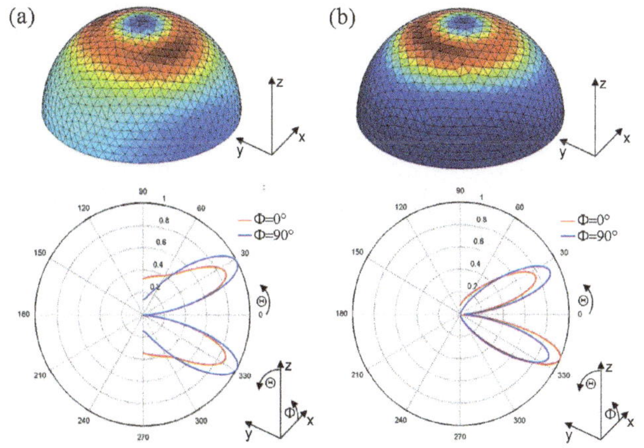

Fig. 6. Radiation patterns and directivity diagrams from analytical computation (a) and simulation (b) based on an eccentricity factor e=0.5.

4 Conclusions

The thought experiment for the eccentric shift of the inner conductor is confirmed by the theoretical analysis, as well as by the simulation results of Ansoft HFSS. Due to the neglect of the higher order modes with the theoretical analysis, the results differ slightly in the near field area. Nevertheless the simulation results of the near field are verified by the theoretical analysis in spite of exclusive consideration of the fundamental TEM mode. It could be shown unambiguously that the position of the maximum temporal average of Poynting's vector, is a function of the inner conductor's eccentricity factor e in the emitter aperture. Depending on the geometry and material dates, an eccentric position of the inner conductor could be determined, so that the maximum of the power density is localized directly above the emitter centre in the near field area. The simulated far field radiation patterns differ from the analytically calculated results mainly in the position of the maxima. A better agreement between the simulation data and the analytical results can only be obtained by an expansion of the proposed analytical solution to consider for the higher order modes in the emitter aperture.

References

Bakhtiari, S., Ganchev, S. I., and Zoughi, R.: Analysis of radiation from open-ended coaxial line into stratified dielectrics, IEEE Transaction on Microwave Theory and Techniques, 42, 1261–1267, 1994.

Kalantaewskaja, K. A.: Morphologie und Physiologie des Hautgewebes, Moskau, Medizin-Verlag, 1972.

Maloney, J. G., Smith, G. S., and Scott, W. R.: Accurate computation of the radiation from simple antennas using the finite-difference time-domain method, IEEE Transactions on Antennas and Propagation, 38, 1059–1068, 1990.

Mosig, J. R., Besson, J. C. E., Gex-Fabry, M., and Gardiol, F. E.: Reflektion of an open-ended coaxial line and applikation to non-destruktive measurement of materials, IEEE Transactions on Instrumentation and Measurement, 30, 46–51, 1981.

Murray, R.: Handbuch für Mathematik, New York, McGraw - Hill, 1979.

Nevels, R. D., and Wheeler, J. E.: Radiation from a dielectric coated hemispherical conductor fed by a coaxial transmission line, IEEE Transactions on Electromagnetic Compatibility, 31, 16–20, 1989.

Rossetto, F., Diederich, C. J., and Stauffer, P.: Thermal an SAR characterization of multielement dual concentric conductor microwave applicators for hyperthermia, a theoretical investigation, Medical Physics, 27, 4, 2000.

Stauffer, P., Rossetto, F., Leoncini, M., and Gentilli, G. B.: Radiation Patterns of Dual Concentric Conductor Microstrip Antennas for Superficial Hyperthermia, IEEE Transaction on Biomedical Engineering, 45, 5, 1998.

Stuchly, M. A. and Stuchly, S. S.: Coaxial line reflection methods for measuring dielectric properties of biological substances at radio and microwave frequencies – a review, IEEE Transactions on Instru-mentation and Measurement, 29, 176–183, 1980.

Stuchly, M. A., Brady, M. B., Stuchly, S. S., and Gajda, G. B.: Equivalent circuit of open-ended coaxial line in a lossy dielectric, IEEE Transactions on Instrumentation and Measurement, 31, 116–119, 1982.

Sumbar, E., Chute, F. S., and Vermeulen, F. E.: Implementation of radiation boundary conditions in the finite element analysis of eletromagnetic wave propagation, IEEE Transactions on Microwave Theory and Techniques, 39, 267–273, 1991.

Multi-frequency sensor for remote measurement of breath and heartbeat

M. Jelen and E. M. Biebl

Fachgebiet Höchstfrequenztechnik, Technische Universität München, Germany

Abstract. Remote measurement of breath and heartbeat is desirable in many situations. It avoids the discomfort resulting from electrodes applied on the skin for long-term patients or during sports acvtivities. Also, surveillance of high security areas or finding survivors of disasters are interesting applications. Common methods identify the movement of heart and thorax by using the range resolution provided by UWB pulse radar systems. In this paper a low-cost approach is presented, that is based on detection of movement by means of Doppler radar sensors. Combining three sensors working in the ISM bands at 433 MHz, 2.4 GHz and 24 GHz, the presence of persons was reliably detected and the frequency of breath and heartbeat was measured.

1 Introduction

Heartbeat and respiration are a common attribute of all humans. They have a characteristic form and frequency which, in case of the heartbeat, can't be affected voluntarily.

In technical systems (e.g. in cars) there is often the need to detect the presence of a human. This task can be fullfilled by a wide variety of sensors, such as mechanical switches (e.g. in the seat) or ultrasonic sensors.

Also the frequencys of pulse and respiration are often measured. One obvious field of application is the medical sector, but numerous other applications (e.g. sport/fitness) can be thougt of. Currently the measurements are performed by sensors which are in contact with the person beeing observed. This may be a serious disadvantage in situations were measurements must be taken over a long period of time (e.g. long term patients in hospital) or were physical contact is not possible.

An approach of detecting humans by means of radar, mainly for security applications, is described in Greneker (1997). Bimpas et al. (2003) covers the use of a radar system for the detection of persons trapped in building ruins after an earthquake.

This article describes a multi-frequency-sensor approach for both the reliable detection of persons and the measurement of heartbeat and respiration rate.

2 Physiological and technical basics

The human heart is approximately the size of a fist. It is located off-center in the thorax, two-third in the left part, one third in the right. In resting state it beats abt. 70 times per minute. With physical effort the activity can increase up to 200 bpm. The heart is covered by several layers, the most important are skin, fat and bone. The thickness of these layers varies from person to person.

A movement detection of the heart by means of radar implies the transmission of electromagnetic waves into the human body as well as reflections at a border. To estimate the amount of reflected energy it is necessary to know the dielectric properties of the different body tissues. A rather complete list can be found in Http://niremf.ifac.cnr.it/tissprop/. For the calculation the well-known equations

$$R[dB] = 10 \log \left(\frac{|\underline{Z}_{F2}| - |\underline{Z}_{F1}|}{|\underline{Z}_{F1}| + |\underline{Z}_{F2}|} \right)^2 \tag{1}$$

and

$$T[dB] = 10 \log \left(\frac{2\,|\underline{Z}_{F2}|}{|\underline{Z}_{F1}| + |\underline{Z}_{F2}|} \right)^2 \frac{|\underline{Z}_{F2}|}{|\underline{Z}_{F1}|} \tag{2}$$

can be used. $\underline{Z}_F$ can be calculated from μ_r and ϵ_r using the following equation:

$$\underline{Z}_F = \sqrt{\frac{\mu_0 \underline{\mu}_r}{\epsilon_0 \underline{\epsilon}_r}} = Z_0 \sqrt{\frac{\underline{\mu}_r}{\underline{\epsilon}_r}} = Z_0 \sqrt{\frac{\mu_r (1 - j \tan \delta_\mu)}{\epsilon_r (1 - j \tan \delta_\epsilon)}} \tag{3}$$

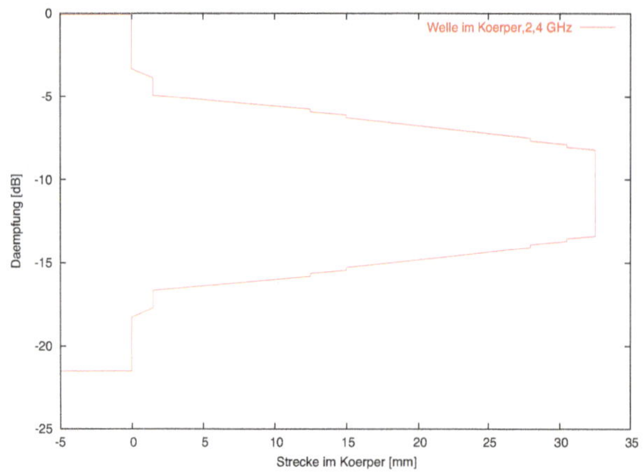

Fig. 1. Propagation of a wave at 440 MHz through the thorax and back.

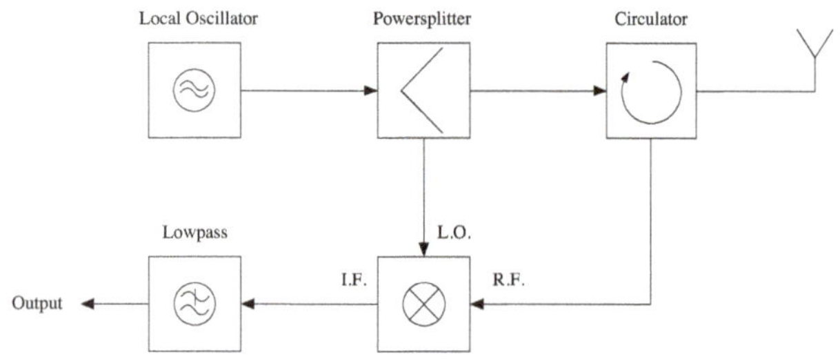

Fig. 2. Simple block diagram of a cw-doppler-radar.

Figure 1 shows an estimation of a wave at 430 MHz being propagated through the body towards the heart and being reflected back on the border fat/heart. The graph shows that a reflected wave is attenuated by approximately 22 dB. The reflection coefficent air-skin is about 2.38 dB at this frequency. Because the surface of the skin is also much larger than the surface of the heart, the difference between the two reflected signals will be at least 30 dB. Those values are very similiar for 2.4 GHz.

The reflection coefficent at the border air/skin at 24 GHz is about -3.2 dB. In good approximation all power which is transmitted into the body can be considered as lost, so that no reflections from the heart can be measured at this frequency.

Two radar principles could be used for this task: ultra-wide-band(UWB)-radar and cw-doppler-radar.

The first one uses short pulses which are reflected at a target. The distance resolution that can be achieved is inverse proportional to the length of the pulses. As the movement of the heart is in the sub-cm-region, very short pulses (and therefore a very high bandwith) would be needed. Despite from frequency regulation issues this would lead to a sophisticated broad-band design, not suitable for a low-cost approach. Staderini covers the use of such a system in medical applications in detail.

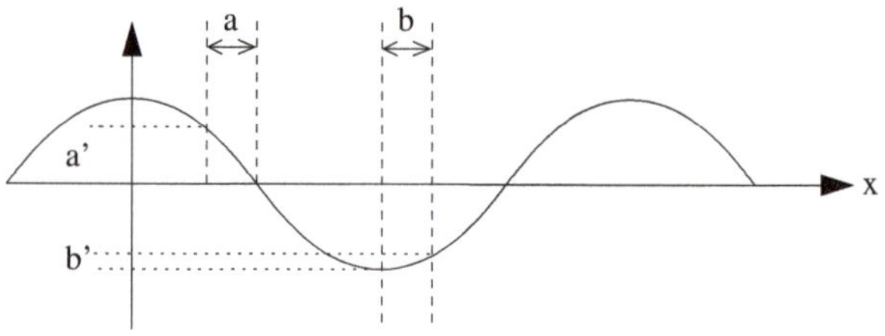

Fig. 3. Influence of phase on mixer output voltage.

The second one is much simpler to design. Figure 2 shows a simple block diagram of a doppler-radar circuit. Contrary to the pulseradar, the signal is sent continously and unmodulated. Therefore only a small bandwith is occupied. The information is contained in amplitude, frequency and phase of the received signal. The signal of the local oscillator is divided in two parts: One part is radiated via the antenna, the second part feeds the LO-port of the receiving mixer. The received signal feeds the IF-port of the mixer. If the frequency of the transmitted signal is ω_{TX}, and the frequency of the received signal is ω_{RX}, and φ_{TX} and φ_{RX} are the corresponding phases, the following signal can be observed at the IF-port of the mixer:

$$\begin{aligned} IF = {} & A_1 \cos[(\omega_{TX} - \omega_{RX})t - (\varphi_{TX} - \varphi_{RX})] + \\ & A_2 \cos[(\omega_{TX} + \omega_{RX})t - (\varphi_{TX} - \varphi_{RX})] + \\ & \ldots \end{aligned} \tag{4}$$

A_1 und A_2 are constants depending on the mixer. If the frequency of the sent and received signals are equal, the equation simplifies to the following:

$$V_{Out} = A_1 \cos(\Delta\varphi) \tag{5}$$

A dc voltage depending on the phase can be observed at the output of the mixer. If the object which reflects the wave moves around a fixed position (as it is the case with the beating heart) the amplitude of this voltage changes. But the amplitude depends also on the absolute position of the target. This is illustrated in Fig. 3. A phase variation due to a movement within ‘a’ causes the voltage change of ‘a’. A phase variation about (the same absolut) amount ‘b’ on another absolut position causes a much smaller change of ‘b’. Because the amplitude of the heartbeat is much smaller than the used wavelengths, this effect is significant for the detection of heartbeats.

To overcome this problem of ‘blind spots’, the approach shown here uses three sensors at three different frequencies. In almost all cases at least one sensor will be able to achieve a meaningfull measurement result.

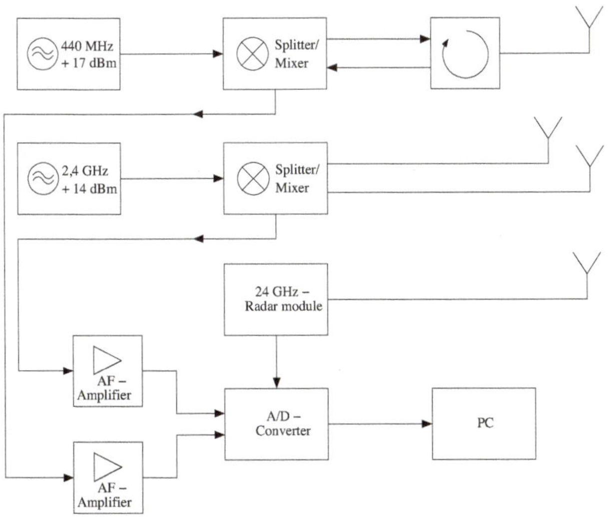

Fig. 4. Block diagram of the system used for the presented measurements.

Table 1. Used frequencies in the ISM range.

Wavelength	Frequency range	Output Power
70 cm	433,05 - 434,79 MHz	0 dBm
12,5 cm	2,4 - 2,5 GHz	0 dBm
1,25 cm	24 - 24,25 GHz	0 dBm

3 Measurement setup

Measurements were performed at three different frequency ranges. To make commercial applications possible, all chosen frequencies are located within the ISM (Industrial, Scientific, Medicine) bands. Table 1 gives an overview over the used frequencies and output powers.

Figure 4 shows a block diagram of the system used for the measurements. For 440 MHz, a circulator was available, thus a system according to Fig. 2 could be set up. The RF was supplied by a commercial signal generator. The maximum available output power was +17 dBm. Because a resistive power splitter was used to split the transmit and reference signal and due to the losses in the circulator, approximately +10 dBm (10 mW) were available at the antenna.

For 2.4 GHz no circulator was available, so two separate antennas were used for both the transmit and the receive path. Both antennas were realised as patch antennas etched on a low-loss substrate. To keep losses low, the power splitter and mixer were also included on the same substrate. Figure 5 shows the system. The power was also generated by a laboratory generator, and also about +10 dBm were fed into the transmit antenna.

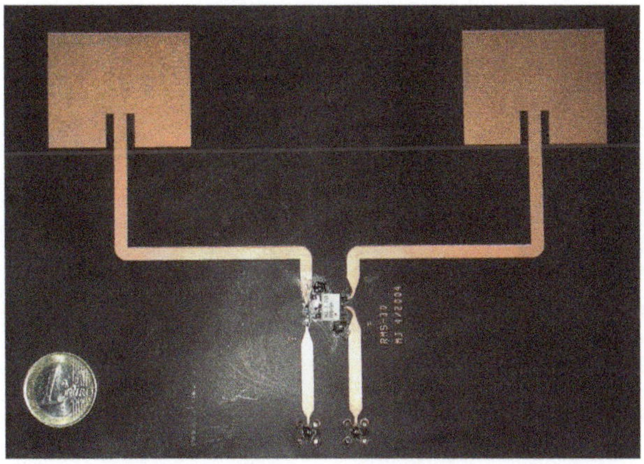

Fig. 5. The sensor used for 2.4 GHz.

The 24 GHz measurements were performed using a commercially available doppler radar module with a horn antenna with about 20 dBi antenna gain. The output of this module was ac-coupled, so the response to very slow movements (e.g. breath movement) was strongly attenuated or couldn't be detected at all. This can be seen in the measurement results.

All outputs were fed into a professional 16 bit A/D converter card connected to a notebook. The captured data was processed using MATLAB. First, the dc-offset was removed from the signal by calculating the mean value and subtracting it from all samples. Then the captured time domain data was windowed by a hanning-window to reduce the leakage-effect and converted to the frequency domain using the fast-fourier-transformation (FFT). There the signal was band-pass-filtered with a passband from 0.2 Hz to 5 Hz. For the time domain representation the filtered signal was transformed back into the time domain.

The achievable frequency resolution Δf is direct proportional to the length of the captured sequence N:

$$\Delta f = \frac{f_s}{N} \tag{6}$$

With a constant sample rate this can be converted into the recording time t_s:

$$\Delta f = \frac{1}{t_s} \tag{7}$$

If a resolution of 0.1 Hz is needed, the capture time must be at least 10 s.

4 Measurement results

For the actual measurements a sample rate $f_s = 8$ kHz was used. For the first measurement a person was seated on a chair and asked to sit quiet without moving and hold the

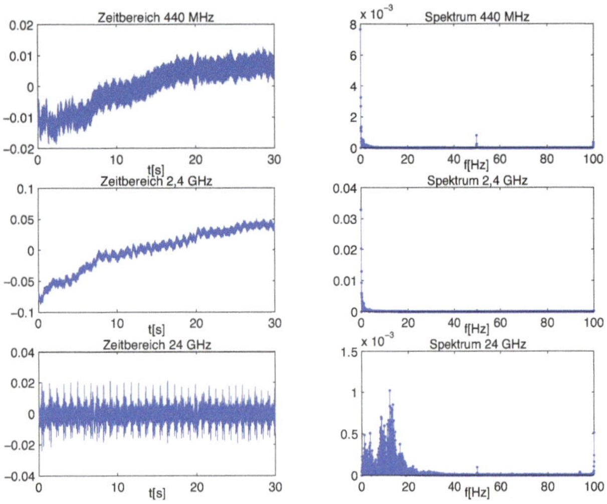

Fig. 6. Raw time- and frequency domain data.

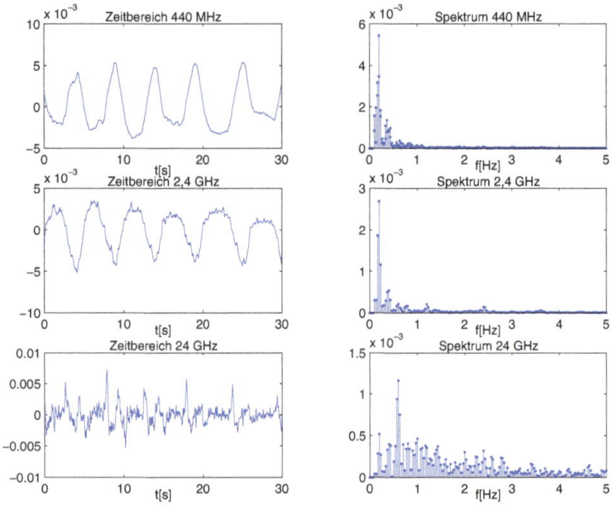

Fig. 8. Breath and heartbeat captured from a distance of 2 m.

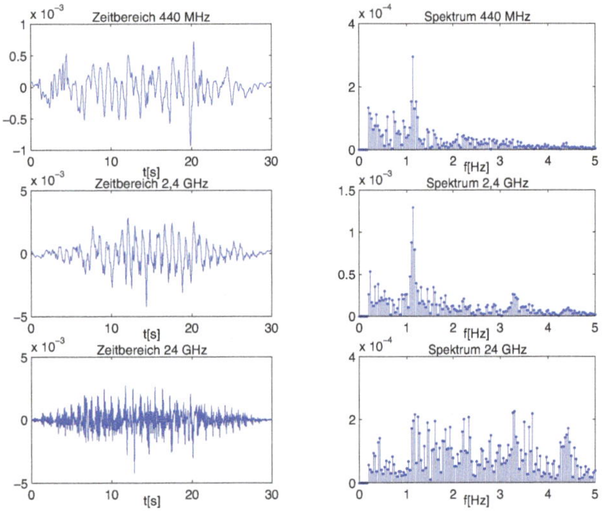

Fig. 7. Filtered time- and frequency domain data.

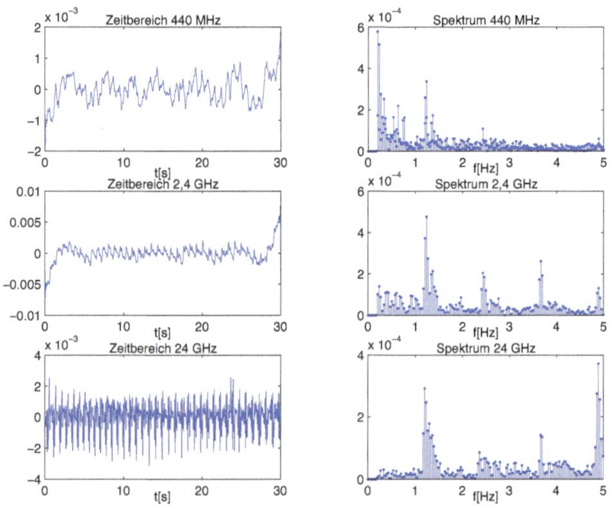

Fig. 9. Heartbeat captured from a distance of 2 m.

breath for 30 s. The data from all three sensors were taken parallel. Figure 6 shows the raw time- and frequency domain data as captured from the sensors, Fig. 7 shows the same data after digital filtering.

As one can see in Fig. 6 the spectral components are limited to very low frequencies. Therefore the maximum frequency range in Fig. 7 was reduced to 5 Hz. The frequency resolution with a measurement time of 30 s is 1/30 Hz. A peak is visible at 1.2 Hz. This equals a heart rate of 72 bpm which was confirmed by a parallel 'classical' measurement.

For the second measurement the person was again seated in the chair about 2 m from the sensors and asked to sit quiet, this time breathing. Figure 8 shows the results. The breath movement is clearly visible in the graphs resulting from the measurements at 440 MHz and 2.4 GHz. As mentioned before, these low frequencies were cut by the hardware in the 24 GHz sensor. The spectrum has a peak at about 0.26 Hz, which equals a breath frequency of about 15.6 per minute. This was in fact the breath frequeny of the test person.

After the measurement the person was asked to stop breathing so that only the heartbeat could be detected. Figure 9 shows the result. Here a peak in the spectrum is visible at all three measurement frequencies at 1.3 Hz. As a reference, the heart rate of the test person was measured with a pulse watch which showed 79 beats per minute. This equals 1.317 Hz.

Comparing the peak powers of breath and heartbeat with

Table 2. Increase of maximum compared to reference measurement.

Increase			
Measurement distance	440 MHz	2,4 GHz	24 GHz
Distance 1 m	> 35 dB	> 35 dB	> 35 dB
Distance 3 m	22,5 dB	23,5 dB	> 40 dB
Distance 3 m + door	16,5 dB	31,1 dB	–
Distance 8 m	21 dB	14 dB	30 dB
Distance 12 m	2,5 dB	15,5 dB	16,9 dB

the equation

$$V = 20log\frac{\text{peak}_{\text{respiration}}}{\text{peak}_{\text{heartbeat}}} \tag{8}$$

a difference of about 16 dB for 2.4 GHz and 20 dB for 440 MHz is found. Assuming that all returning power is reflected at the border air/skin at the breath measurement, these values are close to those predicted before. This may be a hint that actually reflections at the heart were measured.

If only the information is needed whether a person is present within the observed range, a threshold in the power spectrum could be defined. This was tested for different constellations. In all cases the persons were succesfully detected if they were in the line of sight of the antennas. Even through obstacles, such as a closed wooden door, reliable recognition of the presence of persons could be achieved. Table 2 gives an overview over the increase of the low-frequency-spectrum compared to a reference measurement in an empty room.

5 Conclusions

CW-doppler-radar sensors at three different frequencies were sucessfully used to detect the presence of persons and to measure heartbeat and respiration rate over a distance of several meters. This was achieved with a low-cost hardware setup which could be easily implemented, for example, in the interior of a car. As expected, 440 MHz and 2.4 GHz show very similiar properties for this application, whereas 24 GHz can only detect body movements due to its very small penetration depth.

The system still has some potential for improvements, for example the sensitivity of the radar sensors could be increased and therefore the output power could be decreased even further. Also only very basic DSP techniques were used for this measurements. A combination of the output from different sensors in combination with more sophisticated DSP algorithms could help to eliminate unwanted responses and overcome the disadvantage of the dependence from the absolute position.

References

Http://niremf.ifac.cnr.it/tissprop/.

Bimpas, M., Nikellis, K., Paraskevopoulos, N., Economou, D., and Uzunoglu, N.: Devlopement and testing of a detector system for trapped humans in building ruins, in: 33rd European Microwave Conf., 999–1002 , 2003.

Greneker, E.: Radar Sensing of Heartbeat and Respiration at a Distance with Security Applications, Proc. of SPIE, 22–27, 1997.

Staderini, E.: UWB Radars in Medicine, IEEE Aerospace and Electronic Systems Magazine, 13–18, 2002.

7

Concept for the controlled plane wave exposure for animal experiments using a parabolic reflector

S. Tejero, S. Schelkshorn, and J. Detlefsen

Institute for High Frequency Engineering, Technische Universität München, 80290 München, Germany

Abstract. In this paper a low-cost concept for the controlled RF plane wave exposure for *in vivo* experiments is presented. The exposure setup is based on the use of a parabolic reflector to convert the incident spherical wavefront emanating from the primary source into a plane wave. The employed paraboloid is a common prime focus paraboloid used in satellite-TV links. The main problems of the focussed approach are identified and a solution based on a defocussed system is introduced. It results in a compact, cost-effective and still power-efficient setup for the RF exposure at microwave frequencies.

Simulation results show a very good performance of the concept achieving a quasi-plane wave incident on the animals with minimum variations of the exposure dose.

---

1 Introduction

Together with the great expansion of mobile communications in the last years, the public concern about possible negative effects of these systems on human beings has grown. Although it is impossible in the strict sense to prove the absence of any effect, to get a strong body of evidence for this absence should be enough. However, a body of evidence can only be considered strongly if it is coherent and comprehensive with respect to the raised health concern and not an accumulation of more or less randomly collected negative findings. For our special case of developing an exposure setup, this means that the exposure conditions must be controlled, well defined, standard and repeatable. In this paper a low-cost setup with these characteristics for a specific exposure case is presented. However, the concept can be applied to other setups with different requirements.

The setup is intended for a long-term, continuous and controlled exposure of living Wistar rats. The animals are divided into three different groups: two of them will be exposed, with a GSM- and a UMTS-signal respectively, and one serves as control group, i.e. the animals will have the same environmental conditions but no exposure at all. The isolation between the three groups must be assured to ensure controlled exposition.

To obtain meaningful conclusions from the statistics the number of exposed animals must be high enough, a minimum of 100 rats per setup should be exposed simultaneously. Other requirements are imposed by the *in vivo* characteristic of the experiment: the exposure chamber must be air-conditioned and the access to the cages must be assured for the daily animal care.

The animals will be exposed by a linear polarized plane wave incident on them. Incident plane wave is the most commonly used exposure condition for *in vivo* experiments. Of course, the variations of the exposure dose must be kept as low as possible.

To achieve the plane wave condition a long distance from the antenna to the observation point is usually required. That allows to consider the spherical wavefront radiated from the source as a plane or quasi plane wave, i.e. a wave where small deviations from the pure plane wave are tolerated. This approach requires a large exposure setup and is also very inefficient concerning exposure energy, as the signal power density decreases with the square of the distance.

Other energy efficient approaches found in the literature are not suitable for long-term continuous exposure (e.g. Schönborn et al., 2004) or are based on the propagation of guided TEM waves and may run into problems when frequency increases, as the dimensions required to avoid propagations of high order modes are too small (e.g. Hansen et al., 1999; Balzano et al., 2000).

The setup presented in this document is in analogy to the "compact range" concept (e.g. Balanis, 1997), which is used in antenna measurement facilities to obtain a plane wave in relative short ranges. The idea behind is to convert a spherical wavefront into a plane one by using parabolic reflectors. The requirements in antenna measurements are very restrictive in terms of error to a pure plane wave and polarization changes, resulting in big reflector sizes, high distance from the source to the quiet zone – the zone where the plane wave condition is best approximated – and very high price. The approach presented here is based on a common parabolic reflector used in satellite-TV links and results in a compact, cost-effective and still power-efficient setup.

Although the setup described in this paper is intended for a specific experiment, our concept can also be used for other kind of plane wave exposure setups.

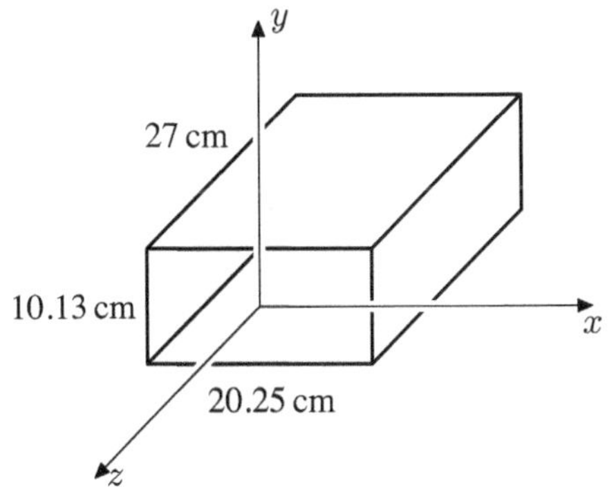

Fig. 1. Open waveguide for the GSM band (925 MHz) operating in the fundamental mode TE_{10} and radiating 1 W.

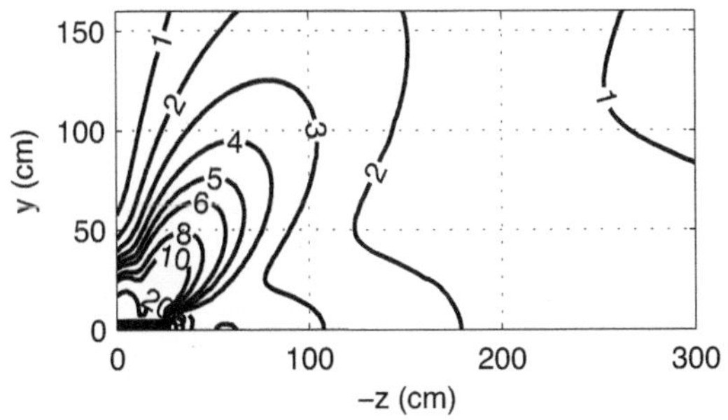

Fig. 2. Field $|E_y|$ (in V/m) of the GSM band (925 MHz) waveguide operating in the fundamental mode TE_{10} and radiating 1 W in the absence of the paraboloidal reflector.

2 Development of the setup

The principal idea of the exposure concept described here is the well known approach of using a parabolic reflector to convert a spherical wavefront emanating from a primary source located at the focal point into a plane wave. The reflector employed is a common prime focus paraboloid with a diameter of 320 cm and a focal distance of 112 cm and originally intended for satellite-TV links. The use of a mass-production good reduces the total cost of the equipment drastically.

Assuming a common prime focus paraboloid illuminated by an ideal source located at its focus, i.e. a point source radiating only towards the reflector and illuminating it uniformly, the reflected wavefront on the paraboloid at the focal plane would be a plane wave with a rather constant electric field and only disturbed by the diffraction caused by the reflector rims. The main problems of this approach are those created by the non-ideal feed. One problem is the spillover effect: a real feed not only radiates towards the reflector, it radiates also towards the cages zone, disturbing the homogeneity of the field distribution. The other main problem is to achieve a uniform illumination of the reflector's surface. An offset paraboloid could be used to avoid these two problems: the exposure zone, located at the focal plane, does not contain the feed and the solid angle covered by the reflector viewed

Table 1. Radiation characteristic of infinitesimally small antennas.

Antenna	Radiation characteristic and 3 dB-beamwidth E-plane		H-plane	
Dipole	$\cos\theta$	90°	1	360°
Aperture in free-space	$(1+\cos\theta)/2$	131°	$(1+\cos\theta)/2$	131°
Aperture on a ground plane	1	360°	$\cos\theta$	90°

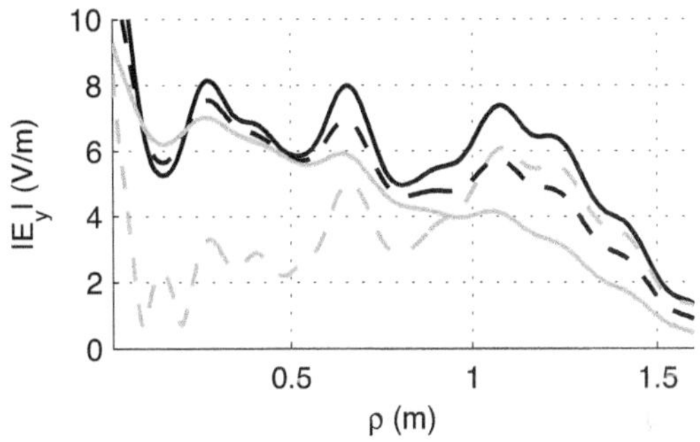

Fig. 3. Electric field at $z=-150$ cm reflected on the paraboloid from Fig. 7 illuminated by a point source located at the focus and radiating in the UMTS band (1966 MHz) with different patterns: uniform (solid black), $(1+\cos\theta)/2$ (dashed black), $\cos\theta$ (solid gray) and $\sin\theta$ (dashed gray).

from the focus is small. Unfortunately, the low-cost commercially available offset paraboloids have diameters lower than 2 m and this is not enough to expose a high number of animals.

2.1 Feed spillover problem

To identify the region where the field radiated by the primary feed can be neglected a typical wide-beam feed, the open waveguide operating in the fundamental mode, is analyzed. The analyzed waveguides (e.g. Fig. 1) radiate 1 W and operate in the fundamental mode TE_{10} in the GSM (925 MHz) and the UMTS (1966 MHz) bands respectively. The simulation for the field is carried out with the *Method of Moments* (MoM). As shown in Fig. 2 for the GSM band case (and the same is valid for the UMTS one), 150 cm behind the aperture ($z<-150$ cm) the field is 10 times lower than in the vicinity of the aperture and also much lower than the field expected from the reflection on the paraboloid. Hence, it can be stated that the spillover effect of the primary feed becomes negligible for $z<-150$ cm. Furthermore, by means of experimental measures, this disturbing field radiated toward the cages by the primary source can be additionally reduced.

Hence, the feed spillover problem will be solved by placing the exposure zone in a region in which the assumption of negligible field of the primary feed is valid: i.e. 2 m behind the feed aperture, in $-250\,\text{cm}\leq z\leq -200$ cm.

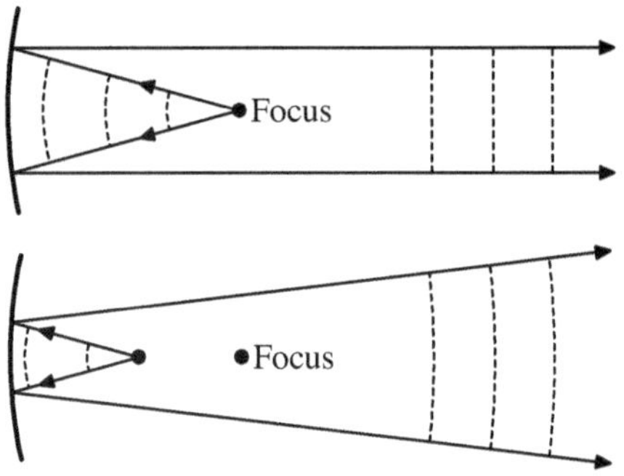

Fig. 4. Reflected wave on a parabolic reflector in a focussed and a defocussed system.

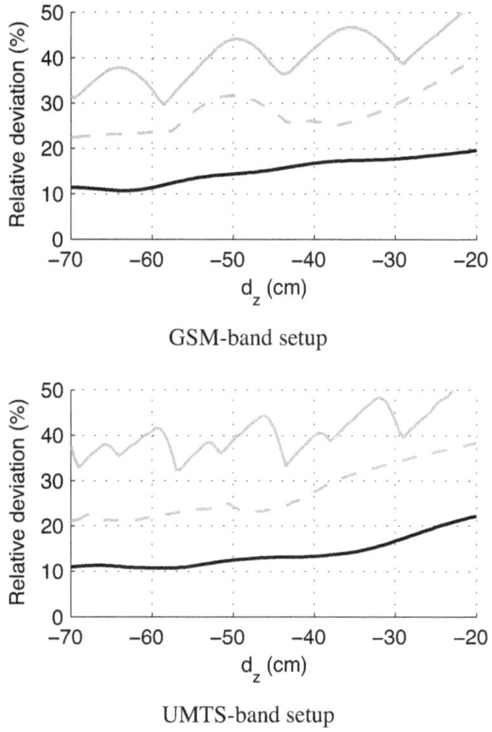

Fig. 5. Standard deviation (black line), maximum positive deviation (gray solid line) and maximum negative deviation (gray dashed line) from the mean value of the reflected field evaluated at the exposure volume as a function of the defocussing distance. The optimum defocussing distances are –59 cm and –56 cm for the GSM- and UMTS-band setups respectively.

2.2 Uniform illumination problem

Another problem of the focused approach is to illuminate the reflector uniformly. As the focal distance-to-diameter ratio is low, i.e. 0.38, the angle which must be uniformly illuminated by the source is very large, i.e. 142°. Comparing this value with the half-power beamwidth of elemental radiators (see Table 1) gives us an idea of the realization difficulty of a source with the desired beamwidth. Also the angle required to illuminate only the region with radius $\rho < 125$ cm (the minimum to expose the required number of animals) is rather large, i.e. 116°.

Figure 3 shows the reflected field on the paraboloid from Fig. 7 illuminated with different radiation patterns. It is observed that to obtain a similar homogeneity as with the uniform illumination, the exposure zone radius must be reduced at least to 1 m, not enough to expose a high number of animals.

Also the presence of ripples is observed in the reflected field. These ripples are caused by abrupt changes in the field distribution at the focal-plane aperture (Boonzaaler and Pistorius, 1988). Illumination patterns with a smooth tapering, like $\cos\theta$, prevent these undesired ripples.

To overcome the illumination problem the reflector will be defocussed. As seen in Fig. 4, in a defocussed configuration the reflected wavefront is curved and as the wave propagates the illuminated area increases. The defocussing together with the placement of the quiet zone far enough from the reflector can be used to reduce the primary source's beamwidth requirements. Even though due to the defocussing the wave is not plane any more, the radius of the wavefront is high enough to consider the wave as quasi plane in the region occupied by a cage and to keep the loss in power efficiency within an acceptable limit.

Furthermore, the defocussing generates a smooth tapering of the field at the focal plane preventing the above mentioned undesired ripples in the reflected field.

To obtain the optimum defocussing distance d_z, some statistical parameters of the electric field in the exposure zone will be analyzed. By means of the *Physical Optics* (PO) the reflected field on the paraboloid is obtained for different d_z values in the volume identified for the exposure, i.e. $-250\,\text{cm} \leq z \leq -200\,\text{cm}$ and $15\,\text{cm} \leq \rho \leq 125\,\text{cm}$, at the points given by a 4 cm rectangular grid. With the 10 192 field samples, the standard deviation and the maximum positive and negative deviations from the mean value are obtained. Analyzing these parameters (shown in Fig. 5), the optimum defocussing distances has been set to –59 cm and –56 cm for the GSM- and UMTS-band setups respectively.

2.3 Proposed approach

Figure 6 shows the final exposure setup which is placed in a metallic shielded chamber. Besides the paraboloid and the defocussed feed, the absorber material on the walls and the cages for the animals are shown.

To avoid the uncontrolled reflection of the waves propagating in the chamber, absorber material is put at the walls. The absorber material is a high loss material with an impedance comparable to the characteristic impedance of free space, i.e. an incident wave is mainly absorbed and only a small amount is reflected. With typical absorber materials the amplitude of the reflected wave is at least 20 dB lower than the incident wave.

The exposure volume is the cylindrical region with radius $\rho < 125$ cm and depth $-250\,\text{cm} \leq z \leq -200\,\text{cm}$.

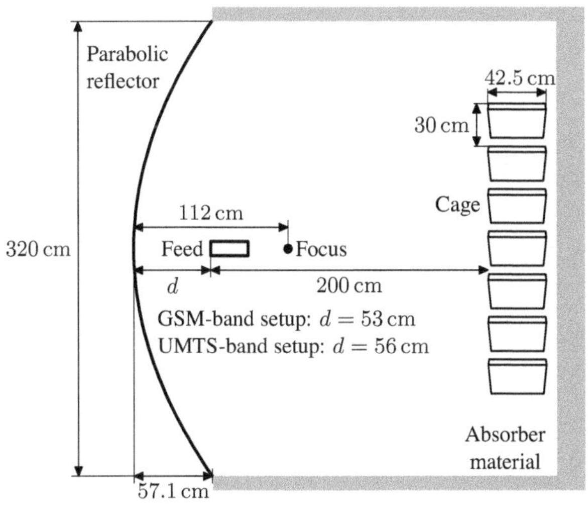

Fig. 6. Side view of the exposure setup.

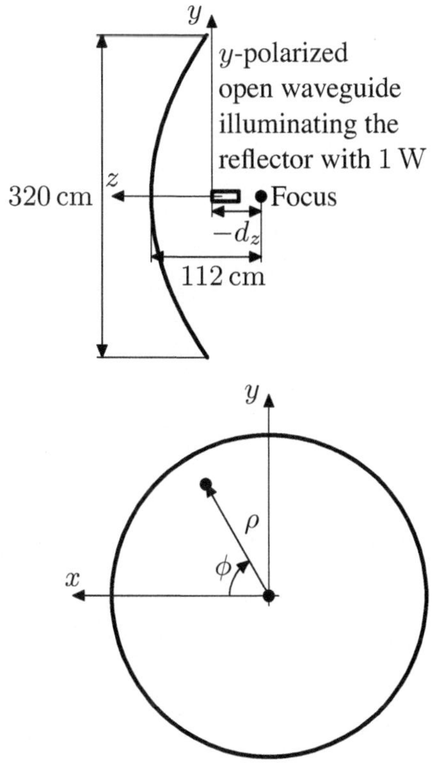

Fig. 7. Simulated exposure setup: d_z=−59 cm for the GSM-band setup and –56 cm for the UMTS-band setup

The animals are inside polysulfan cages with size ($H \times W \times L$) 24×28.5×42.5 cm^3. The cages are placed on wood slats with the cage frontal looking toward the reflector. The horizontal distance between the cages is 30 cm and between the slats also 30 cm. As the diameter of the cross section of the quiet zone is planned to be 2.5 m, a large number of cages (approx. 40) can be placed in the setup's quiet zone.

Table 2. Simulation results for the field distribution in the exposure volume.

	GSM Band	UMTS Band
Mean value	7.6 V/m	7.6 V/m
Standard deviation	11.2 %	10.7 %
Maximum positive deviation	33 %	31 %
Maximum negative deviation	−23 %	−23 %
Relative frequency of points with a deviation under 15 %	76 %	83 %
Relative frequency of points with a deviation over 25 %	0.5 %	0.5 %
Maximum phase deviation from a plane wavefront	±12°	±12°
Power efficiency	34 %	34 %

3 Simulation results

The simulation for the electric field in the setup is carried out using a combination of two numerical approaches: the *Method of Moments* (MoM) and the *Physical Optics* (PO) approximation. The MoM, typically used for electrically small problems, is used to simulate the feed. The PO approximation, a technique to model the electric currents in electrically large objects, is employed for the parabolic reflector.

The final setup's model is shown in Fig. 7: a defocussed paraboloid – with defocussing distance d_z=−59 cm for the GSM-band setup and –56 cm for the UMTS-band setup – with a diameter of 320 cm and a focal distance of 112 cm illuminated by a y-polarized open waveguide operating in the fundamental mode (TE_{10}) and radiating 1 W (see Fig. 1). The field distribution has been calculated at two different frequencies: at 925 MHz (GSM band) and at 1966 MHz (UMTS band).

As stated in Sect. 2, the near field of the open waveguide can be neglected for $z < -150$ cm. Therefore, only the reflected component of the total field (the PO component) is investigated.

To analyze the statistics of the field in the exposure zone, the reflected field on the paraboloid is calculated in the exposure volume, –250 cm$\leq z \leq -200$ cm and 15 cm$\leq \rho \leq$125 cm, at the points given by a 2 cm rectangular grid. With the 80 106 field samples a statistical analysis is done. The standard deviation is 11.19% for the GSM-band setup and 10.70% for the UMTS-band setup.

The power efficiency – defined as the quotient between the power flowing through the frontal exposure surface and the power fed into the primary feed – is 34.4% for both setups. Assuming a mean whole-body *Specific Absorption Rate* (SAR) of 0.025$\frac{\mathrm{W/Kg}}{\mathrm{W/m^2}}$ for a Wistar rat (Fröhlich et al., 2003) the efficiency in terms of mean whole body SAR per unit of input power (SAR efficiency) is estimated to be 0.002 $\frac{\mathrm{W/Kg}}{\mathrm{W}}$. Assuming that 40 cages with 3 rats each cage will be exposed, a total of 120 rats with an estimated weight

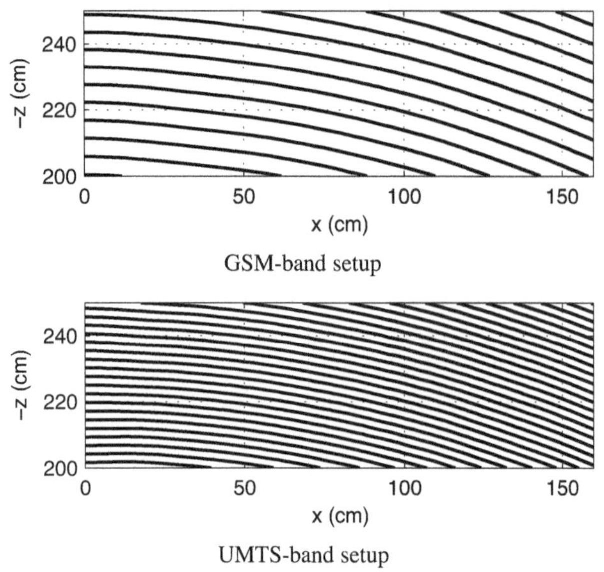

Fig. 8. Phase fronts ($\Phi(E_y)=n\cdot 60^\circ$, n=0, 1, 2, ...) at the xz-plane of the electric field reflected on the paraboloid of Fig. 7. In the region occupied by a cage ($W\times H\times L$=24×28.5×42.5 cm^3) the maximum deviation from a plane phase front is lower than ±12°.

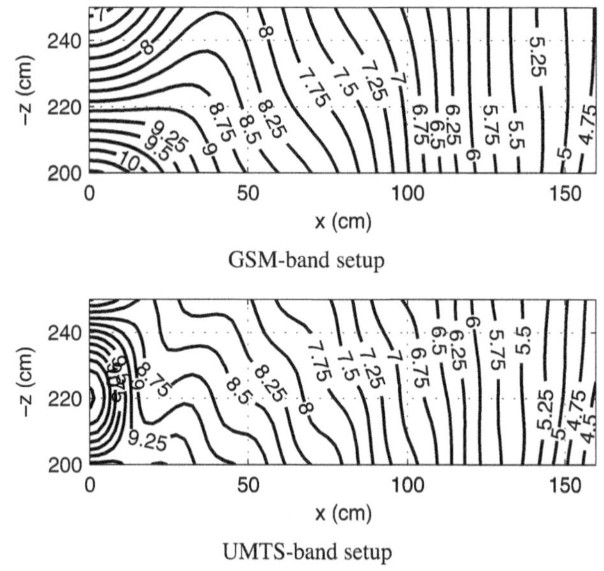

Fig. 9. Absolute value of the electric field (in V/m) reflected on the paraboloid of Fig. 7 illuminated with 1 W. In the exposure volume the standard deviation is 11%.

of 275 g will absorb 6.5% of the power radiated by the primary source.

In Fig. 8 the phase fronts at the xz-plane are presented. In the region occupied by a cage with dimensions $W\times H\times L$=24×28.5×42.5 cm^3 the maximum deviation from a plane phase front is lower than ±12° for both setups.

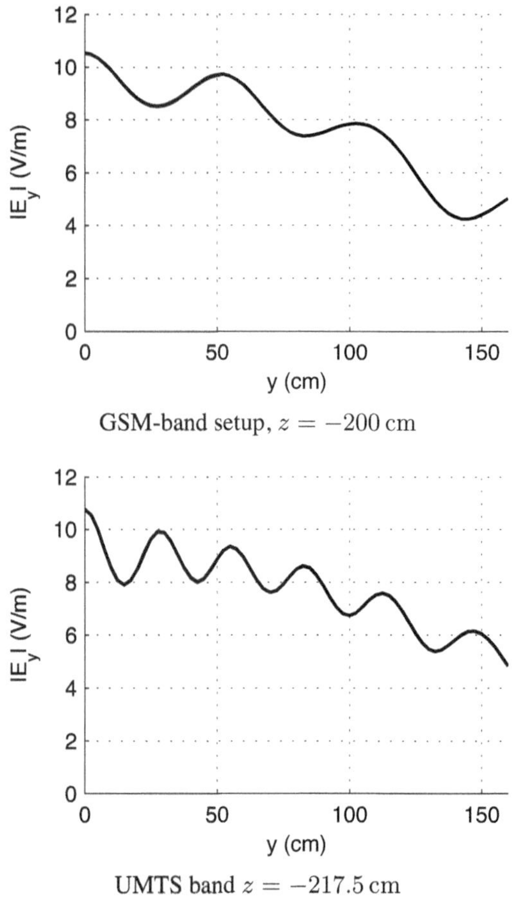

Fig. 10. Absolute value along the worst case line of the electric field reflected on the paraboloid of Fig.7 illuminated with 1 W. The maximum deviations from the mean value are +33%/–23.50% for the GSM-band setup and +31%/–23.50% for the UMTS-band setup.

In Fig. 9 contour plots of the absolute value of the electric field are shown. Figure 10 presents the field along the worst case line, where the field presents maximum variations. The maximum deviations from the mean value are +33%/–23.50% for the GSM-band setup and +31%/–23.50% for the UMTS-band setup.

Figure 11 shows the polarization ellipses at the plane z=−200 cm. In the region with radius ρ<125 cm the maximum inclination of the main axis is lower than 3° for both setups and the axis relation greater than 10 and 17 for the GSM- and UMTS-band setups respectively. Hence, the resulting polarization can be considered linear.

4 Conclusions and outlook

A RF exposure setup based on a parabolic reflector is presented. The problems related to a common prime focus paraboloid are solved by defocussing the system and placing the quiet zone away from the feed. A quasi plane wave is generated in a cylindrical region with depth −250 cm≤z≤−200 cm and radius 15 cm<ρ<125 cm with a

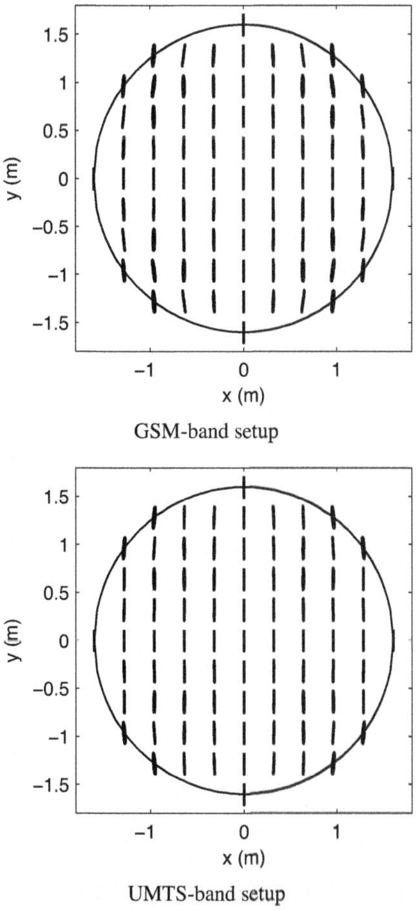

Fig. 11. Polarization ellipses at the plane $z=-200$ cm of the electric field reflected on the paraboloid of Fig. 7. In the exposure region the maximum inclination of the lines is lower than 3° and the axis relation greater than 10 and 17 for the GSM- and UMTS-band setups respectively. The polarization can be considered linear.

maximum phase error of $\pm 12°$ in the volume of a cage and an absolute value standard deviation of 11.2% and 10.7% for the GSM- and UMTS-band setups respectively. Within this volume a total of 40 cages with frontal dimensions of 30 cm×30 cm can be homogeneously exposed. The power efficiency is approximately 34% and the SAR efficiency approximately $0.002\,\frac{\mathrm{W/Kg}}{\mathrm{W}}$. The advantages of this concept are the small size of the equipment – a total volume of $3.2\times3.2\times3.5\,\mathrm{m}^3$ is occupied – and the use of a mass-production reflector, which will reduce the total cost of the equipment drastically.

In this paper waves reflected by the absorber material have been neglected and the configuration is still considered without racks, cages and animals, that means without taking into account the mutual effects generated by these objects. Even though the results presented here are expected to be in good agreement with the real field distribution, the final assessment of the exposure will have to take into account all these influences. It is expected that the limits recommended in the literature (Kuster and Schönborn, 2000) will be fulfilled.

References

Balanis, C. A.: Antenna Theory Analysis and Design, John Wiley & Sons, New York, 2. edn., 1997.

Balzano, Q., Chou, C.-K., Cicchetti, R., Faraone, A., and Tay, R. Y.-S.: An Efficient RF Exposure System with Precise Whole-Body Average SAR Determination for *in vivo* Animal Studies at 900 MHz, IEEE Transactions on Microwave Theory and Techniques, 48, 2040–2049, 2000.

Boonzaaler, J. J. and Pistorius, C. W. I.: Degradation of a Plane Wave in a Compact Range, in Antennas and Propagation Society International Symposium, vol. 3, Syracuse, New York, USA, 1402–1405, 1988.

Fröhlich, J., Berdiñas Torres, V., Klingenböck, A., Nikoloski, N., and Kuster, N.: Endbericht Projekt St.Sch4209 "Untersuchung der SAR-Verteilung elektromagnetisch exponierter Versuchstiere", Tech. rep., Bundesamt für Strahlenschutz, 2003.

Hansen, V. W., Bitz, A. K., and Streckert, J. R.: RF Exposure of Biological Systems in Radial Waveguides, IEEE Transactions on Electromagnetic Compatibility, 41, 487–493, 1999.

Kuster, N. and Schönborn, F.: Recommended Minimal Requirements and Development Guidelines for Exposure Setups of Bio-Experiments Addressing the Health Risk Concern of Wireless Communications, Bioelectromagnetics, 21, 508–514, 2000.

Schönborn, F., Poković, K., and Kuster, N.: Dosimetric Analysis of the Carousel Setup for the Exposure of Rats at 1.62 GHz, Bioelectromagnetics, 25, 16–26, 2004.

8

Computation of currents induced by ELF electric fields in anisotropic human tissues using the Finite Integration Technique (FIT)

V. C. Motrescu and U. van Rienen

Institute of General Electrical Engineering, University of Rostock, Albert-Einstein-Str. 2, 18059 Rostock, Germany

Abstract. In the recent years, the task of estimating the currents induced within the human body by environmental electromagnetic fields has received increased attention from scientists around the world. While important progress was made in this direction, the unpredictable behaviour of living biological tissue made it difficult to quantify its reaction to electromagnetic fields and has kept the problem open. A successful alternative to the very difficult one of performing measurements is that of computing the fields within a human body model using numerical methods implemented in a software code. One of the difficulties is represented by the fact that some tissue types exhibit an anisotropic character with respect to their dielectric properties. Our work consists of computing currents induced by extremely low frequency (ELF) electric fields in anisotropic muscle tissues using in this respect, a human body model extended with muscle fibre orientations as well as an extended version of the Finite Integration Technique (FIT) able to compute fully anisotropic dielectric properties.

1 Introduction

Simulations that take into account the anisotropy of muscle tissue were already performed by (Dawson and Stuchly, 1998). This paper reports a number of other authors that earlier incorporated anisotropic tissue properties within numerical methods like FD, FDTD and FEM. Dealing with SPFD and magnetically induced fields and currents, the paper of (Dawson and Stuchly, 1998) presents plenty of results for miscellaneous exposure conditions under the limitation of considering all the muscle fibers oriented parallel or perpendicular to the longitudinal axis of the body. These hypotheses can offer an image of the limits of induced current densities as they represent extreme cases, however as mentioned in the paper, the anisotropy of muscle tissue changes the current paths giving thus importance to the real fiber orientation especially for organ dosimetry.

In order to present our method for computation of electrically induced currents in the human body accounting for the anisotropy of skeletal muscles, we perform some simulations with a realistic human body model standing under a transmission power line employing the algorithm of the Finite Integration Technique (FIT) – a numerical method which discretizes Maxwell's equations on a grid pair preserving the analytical properties of the original equations (Weiland, 1977).

The human body model known as HUGO is based on the cryosection images provided by the Visual Human Project of the National Library of Medicine, Maryland, USA, which were assembled in a 3-D voxel model and then extended with muscle fiber orientations by a working group at the Institute of Biomedical Engineering, University of Karlsruhe, Germany (Sachse, 1998). It is known that muscle tissues present a transversal isotropic anisotropy with regard to their dielectric properties (Sachse, 1997). This property leads to diagonal material tensors in a local coordinate system aligned to the main directions of the fibers, while in a global coordinate system the anisotropy of muscle tissue is depicted by a full and symmetric tensor.

In a previous paper (Motrescu and van Rienen, 2004) we published details about the human body model, dielectric properties of biological tissues, and the extension of the Finite Integration Technique for computing anisotropic tissues. At that time, we performed simple electrostatic simulations with anisotropic muscle tissues considering different anisotropic ratios for the validation of our method. Comparison to the isotropic case has shown that the anisotropy of muscles changes the distribution of the electric potential inside the muscles and consequently the displacement current computed with this potential.

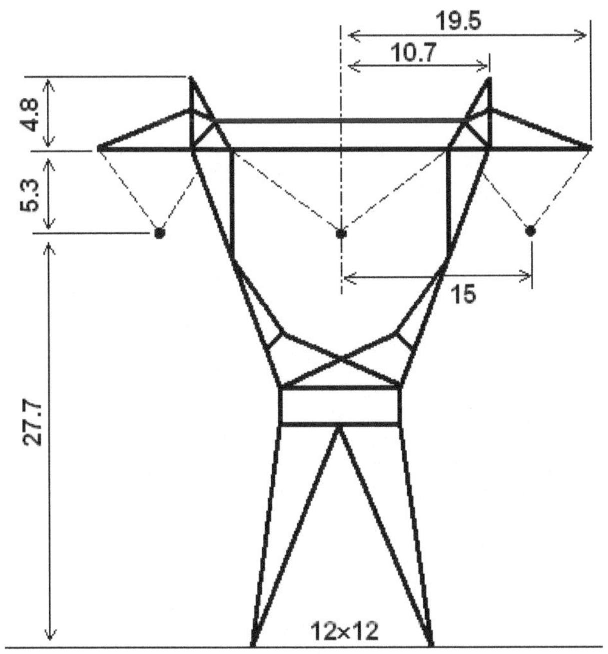

Fig. 1. Tower geometry (All dimensions are given in meters).

In this paper we publish results from electro-quasistatic simulations with a model which includes the anisotropy of both conductivity and permittivity, the latter expressing the capacitive effect at the separation plane between different tissue types.

2 Materials

As already mentioned, the electric field source chosen for our simulations is represented by a high voltage transmission power line operating at 765 kV and 50 Hz with parameters taken from EPRI (1982). Within an electromagnetic simulation software package based on the Finite Integration Technique (CST EMStudio^{TM}), a high voltage tower of lattice type, made of steel, for a single horizontal circuit, was modelled with the main dimensions given in Fig. 1.

The power line consists of two such towers situated at a distance of 408 m from each other such that the conductors in between have a length of 396 m corresponding to the average span length provided in EPRI (1982). For each phase consisting of four conductors with a diameter of 4 cm, spaced 60.96 cm from each other, we calculated an equivalent diameter of 43.63 cm with the formula given in Eq. (1) which is valid for conductors structured in regular bundles and calculation of fields that are far from the conductor surface (Deno and Zaffanella, 1982).

$$d_{eq} = D \sqrt[n]{\frac{nd}{D}} \quad (1)$$

In Eq. (1), D is the bundle diameter, n is the number of subconductors and d is the diameter of the subconductors.

Table 1. Minimum clearances for the 765 kV transmission power line.

Distance	Value in meters
Phase to tower	4.2
Phase to ground	16.2
Phase to phase	13.8

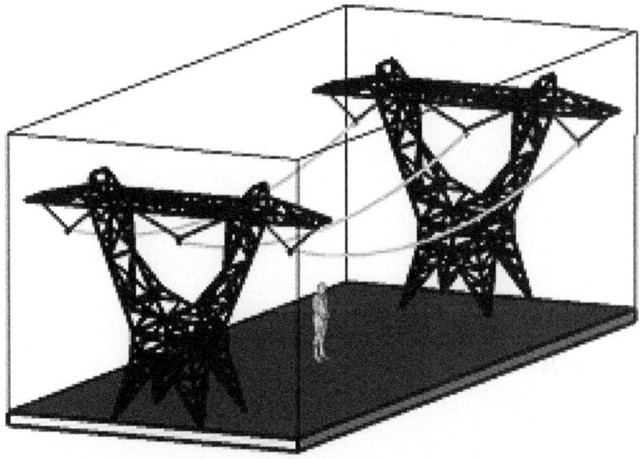

Fig. 2. Human body standing under the power line. Perspective view of the computational domain.

Shield wires were omitted considering that because they are positioned above the phase conductors, their influence is very small near the ground level. All the insulators were modeled as simple insulating rods while the ground was assigned to a perfect electric conductor (PEC) layer in galvanic contact with both towers assigned to the same layer. The ground was assumed to be perfect electric conductor based on the fact that earth surface charges redistribute very fast (0.1 to 100 ns) under a change in the applied field compared to the period of power frequency (Deno and Zaffanella, 1982). The power line configuration does not exceed minimum clearances recommended by the designers and provided in Table 1. At midspan the conductors reach the level of 16.2 m above ground which is equivalent to the minimum phase to ground clearance. The entire computational domain shown in Fig. 2 where the human body is standing under the power line, has the following dimensions: 414 m×52 m×40 m.

3 Methods

Due to the large sizes of the transmission line, we perform the simulation in two steps. In a first step the human body is removed from the computational domain and the power line, discretized on a grid of 6.4 million nodes, is simulated within the CST EMStudio^{TM} software package as an electrostatic problem with open boundary conditions.

The potential value found at the level of two meters above ground, is imposed in a second step as a boundary condition in an electro-quasistatic simulation with a much finer grid and fully anisotropic material properties, with the computational domain consisting of the human body surrounded by air. The second step computation is performed within a C^{++} software code (implemented by us for this purpose) that uses the CST EMStudio^{TM} only for the visualization of the computed fields.

3.1 Electrostatic simulation

Since electrostatic fields exist only in non-conductive media, there are no conduction, convection, or impressed currents in the mathematical model of electrostatics (van Rienen, 2001). The electric fields and displacement currents are calculated by solving in conjunction with the potential theory, a Poisson type equation of the form:

$$div\,(\underline{\underline{\varepsilon}}\; grad\;\varphi) = -\rho \tag{2}$$

where $\underline{\underline{\varepsilon}}$ is the rank two tensor of dielectric permittivity, φ is the electric potential and ρ the charge density. When discretized with FIT, the formula in Eq. (2) becomes:

$$\tilde{\mathbf{S}}\mathbf{M}_{\varepsilon}\tilde{\mathbf{S}}^{T}\,\Phi_E = \boldsymbol{q} \tag{3}$$

with the discrete divergence operator $\tilde{\mathbf{S}}$, the material operator $\mathbf{M}_{\varepsilon}$ and the discrete gradient operator $\mathbf{G}{=}-\tilde{\mathbf{S}}^{T}$. The discrete potential is denoted by Φ_E and the discrete charge density by $\boldsymbol{q}$.

The electric fields arising from a power line over a flat ground are practically vertical near the ground level (Deno and Zaffanella, 1982) reaching the maximum values at midspan. Figure 3 shows the lateral profile of the power line taken at two meters above the ground level in the midspan. Maximum values of both, potential (11.230 kV) and electric field (5.796 kV/m) were found under the midphase. This conforms to the potentials allocated on the phase conductors i.e. 441.672 kV on the midphase B and a negative potential of –220.836 kV on both lateral phases, A and C. The situation corresponds to the time t=0, (origin of the time axis) in Fig. 4 where the three-phase harmonic voltage system was represented. The phases were considered in a "wye" configuration. The ground potential was set to zero volt.

3.2 Electro-quasistatic simulation

At the frequency of 50 Hz the electromagnetic field has a wavelength of 6000 km which is very large compared to the size of a person so that the quasistatic assumption is fulfilled. For a predominantly electric field (electro-quasistatic, EQS), the time-derivative of the magnetic flux is negligible while the displacement currents have to be taken into account. Regarding the time harmonic EQS fields, these assumptions reduce the Maxwell's equations to the following complex Poisson type equation:

$$div\,((i\omega\underline{\underline{\varepsilon}} + \underline{\underline{\sigma}})grad\;\underline{\varphi}) = div(\underline{\boldsymbol{J}}_i) \tag{4}$$

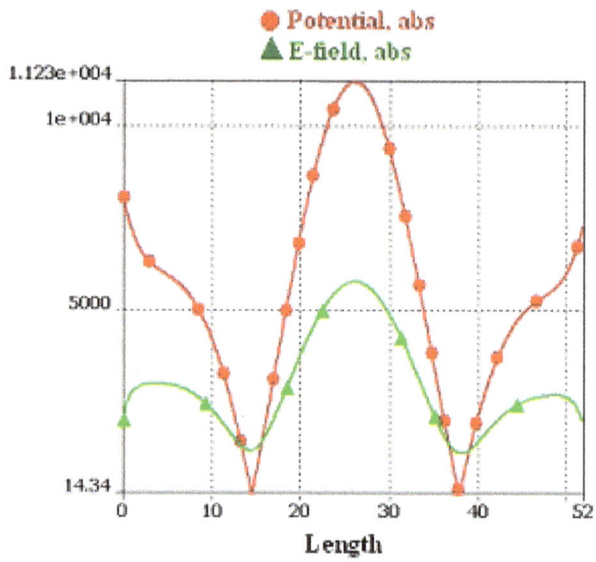

Fig. 3. Lateral profile of the power line at the level of two meters above ground (Markers have only an identification role).

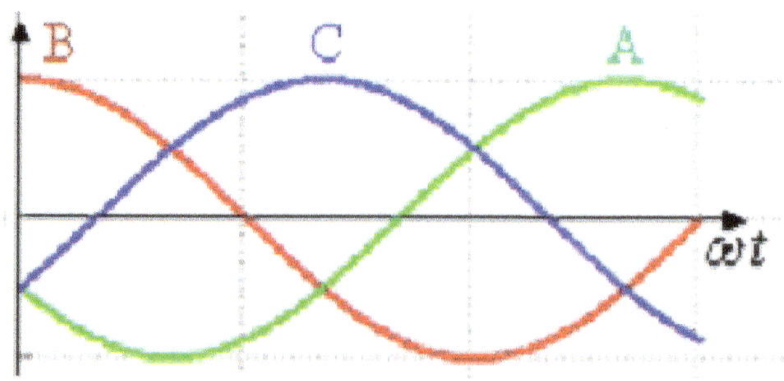

Fig. 4. Three-phase system of 50 Hz voltages.

The propagation medium is considered anisotropic described by the rank two symmetric tensors $\underline{\underline{\sigma}}$ and $\underline{\underline{\varepsilon}}$ having no electric charges in motion. The complex amplitude of the electric field is defined as the gradient of the complex scalar potential $\underline{\boldsymbol{E}}{=}-grad\;\underline{\varphi}$. The complex impressed current density is denoted by $\underline{\boldsymbol{J}}_i$.

With FIT, the continuous equation in Eq. (4) is discretized into the following one:

$$\tilde{\mathbf{S}}(i\omega\mathbf{M}_{\varepsilon} + \mathbf{M}_{\sigma})\tilde{\mathbf{S}}^{T}\,\underline{\boldsymbol{\Phi}}_E = \tilde{\mathbf{S}}\underline{\boldsymbol{j}}_i \tag{5}$$

with the discrete divergence operator $\tilde{\mathbf{S}}$, the material operators $\mathbf{M}_{\varepsilon}$, $\mathbf{M}_{\sigma}$ and the discrete gradient operator $\mathbf{G}{=}-\tilde{\mathbf{S}}^{\mathbf{T}}$. For the electro-quasistatic computations, the algorithm of FIT extended with fully anisotropic material properties was implemented in a software code which uses solvers from PETSc (Balay et al., 1997) for the solution of linear algebraic systems. The human body model of 8 mm resolution consisting of cubic voxels and extended with muscle fibre orientation was imported within the software code in a box with sizes: 592 mm×336 mm×1872 mm. At left, right, front and back we added on each side 72 mm of air. On the top-bottom direction, the air was added such that for different positions of the human body on this direction, the computational domain has always a height of two meters. This restriction comes from the fact that at the top side we impose through the Dirichlet boundary conditions, the maximum value of the unperturbed potential computed within the electrostatic simulation at the level of two meters above the ground, i.e. 11.230 kV. At bottom side we impose the corresponding ground potential from the electrostatic simulation i.e. zero volt. The left-right and front-back sides were treated conveniently with Neumann boundary conditions. This set of boundary conditions enforces the fields to be parallel to the longitudinal axis of the human body keeping thus the direction of the unperturbed electric fields under the power line near the ground level. The entire domain is discretized with a number of approximately 1.4 million mesh-nodes when the mesh is regular having the same size with the human body model resolution i.e. 8 mm. With respect to the position of the human body relative to the bottom boundary of the computational domain, we imagined three situations:

- Both feet touching the ground;
- Only one foot touching the ground;
- Both feet in the air.

With respect to the dielectric properties of tissues we imagined the following five situations:

- All tissues are isotropic;
- Anisotropic muscle with realistic fibre orientation;
- Anisotropic muscle with all fibres oriented on the top-bottom direction of the body;
- Anisotropic muscle with all fibres oriented on the left-right direction of the body;
- Anisotropic muscle with all fibres oriented on the front-back direction of the body.

Combining the two sets from above, we run a number of fifteen simulations. The dielectric properties of the isotropic tissues used in simulations were taken mainly from (Andreuccetti et al., 1997) which is a web server application calculating the dielectric properties of human body tissues using the parameters published by (Gabriel et al., 1996a). The anisotropic properties of muscles, in a ratio of 1.17 for the conductivity, and 2.27 for the permittivity, were taken directly from (Gabriel et al., 1996b).

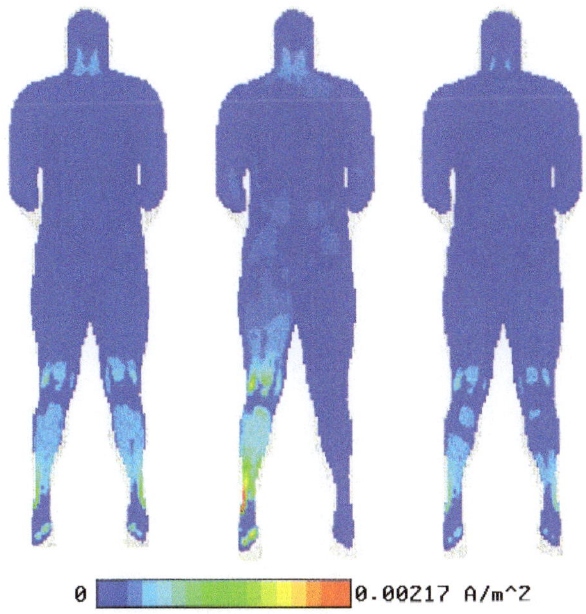

Fig. 5. Differences between real parts of the induced current densities in a section of the body computed with isotropic and anisotropic (natural orientation) muscles for the cases: left – both feet touch the ground, middle – only one foot touches the ground, right – both feet isolated from ground.

4 Results

As results from the electro-quasistatic simulations we provide a scalar plot of the difference between real parts of the induced current densities in a section of the body computed with isotropic and anisotropic (natural orientation) muscles for the cases when both feet touch the ground, only one foot touches the ground, and both feet isolated from ground.

Maximum values of the current density in different parts of the body are tabulated in Appendix A.

5 Discussions

The current density values from Table A1 show that the basic restriction for human exposure to ELF electric fields, of 2 mA/m^2 (for the general public, in head, neck and trunk) established by ICNIRP (1998) are not exceeded under the power lines.

6 Conclusion

In this paper we investigated the induced current densities in a human body standing under a transmission power line considering the anisotropy of muscle tissues and different exposure conditions. The results were presented in a plot as well as tabulated.

Appendix A

Table A1. Maximum current densites found in different parts of the body expressed as rms values in [mA/m^2] for the situation when the human model touches the ground with both feet, with only one foot or is isolated from ground and muscle tissues are considered isotropic or anisotropic with fibres orientation specified in the left column.

	Ankles	Head	Neck	Trunk
Both feet touching the ground				
Isotropic	2.296	0.779	0.463	0.485
Anisotropic natural	3.682	0.845	0.679	0.587
Anisotropic front-back	2.315	0.887	0.454	0.483
Anisotropic left-right	2.421	0.819	0.492	0.492
Anisotropic top-bottom	3.690	0.849	0.671	0.605
One foot touching the ground				
Isotropic	4.590	0.851	0.505	0.512
Anisotropic natural	7.106	0.923	0.741	0.615
Anisotropic front-back	4.733	0.969	0.496	0.568
Anisotropic left-right	5.601	0.895	0.537	5.385
Anisotropic top-bottom	7.100	0.928	0.733	0.646
Both feet in the air				
Isotropic	1.553	0.649	0.385	0.399
Anisotropic natural	2.444	0.705	0.565	0.484
Anisotropic front-back	1.604	0.740	0.378	0.397
Anisotropic left-right	1.503	0.683	0.409	0.410
Anisotropic top-bottom	1.553	0.649	0.385	0.399

Acknowledgements. The authors are grateful to M. Clemens (Helmut-Schmidt-Universität der Bundeswehr Hamburg) as well as to H. De Gersem (TU Darmstadt) for many useful discussions.
V. Motrescu was supported by DFG under the contract (RI 814/12-1) and by CST GmbH.

References

Andreuccetti, D., Fossi, R., and Petrucci, C.: An Internet resource for the calculation of the dielectric properties of body tissues in the frequency range 10 Hz–100 GHz, Insitute for applied physics "Nello Carrara", Florence, Italy, http://niremf.ifac.cnr.it/tissprop/, 1997.

Balay, S., Gropp, W. D., McInnes, L. C., and Smith, B.: PETSc – Efficient Management of Parallelism in Object Oriented Numerical Software Libraries, Modern Software Tools in Scientific Computing, Birkhauser Press, 163–202, 1997.

CST EMStudio™, CST GmbH, Bad Neuheimer Str. 19, D-64289, Darmstadt, Germany.

Dawson, T. W. and Stuchly, M.: Effects of skeletal muscle anisotropy on human organ dosimetry under 60 Hz uniform magnetic field exposure, Phys. Med. Biol. 43, 1059–1074, 1998.

Deno, D. and Zaffanella, L.: Field effects of overhead transmission lines and stations, in: Transmission Line Reference Book: 345 kV and Above, 2nd edn., (Palo Alto, CA: EPRI), 365–79, 1982.

EPRI: EHV tower geometries and line characteristics, in: Transmission Line Reference Book: 345 kV and Above, 2nd edn., (Palo Alto, CA: EPRI), 30–168, 1982.

Gabriel, S., Lau, R. W., and Gabriel, C.: The dielectric properties of biological tissues: II. Measurements in the frequency range 10 Hz to 20 GHz., Phys. Med. Biol. 41, 2251–2269, 1996a.

Gabriel, S., Lau, R. W., and Gabriel, C.: The dielectric properties of biological tissues: III. Parametric models for the dielectric spectrum of tissues, Phys. Med. Biol. 41, 2271–2293, 1996b.

ICNIRP: Guidelines for limiting exposure to time-varying electric, magnetic, and electromagnetic fields (up to 300 GHz), Health Physics Vol. 74, No 4, 494–522, 1998.

Motrescu, V. C. and van Rienen, U.: Computation of electrostatic fields in anisotropic human body tissues using the Finite Integration Technique (FIT), Kleinheubacher Berichte, Ad. Radio Sci., 2, URSI, 309–313, 2004.

Sachse, F. B., Werner, C., Meyer-Waarden, K., and Dössel, O.: Comparison of Solutions to the Forward Problem in Electrophysiology with Homogeneous, Heterogeneous and Anisotropic Impedance Models, Biomedizinische Technik, vol. 42, 277–280, 1997.

Sachse, F. B., Wolf, M., Werner, C., and Meyer-Waarden, K.: Extension of Anatomical Models of the Human Body: Three-Dimensional Interpolation of Muscle Fiber Orientation Based on Restrictions, Journal of Computing and Information Technology, vol. 6 no. 1, 95–101, 1998.

van Rienen, U.: Numerical Methods in Computational Electrodynamics, Linear Systems in Practical Application, Lecture Notes in Computational Science and Engineering, Vol. 12, Springer Verlag, Berlin Heidelberg, 2001.

Weiland, T.: Eine Methode zur Lösung der Maxwellschen Gleichungen für sechskomponentige Felder auf discreter Basis, AEÜ, 31, 116–120, 1977.

9

A comparision of Hodgkin-Huxley and soliton neural theories

R. Appali, S. Petersen, and U. van Rienen

Institute of General Electrical Engineering, Chair of Electromagnetic Field Theory, University of Rostock, Justus-von-Liebig-Weg 2, 18059 Rostock, Germany

Abstract. Hodgkin and Huxley were the pioneers to abstract biological neuron as an electric circuit and nerve signal as the voltage impulse. The Hodgkin-Huxley theory (Hodgkin and Huxley, 1952) has set the direction and defined the goals for much of the ensuing research in biophysics. However, in 2005, T. Heimburg and A. D. Jackson, biophysicists from Copenhagen proposed a new neural theory called Soliton theory (Heimburg and Jackson, 2005). In this theory, the nerve conduction is proposed as a density wave.

In this paper, Hodgkin-Huxley and Soliton theories are described and a theoretical comparison has been carried out throughout the analysis of the theories and models.

1 Introduction

Nerves are the information transmitter lines from the brain to organs. The nerves are described as the bundle of discrete individual cells called neurons. These individual cells transmit the information as signals.

The main parts of a neuron are the cell body (*soma*), signal receiving ends (*dendrites*), signal transmitter (*axon*), and connecting ends to other neurons or glands (*synapses*). The scientific model of the neuron was first given in electrical terms by Hodgkin and Huxley (Hodgkin and Huxley, 1952). Hodgkin and Huxley conducted the experiments on non-myelinated giant axon of squid to study the neuronal properties. The Hodgkin-Huxley equations (Hodgkin and Huxley, 1952) are the starting point for detailed neuronal models.

2 The Hodgkin-Huxley theory

The theory was the summation of active experimental and theoretical association of Hodgkin and Huxley. Their five publications in 1952 in the Journal of Physiology are quite renowned. The four key research developments which led to the theory were: (Häusser, 2000)

- Firstly, Cole and Curtis demonstrated that the action potential is associated with a large increase in membrane conductance (Cole and Curtis, 1939).
- Secondly, Hodgkin and Huxley made the first intracellular recording of an action potential (Hodgkin and Huxley, 1952). This demonstrated that the action potential exceeds zero mV, denying Bernstein's hypothesis (Bernstein, 1902) that the underlying increase in membrane permeability is non-selective.
- Hodgkin and Katz (Hodgkin and Katz, 1949) explained the overshooting action potential as the result of an increase in sodium permeability approving the work of Overton (Overton, 1902).
- Finally, Hodgkin, Huxley and Katz developed the voltage-clamp circuit to facilitate quantitative measurement of ionic currents from squid axon.

The step depolarization of squid axon triggering an inward current followed by an outward current was then proved by Hodgkin and Huxley. With the aid of ionic substitution, they demonstrated that this net current could be separated into two distinct components, a fast inward current carried by Na^+ ions, and a more slowly activating outward current carried by K^+ ions. Using ingenious voltage-clamp protocols, they concluded that these two currents result from independent permeability mechanisms for Na^+ and K^+ with conductance changing as a function of time and membrane potential. This striking conceptual breakthrough was then termed as the "ionic hypothesis" (Häusser, 2000).

The three different types of ion current, viz., sodium, potassium, and a leak current that consists mainly of Cl^- ions contribute to the voltage signal in the neuron. The flow of these ions through the cell membrane is controlled by their respective voltage-dependent ion channels. The leak current takes care of other channel types which are not described in particular. The most remarkable achievement of this theory

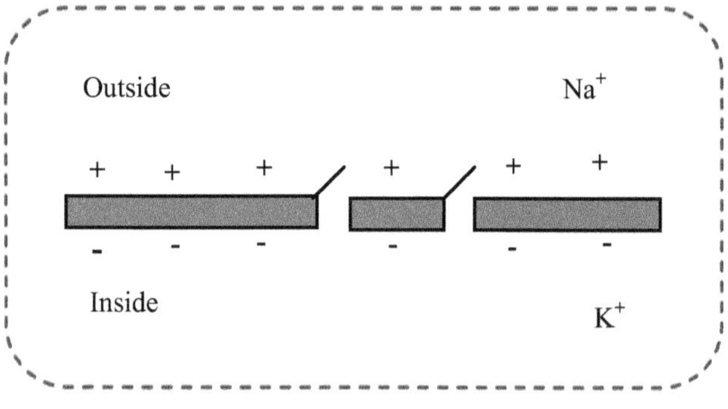

Fig. 1. Schematic model of Hodgkin-Huxley theory (modified from Gerstner et al., 2002).

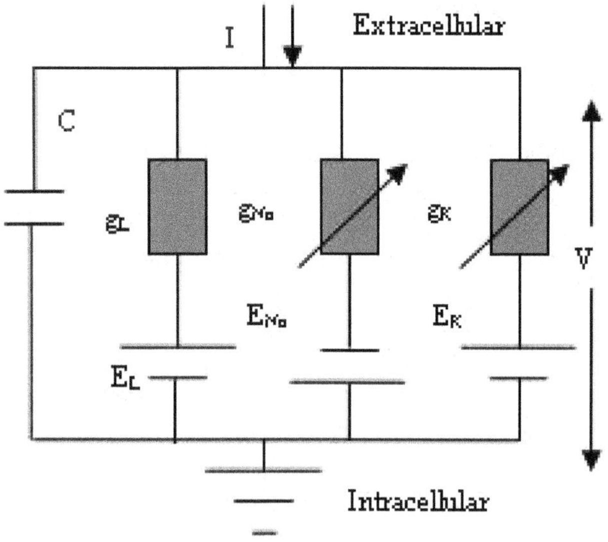

Fig. 2. Equivalent circuit of H-H model (Hodgkin and Huxley, 1952).

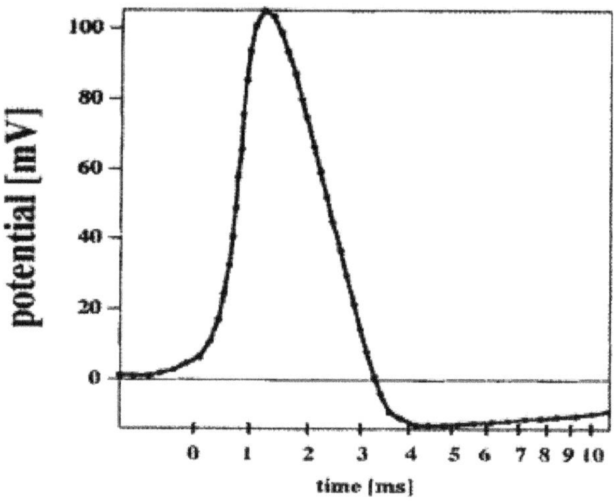

Fig. 3. Action potential adapted from the original paper of Hodgkin and Huxley.

was the empirical representation of the experimental data in a quantitative model (Hodgkin and Huxley, 1952), the very first complete description of the excitability of single neuron.

2.1 The model

The Hodgkin-Huxley model (H-H model) can be explained with the help of Fig. 1. The semi-permeable cell membrane separates the interior of the cell from the extracellular liquid and acts as a capacitor. If an input current $I(t)$ is injected into the cell, it may add further charge on the capacitor C, or leak through the channels in the cell membrane. Because of active ion transport through the cell membrane, the ion concentration inside the cell is different from that of the ion concentration in extracellular liquid. The Nernst potential generated by the difference in ion concentrations is represented by a battery. The equivalent circuit is shown in Fig. 2.

As mentioned above, the Hodgkin-Huxley model describes three types of channels. All channels may be characterized by their resistance or, equivalently, by their conductance. The leakage channel is described by a voltage-independent conductance and the conductance of the other ion channels is voltage and time dependent.

If all channels are open, they transmit currents with a maximum conductance g_{Na} or g_K, respectively. Normally, however, some of the channels are blocked. The probability that a channel is open is described by additional variables m, n, and h. The combined action of m and h controls the Na^+ channels. The K^+ gates are controlled by n. Hodgkin and Huxley formulated the three ionic components as

$$C\frac{du}{dt} = g_{Na}m^3h(u - E_{Na}) + g_K n^4(u - E_K) + g_L(u - E_L) + I(t). \tag{1}$$

The parameters E_{Na}, E_K, and E_L are the reversal potentials. Reversal potentials and conductances are empirical parameters. So, it is demonstrated that the neurons transmit information by firing and propagating electrical action potentials along their axons. Usually, motor neurons possess a myelin sheath around their axons acting as an insulator and enhancing the speed of signal transmission.

The model can reproduce and explain a wide range of data from squid axon like the shape and propagation of the action potential (see Fig. 3), its sharp threshold, refractory period, anode-break excitation, accommodation and sub threshold oscillations. The model can also describe many channel types with small parameter changes indicating the generality of the approach. Thus, the H-H model links the microscopic level of ion channels to the macroscopic level of currents and action potentials (see Fig. 4).

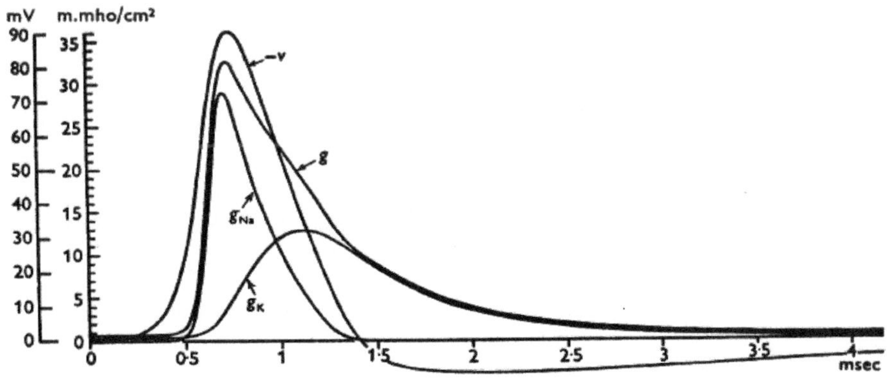

Fig. 4. Separation of ionic conductances underlying the action potential in H-H model. Modified from Hodgkin and Huxley (1952).

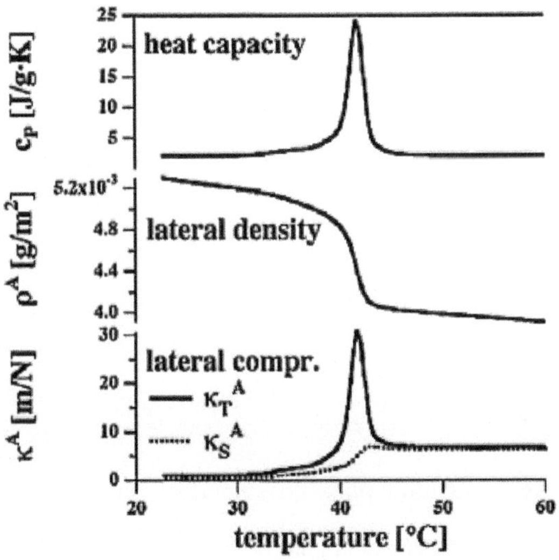

Fig. 5. Heat capacity, lateral density and lateral compressibility as a function of temperature (adapted from Heimburg and Jackson, 2005).

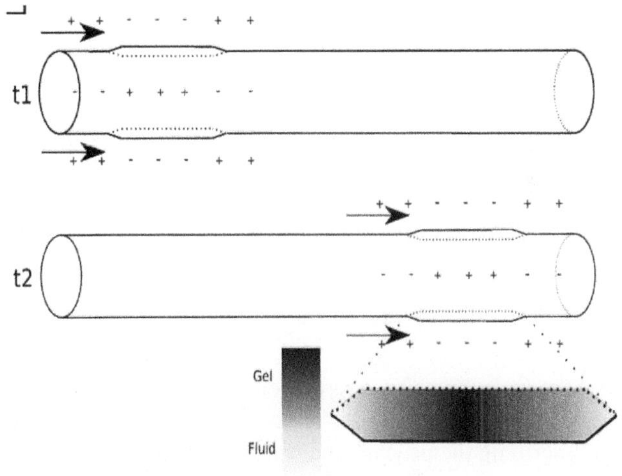

Fig. 6. Schematic model of membrane lipid thermodynamic theory (adapted from Anderson et al., 2009).

3 Soliton theory

T. Heimburg and A. D. Jackson proposed an alternative theory in 2005. The theory is a thermodynamic theory of nerve pulse propagation. This theory is based on the lipid characteristic of transition from a fluid to a gel state over a range of several degrees, slightly below the body temperature (Heimburg and Jackson, 2005, and references there in). That is, when a biological membrane is compressed, the fluid membrane gradually reaches to a denser gel state. Now, a sufficient pressure/ compression either in membrane area or volume will bring a transition to the gel state (Heimburg, 2007a). This transition is associated with the release of heat (Heimburg, 2007a; Heimburg, 2007b). The relation between heat capacity, density, compression and temperature are shown in Fig. 5.

Lipid membranes have the properties required for the generation and propagation of solitons. Pure lipid membranes exhibit "non-linearity" (maximally compressible close to the transition and decreased compressibility with increasing density or pressure) and "dispersion" (lipid compressibility is frequency dependent both for area and volume compressibility) (Heimburg and Jackson, 2005; Heimburg and Jackson, 2007a).

Both the effects of non-linearity and dispersion produce a self-sustaining and localized density pulse with a moving segment of the nerve membrane in the gel state (see Fig. 6). This pulse is called a soliton (Heimburg and Jackson, 2007a). The solitary wave maintains its shape while it travels at a constant speed less than the sound velocity in the lipid membrane. Solitons can propagate over long distances without loss of energy (Heimburg and Jackson, 2005). The pulse is also called as adiabatic pulse, because no energy is lost to the environment during propagation (Heimburg and Jackson, 2007b). The soliton mode of pulse propagation is basically different from the AP given by the H-H model (Anderson et al., 2009).

The solitary wave is associated with a transient voltage change due to the transient change in state of membrane. This affects the membrane potential difference. Especially, it changes both the area and thickness of membrane and thereby capacitance (Heimburg and Jackson, 2007a).

These changes seem to be as a voltage pulse and leads to a capacitive current due to the unbalanced charge of membrane. The reversible heat observed during action potential is proposed as a result of lipid melting enthalpy i.e. production of latent heat during lipid transition from fluid to gel state and the reabsorption of heat as the system returns to the fluid state (Anderson et al., 2009).

Thus the heat flow is correlated with the changes in membrane density and potential.

3.1 Soliton equation

Hydrodynamic equation for the propagation of a density pulse in the presence of dispersion for a cylindrical membrane along x-axis is (Heimburg and Jackson, 2005) given by

$$\frac{\partial^2}{\partial t^2}\Delta\rho^A = \frac{\partial}{\partial x}\left[c^2 \partial/\partial_x \Delta\rho^A\right] - h\frac{\partial^4}{\partial x^4}\Delta\rho^A \tag{2}$$

After the co-ordinate transformation ($z = x - v \cdot t$), by introducing the propagation velocity, the time independent equation describing the propagation density excitation is given as (Heimburg and Jackson, 2007a)

$$v^2\frac{\partial^2}{\partial z^2}\Delta\rho^A = \frac{\partial}{\partial z}\left[\left(c0^2 + p\Delta\rho^A + q(\Delta\rho^A)^2 + \ldots\right)\partial/\partial_z\Delta\rho^A\right] - h\frac{\partial^4}{\partial z^4}\Delta\rho^A \tag{3}$$

Where

v is propagation velocity

ρ^A is lateral membrane density

c is sound velocity in fluid phase of membrane

p and q are the parameters determined from sound velocity and density dependence

h is the parameter to set the linear scale of propagating pulse.

This theory also explains the effect of anesthetics on the nerve pulse propagation. According to this theory, the anesthetics in the membrane lower the fluid-gel transition temperature, a phenomenon known as freezing point depression (Heimburg and Jackson, 2007c). So it is difficult to force the membrane through the fluid-gel transition, thereby the action potential generation is suppressed. It is noted that freezing point depression is a general thermodynamic phenomenon and not chemical in nature (Anderson et al., 2009).

4 Comparison

The major differences between both the neural theories are (Hodgkin and Huxley, 1952; Heimburg and Jackson, 2005; Heimburg and Jackson, 2007a; Anderson et al., 2009):

H-H The action potential is due to the electric current conducted across the membrane by ion channels that act as resistors.

S The nerve pulse is a solitary wave generated by the lipid transition in the membrane.

H-H Nerve pulse propagation is then explained as a moving self regenerating pulse of local capacitive discharges through the resistors followed by capacitor recharging.

S The propagation is due to the combined effect of nonlinearity and dispersion.

H-H The theory is pure electrical with ionic hypothesis.

S The theory is physical/ mechanical i.e. based on thermodynamics.

H-H The nerve signal or action potential is electrical pulse.

S The nerve signal is propagating adiabatic electromechanical pulse.

H-H The pulse propagation dissipates energy in the form of heat.

S The propagating pulse does not dissipate heat energy.

H-H The nerve's physical changes are not explained.

S The nerve changes dimensions and gets thicker during nerve pulse. The voltage pulse accompanies the propagating signal, as a result of these transient geometric changes of the membrane.

Both the theories explain the existence of voltage pulse in nerve pulse propagation. However, as an information transmitting pulse in H-H model and as a by-product signal in soliton theory.

5 Discussion

Hodgkin-Huxley model of nerve pulse propagation is undoubtedly a remarkable achievement in the field of physiology. However, the theory could not explain the physical phenomena such as reversible heat changes, density changes and geometrical changes observed in the experiments (Iwasa et al., 1980; Tasaki, 1982, 1999; Tasaki et al., 1989, Tasaki and Byrne, 1992).

Soliton model is an alternative model, convincingly explaining some of the unexplained observations in the experiments. However, there are several other questions that this has to answer like ion flow involvement in nerve signal propagation as stated by H-H model and also the faster propagation in myelinated nerves than in unmyelinated.

6 Context

This is a theoretical comparison of both the theories. This has been carried out to analyze and choose a detailed neuron model which fits in for the goals of our project. We are examining the soliton theory to model and simulate the action potential and the coupling of electrode for a detailed study of the neuron.

Acknowledgements. We are grateful to DFG (German Science Foundation) for funding our project in the Research Training Group 1505/1.

References

Andersen, S., Jackson, A. D., and Heimburg, T.: Towards a thermodynamic theory of nerve pulse propagation, Progr. Neurobiol., 88, 104–113, 2009.

Bernstein, J.: Studies on the thermodynamics of bioelectric currents, Pflügers Arch. Ges. Physiol., 92, 521–562, 1902.

Cole, K. S. and Curtis, H. J.: Electric impedance of the squid giant axon during activity, J. Gen. Physiol., 22, 649–670, 1939.

Gerstner, W. and Kistler, M. W.: Spiking neuron models. Single neurons, populations, plasticity, Cambridge Univ. Press, p. 34, 2002.

Häusser, M.: The Hodgkin-Huxley theory of the action potential, Nature neuroscience, 3, 1165, doi:10.1038/81426, 2000.

Heimburg, T. and Jackson, A. D.: On soliton propagation in biomembranes and nerves, Proc. Natl. Acad. Sci. USA., 102, 9790–9795, 2005.

Heimburg, T.: Chapter 6: lipid melting, in: Thermal Biophysics of Membranes, Wiley–VCH, pp. 75–97, 2007a.

Heimburg, T. and Jackson, A. D.: On the action potential as a propagating density pulse and the role of anesthetics, Biophys. Rev. Lett., 2, 57–78, 2007a.

Heimburg, T.: Chapter 7: phase diagrams, in: Thermal Biophysics of Membranes, Wiley–VCH, pp. 99–122, 2007b.

Heimburg, T. and Jackson, A. D.: Thermodynamics of the nervous impulse. In: Nag, K. (Ed.), Structure and Function of Membranous Interfaces, Wiley and Sons, 2007b.

Heimburg, T. and Jackson, A. D.: The thermodynamics of general anesthesia Biophys. J., 92, 3159–3165, 2007c.

Hodgkin, A. L. and Huxley, A. F.: A quantitative description of membrane current and its application to conduction and excitation in nerve, J. Physiol., 117, 500–544, 1952.

Hodgkin, A. L.: The ionic basis of nervous conduction, Nobel Lectures, Physiology or Medicine, 1963–1970.

Hodgkin, A. L. and Katz, B.: The effect of sodium ions on the electrical activity of the giant axon of the squid, J. Physiol. (Lond.), 108, 37–77, 1949.

Iwasa, K., Tasaki, I., and Gibbons, R. C.: Swelling of nerve fibers associated with action potentials, Science, 210, 338–339, 1980.

Overton, E.: Contributions to general muscle and nerve physiology II. Indispensability of sodium (or lithium) ions for muscle contraction, Pflügers Arch. Ges. Physiol., 92, 346–386, 1902.

Tasaki, I.: Physiology and Electrochemistry of Nerve Fibers, Academic Press, New York, 1982.

Tasaki, I.: Evidence for phase transitions in nerve fibers, cells and synapses, Ferroelectrics, 220, 305–316, 1999.

Tasaki, I. and Byrne, P. M.: Heat production associated with a propagated impulse in bullfrog myelinated nerve fibers, Jpn. J. Physiol., 42, 805–813, 1992.

Tasaki, I., Kusano, K., and Byrne, P. M.: Rapid mechanical and thermal changes in the garfish olfactory nerve associated with a propagated impulse, Biophys. J., 55, 1033–1040, 1989.

10

Efficient $\omega - k$-algorithm for circular SAR and cylindrical reconstruction areas

A. Dallinger, S. Schelkshorn, and J. Detlefsen

Technische Universität München, Lehrstuhl für Hochfrequenztechnik, Fachgebiet Hochfrequente Felder und Schaltungen, D-80333 München, Germany

Abstract. We present a novel reconstruction algorithm of $\omega - k$ type which suits for wideband circular synthetic aperture data taken in stripmap mode. The proposed algorithm allows to reconstruct an image on a cylindrical surface. The range trajectory is approximated by Taylor Series expansion using only the quadratic terms which limits the angular reconstruction range (cross range). In our case this is not a restriction for the application. Wider areas with respect to cross range can be realized by joining several reconstructed images side by side to build a wider image by means of digital spotlighting.

1 Introduction

Broadband imaging in three dimensions is feasible using the scattered field distributions measured along curved lines. As a special case also circular apertures can provide 3-D spatial resolution (Ishimaru et al., 1998). A circular aperture can be realized by either using an array of sensors or by moving one sensor on a circular trajectory generating a synthetic aperture of circular shape.

In general holographic data can be focused for a known aperture and reconstruction area geometry as long as the sampling constraints are met. A direct inversion scheme of the linearized inverse scattering problem can be realized e.g. by means of correlation or backprojection algorithms. These algorithms are not optimum with respect to computational efficiency but have the advantage to work easily with arbitrary geometries.

Fast and efficient algorithms ($\omega - k$ type) exist for focusing spotlight circular synthetic aperture data on planes parallel to the circular aperture plane (Soumekh, 1996). In the context of imaging of humans it would be desirable to reconstruct scattering centers primarily on cylindrical surfaces. Therefore we present an efficient $\omega - k$ stripmap reconstruction algorithm adapted for focusing holographic data, taken by a monostatic circular aperture, onto a cylindrical reconstruction area. This type of imaging is related to linear stripmap slant range imaging (Soumekh, 1999), but the derivation of the spectral support function of the scattering equation relies on an approximation of the range trajectory. This limits the length of the circular aperture and the size of the reconstructed area. In our case this is not a severe restriction because target signatures usually cannot be observed over wider angular regions compared to the limits given by the approximation. If wider angular regions have to imaged so-called digital spotlighting can be used in order to join several small stripmap images side by side. Also reconstruction areas lying within a widespread scattering area can be imaged using this technique.

2 Derivation of the algorithm

2.1 Geometrical definitions

Figure 1 shows the geometrical setup. A monostatic broadband imaging sensor is moved around the object on a circular path. The angular path variable is φ_a. For convenience we assume that the range of φ_a is symmetric with respect to the $x - z$-plane, i.e. $\varphi_a \in [-\varphi_{a,max} \dots \varphi_{a,max}]$. The path length in angular coordinates is $L_{\varphi_a} = 2\,\varphi_{a,max}$ and the radius ρ_a. We also assume that the sensor illuminates the object homogeneously, i.e. the object resides within the [3]dB antenna beamwidth. The scattered field is sampled on the discretized circular path with an angular increment of $\Delta\varphi_a$. The object is reconstructed on a cylindrical surface with radius ρ_o and an angular extent of $L_{\varphi_o} = 2\,\varphi_{o,max}$ with $\varphi_o \in [-\varphi_{o,max} \dots \varphi_{o,max}]$.

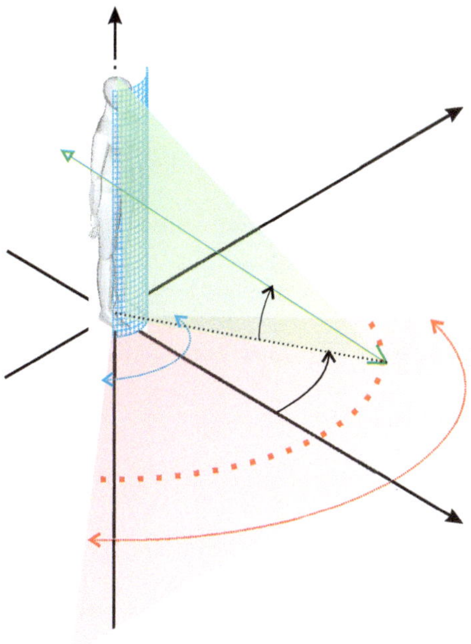

Fig. 1. SAR imaging geometry: The imaging sensor is moved on a circular path around the person. It is assumed that the person is illuminated homogeneously. The imaging area is located on a cylindrical surface.

Both the aperture and the reconstruction area are described in polar coordinates, that is for the aperture $\boldsymbol{r}_\mathrm{a} = (\rho_\mathrm{a}\cos\varphi_\mathrm{a}, \rho_\mathrm{a}\sin\varphi_\mathrm{a}, z_\mathrm{a})^\top$ and for the reconstruction area $\boldsymbol{r}_\mathrm{o} = (\rho_\mathrm{o}\cos\varphi_\mathrm{o}, \rho_\mathrm{o}\sin\varphi_\mathrm{o}, z_\mathrm{o})^\top$.

2.2 Linearized inverse scattering

The first order Born approximation (Morse and Feshbach, 1953) is used in order to linearize the inverse scattering problem. The description of the object may be further simplified by assuming distributed point sources which do not interact with each other in an electromagnetic sense. Thus the scattering of the incident wave reduces to a single bounce, line of sight (direct path between the transmitter/receiver and the point scatterer) process, which is linear. Hence contributions from each point scatterer superimpose coherently. The scattered field (monostatic case, i.e. the incident wavevector $\boldsymbol{k}_\mathrm{i}$ is equal to the scattered wavevector $-\boldsymbol{k}_\mathrm{s}$) is found to be

$$u_\mathrm{s}(\boldsymbol{r}_\mathrm{a}, k) = \int o(\boldsymbol{r}_\mathrm{o}) \exp[-j2kr(\boldsymbol{r}_\mathrm{a}|\boldsymbol{r}_\mathrm{o})]\mathrm{d}V_\mathrm{o} \quad . \tag{1}$$

The object itself is described by an object function $o(\boldsymbol{r}_\mathrm{o})$ which is assumed to be a superposition of distributed point sources, i.e. $o(\boldsymbol{r}_\mathrm{o}) = \sum_{n=1}^{N} \delta_n(\boldsymbol{r}_{\mathrm{o},n})$. k stands for the wavevector of the propagation medium. For the monostatic case we have $k = |\boldsymbol{k}| = |\boldsymbol{k}_\mathrm{i}| = |-\boldsymbol{k}_\mathrm{s}| = 2\pi/\lambda = 2\pi f/c_0$. The support of k is given by the bandwidth B of the system, i.e. $k \in [2\pi f_{\mathrm{min}}/c_0 \ldots 2\pi f_{\mathrm{max}}/c_0]$.

The distance from the aperture to a point of the reconstruction area can be determined by

$$r(\boldsymbol{r}_\mathrm{a}|\boldsymbol{r}_\mathrm{o}) \\ \sqrt{{\rho_\mathrm{a}}^2 + {\rho_\mathrm{o}}^2 - 2\rho_\mathrm{a}\rho_\mathrm{o}\cos(\varphi_\mathrm{o} - \varphi_\mathrm{a}) + (z_\mathrm{o} - z_\mathrm{a})^2}. \tag{2}$$

2.3 Angular spectral domain of a point scatterer

The inversion of Eq. 1 leads to the unknown object function $o(\boldsymbol{r}_\mathrm{o})$. For an $\omega - k$ type inversion one has to find the spectral support function of a point scatterer with respect to the path variable. Therefore we approximate Eq. 2 by using a Taylor series expansion of the cosine term (i.e. $\cos x = 1 - \frac{x^2}{2!} + \ldots$). By neglecting higher order terms the distance becomes

$$r(\boldsymbol{r}_\mathrm{a}|\boldsymbol{r}_\mathrm{o}) = \sqrt{\alpha + \beta(\varphi_\mathrm{o} - \varphi_\mathrm{a})^2} \tag{3}$$

with $\alpha = (\rho_\mathrm{o} - \rho_\mathrm{a})^2 + (z_\mathrm{o} - z_\mathrm{a})^2$ and $\beta = \rho_\mathrm{a}\rho_\mathrm{o}$. The scattered signal measured at the receiver is

$$u_s(\boldsymbol{r}_\mathrm{a}, k) = u_s(\varphi_\mathrm{a}, k) = e^{-j2k\sqrt{\alpha+\beta(\varphi_\mathrm{o}-\varphi_\mathrm{a})^2}} \quad . \tag{4}$$

$\boldsymbol{r}_\mathrm{a}$ depends in this case only on the cross range angle φ_a (ρ_a and z_a are constant). By taking the Fourier transform of both sides of Eq. 4 with respect to the variable φ_a the angular spectral function is obtained to

$$u_s(k_{\varphi_\mathrm{a}}, k) = \mathcal{F}_{\varphi_\mathrm{a}}\{u_s(\varphi_\mathrm{a}, k)\} \\ = \int_{-\pi}^{\pi} e^{-j2kr} e^{-jk_{\varphi_\mathrm{a}}\varphi_\mathrm{a}} d\varphi_\mathrm{a} \quad . \tag{5}$$

The integral can be evaluated by the method of stationary phase. The general result is given by

$$\int_{-\infty}^{\infty} g(\chi)e^{j\phi(\chi)}d\chi = \sqrt{\frac{j2\pi}{\phi''(\chi^*)}} g(\chi^*)e^{j\phi(\chi^*)} \quad . \tag{6}$$

$\chi^* \in \chi$ is called the stationary point and is determined by the solution of $\phi'(\chi^*) = 0$. The stationary point φ_a^* of Eq. 5 can be determined to be

$$\varphi_\mathrm{a}^* = \varphi_\mathrm{o} - \frac{\sqrt{\alpha}k_{\varphi_\mathrm{a}}/\beta}{\sqrt{4k^2 - k_{\varphi_\mathrm{a}}^2/\beta}} \quad . \tag{7}$$

The angular spectrum follows to

$$u_s(k_{\varphi_\mathrm{a}}, k) = e^{-j\sqrt{4k^2 - k_{\varphi_\mathrm{a}}^2/\beta}\sqrt{\alpha} - jk_{\varphi_\mathrm{a}}\varphi_\mathrm{o}} \tag{8}$$

neglecting all changes of amplitude terms.

2.4 Inversion scheme

The term $\sqrt{\alpha}$ in Eq. 8 is the distance of closest approach from the sensor to a point scatterer ($\varphi_\mathrm{a} = \varphi_\mathrm{o}$). It is used as the range axis r_τ (in terms of propagation delay $\tau = 2r/c_0$).

For the reconstructed image the range and cross range variables, r_τ and φ, respectively, are used (whereas $\varphi = \varphi_a$). The spacial spectrum of point scatterer n at position $(r_{\tau,n}, \varphi_{o,n})$ can be described by

$$\mathcal{F}_{\varphi,r_\tau}\left\{\delta(\varphi - \varphi_{o,n}, r_\tau - r_{\tau,n})\right\} = e^{-jk_{r_\tau} r_{\tau,n} - jk_\varphi \varphi_{o,n}} \quad . \tag{9}$$

Comparing Eqs. 8 and 9 and taking into account Eq. 1 the inversion scheme results in

$$o(r_\tau, \varphi) = \mathcal{F}^{-1}_{k_\varphi,k_{r_\tau}}\left\{\text{Stolt Mapping}\left[\mathcal{F}_{\varphi_a}\{u_s(\varphi_a, k)\}\right]\right\}. \tag{10}$$

The so called Stolt Mapping is a variable transformation which is described by

$$k_{r_\tau} = \sqrt{4k^2 - k_{\varphi_a}{}^2/\beta}$$
$$k_\varphi = k_{\varphi_a}\,. \tag{11}$$

After having performed the Stolt Mapping an interpolation of the data, which is distributed on a non-linear grid $(k_{r_\tau},\ k_\varphi)$ is necessary. Methods like the Nonuniform Fourier transform (NUFFT), (Fessler and Sutton, 2003) or Gridding (Jackson et al., 1991; Wajer et al., 2000) can be applied. We use NUFFT because precomputed interpolation matrices can be used which makes the computational process very fast.

The inversion scheme is straight forward and has close similarities to the case when a linear monostatic aperture is used (Soumekh, 1999; Carrara et al., 1995).

2.5 Angular spectral support

The support of the angular spectrum can be estimated by calculating the angular spectral variable k_{φ_a} which is

$$k_{\varphi_a} = \frac{d\angle\{u_s(\varphi_a, k)\}}{d\varphi_a}$$
$$= \frac{2k\rho_a\rho_o \sin(\varphi_o - \varphi_a)}{\sqrt{\rho_a{}^2 + \rho_o{}^2 - 2\rho_a\rho_o\cos(\varphi_o - \varphi_a) + (z_o - z_a)^2}}. \tag{12}$$

The angular spectral support band is determined by the angular extent of the reconstruction area and the length of the aperture. Assuming that $L_{\varphi_a} = L_{\varphi_o}$ the boundaries of the support band can be estimated from Eq. 12 which results in

$$|k_{\varphi_a}| \le \frac{2k_{\max}\rho_a\rho_o \sin(L_{\varphi_a})}{\sqrt{\rho_a{}^2 + \rho_o{}^2 - 2\rho_a\rho_o\cos(L_{\varphi_a}) + (z_o - z_a)^2}}\,. \tag{13}$$

Due to the strong chirp character of the received signal the angular spectrum of a point scatterer is a bandpass function. The boundaries are calculated by using Eq. 12. Assuming a point scatterer at position $\varphi_o = 0$ its angular spectral support is given by

$$|k_{\varphi_a,n}|$$
$$\le \frac{2k_{\max}\rho_a\rho_o \sin(L_{\varphi_a}/2)}{\sqrt{\rho_a{}^2 + \rho_o{}^2 - 2\rho_a\rho_o\cos(L_{\varphi_a}/2) + (z_o - z_a)^2}}, \tag{14}$$

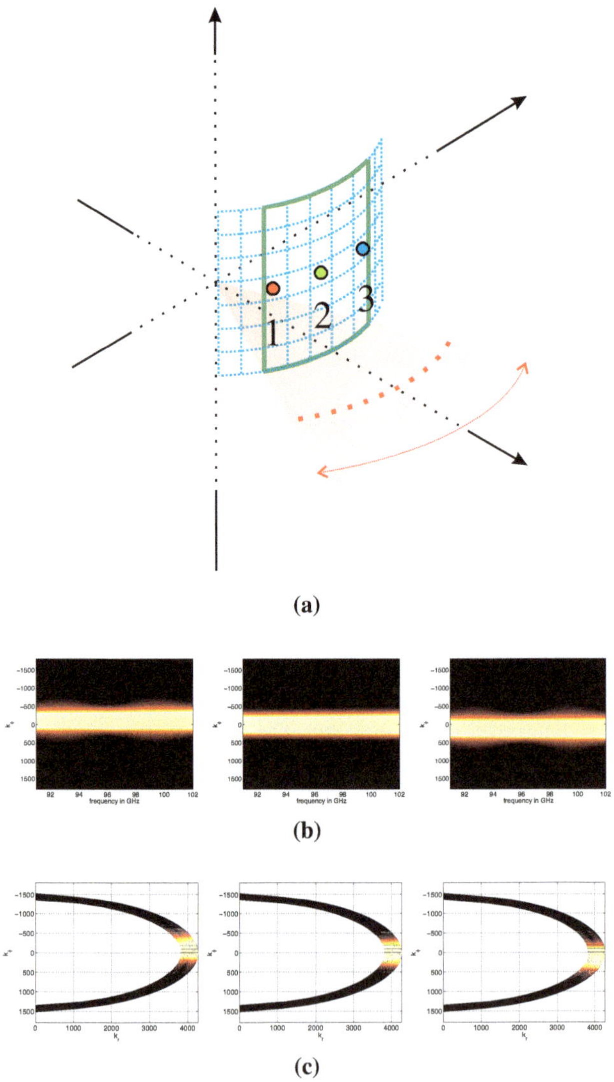

Fig. 2. The spectrum of point scatterers with different angular positions ($L_\varphi = [40]^\circ$, $\Delta\varphi = [0.1]^\circ$, $\rho_a/\rho_o = 3$, $|z_a - z_o| = [0.5]m$, $B = [11]GHz$)
(a) setup of three point scatterers (numbered from 1 to 3) placed on a cylindrical surface (blue grid) within the reconstruction area (green cylindrical boundary), **(b)** angular spectrum vs. frequency $u_s(k_{\varphi_a}, k)$, **(c)** angular spectrum vs. the spectral range domain $u_s(k_{\varphi_a}, k_{r_\tau})$
The left column belongs to point scatterer 1 ($\varphi_o = [-10]^\circ$), the middle column to scatterer 2 ($\varphi_o = [0]^\circ$) and the right column to scatterer 3 ($\varphi_o = [10]^\circ$).

which can be used to calculate the angular resolution (see Sect. 2.6.2).

The center of the angular spectrum is depending on the actual position of the point scatterer with respect to its angular position. Figure 2 shows the spectrum $u_s(k_{\varphi_a}, k)$ of three point scatterers placed on a cylindrical surface but with different angular positions.

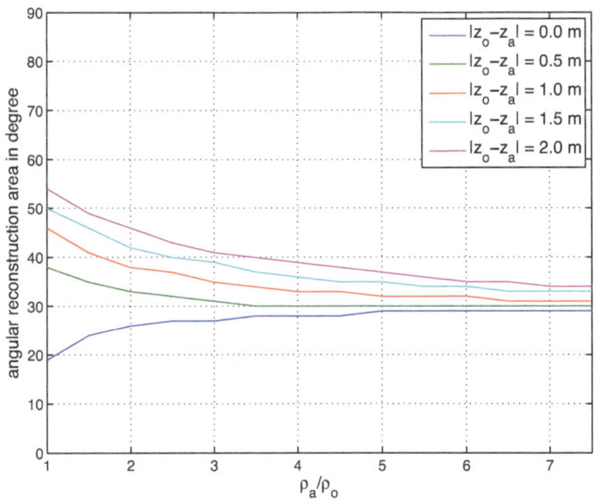

Fig. 3. Estimation of the valid angular reconstruction area respective the normalized radius ρ_a/ρ_o. The tolerated phase error is $[10]°$.

2.6 Resolution

2.6.1 Range resolution

The range resolution Δr_τ is given by the bandwidth B of the radar system and is calculated by $\Delta r_\tau = c_0/2B$, with c_0 the free space propagation velocity. The resolution in z-direction follows to be $\Delta z = \Delta r_\tau / \sin\theta$, θ is the inclination angle of the antenna (see Fig. 1).

2.6.2 Cross range resolution

The cross range resolution can be estimated from the extension of the angular spectral support band of a single point scatterer Eq. 14, i.e. $\Delta\varphi = \pi/|k_{\varphi_a,n}|$.

3 Validity of the algorithm respective the cross range approximation

Due to the approximation of the cosine term in Eq. 3 the expression for the angular spectrum is only valid within a limited angular range. Figure 3 shows the valid region for a cylindrical reconstruction area versus the normalized distance ρ_a/ρ_o assuming the phase difference of the approximated signal Eq. 3 with respect to the ideal signal Eq. 1 to be smaller than ten degrees.

Since the approximation gets worse with increasing angular domain the quality of the image will gradually degrade as the distance to the center of the image plane is increased. This effect is illustrated in Fig. 4 where an array of point scatterers is reconstructed.

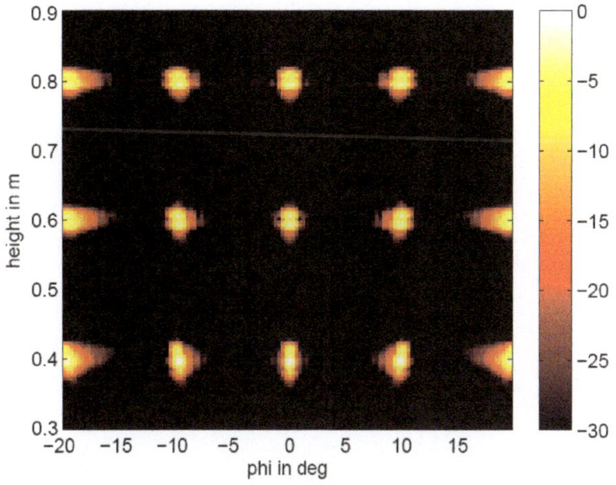

Fig. 4. Reconstructed image of point scatterers in log. scale ($L_\varphi = [40]°$, $\Delta\varphi = [0.1]°$, $\rho_a/\rho_o = 3$, $B = [11]GHz$): the angular resolution gets worse with increasing angular offset to the center of the image (the range resolution enhances with increasing range which is directly related to the height z_o of the point scatterer placed on a cylindrical surface)

4 Digital spotlighting

4.1 Circular convolution characteristics

Scatterers placed outside the unambiguous range are reconstructed as ghost targets. This effect can be attributed to the circular convolution characteristics of the discrete Fourier transform.

The same occurs in cross range direction. Since the size of the synthetic aperture is L_{φ_a} the sample spacing in the angular spectral domain is $\Delta k_{\varphi_a} = 2\pi/L_{\varphi_a}$. The cross range mapping in the spatial frequency domain is defined by $k_\varphi = k_{\varphi_a}$, i.e. the resultant samples of the angular spectrum of the object function $o(k_\varphi)$ are also separated by $\Delta k_{\varphi_a} = 2\pi/L_{\varphi_a}$. The angular length of the illuminated target area may be $L_{\varphi_o}{}'$ (including also scatterers lying outside the reconstruction area). To avoid aliasing in the φ domain, the sample spacing in the k_φ domain should satisfy the Nyquist rate $\Delta k_\varphi \leq 2\pi/L_{\varphi_o}{}'$. If the size of the target area is greater than the synthetic aperture length, i.e. $L_{\varphi_o}{}' > L_{\varphi_a}$ than $o(k_\varphi)$ will be aliased. Scatterers placed outside the angular reconstruction boundary as illustrated in Fig. 3 show up as ghost targets, see Fig. 3.

4.2 Zero-padding in angular direction

By increasing the length of the aperture L_{φ_a} by means of zero-padding $u_s(\varphi_a, f)$ with respect to the angular variable φ_a an effective aperture with length $L_{\varphi_a}{}' = L_{\varphi_o}{}'$ is obtained and $\Delta k_{\varphi_a} = \Delta k_\varphi$ which means that $o(k_\varphi)$ will not be aliased. This operation has no influence on the resolution of the image but increases the computational load which can be reduced with the following method.

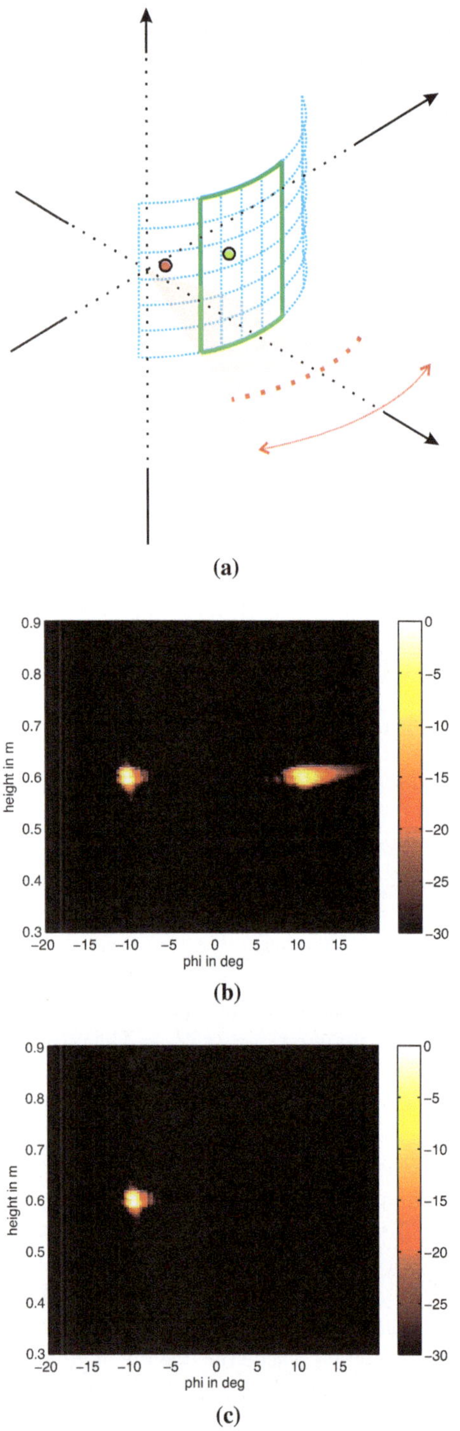

Fig. 5. Digital Spotlighting ($L_\varphi = [40]°$, $\Delta\varphi = 0.1°$, $\rho_a/\rho_o = 3$, $B = [11]GHz$): **(a)** two point scatterers placed on a cylindrical surface: scatterer 1 lies in the reconstruction area at ($\varphi_{o,1} = [-10]°$, $z_{o,1} = [0.6]m$), scatterer 2 at ($\varphi_{o,2} = [-30]°$, $z_{o,2} = [0.6]m$), **(b)** scatterer 2 shows up as a ghost target in the reconstructed image, **(c)** zero padding in the angular direction increases the unambiguous angular region, the ghost target is gone

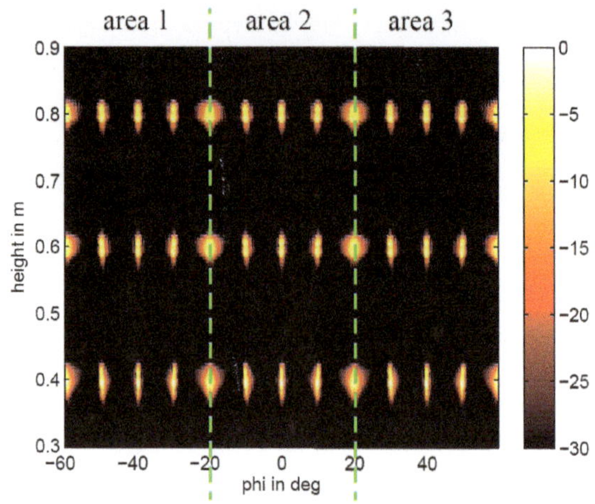

Fig. 6. Three reconstruction areas joined side by side to obtain a wide cross range image ($L_\varphi = 3 \times [40]°$, $\Delta\varphi = [0.1]°$, $\rho_a/\rho_o = 3$, $B = [11]GHz$). The SAR data $u_s(\varphi_a, f)$ is divided in 3 parts of equal length and zero-padded by a factor of two. Additionally a rectangular bandpass filters out the appropriate spectral support region

4.3 Filtering in angular spectral domain

To image a reconstruction area embedded inside a wider scattering area (digital spotlighting) one method is to apply a bandpass filter in the angular spectral domain. According to Sect. 2.5 the boundaries of the angular spectral support of the reconstruction area can be calculated by Eq. 12. The spectral support of a point scatterer is a bandpass function. The spectral overlap of the reconstruction area and the spectrum of the illuminated scatterer lying outside is determined by the angular position of these scatterer. A bandpass operation which filters out the spectral support of the reconstruction area will attenuate the ghost target but also slightly degrade the image quality.

In combination with the method introduced in Sect. 4.2 the filter-width can be wider and an optimum solution respective the digital spotlighting task can be found which does not lead to an image degradation.

In reality the beamwidth of the antenna pattern also helps to filter out the area of interest.

4.4 Subaperture imaging and image joining

The angular approximation used to derive the angular spectral support function limits the imaging area in angular direction. Using subaperture digital spotlighting techniques smaller angular segments can be processed and the results can be joined side by side in order to obtain a wider imaging area with approx. a constant resolution all over the total image.

Figure 6 shows an image of a wide angular reconstruction area. Three areas have been processed by subaperture digital spotlighting and joined side by side.

Table 1. Parameter settings of data acquisition using a homodyne FMCW radar

frequency range	$f_{\text{min}} \cdots f_{\text{max}}$	[91]GHz…[102]GHz
bandwidth	B	[11]GHz
range resolution	Δr	[1.36]mm
sweeptime	T	[2]ms
TX power	P_{out}	[20]dBm
RX sampling rate	f_s	[500]kHz
aperture radius	ρ_a	[0.60]m
aperture height	z_a	[0]m
aperture length	L_{φ_a}	$[-90]^\circ \ldots [90]^\circ$
angular increment	$\Delta\varphi_a$	$[0.1]^\circ$
angular image length	L_φ	$[40]^\circ$
radius cyl. imaging area	ρ_o	[21.5]cm
resolution z direction	Δz	[15.7]mm
resolution cross range	$\Delta\varphi$	$\approx [2.5]mm$

5 Imaging results

In order to validate the derived algorithm and the related techniques an ultrawideband FMCW homodyne radar has been used for data acquisition. The parameter settings are summarized in Table 1.

For validation an artificial object consisting of several small metallic spheres (radius [2]mm) as shown in Fig. 7 has been imaged. The point scatterers have been placed on a PVC cylinder with radius $\rho_o = [21.5]cm$. The reconstruction area has been set to $\varphi_o = [-100]^\circ \ldots [100]^\circ$ and $z_o = [0]m \ldots [0.6]m$. The same object also has been imaged with a backprojection algorithm in order to have a reference to compare with.

6 Conclusions

We have developed a $\omega - k$ algorithm adapted to circular apertures and cylindrical reconstruction areas. Within certain angular limitations an approximation to the exact spectral support function has been derived which is suitable for broadband imaging applications. We have validated the algorithms both with synthetic and measured real world data obtained with a ultrawideband FMCW radar. No major degradation of the image quality is observed when compared to a reference backprojection algorithm. However the computational efficiency is several orders better and the image shown in Fig. 7b (640 × 128 pixels) is computed in about 5 s (on a P4, 1.5 GHz).

(a) Photo of the object

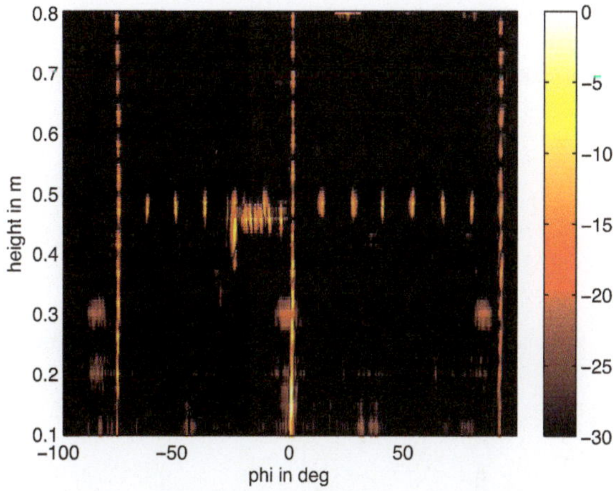

(b) $\omega - k$: reconstructed image consisting of 5 joined angular areas

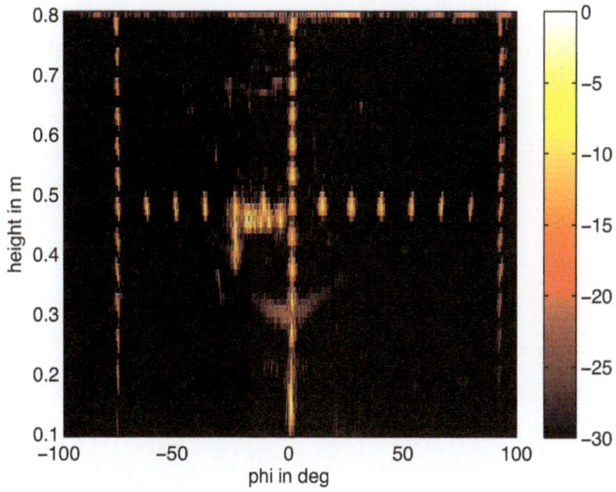

(c) backprojection: reconstructed image

Fig. 7. Small metallic spheres with a radius of [2]mm and a spacing of [5]cm. This structure has been placed on PVC cylinder with radius $\rho_a = [22.5]cm$

References

Carrara, W. G., Goodman, R. S., and Majewski, R. M.: Spotlight Synthetic Aperture Radar Signal Processing Algorithms, Artech House, 1995.

Fessler, J. A. and Sutton, B. P.: Nonuniform fast Fourier transforms using min-max interpolation, IEEE Transactions on Signal Processing, 51, 560–574, 2003.

Ishimaru, A., Chan, T.-K., and Kuga, Y.: An Imaging Technique Using Confocal Circular Synthetic Aperture Radar, IEEE Transactions on Geoscience and Remote Sensing, 36, 1524–1530, 1998.

Jackson, J. I., Meyer, C. H., Nishimura, D. G., and Macovski, A.: Selection of a convolution function for Fourier inversion using gridding [computerised tomography application], Medical Imaging, IEEE Transactions on, 10, 473–478, 1991.

Morse, P. M. and Feshbach, H.: Methods of Theoretical Physics, vol. Part II, McGraw-Hill Book Company, Inc., 1953.

Soumekh, M.: Reconnaissance with Slant Plane Circular SAR Imaging, IEEE Transactions on Image Processing, 5, 1252–1265, 1996.

Soumekh, M.: Synthetic Aperture Radar Signal Processing with Matlab Algorithms, John Wiley and Sons, Inc., 1999.

Wajer, F., Lethmate, R., van Osch, J., Graveron-Demilly, D., Fuderer, M., and van Ormondt, D.: Interpolation from Arbitrary to Cartesian Sample Positions: Gridding, in: ProRISC/IEEE workshop, 571–577, 2000.

11

Microwave sensors for detection of wild animals during pasture mowing

A. Patrovsky[1,2] **and E. M. Biebl**[2]

[1]Technische Universität München, Fachgebiet Höchstfrequenztechnik, Arcisstraße 21, 80333 München, Germany
[2]now at: Ecole Polytechnique, University of Montreal, C.P. 6079 Montreal, H3C 3A7 Canada

Abstract. More than 400 000 wild animals are killed or severely injured every year during spring time pasture mowing. Conventional methods for detection and removal or expulsion of animals before mowing are either inefficient or very time-consuming. The first really working method is based on a pyro-detector which senses the temperature contrast between the animals body and the surrounding pasture. Unfortunately, the detection reliability of this sensor decreases with increasing ambient temperature and strong sunlight, i.e. for typical weather conditions, when pasture is mowed, especially around noon.

In this paper, a detector is presented that exhibits complementary behaviour. It works best during dry conditions (i.e. around noon), but has a tendency to false alarms when dew is present (i.e. morning and evening). The sensor is based on a commercial, low-cost Doppler module at 24 GHz. It senses the difference of radar cross section between the animals body (high water content, specular reflection) and the pasture (low water content, diffuse reflection). The signal is analysed by means of a non-linear Wigner time-frequency transformation. Experimental results are presented for a laboratory setup as well as for measurement in actual springtime pasture. The results prove that a microwave sensor is capable of reliably detecting animals of the size of a fawn even if it is covered by a layer of pasture.

1 Introduction

More than 400 000 wild animals are killed or severely injured every year during spring time pasture mowing for hay and silage production in the western part of Germany only. Fawns, leverets (young hares) and clutches of birds breeding in the meadows are hidden in the pasture and cannot be detected visually by an operator of a mowing-machine. Thus, several strategies and methods have been developed in the past to find and remove or expel these animals before and during mowing:

- *Scaring away:* One day before mowing the meadow some unusual disturbance is made, e.g. by loud noise, scarecrows or dogs. During night the roe deer leads her fawns hidden in this meadow into the forest or another meadow. This method is not very reliable (especially in densely populated areas wild animals are very tolerant to disturbance) and requires careful planning of the mowing. Obviously, this method does not prevent the destruction of clutches.

- *Searching with a hunting dog:* In the morning before mowing the meadow is searched by use of a well-trained hunting dog. This method is extremely time-consuming requiring much more time than the mowing. Thus, it is restricted to relatively small meadows. Clutches are easily removed, but it is difficult to prevent the return of found and removed fawns into the meadow.

- *Mechanical solutions:* Several kinds of mechanical protection systems like large combs in front of the mower have been developed and investigated. Unfortunately, these devices tend to impede the mowing machine or to pass over the cowering animal.

In conclusion, these conventional methods are very time-consuming or ineffective. Recently, a system based on thermal sensors (Fig. 1) has been developed at the German Aerospace Center (DLR, Oberpfaffenhofen, Germany) in cooperation with the company ISA, Weiden, Germany (Haschberger, 1996) . At a transverse aluminium bar carried by a person ten pyro-sensors are mounted. Each sensor scans a stripe of the meadow (about 80 cm wide) and detects the thermal contrast between the animal's body or the clutch and the surrounding pasture. Many of these systems are operated since years and exhibit a high, documented detection rate. However, since pasture is usually mowed during sunshine, the thermal contrast decreases during the course of a day leading to a significant number of false alarms. For an automated detection system, e.g. mounted at a mowing machine, only a very small rate of false alarms can be tolerated.

Thus, an additional microwave sensor based on a commercial 24 GHz Doppler radar module has been developed that detects the contrast regarding reflectivity and radar cross

Fig. 1. Portable detection system based on eight pyro-sensors.

section between the animal's body and the pasture. The typical radar response of the convex, due to the water content highly reflective body differs significantly from the "noisy" response of the pasture. Investigation of the time-frequency spectrum of the Doppler signal showed that moving the sensor over the body results in a chirped Doppler signal that can reliably be detected by use a nonlinear Wigner transformation. The microwave wildlife sensor was successfully tested in the laboratory as well as in real pasture.

2 Doppler radar sensor

Figure 2 shows the proposed setup of an automated detection system mounted at a mowing machine. A set of microwave sensors is mounted at a standard tool bar (e.g. carrier of a fertilizer spreader) scanning the stripe of pasture to be mowed next. If an animal is detected, the operator is noticed and the position is marked on a GPS based map. When the next stripe is mowed, the operator can either remove the animal or simply go around leaving out a small spot of pasture.

In the past it has been proven that even simple Doppler sensors are capable of performing complex sensing task, if suitable signal processing is applied to the Doppler signal (e.g. Biebl, 1999; Rasshofer, 1999). The Doppler radar sensor employed in our investigations is a commercial, low-cost module based on hollow-guide technology. The signal of a 24 GHz Gunn oscillator is partly radiated via a horn antenna (about 5 mW of output power) and partly used as LO of a Schottky diode mixer (see Fig. 3). Using this mixer the received Doppler-shifted signal is down-converted to baseband. The down-converted signal is filtered by a band-pass (10 Hz–1 kHz), amplified by 70 dB and fed into a standard PC soundcard for A/D-conversion and processing. Doppler-module, analogue processing (filter and amplifier) and some circuitry for power supply was assembled in a rigid housing

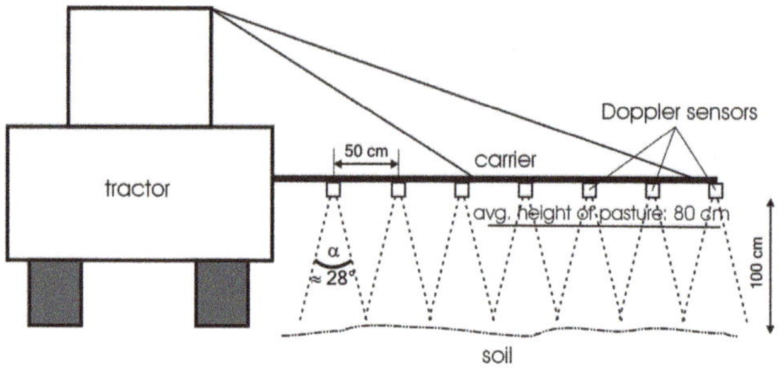

Fig. 2. A proposed microwave detection system mounted at a mowing machine.

including a protective dielectric cover for the horn antenna as shown in Fig. 4.

The geometry of the horn antenna leads to a beam-width of 25° and 28° in E- and H-plane, respectively. We obtained best results when the axis of the horn antenna was inclined to an angle of 20° with respect to the vertical direction (see Fig. 3). Our experiments showed that a reliable recognition of an animal's body can be obtained by means of the typical sequence of the specular reflection of the radar signal from the body. When the sensor goes over the body at constant speed it will initially detect diffuse reflection. In this case, the Doppler signal look like noise with relatively small amplitude. As soon as the beam hits the animal's body specular reflection results in a sharp increase of the amplitude. According to the radiation pattern the amplitude increases further, then decreases and, eventually, only noise from the diffuse reflection is detected. In addition, the frequency of the Doppler signal changes in time: Since the Doppler frequency depends of the effective inclination angle, i.e. the angle of the line between the point of specular reflection and the sensor's phase centre, the Doppler frequency decreases during the movement of the sensor of the animal's body. In other words, a down-chirped time-frequency spectrum of the Doppler signal is expected.

3 Signal analysis

Several possibilities can be taken into account in order to analyze the Doppler signal caught by the radar sensor.

An evaluation in the time domain by cross correlation with sample signal sequences has been investigated in the laboratory. Results showed however that this method was too sensitive on specific objects. Analysis exclusively in the frequency domain is not sufficient, because time information is needed for location of the object to be detected. So-called joint time-frequency analysis (Cohen, 1995; Quian et al., 1996) is used in these cases where time varying spectra emerge.

As it is known from the uncertainty principle, time and frequency resolution cannot be infinite at the same time. The time-bandwidth-product is a measure for the overall resolution if a signal or signal section in the time-frequency-plane.

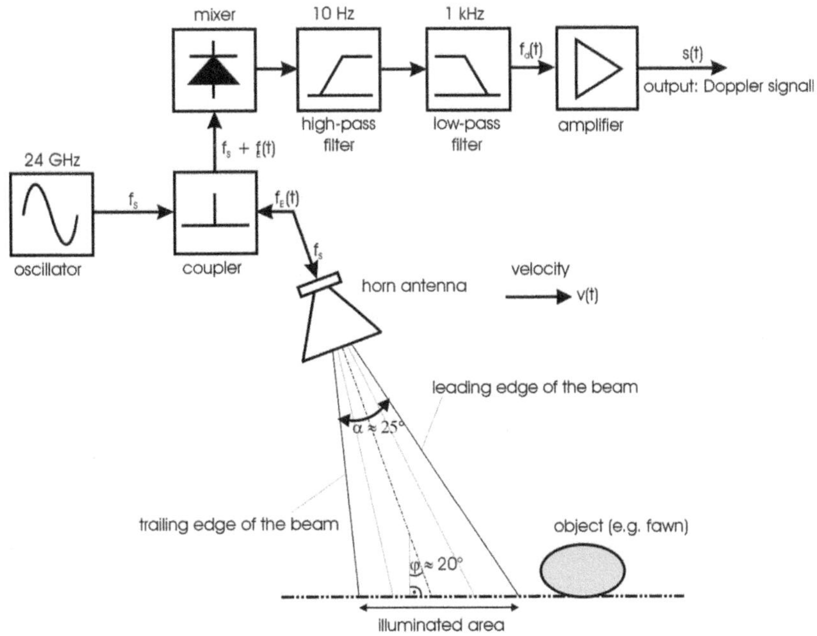

Fig. 3. Doppler radar front-end of the microwave sensor.

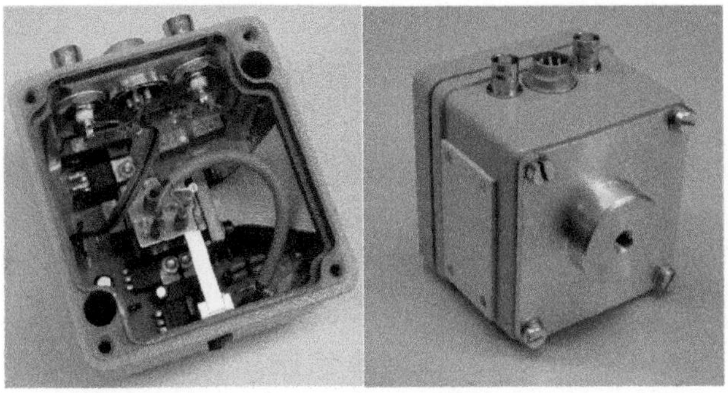

Fig. 4. Photographs of the microwave sensor.

It should be mentioned that the time-frequency characteristics of a signal section differs from these of the entire signal.

The simplest joint time-frequency analysis method is the short time Fourier transformation (STFT) which is basically a Fourier transformation of consecutive sections of a time signal limited by a time window. The shape of this window has to be chosen carefully in order to get good confinement in the frequency domain. The Hanning window was used for further investigations. A compromise has to be found for the window width, which determines the resolution in the time and frequency domain.

Mathematically, the STFT is given by

$$STFT_x(t,f) = \int_{-\infty}^{+\infty} \gamma(t'-t)x(t')\,e^{-j2\pi f t'}dt' \qquad (1)$$

where $x(t)$ is a real time signal and $\gamma(t'-t)$ is the time window around t.

Advantages of the STFT are simplicity and linearity as well as the availability of efficient algorithms, namely the Fast Fourier Transform (FFT). However, by applying a window on the time signal, the time-bandwidth-product of the resulting signal section degrades dramatically, so that the overall resolution was found not to be sufficient for radar detection of animals.

Wavelet transform has been considered as well, which uses scaled, time-limited wavelets as basis functions instead of sine or cosine functions. As a result of the scaling, the frequency coefficients are calculated more often, but the absolute bandwidth rises with frequency. This may be an advantage for broadband signals, but if we assume a quasi-constant velocity of the vehicle, the Doppler signal bandwidth is rather narrow and convincing improvements cannot be expected from this technique.

A nonlinear transformation, the so-called Wigner-distribution (WD), showed to have favorable properties concerning the resolution problem.

Detailed motives and derivation of the WD can be found in (Quian, 1996). An intermediate step in the derivation is the ambiguity function, which can be regarded as a parallel autocorrelation function in time and frequency domain. It represents an important valuation criterion in radar engineering. After transfer from the relative to the absolute time-frequency-plane, the Wigner distribution (WD) of a time signal $x(t)$ is written as

$$W_{xx}(t,f) = \int_{-\infty}^{+\infty} x^*\left(t-\frac{\tau}{2}\right)\cdot x\left(t+\frac{\tau}{2}\right)\,e^{-j2\pi f\tau}\,d\tau. \quad (2)$$

The xx index is added in order to distinguish the auto-WD from the likewise existing cross-WD of two different

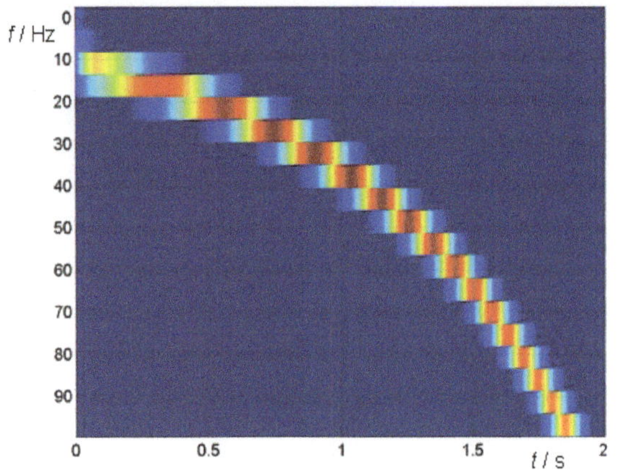

Fig. 5. Time-frequency distribution $|STFT|^2$of a frequency chirp calculated by means of short-time Fourier transform (frequency resolution: 5.4 Hz; time resolution: 0.186 s).

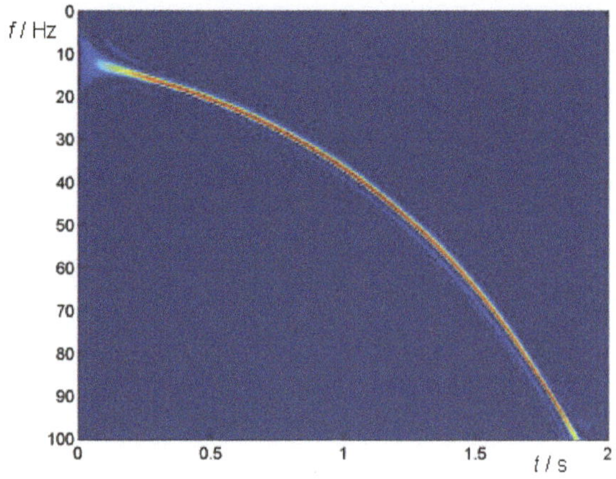

Fig. 6. Time-frequency distribution of a frequency chirp calculated by means of pseudo-Wigner-Ville-transform PWVT (frequency resolution: 0.49 Hz).

signals. Even if the conjugate complex signal is not needed for real signals, it is added here for later considerations. Some characteristics of the WD are listed in (Quian, 1996). It can be seen that the WD is a nonlinear, or more precisely, a bilinear transformation. Of course, as every nonlinear transformation, it will exhibit the really existing peaks in the time-frequency-plane, but also sum and difference terms. If for instance the time-frequency distribution of a signal with frequency components f_1 and f_2 is needed, the WD will also produce peaks at $(f_1-f_2)/2$ and $(f_1+f_2)/2$. Ville proposed to use the analytic signal to get rid of the $(f_1-f_2)/2$cross-term at low frequencies. The analytic signal is mathematically described by the Hilbert-transformation and serves to eliminate the negative frequency components of a real time signal. The Wigner-transformation of the analytic signal is called Wigner-Ville-distribution (WVD).

Up to now the transformation is carried out over an infinite time interval. It can be shown that, in this case, the result has the same time-bandwidth-product as the original signal. Due to limited computational resources, the time interval must be limited by a time window. Furthermore, this implicates reduced cross-terms, which arise from non-linearity. In comparison to the STFT, however, the length of the window can be chosen freely without reducing the time resolution. With a window applied, the transformation is called Pseudo-Wigner-Ville-transformation (PVWT).

$$PWV_{xx}(t,f)=\int_{-\infty}^{+\infty}\gamma(\tau)\cdot x_a^*\left(t-\frac{\tau}{2}\right)\cdot x_a\left(t+\frac{\tau}{2}\right)e^{-j2\pi f\tau}d\tau \quad (3)$$

$\gamma(\tau)$ represents the time window function and x_a the analytic signal.

In order to become aware of the benefits of the PWVT, it was compared to the STFT on the basis of a down-chirped signal from an audio signal generator. Figure 5 shows the course of the spectral peak for the STFT. The frequency resolution is noticeably bad and the time window cannot be chosen narrower as this would result in a too heavily blurred spectrum. The course of the spectrum peak calculated by the PWVT is presented in Fig. 6. It is clearly superior in terms of frequency resolution and the noisy curve before and after the object slope also indicates an improved time resolution. A disadvantageous property of the Wigner transformation is the generation of spurious cross-terms next to the proper energy distribution, which are a consequence of the nonlinear operation. These can cause severe perturbations during radar detection, if reflections from the object are weak. Another drawback is the missing possibility of compensation for mechanical oscillations after the transformation. In contrast to linear transformations it is not feasible to use an extra sensor to catch only the oscillation and to subtract this portion of the Doppler signal afterwards.

4 Evaluation in the laboratory

The sensor hardware and the analogue and digital signal processing was developed and evaluated in a laboratory setup shown in Fig. 7. The sensor was mounted beneath a large model train running on wooden bar forth and back at constant, but adjustable speed. As a lab-model of pasture we grew a stripe of grass and as a lab model of a fawn we used a fabric-covered hot-water bottle made of rubber. Comparison with a real animal of similar size showed that we obtain from these models very similar signals for both the pyro-sensor and the microwave sensor.

The Doppler baseband signal, obtained from the radar sensor, lies in a frequency range from 10 Hz to 500 Hz, depending on the antenna elevation angle and velocity of the vehicle. For this reason an ordinary PC sound card was used as A/D converter. The sampling rate should not be chosen

Fig. 7. Laboratory setup for evaluation of the performance of microwave sensor and signal processing.

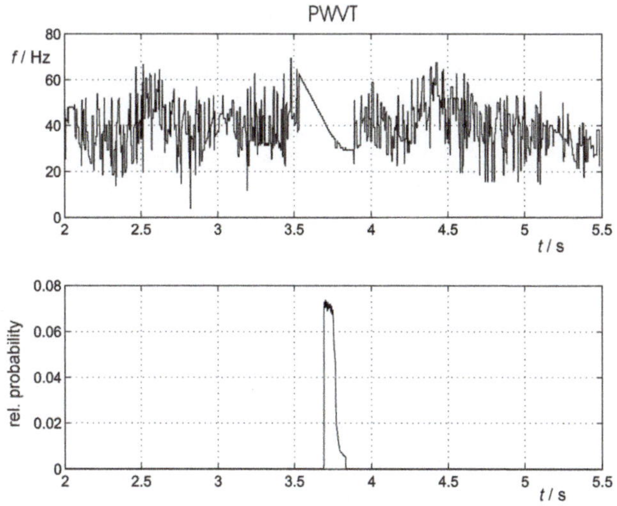

Fig. 8. Typical time-frequency and detection probability response obtained from the hot-water bottle (lab model of a fawn).

higher than necessary in order to economize computational resources for real-time analysis. Once the course of the spectral peak over time is calculated, as shown in Fig. 8, the detection algorithm consists of finding the characteristic frequency slope. A straight line is compared to the slope and

Fig. 9. Microwave sensor mounted at a tractor scanning the natural cover next to a dirt road.

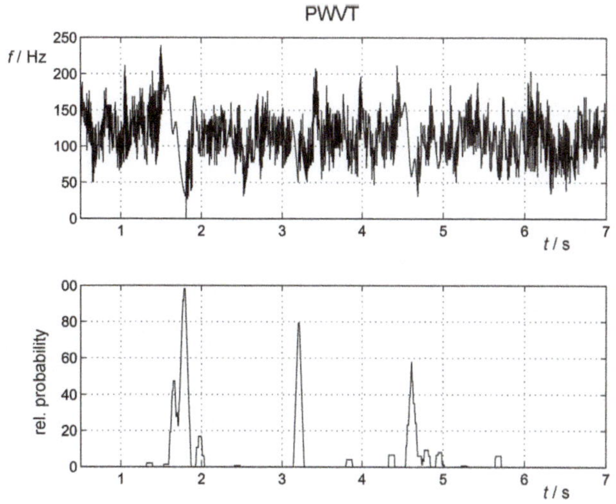

Fig. 10. Time-frequency spectrum obtained from a measurement run over two corner reflectors (at t=1.8 s and t=4.6 s) and a hare (at t=3.2 s).

the least square error is calculated. The slope is not always as perfect as shown in Fig. 8. It can be superposed with noise or oscillations, resulting in a big error. If the error is under a predefined limit, the system will indicate the detection of an animal. This error sum and the length of the line, as well as possible limits for its declination, determine the probability of detection or error, respectively.

5 Evaluation in the field

The microwave detection system was evaluated during two measurement campaigns. In late summer, the sensor was mounted at a tractor scanning the sparse natural cover next to a dirt road (Fig. 9). A typical result for detection of a hare is shown in Fig. 10. Two corner reflectors provided reference

Fig. 11. Measurement of a target in spring-time pasture: **(a)** Photograph of the sensor above the pasture, **(b)** measured time-frequency spectrum.

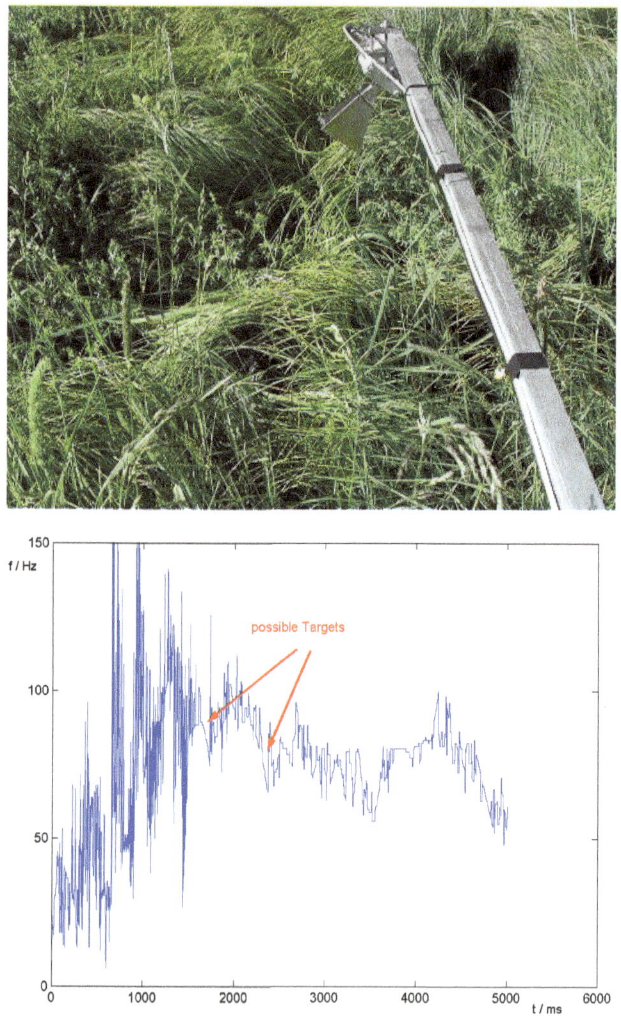

Fig. 12. Measurement of a target beneath a thick cover of herbaceous plants: **(a)** Photograph of the sensor above the pasture, **(b)** two measured time-frequency spectra.

positions. As can be seen, the animal provides a very clear radar response. Due to unevenness of the road, the elasticity of the aluminium bar and the sensor's weight sometimes led to vertical oscillations of the sensor resulting in sinusoidal, low-frequency Doppler signals that disturbed the detection significantly.

During the following spring measurements were made in actual spring-time pasture. In order to prevent the pasture from too much damage the sensor was mounted onto a rotary base. The resulting curved trajectory does not much affect the Doppler signal. Since no fawn could be found in the pasture the hot-water bottle was used as a model. Figure 11 shows the picture of typical spring-time pasture and the respective measured time-frequency spectrum. As can be seen, the target is reliably detected. As similar result was obtained for a target completely covered by the pasture, i.e. invisible for all optical detectors.

Finally, the target was placed beneath a dense cover of thick, herbaceous plants (Fig. 12). Even under this "worst-case" condition, a detection reliability of about 50% was obtained.

6 Conclusion

A microwave sensor based on a commercial low-cost 24 GHz Doppler radar module was employed for detection of animals in spring-time pasture. The detection algorithm is based on a nonlinear pseudo-Wigner-Ville-transformation. A very high reliability of detection was achieved in the laboratory as well as in actual spring-time pasture. Even heavily covered targets could be detected with satisfactory probability. The measurements showed that the coherent detection method (i.e. evaluation of the Doppler signal) is sensitive to vertical oscillations that cannot completely avoided when mounted at a mowing machine. One solution would be to measure these oscillations and to pre-process the signal before performing the nonlinear transformation or to use a non-coherent detection principle like reflectometry or radiometry.

Acknowledgements. The authors appreciate the help and continuing support by R. Nitsche, ISA GmbH, Weiden, and V. Tank, DLR, Oberpfaffenhofen, Germany.

References

Biebl, E. M.: Millimeter Wave systems based on SIMMWICs, in: 1999 URSI General Assembly Digest, (Toronto, Canada), 235, 13.–21. Aug. 1999.

Cohen, L.: Time-Frequency Analysis, New Jersey: Prentice-Hall, ch. 4, 1995.

Haschberger, P., Bundschuh, M., and Tank, V.: Infrared sensor for the detection and protection of wildlife, Opt. Eng. 35(3), 882–889, 1996.

Rasshofer, R. H. and Biebl, E. M.: Advanced millimeterwave speed sensing system based on low-cost active integrated antennas, in: 1999 IEEE MTT–S Int. Microwave Symp. Digest, (Anaheim, CA), 285– 288, 13.–19. June 1999.

Quian, S. and Chen, D.: Joint Time-Frequency Analysis, New Jersey, Prentice-Hall, 1996.

EMC studies on systems with hybrid filter circuits for modern aircraft applications

T. Kut[1], B. Chand[1], A. Lücken[2], S. Dickmann[1], and D. Schulz[2]

[1]Institute of Fundamentals of Electrical Engineering, Helmut Schmidt University, Hamburg, Germany
[2]Institute of Electrical Power Systems, Helmut Schmidt University, Hamburg, Germany

Correspondence to: T. Kut (tobias.kut@hsu-hh.de)

Abstract. In this contribution, a new approach for EMC-filter design is presented. Due to the increasing electrification of modern aircraft, as a result of the More Electric Aircraft concept, new strategies and approaches are required to fulfill the strict EMC aircraft standards (DO-160/ED-14 – Sec. 20). Consequently the weight and volume of the used filter components can be reduced. A promising approach could be a combination of passive and active filters. For the same attenuation effect, so-called hybrid filters achieve either savings in weight and volume, or can obtain an additional filtering effect with minimal weight increase of an existing system. In this paper, the underlying theory is explained in detail, carried out in a simulation tool and the gained insight is demonstrated with a sample measurement.

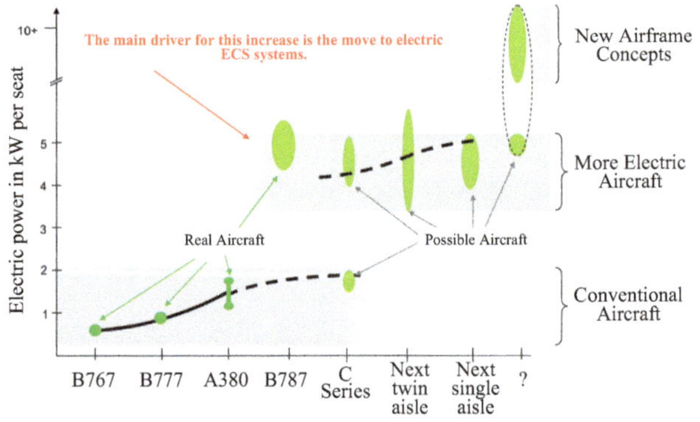

Fig. 1. Increasing electrical power demand in civil aerospace (McLoughlin, 2009).

1 Introduction

In order to reduce the environmental impact of aircraft and to improve the overall system efficiency, global efforts for reducing the aircraft weight are underway. One key issue is the wide application of electric system components instead of heavy mechanical, pneumatic and hydraulic systems (Fig. 1). This approach to change the power supply structure of an aircraft is already known as More Electric Aircraft (MEA) concept. Within this concept the traditional Auxiliary Power Unit (APU) could be replaced by a Multifunctional Fuel Cell System (MFFCS). Therefore, it is necessary to transform the load-dependent fuel cell output voltage by DC/DC converters. To avoid exceeding the EMI limits for power electronics, filter circuits are necessary. These passive filter elements take up a relatively large part of the overall system weight and can be reduced in size only by increasing the switching frequency. In some applications, the EMC filters represent more than one third of the total power supply volume or weight. For this reason some electromagnetically integrated EMC filter structures (van Wyk et al., 2005; Waffenschmidt et al., 2005) and also active EMI filters for printed circuit boards (PCBs) and low power applications (Biela et al., 2006; Son and Sul, 2003) have been proposed. In this paper a new EMI filter structure for DC aircraft applications is presented, which allows a significant reduction of the filter weight or an increasing filter attenuation with minimal additional weight.

2 Hybrid filter topology and its basic operating principle

The proposed concept for the size and weight reduction of EMI filters is the use of both active and passive filters in one system, commonly called hybrid filter, instead of fully passive filters. High frequency active filters are broadly employed in signal processing and communications engineering. However, those applications operate with low current

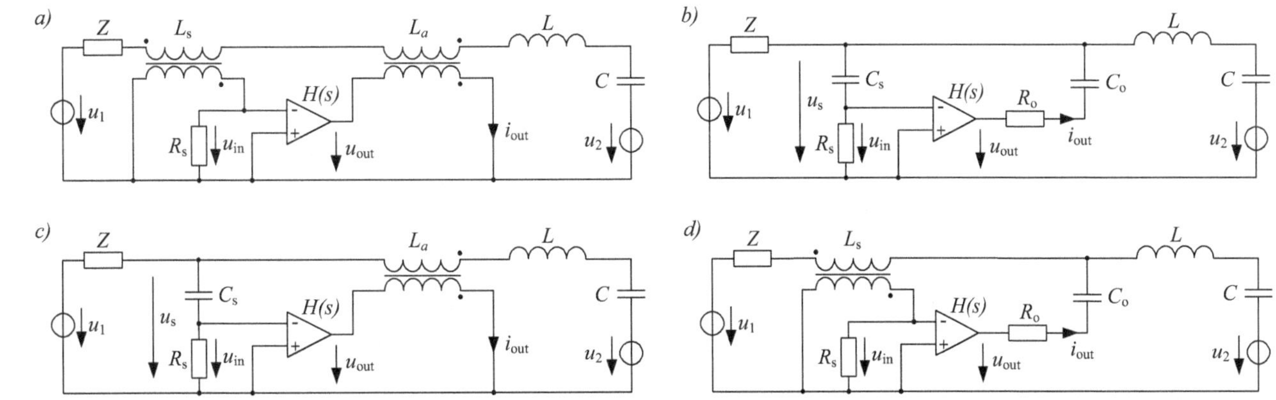

Fig. 2. Different active filter structures for single phase circuits as basic design example.

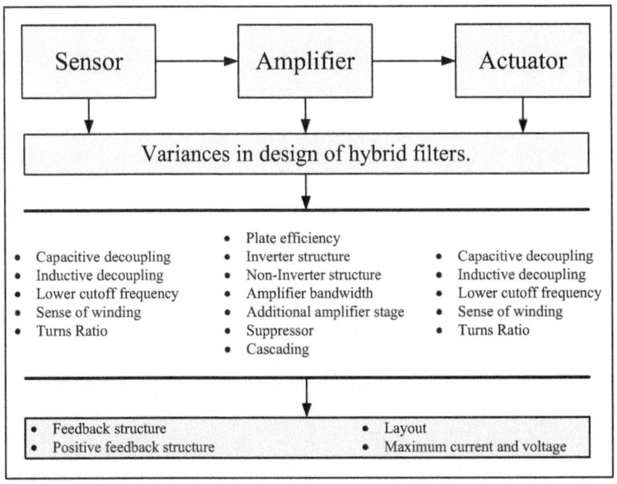

Fig. 3. Basic structure and possible variances in design of hybrid filters.

and voltage levels. For higher power levels, as in aircraft applications, the construction of such systems becomes critical. Wide-band amplifiers have to be used, which should be able to handle high current and voltage levels. This challenge motivates research in this field and is addressed in the following.

For the design of a hybrid EMC filter the knowledge about the different possible active filter structures is essential. Basically four types of circuits (Heldwein et al., 2006) are known in the literature and shown in Fig. 2. The basic strategies rely on the measurement of either current (Fig. 2a and d) or voltage (Fig. 2b and c) (Larsen, 1998; Hironobu, 2003; Honda, 2003; Pelly, 2009) and in the injection of a current (Fig. 2a and c) or a voltage (Fig. 2b and d) to the DC lines. Here, it is obvious that for the integration of active filters into an existing filter network three additional elements must be taken into account. The main task is to decouple the interference signal with a sensor (capacitive or inductive). Subsequently, the signal is rotated by 180 degrees in phase and amplified in the amplitude. This signal is fed back via an actuator (capacitive or inductive) into the existing system. Another degree of freedom (Fig. 3) is, the phase shift is achieved by an inverter or the inversely arranged and coupled inductors. The voltage/current cancelation operates in the limited frequency range from approximately 10 kHz (low pass character at decoupling) up to 5 MHz (bandwidth of the amplifier). The limited bandwidth of the amplifier requires a reduction of the gain at higher frequencies in order to guarantee stability of the system.

3 Simulation results

The basic topology of Fig. 2c was converted into a simulation model. To demonstrate that this approach has advantages over a conventional purely passive filter architecture, a hybrid filter architecture in the simulation was examined more closely. A promising approach (Rehfeld, 2012) for the integration of this active filter section is shown in Fig. 4. It is a simple common-mode filter, consisting of Y-capacitors, and a common-mode choke, supplemented by the active part and excited with a common mode interference. The network is terminated by a conventional Line Impedance Stabilization Network (LISN). By measuring the voltages across the resistors R_m (U_{out}) and R (U_{in}) and the subsequent forming of the quotient (U_{out}/U_{in}) of these two voltages, the transfer function $Z(s)$ of each applied filter circuit can now be calculated. These are presented in comparison in Fig. 5. Here, it becomes clearly visible that the hybrid filter structure for the entire kHz-range has significant benefits over the passive filter structure. This frequency range is particularly important and interesting because the passive components designed for this purpose have a high weight and volume. As described in the previous section, the active filter characteristic is in the lower frequency range (1–10 MHz) affected by the coupling characteristic and the upper frequency range is limited (1–10 MHz) by the respective bandwidth of the operational amplifier. Nevertheless, it is clear that through the implementation of the hybrid filter design advantages in order to save

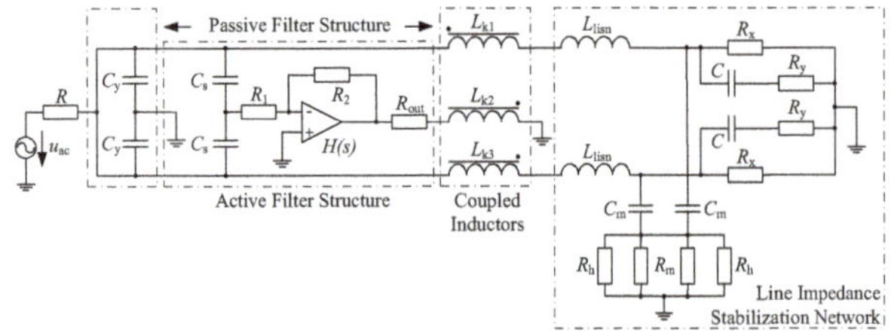

Fig. 4. Schematic of the simulated hybrid filter structure.

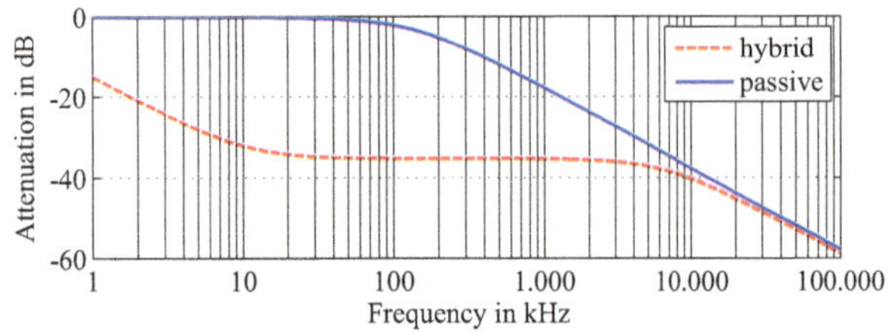

Fig. 5. Comparison of passive and hybrid filter structure simulation results.

weight and/or to increase the available filter attenuation can be achieved. Here, the next step is the consideration of parasitic elements which should be determined from practical measurements and/or an evaluation of mathematical models.

4 Transfer function of hybrid filters

To describe the frequency characteristic of a filter circuit, it is possible to determine it using the method of transfer functions. At higher order filters (series connection of single filters), this method becomes very complex. Another possibility is described below.

4.1 Mathematical description

The $\underline{\mathbf{A}}$ matrix method (Fig. 6) is another way to determine the transfer function. It is based on the fact that each two-port network, which is an electrical system, is described by a matrix.

$$\begin{pmatrix} \underline{U}_{\text{in}} \\ \underline{I}_{\text{in}} \end{pmatrix} = \underline{\mathbf{A}} \cdot \begin{pmatrix} \underline{U}_{\text{out}} \\ \underline{I}_{\text{out}} \end{pmatrix} \tag{1}$$

$$\underline{\mathbf{A}} = \begin{pmatrix} \underline{a}_{11} = \left.\frac{\partial \underline{U}_{\text{in}}}{\partial \underline{U}_{\text{out}}}\right|_{\underline{I}_{\text{out}}=0} & \underline{a}_{12} = \left.\frac{\partial \underline{U}_{\text{in}}}{\partial \underline{I}_{\text{out}}}\right|_{\underline{U}_{\text{out}}=0} \\ \underline{a}_{21} = \left.\frac{\partial \underline{I}_{\text{in}}}{\partial \underline{U}_{\text{out}}}\right|_{\underline{I}_{\text{out}}=0} & \underline{a}_{22} = \left.\frac{\partial \underline{I}_{\text{in}}}{\partial \underline{I}_{\text{out}}}\right|_{\underline{U}_{\text{out}}=0} \end{pmatrix} \tag{2}$$

The advantage for higher order filters is that the matrices of the single filters can easily be multiplied together to obtain a matrix for the overall system – $\underline{\mathbf{A}}_1 \cdot \underline{\mathbf{A}}_2 \cdot \ldots \cdot \underline{\mathbf{A}}_n = \underline{\mathbf{A}}_{\text{total}}$. From this, the transfer function can be determined as the inverse calculation of the matrix element $\underline{a}_{11}$.

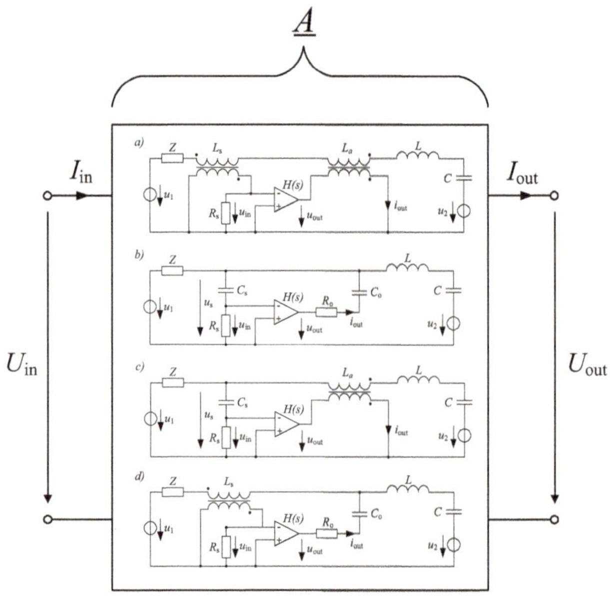

Fig. 6. Mathematical description of filters using $\underline{\mathbf{A}}$ matrices.

The calculation of the matrix element $\underline{a}_{11}$, for example for the circuits presented in Fig. 2b and d yields to the following transfer functions. These two filters have the best filter characteristic for this DC aircraft application here. To simplify this, the following assumptions have been made: the power supply as well as the operational amplifier are ideal. The variables T_1 to T_4 describe the input and output impedances, consisting of C_s/R_s and C_o/R_o, κ describes the gain of the operational amplifier. In order to distinguish the individual elements they are described better by their input and output characteristics. This means for example that the element $\underline{a}_{11-\text{LC}}$ is coupled via an inductor and the amplified signal is fed back through a capacitor.

$$\underline{a}_{11-\text{CC}} = \frac{s^2 T_3^2 + sT_4 + 1}{s^2 \kappa T_1^2 + sT_2 + 1} \tag{3}$$

$$\underline{a}_{11-\text{LC}} = \frac{s^2 T_2^2 + sT_3 + 1}{sT_1 + 1} \tag{4}$$

Based on this filter transfer functions, the matrices were calculated and carried out in a filter simulation tool.

4.2 Implementation in a filter design tool

For various filter circuits (active and passive) these matrices are integrated into a filter design tool (Rehfeld, 2012; Kut et al., 2012). Using a graphical user interface boundary conditions of the filter can be defined. These are the minimum and maximum values of the used inductors, capacitors and resistors. Other input parameters are the attenuation within a given frequency band, the input and output impedance and

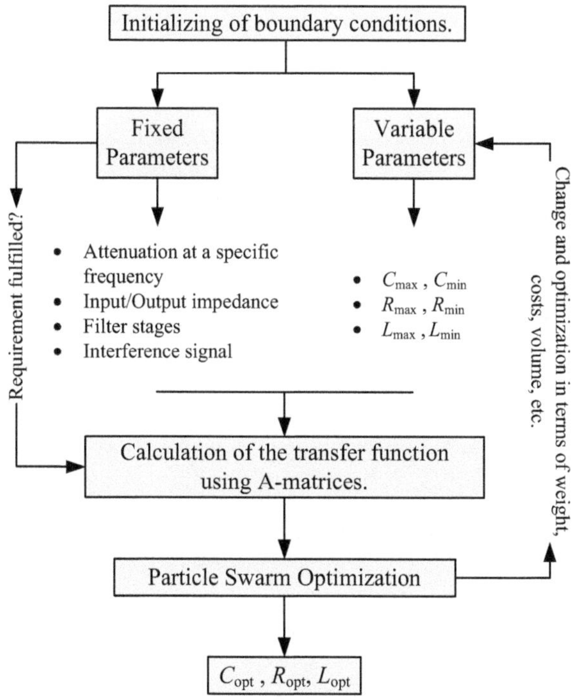

Fig. 7. Flowchart of the filter design procedure.

the number of filter stages (Fig. 7). The optimization in terms of weight, costs, volume, etc. is done by using an algorithm, which is called Particle Swarm Optimization (PSO). The output parameters of the script are the calculated filter elements for the practical implementation (C_{opt}, R_{opt}, L_{opt}).

5 Practical implementation of the analyzed filter

Based on the results of the filter design tool, various filter circuits have been selected, designed, developed and measured. Then, these filter topologies were tested for two different DC/DC converters (Fig. 8b and c). Firstly, this filter principle has been tested on a 2 kW-DC/DC converter and secondly on a 10 kW-DC/DC converter. Here, the conducted interference in the frequency range of 1 kHz to 30 MHz were measured. The two used DC/DC converters are designed as interleaved converters (Fig. 8a). This means that several stages are connected in parallel. The respective semiconductors are switched on and off at different time steps. This has the advantage that the storage throttle can actually be designed for a smaller current. In this respect the current ripple can also be reduced. This decreases the overall weight and there are less electromagnetic interference.

Previous studies (Rehfeld, 2012; Wagner, 2012) have shown that a hybrid filter with an inductive decoupling and capacitive coupling (Fig. 2c) has the best filter characteristic for this application here. Therefore, this filter is used.

Table 1. Comparison of passive and hybrid amplitudes.

	Frequency	Passive	Hybrid	Gain
1	300 kHz	71 dB	52 dB	19 dB
2	600 kHz	54 dB	40 dB	14 dB
3	900 kHz	37 dB	27 dB	10 dB

6 Measurement results

In order to check whether these benefits of the hybrid filter structure can be confirmed by practical measurements, an EMC measurement was performed with DC/DC converter presented in the previous section. Except for the spectrum analyzer the total test setup was integrated into an EMC anechoic chamber to minimize interference from outside. The test setup inside the anechoic chamber is shown in Fig. 9.

Here, a DC source is designed so that it acts as a fuel cell and the electronic load is configured so that it operates similarly as consumer in an electrical aircraft grid. Between these two components the power electronics will be installed with the additional hybrid filter. To differentiate the source and the load, LISNs were installed on both sides. Furthermore, a monitoring unit for controlling and additional auxiliary power supplies are also installed. The interference, recorded with the spectrum analyzer are measured directly on the 50 Ω output of the LISN.

Firstly, the disturbance of the DC/DC converter with a conventional passive filter and secondly with the new hybrid filter structure was measured. The results are shown in Fig. 10. There, once again it shows that with the use of a hybrid filter architecture, the filter attenuation could also be increased and measured in practical applications. The increase of the filter attenuation at the characteristically designed switching frequency (4 × 75 kHz) and its harmonics is compared for discrete frequency points in Table 1.

These findings show that this hybrid filter concept presented here has advantages not only in the simulation, but can also lead to benefits in practical measurements in the filtering of power electronic circuits. As inverse conclusion can be stated that the possible increase of the filter attenuation can be certainly used to design the corresponding passive filters smaller and to save finally weight and volume. The 10 kW-DC/DC converter (PCB, components, heat sink, control, etc.) weighs about 2230 g with passive filters and with hybrid filter with similar filter characteristics (mainly through downsizing and weight reduction of the storage throttle) about 2000 g. In this case, savings of around 10 % are possible.

7 Conclusions

The installation of electric system components instead of heavy mechanical, pneumatic and hydraulic systems in

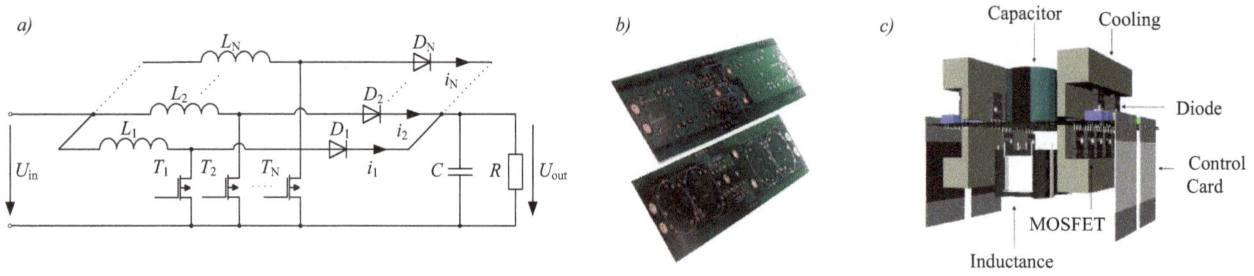

Fig. 8. **(a)** Principle of an interleaved converter as boost, **(b)** Printed circuit board of a 2 kW-DC/DC converter and **(c)** 3-D model of a 4-stage 10 kW-DC/DC converter.

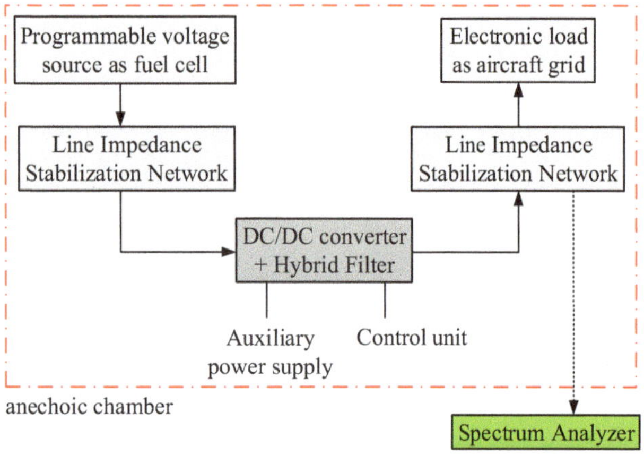

Fig. 9. Measurement setup for comparison of filter structures.

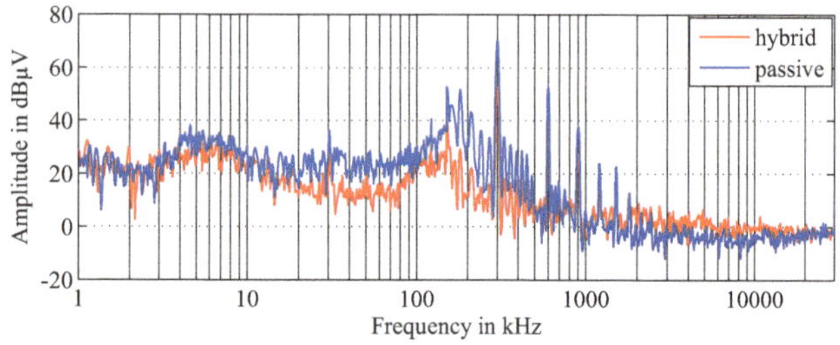

Fig. 10. Comparison of passive and hybrid filter structure meausurement results.

modern aircraft firstly lead to weight saving effects and secondly EMC requirements (DO-160/ED-14) increased as a result of the required DC/DC converter at the output of a possible fuel cell. To make sure, that the achieved weight savings will not be compensated by an increased filter weight of the DC/DC converter, in this paper a new approach of filter topology was introduced for aircraft applications. This method relies on a combination of active and passive filter topologies. This so-called hybrid filter achieves either savings in weight and volume for the same attenuation effect, or can obtain an additional filtering effect with minimal weight increase of an existing system. Starting from the basic topologies and their operating principles, a simulation was performed which shows the advantage of such a topology. Furthermore, various filter structures were carried out in a filter design tool and described by their $\underline{\mathbf{A}}$ matrix with the aim of an automated design process. Subsequently, based on these findings various filter circuits have been selected and tested on two DC/DC converters. The results show that this hybrid filter structure has advantages not only in the simulation, but also can lead to a higher filter attenuation in order to fulfill the strict EMC aircraft standards of DC/DC converters in modern aircraft.

References

Biela, J., Wirthmueller, A., Waespe, R., Heldwein, M., Kolar, J., and Waffenschmidt, E.: Passive and active hybrid integrated EMI filters, in: Applied Power Electronics Conference and Exposition, 2006, APEC '06, Twenty-First Annual IEEE, 7 pp., doi:10.1109/APEC.2006.1620688, 2006.

Heldwein, M., Ertl, H., Biela, J., and Kolar, J.: Implementation of a transformer-less common mode active filter for off-line converter systems, in: Applied Power Electronics Conference and Exposition, 2006, APEC '06, Twenty-First Annual IEEE, 7 pp., doi:10.1109/APEC.2006.1620696, 2006.

Hironobu, K.: Active Filter, Patent, JP 087973, 2003.

Honda, J.: Active EMI Filter, Patent Application, http://www.patentlens.net/patentlens/patent/WO_2004_001927_A3R4/en/, WO 2004/001927 A3R4, 2003.

Kut, T., Chand, B., Lücken, A., Dickmann, S., and Schulz, D.: Optimierte EMV-Filteranalyse für den Einsatz von brennstoffzellengespeisten Gleichspannungswandlern in modernen Verkehrsflugzeugen, in: Deutscher Luft- und Raumfahrtkongress 2012, Berlin, Germany, online available at: http://www.dglr.de/publikationen/2012/281288.pdf, 2012.

Larsen, E. V.: Hybrid active power filter with programmed impedance characteristics, Patent, http://www.patentlens.net/patentlens/patent/US_5737198/en/, US 5737198, 1998.

McLoughlin, A.: More electric – Ready for take off?, in: Power Electronics and Applications, 2009, EPE '09, 13th European Conference on Power Electronics and Applications, 1–7, 2009.

Pelly, B. R.: Active filter for reduction of common mode current, Patent, http://www.patentlens.net/patentlens/patent/US_7583136/en/, US 7583136, 2009.

Rehfeld, M.: EMC-Studies for optimal filter design in aicraft and automotive applications, Master's thesis, Helmut Schmidt University, University of the Federal Armed Forces Hamburg, Germany, 2012.

Son, Y.-C. and Sul, S.-K.: A new active common-mode EMI filter for PWM inverter, IEEE T. Power Electr., 18, 1309–1314, doi:10.1109/TPEL.2003.818829, 2003.

van Wyk, J., Lee, F., Liang, Z., Chen, R., Wang, S., and Lu, B.: Integrating active, passive and EMI-filter functions in power electronics systems:a case study of some technologies, IEEE T. Power Electr., 20, 523–536, doi:10.1109/TPEL.2005.846553, 2005.

Waffenschmidt, E., Ackermann, B., and Ferreira, J.: Design method and material technologies for passives in printed circuit Board Embedded circuits, IEEE T. Power Electr., 20, 576–584, doi:10.1109/TPEL.2005.846530, 2005.

Wagner, R.: Design of a 4-level interleaved converter for the use of fuel cells in modern aircraft, Master's thesis, Helmut Schmidt University, University of the Federal Armed Forces Hamburg, Germany, 2012.

13

Statistical multipole formulations for shielding problems

K. Körber and L. Klinkenbusch

Christian-Albrechts-Universität zu Kiel, Computational Electromagnetics Group, Kaiserstr. 2, 24143 Kiel, Germany

Correspondence to: K. Körber (kko@tf.uni-kiel.de)

Abstract. A multipole-based method is presented for modelling an electromagnetic field with small statistical variations inside an arbitrary enclosure. The accurate computation of the statistics of the field components from the statistical moments of the multipole amplitudes is demonstrated for two- and three-dimensional examples. To obtain the statistics of quantities which depend non-linearly on the field components, higher-order statistical moments of the latter are required.

1 Introduction

Recent research in the field of statistical electromagnetics in general and statistical EMC in particular seems to focus on extremely non-deterministic fields as observed in a reverberation chamber. Much progress has been achieved in this area over the last decades, leading to very general statistical power distributions largely independent of the chamber geometry (Holland, 1999). Relatively little work has been done on the topic of small statistical variations of an otherwise known electromagnetic field. A related example relevant to EMC has been presented in (Ajayi et al., 2008).

In this paper we discuss the case of small statistical variations in various parameters of a shielding problem, e.g. the angle of incidence of a plane wave impinging on a shield. The varying parameters are given by the first few statistical moments of their distributions. From these the statistical moments of the amplitudes of a spherical-multipole expansion are derived. The statistics of the multipole amplitudes then characterizes the statistical properties of the electromagnetic field not only in a single point but in a spherical region around the arbitrarily chosen center of the expansion. In practice, this spherical region can be part of the enclosed volume of any shielding structure. For non-canonical problems, where the calculation of the multipole amplitudes cannot be achieved analytically, data from numerical simulations or measurements are needed to estimate the statistical moments of the multipole amplitudes.

2 Multipole expansion

Consider an arbitrary shielding structure as depicted in Fig. 1.

The electromagnetic field in any linear, homogeneous, and source-free spherical sub-domain of the enclosure can be expanded by means of a spherical-multipole expansion:

$$\boldsymbol{E} = \sum_{n=1}^{\infty}\sum_{m=-n}^{n} A_{n,m}\boldsymbol{N}_{n,m} + \frac{Z}{j}\sum_{n=1}^{\infty}\sum_{m=-n}^{n} B_{n,m}\boldsymbol{M}_{n,m} \tag{1}$$

$$\boldsymbol{H} = \frac{j}{Z}\sum_{n=1}^{\infty}\sum_{m=-n}^{n} A_{n,m}\boldsymbol{M}_{n,m} + \sum_{n=1}^{\infty}\sum_{m=-n}^{n} B_{n,m}\boldsymbol{N}_{n,m}. \tag{2}$$

The coefficients $A_{n,m}$ and $B_{n,m}$ are referred to as the electric and magnetic multipole amplitudes, respectively, $Z = \sqrt{\mu/\varepsilon}$ is the wave impedance in the spherical sub-domain and $\boldsymbol{M}_{n,m}$, $\boldsymbol{N}_{n,m}$ represent the spherical-multipole functions defined by

$$\boldsymbol{M}_{n,m} = (\boldsymbol{r}\times\nabla)\, j_n(\kappa r) Y_{n,m}(\vartheta,\varphi) \tag{3}$$

$$= j_n(\kappa r)\boldsymbol{m}_{n,m}(\vartheta,\varphi) \tag{4}$$

$$\boldsymbol{N}_{n,m} = \left[\frac{1}{\kappa}\nabla\times(\boldsymbol{r}\times\nabla)\right] j_n(\kappa r) Y_{n,m} \tag{5}$$

$$= -\frac{j_n(\kappa r)}{\kappa r} n(n+1) Y_{n,m}\hat{\boldsymbol{r}} - \frac{1}{\kappa r}\frac{d}{dr}[r j_n(\kappa r)]\boldsymbol{n}_{n,m}. \tag{6}$$

Here, $j_n(\kappa r)$ denote spherical Bessel functions of the 1st kind, $\hat{\boldsymbol{r}}$ is the radial unit vector, and the vector wave functions are found as:

$$\boldsymbol{m}_{n,m}(\vartheta,\varphi) = -\frac{1}{\sin\vartheta}\frac{\partial Y_{n,m}}{\partial\varphi}\hat{\boldsymbol{\vartheta}} + \frac{\partial Y_{n,m}}{\partial\vartheta}\hat{\boldsymbol{\varphi}} \tag{7}$$

$$\boldsymbol{n}_{n,m}(\vartheta,\varphi) = +\frac{\partial Y_{n,m}}{\partial\vartheta}\hat{\boldsymbol{\vartheta}} + \frac{1}{\sin\vartheta}\frac{\partial Y_{n,m}}{\partial\varphi}\hat{\boldsymbol{\varphi}}. \tag{8}$$

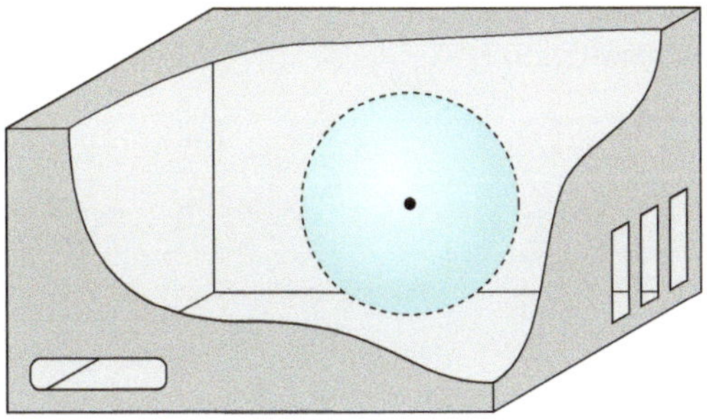

Fig. 1. Spherical domain inside an arbitrary shielding enclosure.

The normalized surface-spherical harmonics are given by

$$Y_{n,m}(\vartheta,\varphi)=\sqrt{\frac{2n+1}{4\pi}\frac{(n-m)!}{(n+m)!}}P_n^m(\cos\vartheta)e^{jm\varphi} \quad (9)$$

with P_n^m being associated Legendre functions of the 1st kind.

For two-dimensional problems, a cylindrical-multipole expansion in plane-polar coordinates (R,φ) consisting of ordinary Bessel-functions J_n in R and harmonic functions in φ can be applied. For the TM$_z$-case this takes the following form (Klinkenbusch, 2005):

$$E_z(R,\varphi) = \sum_{n=-\infty}^{\infty} a_n J_n(\kappa R)e^{jn\varphi} \quad (10)$$

$$H_R(R,\varphi) = \frac{j}{Z}\frac{1}{\kappa R}\frac{\partial}{\partial\varphi}E_z(R,\varphi) \quad (11)$$

$$H_\varphi(R,\varphi) = -\frac{j}{Z}\frac{1}{\kappa}\frac{\partial}{\partial R}E_z(R,\varphi). \quad (12)$$

In any of the described cases, the complete information about the field is contained in the complex-valued multipole amplitudes $(A_{n,m},\ B_{n,m};\ a_n)$. Consequently, the statistics of an electromagnetic field inside the enclosure can be reduced to the statistics of the corresponding multipole amplitudes.

3 Statistical formulation

To describe a statistically varying electromagnetic field we assume its multipole amplitudes to be random variables. For a spherical-multipole expansion truncated at $n=n_{\max}$ there are $n_{\max}(n_{\max}+2)$ index pairs (n,m). For each (n,m) four real random variables need to be considered:

$$\mathrm{Re}\{A_{n,m}\},\mathrm{Im}\{A_{n,m}\},\mathrm{Re}\{B_{n,m}\},\mathrm{Im}\{B_{n,m}\}. \quad (13)$$

We introduce the multipole-amplitude random vector $\boldsymbol{V}$ consisting of all multipole amplitudes considered. It thus has the length $i_{\max}=4n_{\max}(n_{\max}+2)$.

This random vector $\boldsymbol{V}$ will be described by its first few moments, particularly by its expectation value $\eta_{\boldsymbol{V}}$ and its covariance matrix $C_{\boldsymbol{V}}$ (Papoulis, 2008).

3.1 Computation of field statistics

Once the multipole-amplitude random vector $\boldsymbol{V}$ is known in terms of its statistical moments, the field statistics at any points in the domain, where the multipole expansion is valid, can be easily calculated. To this end, we first construct a new field-component random vector $\boldsymbol{W}$ consisting of all real and imaginary parts of all field components at any desired point in the domain. Because of the linearity of the multipole expansion (Eqs. 1–2) the field-component random vector $\boldsymbol{W}=g(\boldsymbol{V})$ depends linearly on the multipole-amplitude random vector $\boldsymbol{V}$, hence the statistical moments of $\boldsymbol{W}$ are easily obtained from the statistical moments of $\boldsymbol{V}$. For the first two moments (expectation value and covariance matrix) this leads to:

$$\eta_{W_k} = g_k(\eta_{\boldsymbol{V}}) \quad (14)$$

$$C_{W_k,W_l} = \sum_i\sum_j C_{V_i,V_j}\left.\frac{\partial g_k}{\partial v_i}\right|_{\eta_V}\left.\frac{\partial g_l}{\partial v_j}\right|_{\eta_V}. \quad (15)$$

Note that for this linear relation n-th order moments of $\boldsymbol{W}$ only depend on n-th order moments of $\boldsymbol{V}$.

This means that the complete field statistics of the whole domain, where the multipole expansion is valid, is contained in the moments of $\boldsymbol{V}$. Usually, the first two moments are sufficient, for arbitrary accuracy higher order moments like skew, curtosis, etc. – as well as their correlations – may have to be considered.

3.2 Compactness of description

The proposed approach of a statistical multipole expansion is a very efficient and systematic way of modelling an electromagnetic field with small variations.

As an example consider the field in a cubic volume of edge length 1.6 m centered around the point of origin at a frequency of 750 MHz (as used in the first examples of Sect. 4). In this case the multipole expansion can be truncated at $n_{\max}\approx 20$ for sufficient accuracy, which leads to a random vector of length $i_{\max}\approx 1760$. On the other hand, to describe the statistics of the field components at each point in space, at least six real random variables would be required per position, i.e. the quadrature components of either the electric or the magnetic field. At a reasonable spatial resolution of at least ten points per wavelength this amounts to a random vector of length $i_{\max}\approx 384000$.

The impact of this difference is even larger when considering that, while the length of the vector of expectation values η is $i_{\max}$, the number of entries in the correlation matrix C is $i_{\max}^2$.

A similar result is obtained for a cylindrical expansion of a square domain of 1.6 m×1.6 m. At 750 MHz the statistical multipole expansion leads to $i_{\max}\approx 194$, the pointwise description of the field components' statistics to $i_{\max}\approx 3200$.

3.3 Sources of statistical variations

To test this new approach we start with a simple example, that is, the electromagnetic field of a homogeneous plane wave where we add a small statistical variation using a single normally distributed real random variable U.

The spherical multipole expansion's (Eqs. 1–2) amplitudes of a plane wave are known (Klinkenbusch, 1996):

$$A_{n,m} = E_0 4\pi j^{n+1} \frac{(-1)^m}{n(n+1)} \left[\boldsymbol{n}_{n,-m}(\vartheta,\varphi) \cdot \hat{\boldsymbol{\xi}} \right]_{\vartheta_0,\varphi_0} \tag{16}$$

$$B_{n,m} = \frac{E_0}{Z} 4\pi j^{n+1} \frac{(-1)^m}{n(n+1)} \left[\boldsymbol{m}_{n,-m}(\vartheta,\varphi) \cdot \hat{\boldsymbol{\xi}} \right]_{\vartheta_0,\varphi_0}. \tag{17}$$

Here, E_0 and $\hat{\boldsymbol{\xi}}$ are amplitude and polarization of the plane wave, while (ϑ_0, φ_0) denote its angle of incidence.

For the two-dimensional $\mathrm{TM_z}$-case, the multipole amplitudes of a plane wave of amplitude E_0 and angle of incidence φ_0 are (Klinkenbusch, 2005):

$$a_n = E_0 j^n e^{-jn\varphi_0}. \tag{18}$$

First we will vary the phase. This can be easily done by multiplying the known amplitudes by a phase factor containing the random variable U as follows:

$$A'_{n,m} = A_{n,m} e^{jU} \tag{19}$$

$$B'_{n,m} = B_{n,m} e^{jU} \tag{20}$$

$$a'_n = a_n e^{jU}. \tag{21}$$

A practically more interesting case is the variation of the angle of incidence. This will be realized by replacing the angle φ_0 in Eqs. (16–18) with a random variable U as described above.

With these relations between the given random variable U and the multipole-amplitude random vector $\boldsymbol{V} = g(U)$, the moments of $\boldsymbol{V}$ can be calculated. Since U is a single real random variable (not a random vector), the expectation values and covariance matrix of $\boldsymbol{V}$ are as follows:

$$\eta_{V_i} = g_i(\eta_U) + \sum_{n=2}^{\infty} \frac{\mu_n(U)}{n!} g_i^{(n)}(\eta_U) \tag{22}$$

$$C_{V_i,V_j} = \sum_{n=2}^{\infty} \frac{\mu_n(U)}{n!} \left(\sum_{k=1}^{n-1} \binom{n}{k} g_i^{(k)}(\eta_U) g_j^{(n-k)}(\eta_U) \right) - \sum_{n=4}^{\infty} \left(\sum_{k=2}^{n-2} \frac{\mu_k(U)\mu_{n-k}(U)}{k!(n-k)!} g_i^{(k)}(\eta_U) g_j^{(n-k)}(\eta_U) \right). \tag{23}$$

Here $\mu_n(U)$ denotes the n-th central moment of U and $g_i^{(k)}$ is the k-th derivative of g_i with respect to its argument.

In this paper we choose U to be normally distributed with zero mean, hence all even moments of U are known in terms of its standard deviation σ_U (Abramowitz, 1972) and all odd moments are zero (see Table 1). For the following calculations the moments of U are considered up to and including eighth order.

Table 1. Central moments of a normally distributed random variable (Abramowitz, 1972).

n	1	2	3	4	5	6	7	8	...
μ_n	0	σ^2	0	$3\sigma^4$	0	$15\sigma^6$	0	$105\sigma^8$	...

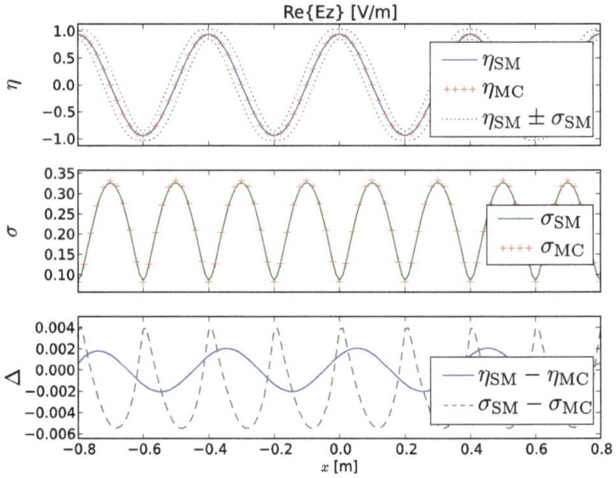

Fig. 2. Expectation value η (upper plot) and standard deviation σ (central plot) of $\mathrm{Re}\{E_z\}$ according to statistical multipole calculation (SM) and Monte Carlo simulation (MC) for a plane wave with varying phase ($\sigma_\psi = 20°$). The lower plot displays the differences between statistical multipole calculation and Monte Carlo simulation.

4 Results

4.1 Plane wave

Figure 2 shows the results for a plane wave with a varying phase ψ of standard deviation $\sigma_\psi = 20°$ at frequency $f = 750$ MHz. The upper diagram represents the expectation value η of the real part of E_z and includes the corresponding standard deviation σ_{SM}. The central diagram shows the standard deviation σ_{SM} alone. Both are calculated by the aforementioned method and – for reference – by a Monte-Carlo simulation (Index $_{\mathrm{MC}}$). The absolute differences between the spherical-multipole calculations are also shown in the lower diagram.

As expected, for a varying phase the standard deviation is largest at the inflection points of the sine curve and minimal at the extreme values. The accordance with Monte Carlo simulations is excellent.

Figure 3 shows the results for a plane wave along the main axis, where the angle of incidence φ_0 is varied according to a standard deviation of $\sigma_{\varphi_0} = 5°$.

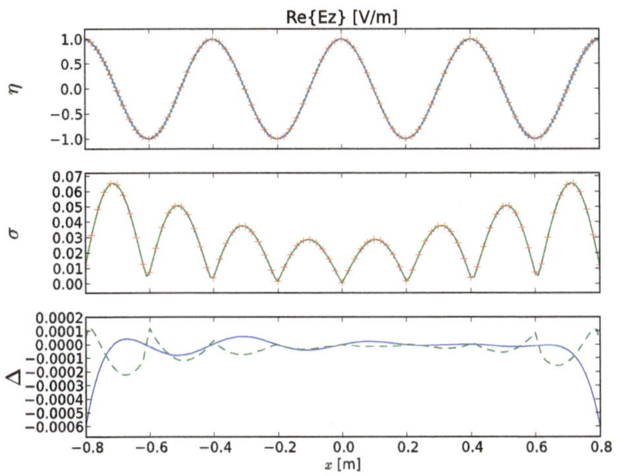

Fig. 3. Expectation value and standard deviation of $\mathrm{Re}\{E_z\}$ according to statistical multipole calculation and Monte Carlo simulation for a plane wave with varying incident angle ($\sigma_{\varphi_0} = 5°$). For a legend see Fig. 2.

Again, the influence of the angle of incidence on the field is largest at the sine curve's inflexion points and minimal at the extreme values, and the agreement with the Monte Carlo simulations is excellent.

4.2 Slitted cylinder

To test our approach with a simple shielding geometry we have used the geometry shown in Fig. 4 representing a thin slitted PEC-cylinder with radius R_S and aperture angle β, illuminated by a plane wave polarized in the z-direction and incident at φ_0. For the deterministic solution of this problem see (Klinkenbusch, 2005).

Figure 5 shows the expectation value of the imaginary part of E_z in the x-y-plane. The shape of the cylinder with $R_0 = 0.5$ m and $\beta = 40°$ is displayed as a black line, the angle of incidence is varying around its expectation value $\eta_{\varphi_0} = 135°$ with standard deviation $\sigma_{\varphi_0} = 5°$. The frequency of the incident field is 750 MHz.

Figure 6 shows in detail the corresponding expectation values, standard deviations and their comparison to Monte Carlo simulations (as in Figs. 2–3) along the x-axis (the dashed line in Fig. 5).

Compared to the previous examples the field structure is much more complex and the standard deviation reaches much larger values. However the multipole expansion again shows excellent correspondence with the Monte Carlo simulation particularly inside the shield.

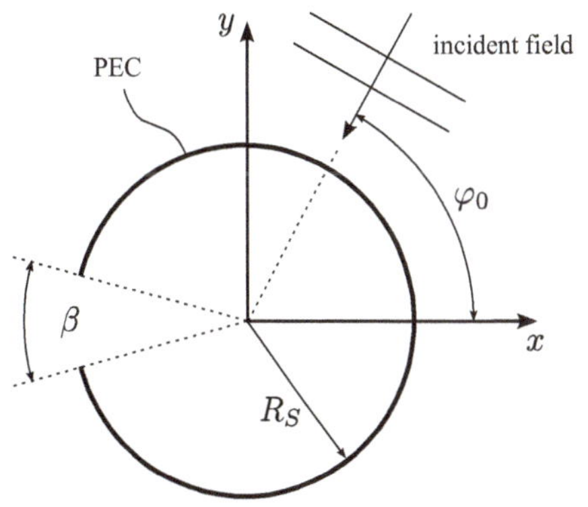

Fig. 4. Geometry of the slitted PEC cylinder in the x-y-plane.

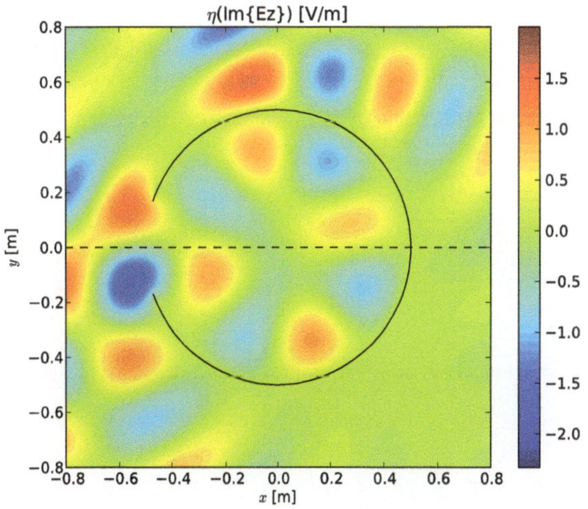

Fig. 5. Expectation value of $\mathrm{Im}\{E_z\}$ according to statistical multipole calculation for a slitted cylinder illuminated by a plane wave with varying angle of incidence ($\sigma_{\varphi_0} = 5°$).

4.3 Shielding effectiveness

The electromagnetic shielding effectiveness has been shown to be an adequate quantity for the characterization of a shield for the high-frequency case (Klinkenbusch, 1996). It is defined as:

$$SE_{em} = 10\log_{10} \frac{2}{\frac{|\boldsymbol{E}^{\mathrm{sh}}|^2}{|\boldsymbol{E}^{\mathrm{un}}|^2} + \frac{|\boldsymbol{H}^{\mathrm{sh}}|^2}{|\boldsymbol{H}^{\mathrm{un}}|^2}}. \tag{24}$$

The quantities marked with ^{sh} are for the shielded case, while those marked ^{un} are for the unshielded case, i.e. the case where no shield is present at all. For the example in Sect. 4.2 the corresponding unshielded fields $|\boldsymbol{E}^{un}|$ and $|\boldsymbol{H}^{un}|$ are those of an undisturbed plane wave and thus are constant. The fields for the shielded case are those of Sect. 4.2 described by the random vector $\boldsymbol{W}$. The random variable for

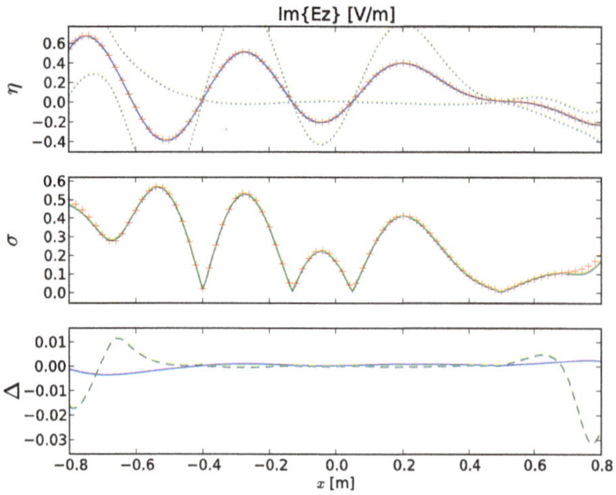

Fig. 6. Expectation value and standard deviation of $\mathrm{Im}\{E_z\}$ according to statistical multipole calculation and Monte Carlo simulation for a slitted cylinder illuminated by a plane wave with varying angle of incidence ($\sigma_{\varphi_0} = 5°$). For a legend see Fig. 2.

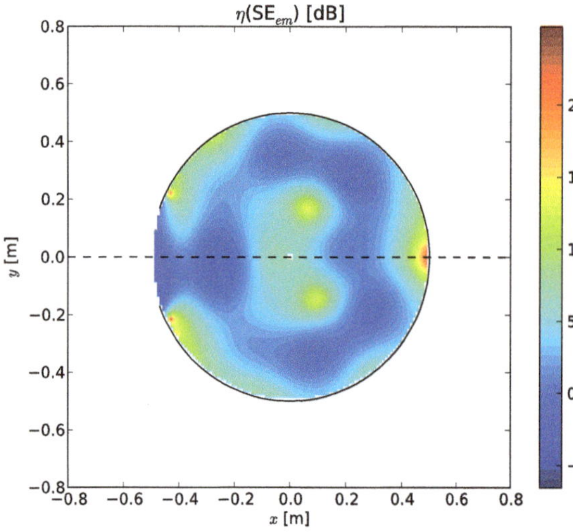

Fig. 7. Expectation value of SE_{em} according to statistical multipole calculation for a slitted cylinder illuminated by a plane wave with varying angle of incidence ($\sigma_{\varphi_0} = 5°$).

the shielding effectiveness will be called X. To compare the results with those ones of the simple cases described before we consider the expectation value and the standard deviation of the shielding effectiveness. These are calculated from the relations

$$\eta_X = g(\eta_{\boldsymbol{W}}) + \sum_i \sum_j \frac{C_{\boldsymbol{W}_i,\boldsymbol{W}_j}}{2} \left. \frac{\partial^2 g}{\partial w_i \partial w_j} \right|_{\eta_{\boldsymbol{W}}} + \ldots \tag{25}$$

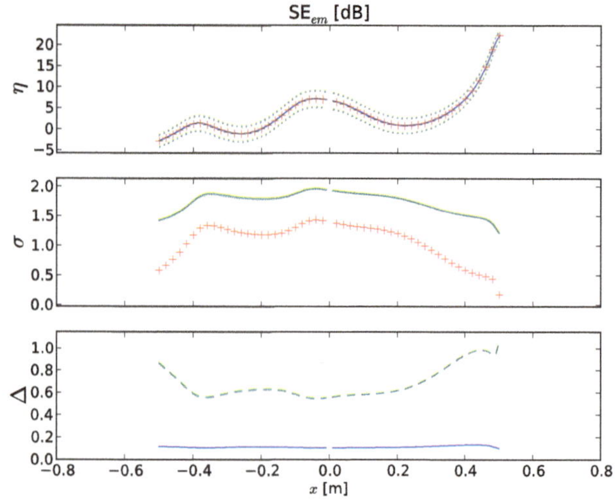

Fig. 8. Expectation value and standard deviation of SE_{em} according to statistical multipole calculation and Monte Carlo simulation for a slitted cylinder illuminated by a plane wave with varying angle of incidence ($\sigma_{\varphi_0} = 5°$). For a legend see Fig. 2.

$$\sigma_X^2 = \sum_i \sum_j C_{\boldsymbol{W}_i,\boldsymbol{W}_j} \left. \frac{\partial g}{\partial w_i} \right|_{\eta_{\boldsymbol{W}}} \left. \frac{\partial g}{\partial w_j} \right|_{\eta_{\boldsymbol{W}}} + \ldots \tag{26}$$

Figure 7 shows the expectation value of SE_{em} in the x–y-plane. Figure 8 shows in detail the corresponding expectation values, standard deviations and their comparison to Monte Carlo simulations along the x-axis. All parameters are identical to those in Figs. 5–6.

While the error of the expectation value is still quite small, the standard deviation derived from the multipole evaluation differs significantly from the Monte Carlo simulation. This can be explained by Eqs. (25–26). The moments of X depend on all moments of $\boldsymbol{W}$, but unlike Eqs. (22–23), where all moments of the normally distributed random variable U were known, in this case only the expectation values and covariance matrix of $\boldsymbol{W}$ have been taken into account for this calculation. Obviously that is not sufficient for a non-linear relation like Eq. (24).

This clearly shows that for the statistics of quantities that depend in a non-linear fashion on the field components, also higher order moments of the multipole amplitudes have to be taken into account.

5 Conclusions

For calculating the statistical moments of the electromagnetic field in case of small variations of various parameters, the statistical-multipole method yields similar results as Monte Carlo simulations with the advantage that the number of quantities to represent the field and its statistics is greatly reduced. However, modeling the multipole-amplitude random vector in terms of only first and second order statistical moments has been shown to be insufficient for the computation of the moments of quantities, which nonlinearly depend on the multipole amplitudes like the electromagnetic shielding effectiveness. In such cases higher-order statistical moments of the statistical multipole expansion have to be considered. However, the basic advantage of representing the field by a minimum number of parameters is still preserved.

Further research includes the application of the method to more general, that is, non-canonical shielding problems, and to characterize the shielding effect by the first few terms of the corresponding multipole expansion and its statistical moments.

References

Ajayi, A., Ingrey, P., Sewell, P., and Christopoulos, C.: Direct Computation of Statistical Variations in Electromagnetic Problems, IEEE T. Electromagn. C., 50, 325–332, 2008.

Abramowitz, M. and Stegun, I. A. (eds.): Handbook of Mathematical Functions with Formulas, Graphs, and Mathematical Tables, Applied mathematics series 55. US Government Printing Office, Washington, DC, USA, Tenth Printing, with corrections edition, 1972.

Holland, R. and St. John, R.: Statistical Electromagnetics, Taylor and Francis, 1999.

Klinkenbusch, L.: Theorie der sphärischen Absorberkammer und des mehrschaligen Kugelschirmes, Habilitationsschrift, Ruhr-Universität Bochum, 1996.

Klinkenbusch, L.: On the shielding effectiveness of enclosures, IEEE T. Electromagn. C., 47, 589–601, 2005.

Papoulis, A. and Pillai, S. U.: Probability, random variables, and stochastic processes, McGraw-Hill series in electrical and computer engineering, McGraw-Hill, Boston, USA, 4th edn., 2008.

An EMC study of an electrical powertrain for transportation systems

B. Chand[1], K. Hähre[2], J. Keghie[1], and S. Dickmann[1]

[1]Fundamentals of Electrical Engineering, Helmut-Schmidt-University/University of the Federal Armed Forces, Hamburg, Germany
[2]Power Electronics, Helmut-Schmidt-University/University of the Federal Armed Forces, Hamburg, Germany

Correspondence to: B. Chand (benian.chand@hsu-hh.de)

Abstract. There are many reasons, why electric drive systems will be used in future more frequently in transportation systems. The electrification allows a flexible use of different energy sources and thus creates a greater independence from fossil energy. However, electric drive systems are a challenge for the manufacturers of transportation systems. The electrical powertrain can not only be exposed to interferences, it may also emit interferences into components and communication cables in the vicinity. This can be a high risk for the drive system and for the passengers. For that reason, an EMC study of an electrical powertrain for transportation systems is presented. There the interference at the unshielded high-voltage cable (HV-cable) is characterized. The analysis of the influence of position variation of the communication cable with respect to the unshielded HV-cables is one main aspect. Different optimizations have been added to the test setup.

1 Introduction

Almost every manufacturer of transportation systems extends his portfolio of products with electric drives. Not only a greater independence from fossil energy, but also more restrictive EMC-standards force the manufacturers to restructure their research and development departments. The electrical powertrain (Fig. 1) can not only be influenced from outside, it can also emit disturbances. These disturbances are critical, when important communication lines are influenced, which could put the drive or the safety of the passengers at risk. A simplified study was done in Weber et al. (2003). An extension to practical cases is done here. In addition, the interferences on the unshielded HV-cables are characterized. In transportation systems the HV-cables are distributed over the whole vehicle. Therefore the position between the HV-cables and existing communication lines can vary. And these different positions yields to different coupling behaviours.

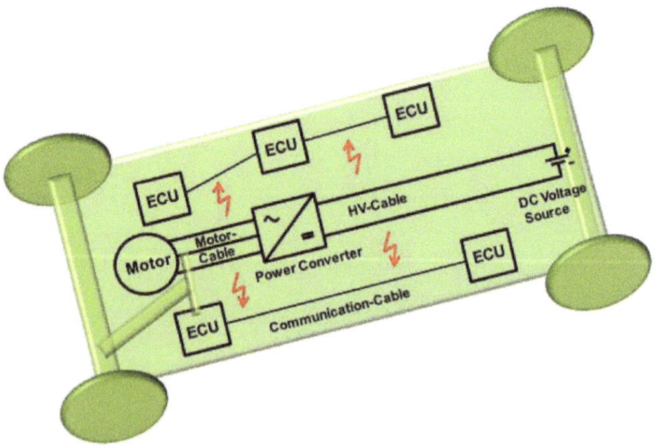

Fig. 1. Electrical powertrain and nearby Electronic Control Units (ECUs) in an automobile.

For this reason, the position of the communication lines to the HV-cables is an essential point in this study. Furthermore, the influence of variations of some parameters is studied. These parameters which are taken into consideration are the motor speed, the PWM-frequency, the terminations and the load. An optimized control unit for the power converter is developed and built up in this study, and optimizations on the measurement setup are done. Different communication cables are used for comparison.

2 Components of the test setup

In Fig. 2 the measurement setup is shown. At first glance one could expect, that the disturbances are only present on the motor cables between the motor and the power converter.

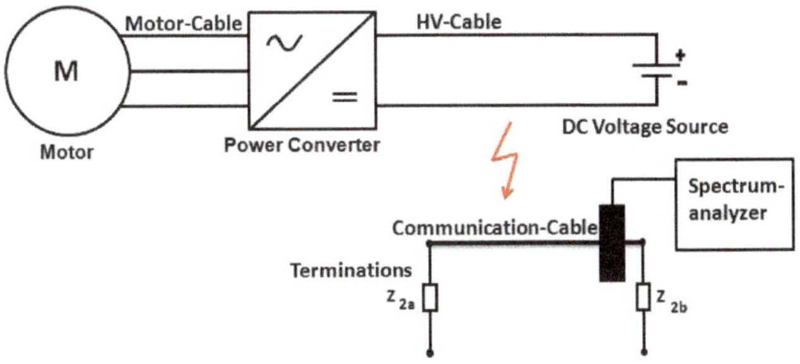

Fig. 2. Measurement setup.

This topic was presented in Kosonen and Ahola (2009). But this contribution shows, that disturbances are also present on the HV-cables at the DC side. Due to packaging reasons, the HV-cables are distributed over the whole vehicle and therefore could influence all components in the vehicle. The motor cables in Fig. 2 are kept very short or inserted into the housing of the motor, to minimize electromagnetic emissions.

2.1 DC voltage source

For the electrification of the powertrains, DC voltages from some hundred Volts to more than 1000 V, depending on the power of the motor, are commonly used. In this study an adjustable source with maximum 350 V and 6.5 A is used.

2.2 Power converter

Power Converters generate from DC voltage/DC current a mono- or multiphase AC voltage/AC current with an adjustable frequency.

The frequency setting is done by a control unit. The schematic design of the inverter is given in Fig. 3.

Two capacitors are used for smoothing and for buffering. The IGBT drivers control the six IGBTs, which are connected to form three half-bridges. At each of the transistors a freewheeling diode is attached, to protect the IGBTs from overvoltage. The inductors describe the motor in star circuit. The branch next to the two capacitors is the chopper. The chopper is used for recovery of the voltage, which is generated due to switching of the IGBTs. By matched time switching of the IGBTs an approximate three-phase sine wave is produced.

2.3 Control unit

A control unit for the inverter has been designed and built up in this study. It includes a 32-bit flash microcontroller, on the base of an ARM Cortex microcontroller architecture. With this control unit it is possible to create a three-phase PWM. Discontinuous step voltage profiles (e.g. rectangular patterns) cause electromagnetic interference, while by using the PWM method, a good approximation of a sine curve

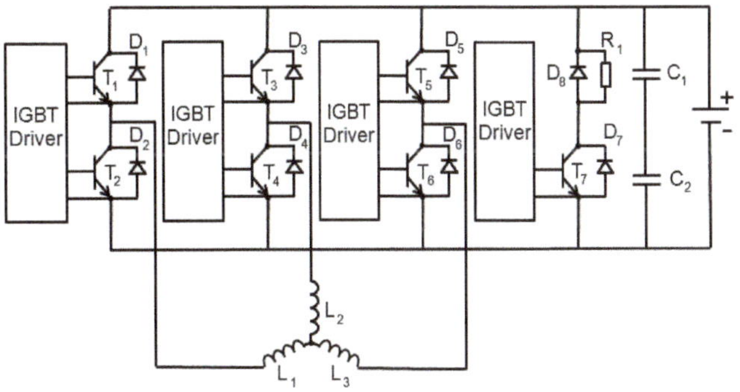

Fig. 3. Equivalent circuit of the power converter.

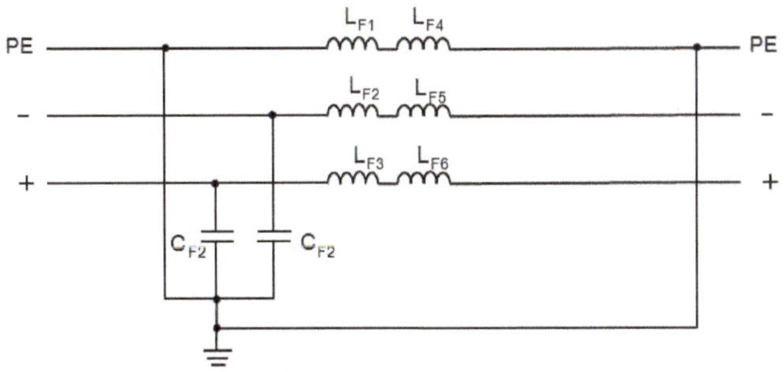

Fig. 4. Equivalent circuit of the DC-filter.

is created. Optocouplers are used for galvanic isolation of the signals. By using this control unit, it is possible to control the motor speed and the PWM-frequency. It should be noted, that the PWM frequency is chosen sufficiently large in comparison to the rotational frequency to obtain a good sine wave, and thus to ensure smooth running of the engine.

2.4 Motor

For the electrification of transportation means, direct current motors, asynchronous motors, synchronous motors and various coupled variants are used. In this study, a 5.5 kW asynchronous motor has been used, since induction motors are very robust and therefore widely used.

2.5 Load

Since the motor without a load only draws a small current, various loads can be used to allow considerable current flow. An elegant way to achieve this, are adjustable resistors, which are connected in parallel in a star circuit to the motor. Thus the current can be drawn from the system according to the values of the resistors.

2.6 Cables

Here unshielded HV-cables of 4 mm^2 copper with a sheath of PVC have been used. For these studies single wire cables and twisted-pair cables as communication cables are used. The

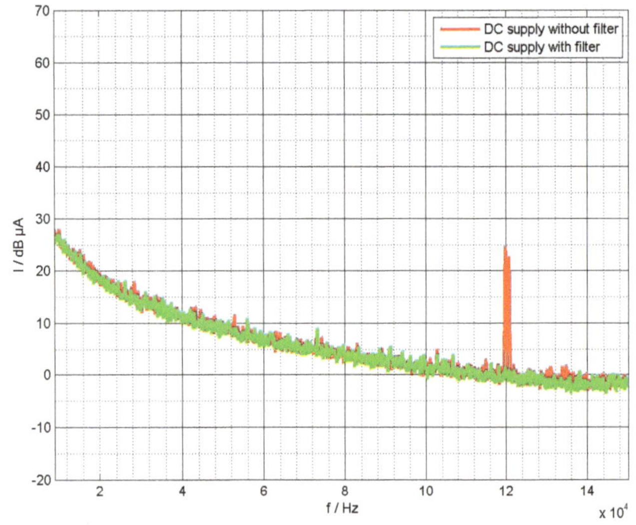

Fig. 5. Comparison of the results with filter and without filter, for the lower frequency range.

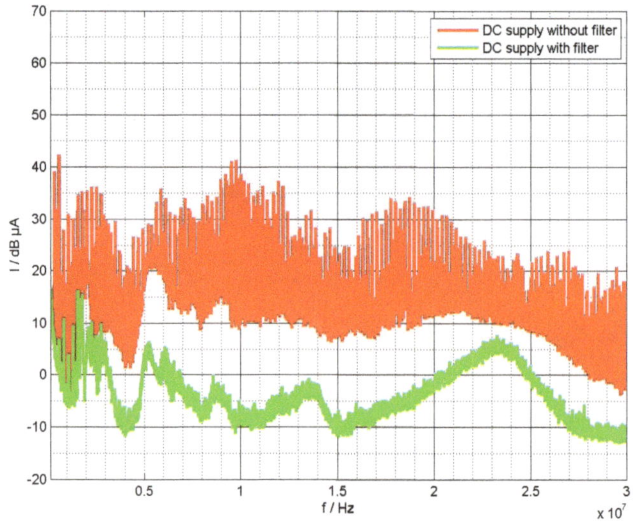

Fig. 6. Comparison of the results with filter and without filter, for the upper frequency range.

single wire is a copper wire with 0.75 mm^2 cross section with a PVC jacket. The twisted-pair cable consists of two twisted wire strands, each 0.25 mm^2, each with a PVC sheath. The single wire represents the communication line for the LIN-Bus. While the bus line for the CAN-Bus and FlexRay-Bus is described by the twisted-pair cable.

2.7 Spectrum analyzer

The measurements are done using a signal analyzer with respect to IEC-CISPR16-2-1 (2008).

3 Optimization of the test setup

The measurements of the HV-cables show strong interferences, which come from the internal pulsed power supplies of the DC source. Since these unwanted interferences would distort the measurement results, a DC-filter (Fig. 4) was constructed and further optimized with simulations for the required frequency range.

The Figs. 5 and 6 show the results with and without filter for both frequency ranges. One can see a good attenuation of interferences from the pulsed DC voltage source. The filter improves the signal from the DC source significantly.

4 Measurements and evaluation

4.1 Optimized measurement setup

The interconnection of the components described in Sects. 2 and 3 yields the test setup shown in Fig. 7. The measurements are carried out in an Absorber Line Shielded Enclosure (ALSE) in order to avoid interferences from outside.

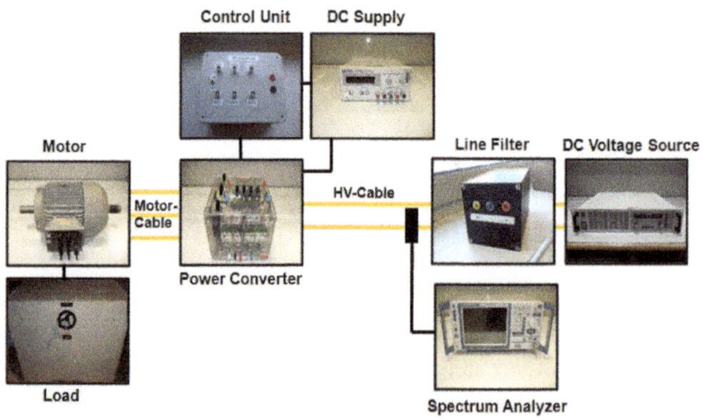

Fig. 7. Components of the whole test setup, connected to one possible measurement setup.

The spectrum has been measured with a current measuring clamp. Since the current through the cables is also of interest, one can calculate it using the frequency-dependent coupling factor k of the clamp (Eq. 1).

$$\frac{I}{\mathrm{d}B\mu A} = \frac{U}{\mathrm{d}B\mu V} + \frac{k}{\mathrm{d}B\frac{1}{\Omega}} \tag{1}$$

4.2 Characterization of the HV-cable

When measuring the current at various positions of both HV-cables, no significant differences in the results of the measurements are seen. On the motor cables, the PWM-frequency and its multiples can be seen clearly. In the higher frequency range the signal form and the magnitude change, like expected, when one compares the current of one cable to all three cables, due to the interaction of the currents on the three cables. The behaviour of the interferences on the

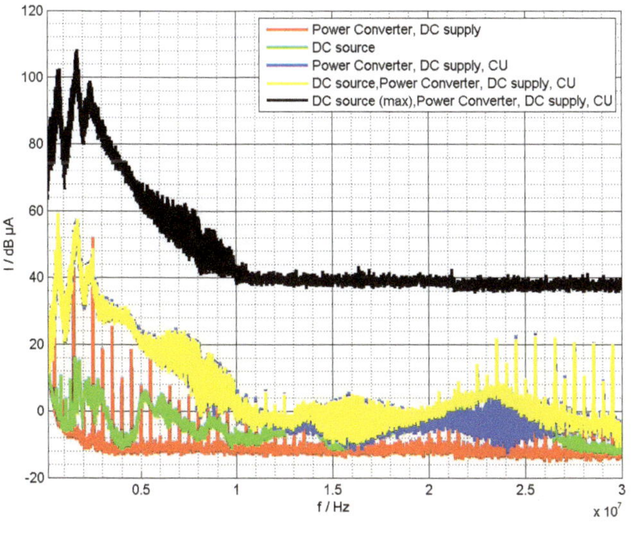

Fig. 8. Composition of the interference.

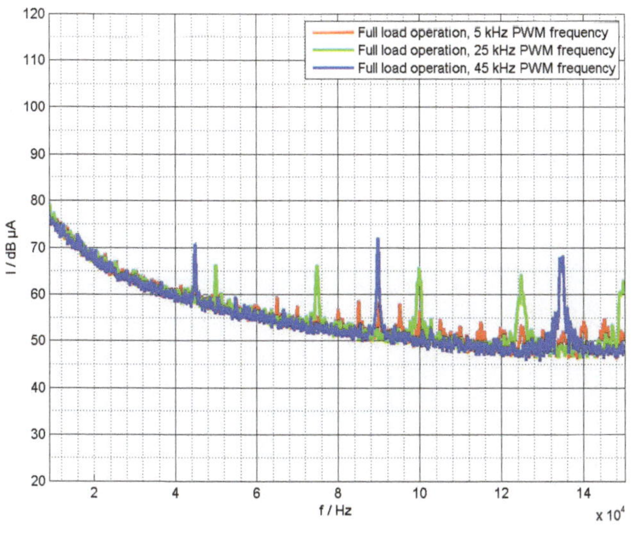

Fig. 9. Comparison of the results for different PWM-frequencies, for the lower frequency range.

HV-cables is composed of the frequency dependent internal interferences of the components, as illustrated by the following results in Fig. 8.

4.3 Influence of the motor speed, the PWM-frequency and the load of the engine

The change of the motor speed respectively the rotational frequency at constant PWM-frequency, does not change the interference behaviour. This is because the resulting short-term disturbances are of transient nature and not considered here, since the frequency is measured under steady state condition. A change in the PWM-frequency at constant motor speed, however, shows a change of gradient. The higher the

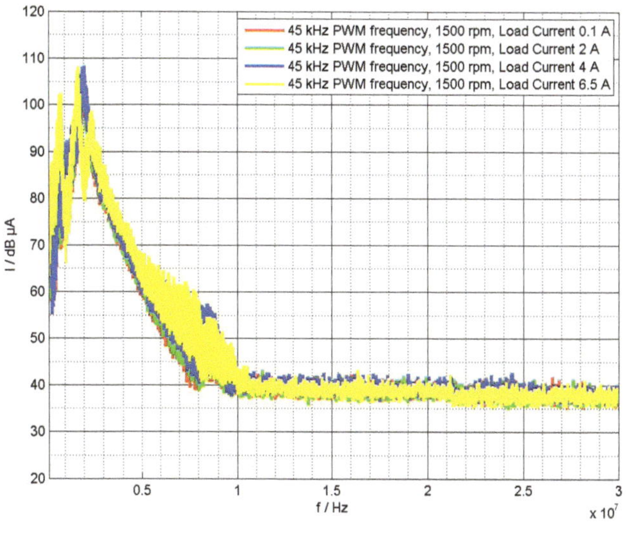

Fig. 10. Comparison of the results for different load currents, for the upper frequency range.

PWM-frequency, the higher the interference currents are and thus more resonances appear. In the lower frequency range, the PWM-frequencies and the multiples are clearly visible, (Fig. 9). Since higher PWM-frequencies lead to higher amplitudes, the PWM-frequencies should be kept as small as possible. But small PWM-frequencies lead to poor quality of the produced sine wave form. For these reasons a compromise between sufficiently high and minimum necessary PWM-frequency have to be found. An increase of the load, which results in an increase of the current, shows no significant differences in the frequency range of 9–150 kHz. In the upper frequency range of 150 kHz–30 MHz, the magnitude changes of course at a higher current, but also the signal form of the interference. The signal form of the interference changes, because for different current/voltage ratio the conducted interference of the components changes internally, as shown in Fig. 10.

4.4 Increasing the distance between the communication cable and the HV-cables

Increasing the distance between the measured cable and the HV-cable in a parallel configuration yields as expected to decreasing magnitude.

In Fig. 11, three different values (5 cm, 10 cm, 30 cm) are used for the parallel distance between the cables. A decreasing of the parallel distance yields further to an increasing of the same ratio of dB of the interference current.

4.5 Change of the height profile of the measured cable with respect to the reference surface

Since the communication lines of transportation systems do not always have a fixed distance from the chassis, the

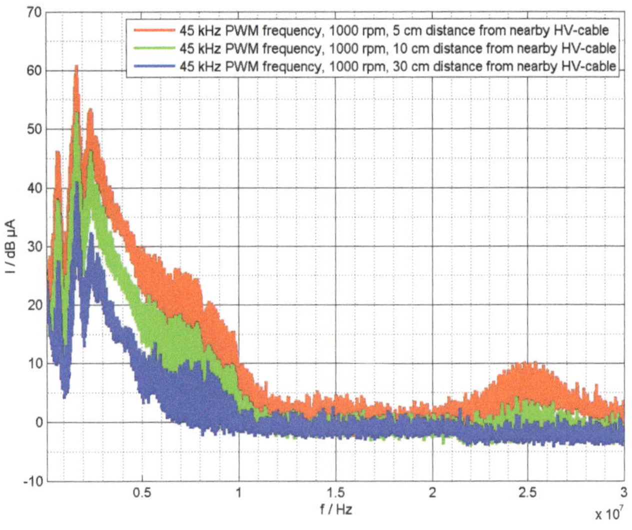

Fig. 11. Influence of position variation of the communication cable with respect to the HV-cables in a parallel configuration, for the upper frequency range.

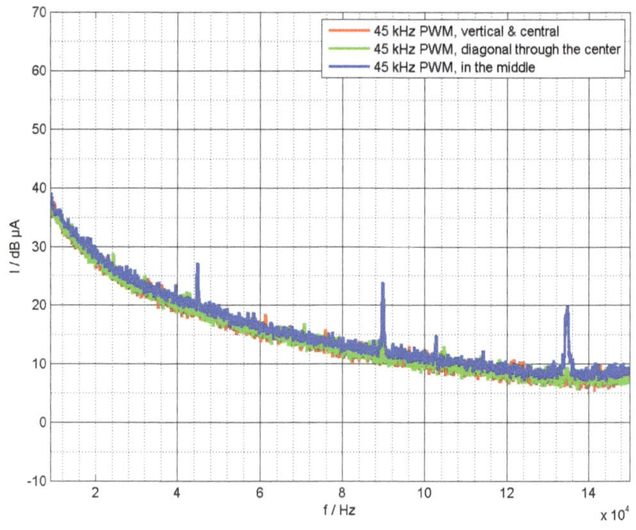

Fig. 12. Influence of position variation of the communication cable with respect to the HV-cables, for the lower frequency range.

influence on communication lines of different heights over a reference surface is studied. One side of the cable is set up 15 cm above the ground surface, the other is 5 cm above the ground plane. The result shows that such variation of the height profile of communication lines over a reference plane has a negligible influence on the interference behaviour for the studied case.

4.6 Communication cable between the HV-cables, diagonally across the HV-cables and perpendicular to the HV-cables

Furthermore, the influences of positioning the communication cable diagonally across the HV-cables, perpendicular to the HV-cables and between the HV-cables are compared. The results in Figs. 12 and 13 shows on the communication cable in the middle of the two HV-cables the most interferences. The inclined position of the measured cable causes a lower coupling of the interference current due to the coupling pattern of the E-and H-fields. The lowest coupling is thus obtained for vertically and centrally located communication cable. The change of the position of the communication cable yields up to 23 dB lower interferences.

4.7 Influence of cable terminations at the communication cable

Another important point is the investigation of the influence of the communication cable terminations. The terminations describe the input resistances in the ECUs in bus systems. The following Fig. 14 shows the results of the measurements with the different terminations.

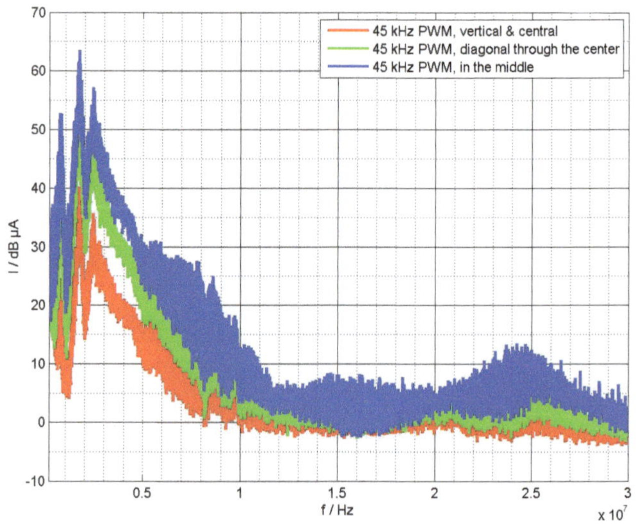

Fig. 13. Influence of position variation of the communication cable with respect to the HV-cables, for the upper frequency range.

It can be seen, that the terminations influence the standing wave patterns on the cables. There is a direct correlation between the choosen terminations and the current magnitude that can be measured. The differences can be up to 40 dB, so it is necessary to consider the termination values well.

4.8 Comparison of single wire and twisted-pair cable

Not only the single-wire measurements are done, also the twisted-pair-cable, which represents the communication link for the CAN-Bus and FlexRay-Bus, is examined. This results for the low frequency range to no differences. For the B-Band (IEC-CISPR16-2-1, 2008) the current in the twisted-pair-cable is twice as large than in the single wire, shown in

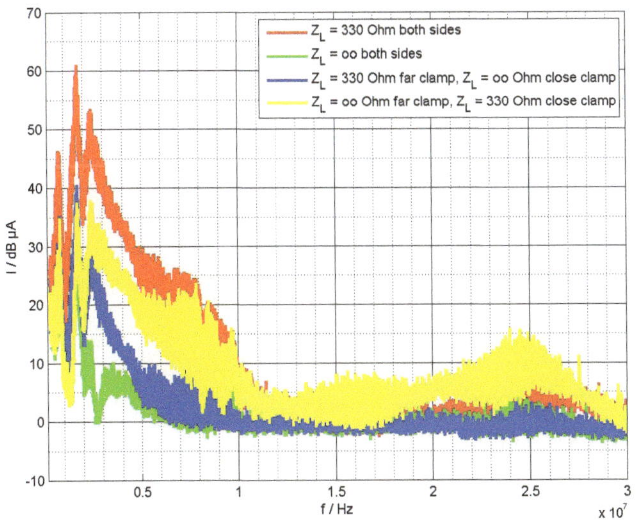

Fig. 14. Comparison of the results for different terminations for both ends of the communication cable, for the upper frequency range.

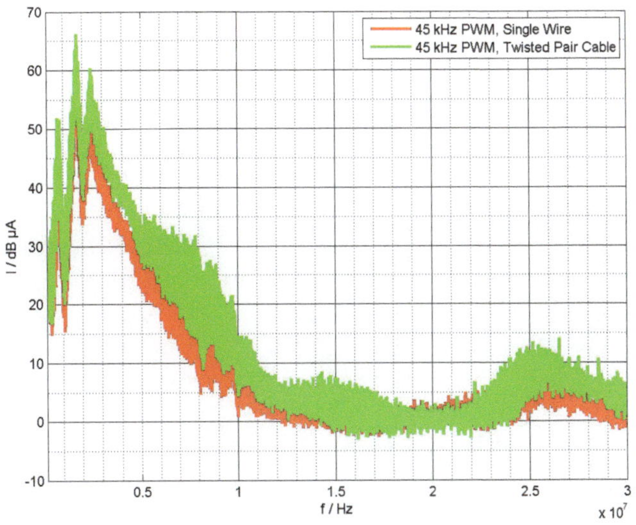

Fig. 15. Comparison of the results for single wire and twisted-pair cable as communication cable, for the B-Band.

Fig. 15. For that reason, as few wires as possible should be used. It should be noted, that the disturbances on the communication lines are common mode interferences.

The resulting currents can yield to load-voltages in Volt-ranges, depending on the terminations and on the voltage and current magnitude of the sources. For that reasons the interferences from the electrical drives to the communication systems should not be neglected, in order to avoid an influence up to a destruction of the communication.

5 Conclusions

In this paper an EMC study of an electrical powertrain for transportation systems is presented. The disturbance at the unshielded HV-cable is characterized. The analysis of the influence of position variation of the communication cable with respect to the unshielded HV-cables is studied. Furthermore, this contribution discussed the influence due to the variation of parameters like the motor speed, the PWM-frequency, the terminations and the load. An optimized control unit for the power converter is developed and built in this study, and optimizations at the test setup are done.

Due to different positions of the communication cables to the HV-cables the results were significantly improved.

To achieve a suppression of interferences on the communication lines, a meaningful positioning of communication cables with respect to HV-cables is encouraged. Using the knowledges gained in this contribution, the safety of the communication links can be increased.

References

IEC-CISPR16-2-1: Specification for radio disturbance and immunity measuring apparatus and methods Part 2-1: Methods of measurement of disturbances and immunity – Conducted disturbance measurements, 2008.

Kosonen, A. and Ahola, J.: Comparison of signal coupling methods for power line communication between a motor and an inverter, 2009, IET 2009, IET Electric Power Applications, 431–440, doi: 10.1049/iet-epa.2009.0114, 2009.

Weber, S., Guttowski, E., Hoene, E., John, W., and Reichl, H.: EMC Issues in Cars with Electric Drives, in: Electromagnetic Compatibility, 2003, EMC 2003, IEEE International Symposium on, 777–782, doi:10.1109/ISEMC.2003.1236706, 2003.

Dead time optimization method for power converter

C. Deselaers[1,*], **U. Bergmann**[2], **and F. Gronwald**[1]

[1]Hamburg University of Technology, 21079 Hamburg, Germany
[2]Jungheinrich AG & Co. KG, 22844 Norderstedt, Germany
[*]now at: DURAG GmbH, 22453 Hamburg, Germany

Correspondence to: F. Gronwald (gronwald@tu-harburg.de)

Abstract. This paper introduces a method for dead time optimization in variable speed motor drive systems. The aim of this method is to reduce the conduction time of the freewheeling diode to a minimum without generation of cross conduction. This results in lower losses, improved EMC, and less overshooting of the phase voltage. The principle of the method is to detect beginning cross currents without adding additional components in the half bridge like resistors or inductances. Only the wave shape of the phase voltage needs to be monitored during switching. This is illustrated by an application of the method to a real power converter.

1 Introduction

Power converter systems are used today in many applications. In particular, in the context of battery supplied vehicles and similar areas the improvement of Electromagnetic Compatibility (EMC) and the reduction of power losses become increasingly important. One approach to improve power converter systems in this respect is to reduce the dead time during switching. To illustrate the concept of dead time, the following Fig. 1 shows a typical half bridge, which is used in power converter systems:

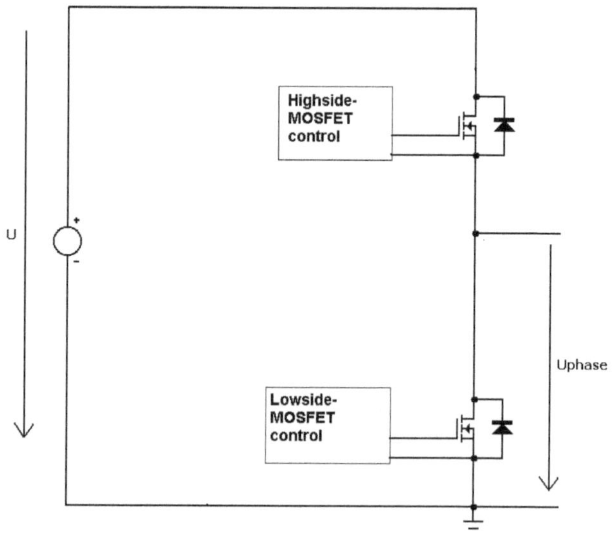

Fig. 1. Principal circuit diagram of a half bridge.

In the operation of the half bridge the Highside- and Lowside-MOSFET are mutually switched. The dead time is here defined as the time delay between the falling edge of the control signal of one MOSFET and the rising edge of the control signal of the opposite MOSFET.

With the reduction of the dead time, the time in which the freewheeling diode conducts and the stored charge in this diode, which is responsible for the Reverse Recovery Effect (Polenov et al., 2009), can be reduced. This effect increases switching losses (Polenov et al., 2009; Reiter et al., 2010) and the snap off of the reverse current results in overvoltage and oscillation of the phase voltage (Semikron, 1998, p. 166). In this context an optimal dead time can prevent the diode from conducting without generating a cross current in the half bridge.

Several methods already have been introduced to reduce the dead time. Many are especially for DC/DC converter systems. For example Reiter et al. (2010) describes an approach which reduces the dead time in one phase and compares after every step the efficiency to a reference phase. An increasing efficiency in one phase will result in an increase of its current. A current balancing controller compensates the resulting deviation of the phase currents. This information can be used for the dead time optimization. As a result, this method uses the existing current balancing controller of a multiphase DC/DC-converter, which does not exist in power converters.

In Yousefzadeh and Maksimović (2005) a method that applies to DC/DC-converters is described which changes the dead time until the duty-cycle command is minimized, corresponding to maximized converter efficiency. This method uses the integrated voltage controller which doesn't exist in that form in a power converter.

Another approach measures the voltage drop across the freewheeling diode (Mappus, 2003). The dead time is reduced until this voltage drop no longer reaches the diodes forward voltage. But measuring this voltage drop is difficult during current commutation because of the influence of induced voltages in the parasitic inductances (Rose et al., 2009).

In addition to these approaches there are several methods which use Zero Voltage Switching (ZVS), compare (Trescaes et al., 2004; Acker et al., 1995; Lau and Sanders, 1997). Most of these methods involve measuring the voltage drop across the switching transistor and the gate-source voltage which is compared to the threshold voltage. However, these approaches can only be used in conjunction with ZVS. In contrast to this, the method described in this paper is focused on a hard switching converter.

Beside the approaches above, there are further methods which apply to hard switching three phase power converters. In Huselstein et al. (1993) an inductance is introduced in the half bridge. This allows detecting an occurring cross current from the change of voltage drop across this inductance: This change can be measured and then the dead time is reduced until a cross current occurs. Here, the main problem is the inclusion of the inductance into the half bridge which can result in a worsening of the switching behavior. An additional method described in Rose et al. (2009) measures the threshold voltage of the MOSFETs involved. The beginning and the end of the current commutation is detected by means of a small inductance introduced in the half bridge. Since the complexity of this method is relatively high, a new designed gate driver is used to reduce the dead time according to this information. This can be disadvantageous because often a simpler and more flexible method is desired which does not require a special Integrated Circuit (IC). Also an additional inductance is needed and it can be assumed that the measurement of the threshold voltage has to be very exact to prevent cross conduction.

In this paper a method is presented which does not require additional components in the half bridge. Only the wave shape of the phase voltage during switching has to be monitored. When a cross conduction occurs, this wave form changes, as it is also stated in (Yousefzadeh and Maksimović, 2005) in view of DC/DC converter systems. These changes have been verified by means of simulation of a half-bridge and measurement in an existing power converter. This power converter is designed for low voltage applications, such that Power-MOSFETs can be used. A measuring circuit has been designed to detect this variation without the need of special or extremely fast components. The measured phase voltage is the basic for the proposed dead time optimization process which successively reduces the dead time until the changes in the wave shape of the phase voltage, that are due to the cross current, are detected.

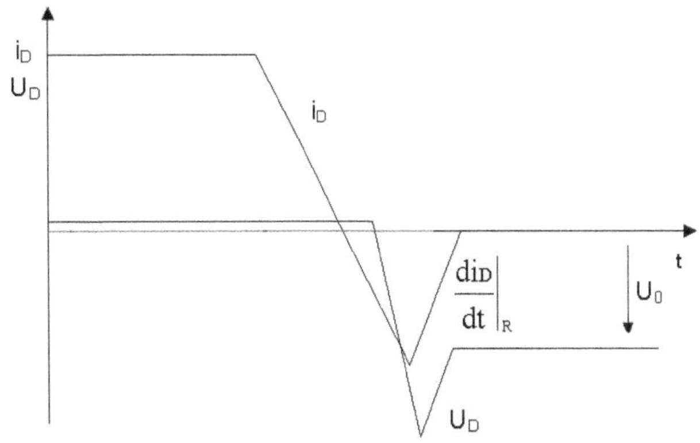

Fig. 2. Typical waveforms of the diode current i_{D} and the voltage drop over the diode u_{D} of a blocking diode, as adapted from (Baliga, 2008). The supply voltage is denoted by U_0.

The structure of the remainder of this paper is as follows: the origin of variation of the phase voltage, which is the basis for the proposed method, is discussed in the following Sect. 2, where also the measuring circuit is introduced and important results of simulations and measurements are presented. In Sect. 3 an actual implementation of the dead time optimization process is described and followed by a conclusion in Sect. 4.

2 Influence of the dead time on the switching behavior

2.1 Reverse Recovery Effect

When the MOSFET takes over the load current from the freewheeling diode of the opposite MOSFET during switch-on, the Reverse Recovery Effect occurs when this diode blocks. The following illustration shows typical current and voltage waveforms during the blocking of the freewheeling diode.

First, the MOSFET starts to take over the load current from the diode during switch-on. Eventually, the MOSFET also has to carry a reverse current of the diode because the stored charge of the diode has to be degraded before it blocks. This results in higher losses in this MOSFET (Baliga, 2008, p. 245). Once the stored charge is degraded the freewheeling diode blocks and the reverse current quickly reduces to zero. This current change induces a voltage in the parasitic inductances of the half bridge (Semikron, 1998, p. 166). If the Reverse Recovery Effect occurs at the transition low-high this voltage results in an overshoot of the phase voltage. The corresponding additional contribution U_{L_σ} to the phase voltage can be characterized by the equation

$$U_{L_\sigma} = L_\sigma \cdot \left.\frac{\mathrm{d}i_D}{\mathrm{d}t}\right|_R \quad (1)$$

Here, the internal inductances of the MOSFETs and of the junction lines are combined in L_σ. The fast current change can also induce oscillations which have a negative impact on EMC (Rose et al., 2009; Semikron, 1998, p. 166).

2.2 Influence of reduced dead time on the Reverse Recovery Effect

The charge Q_R stored in the freewheeling diode can be reduced by decreasing the dead time because this leads to a decreased conducting time of the freewheeling diode (Polenov et al., 2009). The peak I_{RM} of the reverse current depends on this charge (Wang et al., 2004).

$$I_{RM} = \sqrt{\frac{2 \cdot \mathrm{d}i_D/\mathrm{d}t \cdot Q_R}{1+S}} \quad (2)$$

where S is the softness factor of the diode. With a reduced peak, the slope of the reduction of the reverse current, i.e. the term $\mathrm{d}i_D/\mathrm{d}t|_R$, can be decreased as well (Baliga, 2008, p. 251). According to (1) this results in a lower voltage overshoot. It is also mentioned that the losses in the MOSFET during switch-on are reduced due to the lower reverse current and faster switching (Polenov et al., 2009).

2.3 Influence of the cross current on the phase voltage

In the same way as the decrease of the reverse current induces an overshoot of the phase voltage, the cross current can initiate an overshoot of the phase voltage, too. Particularly, the reduction of the cross current is directed in a way such that the voltage induced in the parasitic inductances adds to the phase voltage.

To investigate this behavior, a half bridge was simulated with the main parasitic inductances and resistors included. An existing power converter served as basis for the simulations and also was used for subsequent measurements described below. Figure 3 shows the simulated wave shape of the phase voltage at a dead time where a cross current occurs.

In Fig. 3, the second maximum is due to a decrease of the cross current, as has been confirmed by both simulation and measurement. The width of the peak that is associated to this second maximum depends on the magnitude of the cross current: a larger magnitude implies a larger width and vice versa. This effect can be used to detect the cross current, as described next.

2.4 Setup of the measurement circuit for detecting cross currents

The effect described in the previous paragraph also appears at the transition from the Highside- to the Lowside-MOSFET.

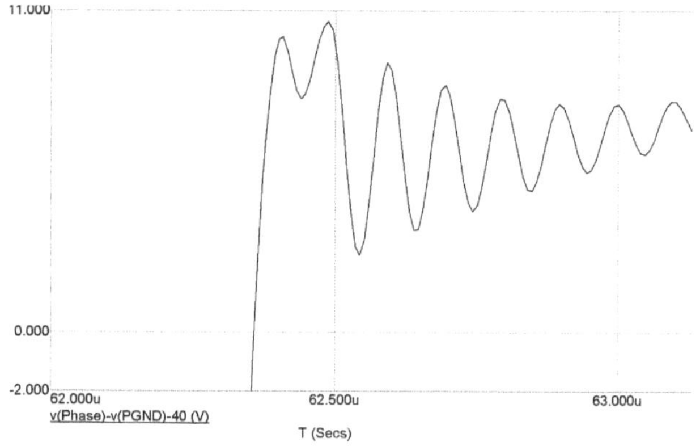

Fig. 3. Waveform of the phase voltage at a comparatively short dead time where a cross current occurs.

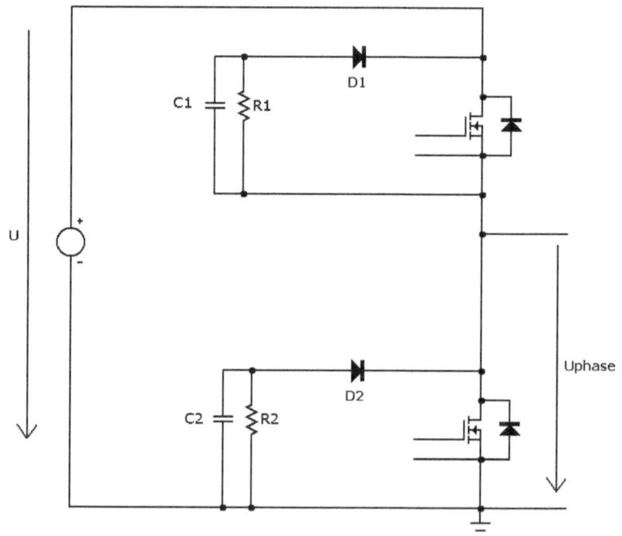

Fig. 4. Measurement circuits for detecting the over- and undershoot of the phase voltage.

The sign of the induced voltage is such that it results in an additional decrease of the phase voltage. This leads to a second undershoot. As a consequence two measurement circuits are necessary: one for detecting the overshoots of the phase voltage and one for detecting the undershoots at the other transition. In Fig. 4 a possible design of these circuits is shown.

First the circuit for detecting the voltage overshoot is described. When the phase voltage becomes larger than the supply voltage plus the forward bias of the diode D1, this diode starts to conduct and the capacitor C1 is charged. To discharge the capacitor for detecting a smaller overshoot after a larger one has occurred, the resistor R1 is connected in parallel. R1 and C1 have to be dimensioned in a way such that the first voltage overshoot is charging the capacitor only to a small fraction and, as a consequence, the second voltage overshoot still can charge the capacitor as well.

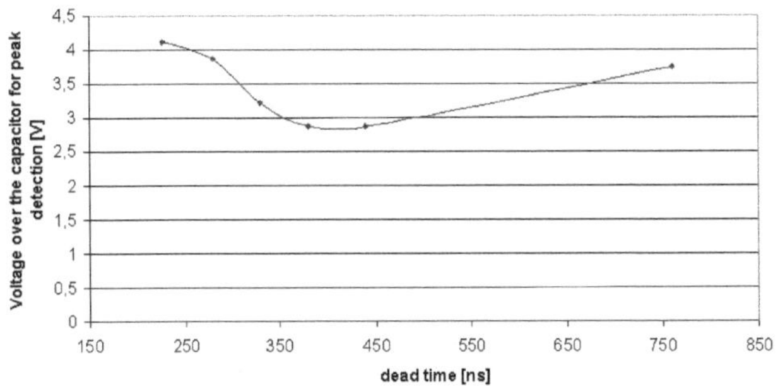

Fig. 5. Voltage across the capacitor for peak detection.

The voltage measured at the capacitor will be characterized by a charging and discharging process and thus approximately have a triangular shape. Then a possibility is to measure only the peak value of this voltage. An option to realize this is to use a capacitor with a diode put in series. This peak detecting capacitor then will be loaded until the maximum voltage will be reached. To discharge this capacitor, a MOSFET put in parallel can be used. This MOSFET can be controlled by the microprocessor that realizes the dead time optimization.

It must be noted that the reference potential of the voltage across the capacitor C1 is given by the phase voltage while the microprocessor who measures this voltage has in most cases a ground reference potential. However, for converting the voltage across the capacitor C1, special ICs can be used, which are e.g. designed for current measurement with a resistor. The peak detection circuit can be set at the output of this IC such that the microprocessor can directly control the discharging MOSFET and measure the voltage over the peak detecting capacitor.

The circuit for detecting undervoltage behaves in the same way. In this case the measurement capacitor is charged when the voltage is below the ground potential plus the forward bias voltage of the diode D2. Here the usage of a peak detection circuit in combination with a voltage transferring IC is also reasonable.

2.5 Influence of the dead time on the measured signal

To investigate the behavior of the measurement circuit of Fig. 4 it is integrated in the half bridge of an actual power converter, which can switch currents up to 400 A. With a successively reduced dead time, the ground referenced voltage across the capacitor for peak detection, as described in the previous section, is measured.

Starting from a large dead time of 750 ns, the absolute value of this voltage decreases because of the reduced Reverse Recovery Effect of the freewheeling diode which leads, according to Eq. (1), to a decreased overvoltage. It is then observed that at a dead time of approximately 350 ns the voltage starts to increase. At this dead time a beginning cross current

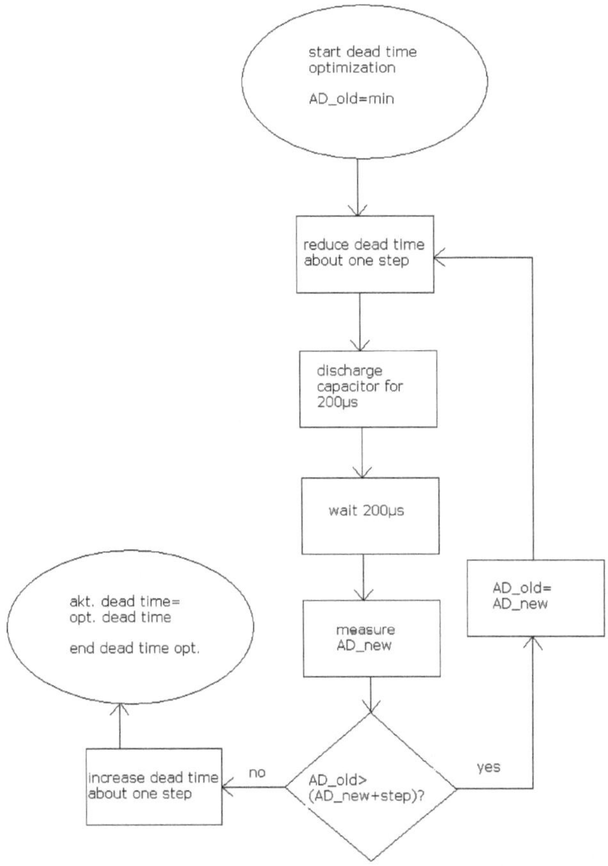

Fig. 6. Principle of the dead time optimization process. The variable AD corresponds to the measured peak voltage.

was detected in several measurements and simulations. Also, beginning from this dead time, increasing losses were measured due to the developing cross current.

At dead times around 350 ns the amplitude of the first overvoltage is strongly reduced in comparison to the larger dead time of 750 ns. But as the simulation result of Fig. 3 shows, the second overshoot becomes larger with increasing cross current. This change of wave shape can also be measured at the half bridge with decreasing dead time. Due to this wider second overshoot the measurement capacitor is stronger charged, which results in the increasing voltage drop. This rise can be detected by a dead time optimization method which is described in the following section.

3 Dead time optimization method

3.1 Process of the dead time optimization method

In general the dead time optimization process only has to be started once at the beginning of the drive. But in some cases it might be useful to repeat this process, e.g. due to an increased

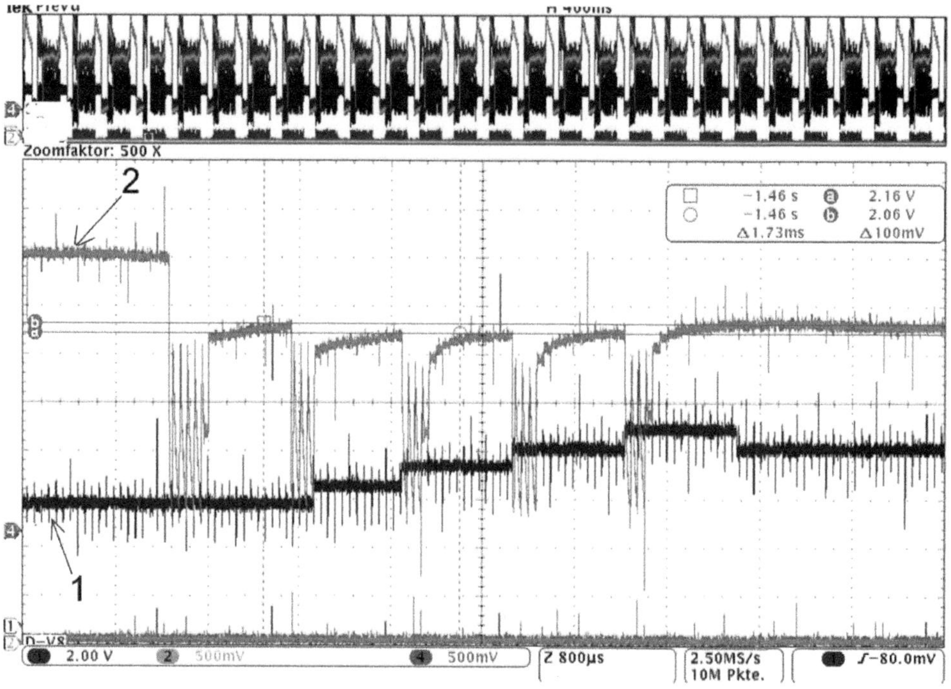

Fig. 7. Dead time optimization process. Curve 1: output of the DA-converter, proportional to dead time: an increase of the DA output voltage corresponds to a decrease of the dead time. Curve 2: measured voltage over the peak detection capacitor C3.

temperature of the converter. This has to be analyzed for a given power converter system.

The PWM-controlling microprocessor has to measure the voltage over the peak detecting capacitors. These signals can directly be transmitted to the AD-converter. It is necessary that the microprocessor can influence the dead time, which is often the case. The dead time is now gradually decreased and after every step the actual voltage over a peak detecting capacitor has to be measured. As an initial condition of this process it is necessary that the capacitor is discharged via the MOSFET which is also controlled by this microprocessor. The discharge has to last until the capacitor C1 or C2 has adapted the new value, which usually takes about 200 μs to 400 μs. After this period, the peak detecting capacitor has to attain the peak voltage. This time depends on the IC used for the voltage transfer. For the cases considered, this time turned out to be about 200 μs.

In the actual process, first the measured peak voltage will decrease as it can be seen in Fig. 5. When the dead time approaches the range where a cross current occurs an increase will be detectable. Figure 6 exemplifies this process.

This process has been integrated in the control of an actual half bridge. For exemplification a voltage was generated via a DA-converter which was chosen to be proportional to the dead time. In Fig. 7 a screenshot of the measured voltages illustrates this process.

First a reduction of the dead time leads to a decrease of the measured voltage. After every dead time reduction, which here is about 50 ns, the peak detection capacitor is discharged until the measurement capacitor C1 has reached the new value. After the fourth step it is seen that the measured voltage starts to increase. This is detected by the controller which then stops the dead time reduction. Since after the fourth step a small cross current exists, the controller increases the dead time about one step. As a result an optimal dead time is found to be about 380 ns.

3.2 Setting of the optimal dead time

Basically four switching transitions have to be distinguished, which are illustrated in Fig. 8.

When the load current is directed out of the half-bridge the Reverse Recovery Effect can be monitored at the transition low-high and in case that the load current runs in the half-bridge the Reverse Recovery Effect can be monitored at the transition high-low. For these two transitions a dead time optimization is useful, as described above, to reduce losses and to improve EMC. The dead time at the other transitions can be chosen relatively high such that no cross current occurs. But it has to be mentioned that the described effect of second overshoot also appears at these transitions. Here it can also be useful to find the optimal, lowest dead time to increase the dynamic range of modulation.

To summarize, four different dead times have been determined. These are, depending on the direction of the load current, two dead times for each switching transition. Therefore the dead times have to be adjusted at each switching tran-

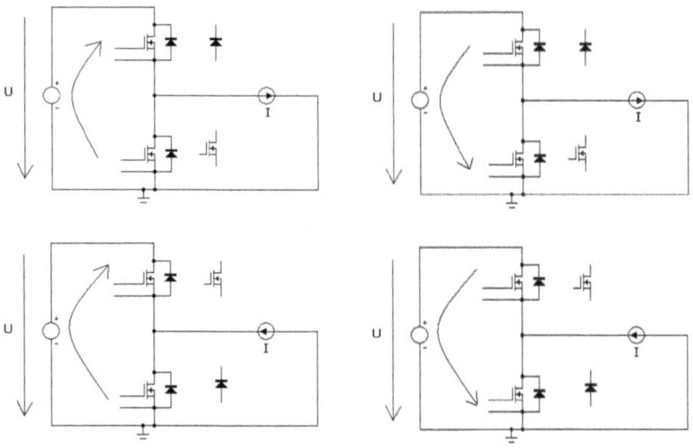

Fig. 8. Switching transitions in a half-bridge, indicated by curved arrows, as adapted from Huselstein et al. (1993).

sition in the zero-crossing of the load current. In many applications the controlling microprocessor measures or calculates the actual current direction, such that this adjustment can easily be achieved.

4 Conclusions

The presented dead time optimization method provides a new alternative to existing approaches. In contrast to the majority of these methods this approach does not need additional components in the main current tracks of the half bridge, which would produce additional losses. A suitable measurement circuit has been proposed to detect the changes in the wave shape of the phase voltage when a cross current occurs. This measured value is the basic variable of the dead time optimization process. The proposed method also has been successfully tested in an actual half-bridge.

With this dead time reduction both the overshoot of the phase voltage and the switching losses are reduced. Also a considerable improvement in the EMC of the power converter has been observed.

So far this method has only been tested with MOSFETs but it should also be applicable to transistors such as IGBTs. Also the application in other contexts, such as the optimization of DC/DC-converters, is conceivable.

References

Acker, B., Sullivan, C. R., and Sanders, S. R.: "Synchronous rectification with adaptive timing control", Proc. IEEE PESC, 1, 88–95, 1995.

Huselstein J.-J., Gauthier, C., and Glaize, C.: Use of the MOSFET channel reverse conduction in an inverter for suppression of the integral diode recovery current, IEEE Power Electronics and Applications, 2, 431–436, 1993.

Jayant Baliga, B.: Fundamentals of Power Semiconductor Devices, Springer Science+Business Media, LLC, Boston, MA, 2008.

Lau, W. and Sanders, S. R.: An integrated controller for a high frequency buck converter, Proc. IEEE PESC, 1, 246–254, 1997.

Mappus, S.: Predictive Gate Drive Boosts DC/DC Power Converter Efficiency, Texas Instruments, Appl. Rep. SLUA281, April 2003.

Polenov, D., Reiter, T., Baburske, R., Probstle, H., and Lutz, J.: The Influence of Turn-Off Dead Time on the Reverse-Recovery Behaviour of Synchronous Rectifiers in Automotive DC/DC-Converters, IEEE EPE'09, 13th European Conference on Power Electronics and Applications, 8 pp., 2009.

Reiter, T., Polenov, D., Probstle, H., and Herzog, H.-G.: PWM Dead Time Optimization Method for Automotive Multiphase DC/DC Converters, IEEE Transaction on Power Electronics, 25, 1604–1614, 2010.

Rose, M., Krupar, J., and Güldner, H.: Zero Body Diode Conduction Time in Variable Speed Drive Converter Systems, IEEE EPE'09, 13th European Conference on Power Electronics and Applications, 10 pp., 2009.

Semikron Application Manual IGBT and MOSFET power modules, 1998.

Trescases, O., Ng, W. T., and Chen, S: Precision gate drive timing in a zero-voltage- switching dc-dc converter, Proceedings IEEE ISPSD'04, 16th International Symposium on Power Semiconductor Devices and ICs, 55–58, 2004.

Wang, Y., Zhang, Q., Ying, J., and Sun, C.: Prediction of PIN Diode Reverse Recovery, IEEE PESC, 4, 2956–2959, 2004.

Yousefzadeh, V. and Maksimovic, D.: Sensorless optimization of dead times in DC-DC converters with synchronous rectification, Proc. IEEE Appl. Power Electron. Conf., 911–917, 2005.

A Low-Cost Versatile Broadband (DC to K-Band) Planar 50 Ω Termination

M. O. Olbrich and E. M. Biebl

Technische Universität München, Fachgebiet Höchstfrequenztechnik, Arcisstr. 21, 80290 München, Germany

Abstract. Numerous standard RF circuits like directional couplers, hybrids, amplitude modulators, oscillators, etc. require a 50 Ω termination. Additionally, in several measurement scenarios, ports have to be terminated. Such terminations are preferable of the same waveguide type as the circuits to avoid degradation of their performance. In general, standard fabrication steps are desired as they are cost-effective. The straightforward solution is a 50 Ω resistor shunted to ground by means of a via. Partial wraparound SMD resistors in a 0402 size have low parasitics making them applicable up to K-band. However, for RF, this combination suffers from the considerable inductive parasitics of the via. Different approaches to avoid the vias, with varying disadvantages exist. We analyzed the approach of broadband matching the non ideal viaresistor combination to 50 Ω by means of a filter structure. The realization uses only standard design and fabrication methods. The design procedure can easily be adapted to arbitrary substrates and fabrication technologies.

The assembled 50 Ω termination was measured using TRL calibration with a VNA. Therefore its characteristics are determined in the intended plane and it can be directly included into a planar 50 Ω circuit. The measured terminations showed a VSWR better than 1.34 within the whole band from DC to K-band, and within 57% of the band (including the 24 GHz ISM band) the VSWR is better than 1.22.

1 Introduction

The objective of the presented design was to gain a standard match for several applications that can directly be included into an actual planar design and that provides its characteristic from DC to K-band. For planar waveguides there is no standard offtheshelf solution as for coaxial cables, for instance. In fact a standard solution for planar terminations would only cover a special substrate material, and furthermore only a special thickness for the respective substrate.

For a few applications, especially at lower frequencies, a coaxial termination can be attached including a coaxial-microstrip transition. Such a solution is not applicable for integrated designs. Various RF applications demand for versatile terminations that are easy to implement onto the respective waveguide plane. A typical example are amplitude modulators. They are often in need of terminating the RF power in the "off" state. Also several measurement scenarios demand for versatile terminations of residual ports.

Generally A VSWR better than 1.22 is considered to represent a perfect match, and a VSWR better than 1.43 provides the desired termination for most applications. The main focus was the 24 GHz ISM band, but the match should be within the above mentioned limits for the whole range from DC to K-band. The design and fabrication procedure should be easily adoptable to different substrates and fabrication technologies. Finally the usual demands for low-costs as well as ease of fabrication and implementation were given.

2 Concept

Several concepts to implement a termination for microstrip lines have been published. Providing a virtual short by replacing a via with a matched open line (Linnér, 1986) reduces the achievable bandwidth. Attenuating the wave by placing lossy material onto the microstrip line (Thompson, 1961) is a difficulttohandle additional fabrication step. The straight forward and low-cost concept is to use standard design and fabrication steps, like vias and discrete SMD components (Lacombe, 1972). We enhance this concept with regard to the frequency range from Ku-band to K-band and in addition we determine the final circuit as well as the component characteristics in the targeted plane to get a definite specification of the termination.

Taconic TLP-5-0100 substrate was taken, the via hole was drilled manually. For the via, a 0.8 mm copper wire was inserted and soldered on both sides. Figure 1 shows a picture of the measured PCB with the via only and the viaresistor combination. The inductive behaviour of the measured via is apparent in the Smith chart of Fig. 2. The IMS RCX 0402PW resistor was selected because of its low parasitics due to its partial wraparound metallization. The unmatched reflection of the via-resistor combination is shown in Fig 3. Here the inductive characteristic of the via is still present and results

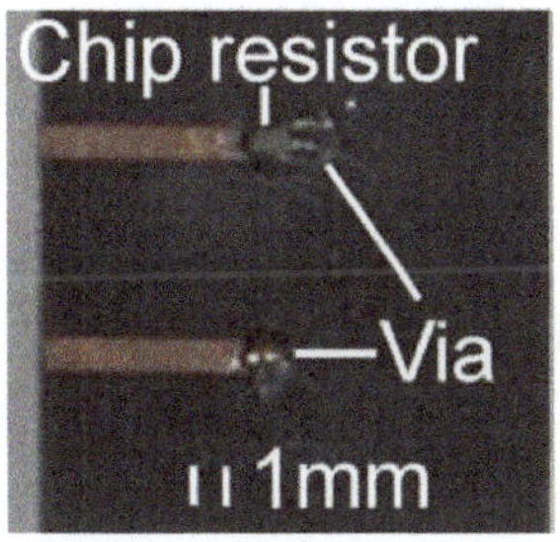

Fig. 1. Picture of the measurement PCB for the via only and the via resistor combination.

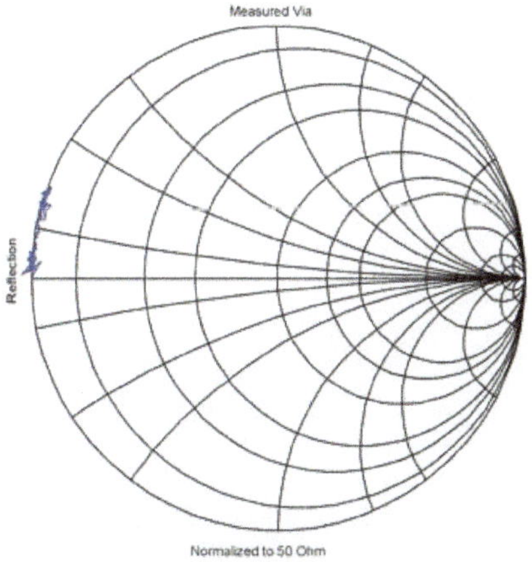

Fig. 2. Smith diagram of the measured via.

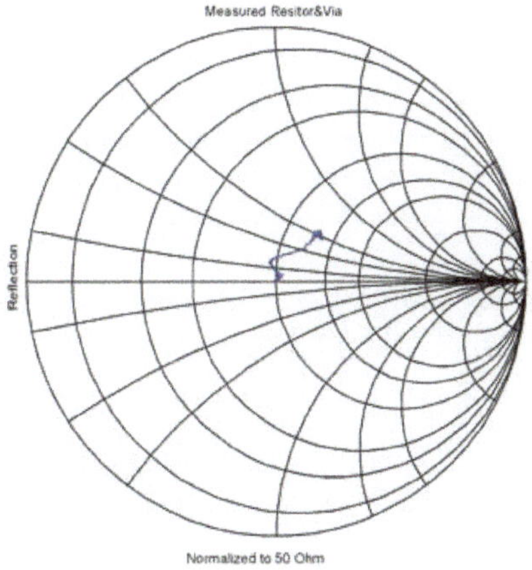

Fig. 3. Smith diagram of the measured via resistor combination.

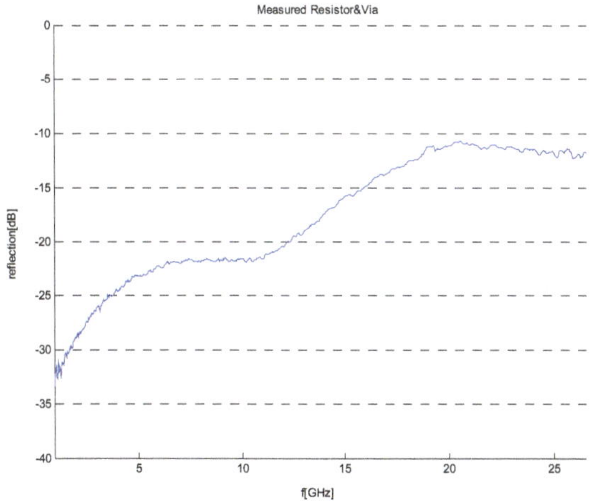

Fig. 4. Magnitude plot of the measured via resistor combination.

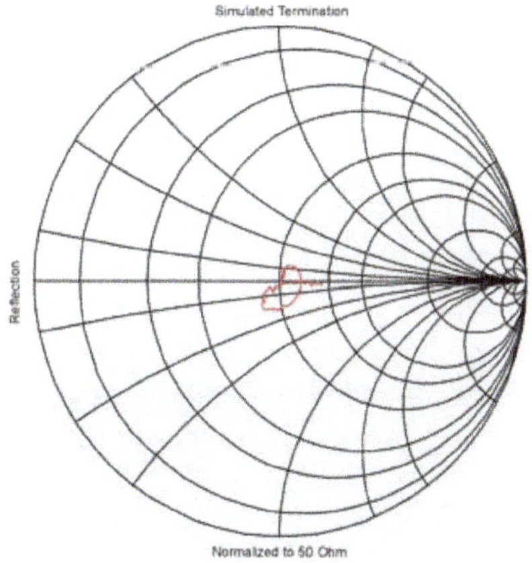

Fig. 5. Smith diagram of the simulated termination.

in the poor matching of a VSWR of more than 1.43 at frequencies above 16 GHz as shown in the magnitude plot in Fig. 4. Whereas at frequencies below 12 GHz, the matching is, as expected, already good. The inductance of the via is the critical parasitic that has to be compensated.

3 Simulation

Generally the parasitics of the via and the resistor limit the performance of the termination. Therefore, the reflection has to be improved by a broadband matching. The broadband matching to 50 Ω was realized by means of a filter structure. Within the desired frequency range discrete components like SMD capacitors or inductors cannot be applied due to their own considerable parasitics. Since its influence is only required at frequencies above 12 GHz, a microstrip filter structure is applicable. Furthermore as the dimensions should be small anyway.

The microstrip filter structure was designed based on the method of moments. Advanced Design System (ADS) by Agilent was used as the simulation tool. The simulation is based on the above mentioned measured reflection coefficient of the via-resistor combination. Starting from an initial dimensioning, the filter was adapted by the cooptimization tool of ADS. Therewith the momentum based filter structure simulation and optimization is combined with the measured viaresistor reflection coefficient.

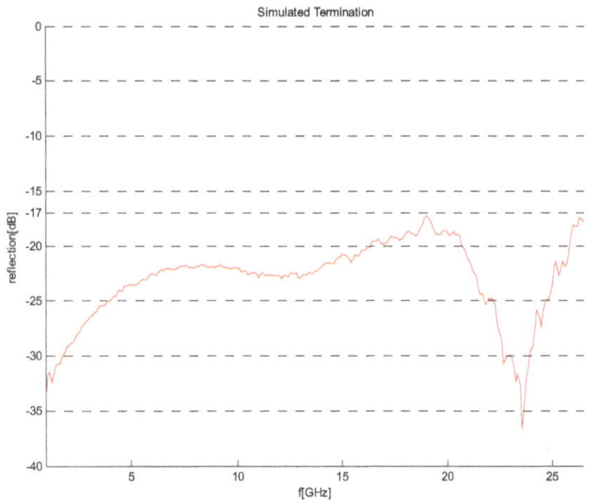

Fig. 6. Magnitude plot of the simulated termination.

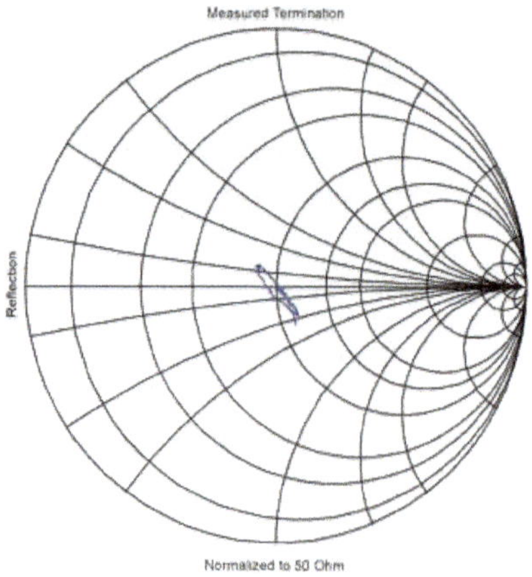

Fig. 8. Scheme of the measurement setup.

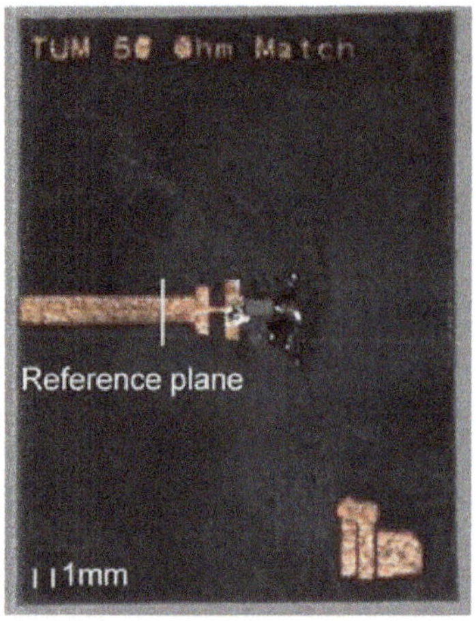

Fig. 7. Picture of the 50 Ω termination.

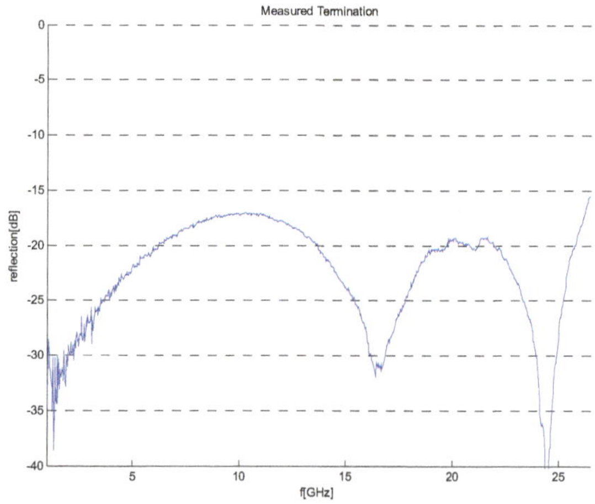

Fig. 9. ADS co optimization schematic.

The order of the filter was given to the simulation tool by the layout of the filter structure. Several simulations with varying filter orders showed that a 3rd order filter provided the desired performance. The simulation results are shown in Fig. 5 in the Smith chart. The magnitude of the reflection coefficient is shown in Fig. 6. Its VSWR is better than 1.33 from DC to K-band.

4 Measurement

The measurement of the components as well as the measurement of the final circuit were performed with a HP 8510 VNA. The TRL calibration method was applied to achieve an onplane specification of the elements. An inhouse developed TRL calkit appropriate for the substrate and frequency range was used. To verify the calibration a passive reference DUT was applied. The onplane specification of the DUT, provided by the TRL calibration, allows the direct use of the measured data for the simulation and the direct connection to a 50 Ω circuit of the final design.

Figure 7 shows a picture of the final circuit consisting of the via, the resistor and the microstrip filter structure. The reference plane of the measurement is set right behind the filter. At this point, the 50 Ω termination can be connected to a 50 Ω microstrip line of an arbitrary circuit.

The measurement showed good correspondence to the simulation results. The best matching is located in the targeted 24 GHz ISM band. The measured data is plotted in Fig. 8 in a Smith chart. Compared to Fig. 3 the inductance of the via is compensated. The magnitude diagram in Fig. 9 shows degradation of at most 5 dB at frequencies below 12 GHz compared to Figs. 4 and 6. The second resonance is about 10 dB more distinctive as compared to the simulation. However, the performance is still within the desired parameters.

5 Conclusion

The presented 50 Ω termination has the advantage of applicability down to DC and a very broadband performance up to K-band compared to other concepts. It is based on standard design and fabrication steps that provide a low-cost and versatile implementation in production determined designs or designs for measurement purposes. The design procedure can easily be adapted to arbitrary substrates and fabrication technologies.

The realized termination shows a VSWR better than 1.34 from DC to K-band and is better than 1.22 within 57% of the range, including the 24 GHz ISM band. Correspondence of the measurement and simulation results verifies the design process. The characteristics were determined in the intended microstrip plane and therefore the layout can be directly attached to a planar 50 Ω circuit.

References

Lacombe, D.: A Multioctave Microstrip 50 Ω Termination, 1972 IEEE Trans. Microwave Theory and Tech., vol. MTT-20. no. 4, 290–291, April 1972.

Linnér, P., Lundén H.: Theorie and Design of Broad-Band Nongrounded Matched Loads for Planar Circuits, 1986 IEEE Trans. Microwave Theory and Tech., vol. MTT-34. no. 8, 892–896, August 1986.

Thompson, G.H.B.: An Easy Method of Matching Microstrip Loads and Attenuators, 1961 IEEE Trans. Microwave Theory and Tech., vol. MTT-9. no. 3, 263–263, May 1961.

17

Recent Applications and Advances of Numerical Modeling and Wavefield Inversion in Nondestructive Testing

R. Marklein[1], K. J. Langenberg[1], K. Mayer[1], J. Miao[1], A. Shlivinski[1], A. Zimmer[1], W. Müller[2], V. Schmitz[2], C. Kohl[3], and U. Mletzko[4]

[1]Fachgebiet Theoretische Elektrotechnik, Fachbereich Elektrotechnik / Informatik, Universität Kassel (UNIK), 34109 Kassel, Germany
[2]Fraunhofer-Institut für Zerstörungsfreie Prüfverfahren (IZFP), Universität des Saarlandes, Gebäude 37, 66123 Saarbrücken, Germany
[3]Bundesanstalt für Materialforschung und Prüfung (BAM), 12200 Berlin, Germany
[4]Materialprüfanstalt (MPA) Stuttgart, Universität Stuttgart, 70511 Stuttgart, Germany

Abstract. This paper presents recent advances and future challenges of the application of different linear and nonlinear inversion algorithms in acoustics, electromagnetics, and elastodynamics. The presented material can be understood as an extension of our previous work on this topic. The inversion methods considered in this presentation vary from linear schemes, like the Synthetic Aperture Radar (SAR) applied electromagnetics and the Synthetic Aperture Focussing Technique (SAFT) as its counterpart in ultrasonics, and the linearized Diffraction Tomography (DT), to nonlinear schemes, like the Contrast Source Inversion (CSI) combined with different regularization approaches. Inversion results of the above mentioned inversion schemes are presented and compared for instance for time-domain ultrasonic data from the Fraunhofer-Institute for Nondestructive Testing (IZFP, Saarbrücken, Germany). Convenient tools for nondestructive evaluation of solids can be electromagnetic and/or elastodynamic waves; since their governing equations, including acoustics, exhibit strong structural similarities, the same inversion concepts apply. In particular, the heuristic SAFT algorithm can be and has been utilized for all kinds of waves, once a scalar approximation can be justified. Relating SAFT to inverse scattering in terms of diffraction tomography, it turns out that linearization is the most stringent inherent approximation. A comparison of the inversion results using the linear time-domain inversion scheme SAFT and well tested nonlinear frequency-domain inversion schemes demonstrates the considerable potential to extend and improve the ultrasonic imaging technique SAFT while consulting the mathematics of wavefield inversion, yet, in particular if the underlying effort is considered, the relatively simple and effective SAFT algorithm works surprisingly well. Since SAFT is a widely accepted imaging tool in ultrasonic NDE it seems worthwhile to check its formal restrictions and assumptions whether they could be overcome and whether they would outperform the standard and original SAFT algorithm.

1 Ultrasonic data: linear and nonlinear inversion results

Convenient tools for NDE of solids can be electromagnetic and/or ultrasonic waves; since their governing equations exhibit strong structural similarities, the same modeling and inversion concepts apply (Marklein, 2002; Marklein et al., 2001, 2002a,b).

In particular, the heuristic synthetic aperture focusing technique (SAFT) algorithm can be – and has been – utilized for all kinds of waves, once a scalar approximation can be justified. Figure 1 shows a typical multi-monostatic (multi-pulse-echo) ultrasonic data set which is processed by the SAFT algorithm. Relating SAFT to inverse scattering in terms of diffraction tomography, it turns out that linearization is the most stringent inherent approximation. Hence, the results of nonlinear inversion schemes such as Contrast Source Inversion (CSI), where CSI has been applied in remote sensing very successfully (van den Berg et al., 2003; Marklein et al., 2001), are compared to the output of SAFT for a carefully designed ultrasonic experiment (Marklein et al., 2002a,b). Here we restrict ourselves to the scalar case, the extension and application of CSI to the vector case in ultrasonic and electromagnetic NDE is a future challenge. First approaches can be found in Abubakar (2000); Pelekanos and Sevroglou (2003). Figure 2 shows the geometries and photographs of the four aluminum samples with two, four, six, and twelve air-filled circular cylinders which are in the experiments embedded in a water tank. Multi-monostatic (multi-pulse-echo) and multi-bistatic (multi-pitch-catch) ultrasonic data sets (Figs. 1 and 4a) have been carefully measured. The SAFT reconstructions are given in Figs. 3 and 4b. Nonlinear CSI inversion results presented in Figs. 4d–f are in general

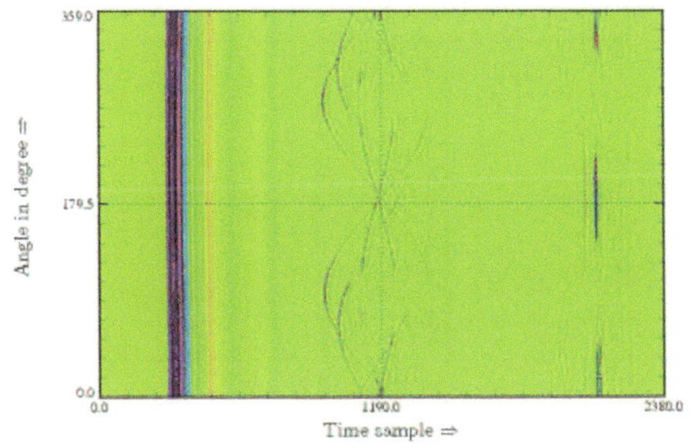

Fig. 1. Measured ultrasonic (elastodynamic) multi-monostatic (multi-pulse-echo) time domain data set for an aluminum cylinder with six air-filled holes.

much better than the SAFT results. Figure 5 shows a comparison of CSI results for experimental and synthetic data. The synthetic data are obtained with a Domain Integral Equation (DIE) solver in the frequency domain. It can be seen that due to phase errors and noise in the measurements of the CSI the reconstructions of the synthetic data are much better.

2 Ultrasonic data: inhomogeneous anisotropic case (InASAFT)

Non-destructive testing of defects in nuclear power plant dissimilar pipe weldings play an important role in safety inspections (see Fig. 6). Traditionally the imaging of such defects is performed using the SAFT algorithm, however since parts of the dissimilar welded structure are made of an anisotropic material, this algorithm may fail to produce correct results.

Here we present results of a modified SAFT algorithm that enables a correct imaging of cracks in complex inhomogeneous anisotropic structures by accounting for the true nature of the wave propagation in such structures. The algorithm is called inhomogeneous anisotropic SAFT (InASAFT) (Shlivinski et al., 2004a,b). The InASAFT algorithm is shown to yield better results over the SAFT algorithm for complex environments, but the InASAFT suffers from the same difficulties of the SAFT algorithm, i.e. "ghost" images and lack of clearly focused images. However these artifacts can be identified through numerical modeling of the wave propagation in the structure, e.g. with the Elastodynamic Finite Integration Technique (EFIT) (Marklein, 2002). Here we consider a realistic dissimilar weld. Figure 6 (left) shows a polished cut of a weld where its modeling, based on the design "blue-prints", can be seen in Fig. 6 (right). Figure 7 (left) shows a synthetic B-scan as a result of the modeling with EFIT (Marklein, 2002), and Fig. 7 (right) displays the InASAFT reconstruction. The crack inside the weld is reconstructed correctly. The experimental B-scan and InASAFT image is given in Fig. 8, where the two focused points in the black ellipse correspond to the crack tips.

3 Electromagnetic data: Ground Penetrating Radar (GPR) application

Figure 9 shows the geometry of a concrete test specimen as a computer model, which was actually realized by the Federal Highway Research Institute in Germany. The specimen contains two circular cylindrical tubes of infinite electrical conductivity with a diameter of 100 mm as an appropriate model for tendon ducts. The reinforcement is perfectly conducting, it exhibits two different mesh sizes of 75 mm and 50 mm. The specimen has the dimensions 2 m×1.5 m×0.7 m, and the concrete material is supposed to be homogeneous and lossless with a relative permittivity $\varepsilon_r=2$. For the 3-D simulation of electromagnetic wave propagation and scattering, a plane wave impinging perpendicularly to the surface is chosen, which has the base band pulse structure shown in Fig. 9 (−3 dB cut-off frequency: $\simeq$0.9 GHz); the electric field strength is linearly polarized parallel to the tendon ducts. The numerical method to solve Maxwell's equations is based on the Finite Integration Technique, and here we use a commercially available implementation (CST, 2004). Figure 10 exhibits 2-D slices out of the 3-D wave field along the line L1 as indicated in Fig. 9, i.e. a line through the coarser reinforcement. Typically, even though the highest frequency wavelength is much larger than the grid spacing, the plane penetrates the reinforcement and leaves it again as a nearly plane wave, nevertheless much weaker in amplitude than before. Hence, the scattering by the tendon ducts is also obvious resulting in a 3-D data field when "scanned" along the surface; these data are 2-D with regard to the scan coordinate, and the third dimension is time. Two slices along the lines L1 and L2 of Fig. 9 of the data are displayed in Fig. 11: The slice along L1 through the coarser grid clearly exhibits the hyperbolic diffraction curves of the tendon ducts, whereas they are not visible along the finer grid. This is confirmed applying a linear diffraction tomographic inverse scattering scheme FT-SAFT (Mayer et al., 1990) to the 3-D data: Fig. 12 clearly shows images of the tendon ducts only below the coarser grid. Thus, simulations of that kind define a limiting grid size for GPR tendon duct imaging, but it turns out that this upper bound cannot yet be met by experiments.

4 Concluding remarks and future work

We have presented latest results of the numerical modeling and linear and nonlinear inversion of electromagnetic and ultrasonic synthetic as well as experimental data. A comparison of the ultrasonic inversion results using the linear time-domain inversion scheme SAFT and well tested nonlinear frequency-domain inversion schemes demonstrates the considerable potential to extend and improve the ultrasonic imaging technique SAFT while consulting the mathematics of wavefield inversion, yet, in particular if the underlying effort is considered, the relatively simple and effective SAFT algorithm works surprisingly well. Even the extension of SAFT to the inhomogeneous anisotropic case, which yields

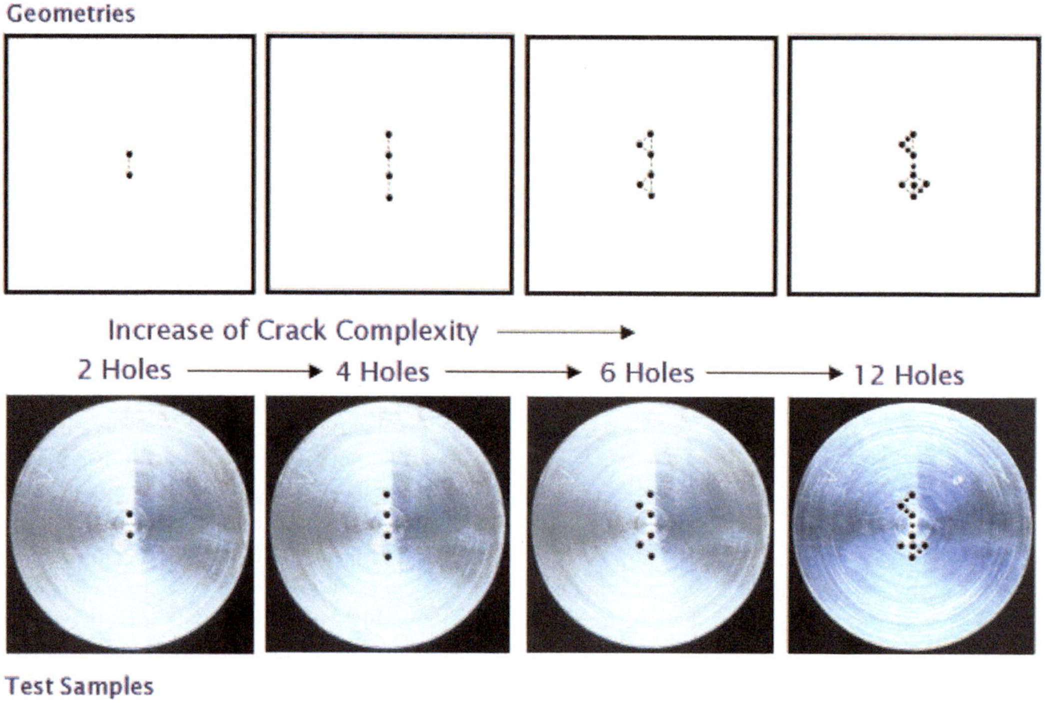

Fig. 2. Stepwise increase of crack complexity: (top) samples geometries of two, four, six and twelve holes; (bottom) aluminum samples.

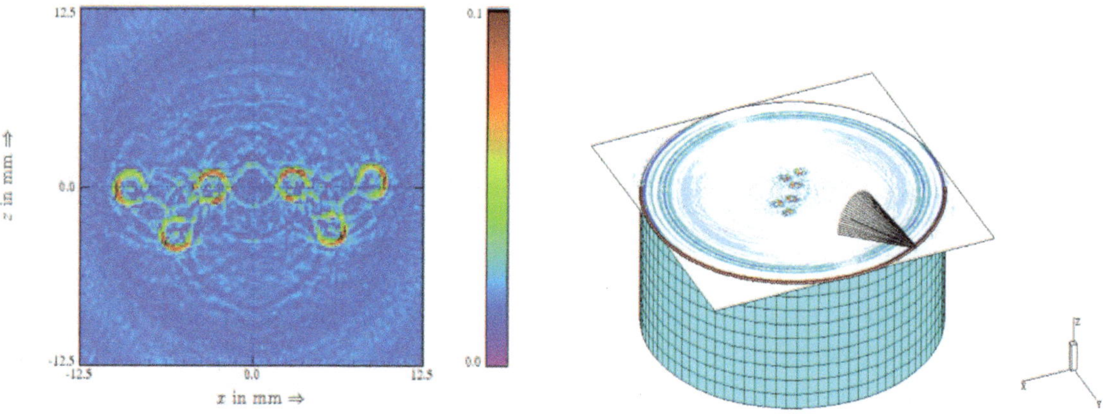

Fig. 3. Left: SAFT reconstruction of the multi-monostatic (multi-pulse-echo) data set; Right: SAFT reconstruction embedded in the 3-D geometry of the test sample.

the InASAFT, is working well and in combination with numerical modeling tools, like EFIT, the effects of linearization, for instance artifacts in the reconstruction, can be interpreted. Obviously, like in the ultrasonic case, simulated electromagnetic data imaging using FT-SAFT leads to an intuitive physical understanding of GPR probing of embedded structures in concrete. The still existing discrepancy between simulation and experiments will be investigated in greater detail in the future, and transmission data will be included as well. Since SAFT/FT-SAFT are a widely accepted imaging tools in ultrasonic NDE it seems worthwhile to check its formal restrictions and assumptions whether they could be overcome and whether they would outperform the standard and original SAFT/FT-SAFT algorithm.

Final conclusions: Not very much remains to be done regarding the theoretical formulation of linear wavefield inversion, yet improvements with nonlinear inversion can certainly be made. Ockham's razor is simple: Will it work in practice? Presently, we are working on the development and implementation of the vector case in electromagnetics and ultrasonics as proposed by Abubakar (2000) and Pelekanos and Sevroglou (2003).

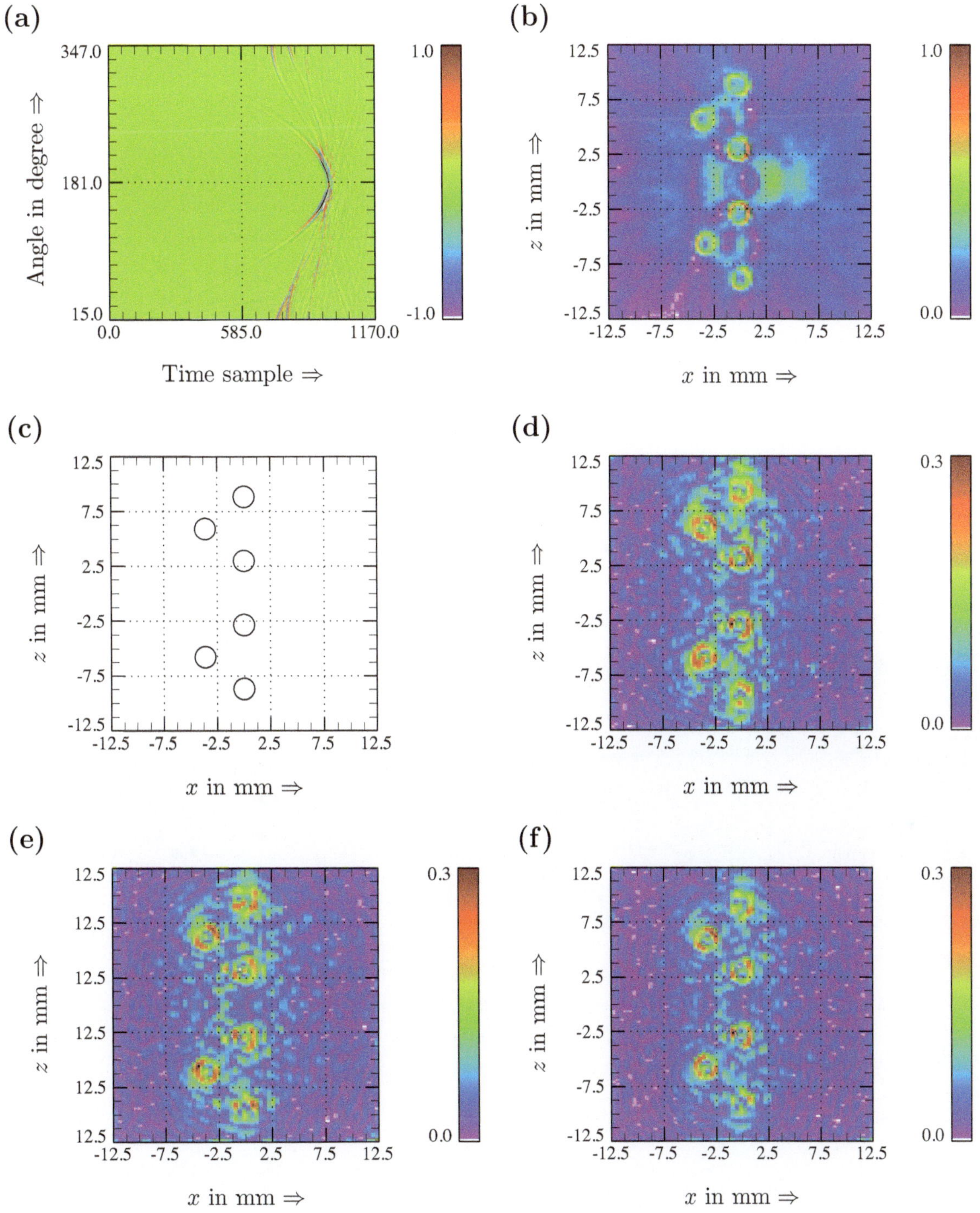

Fig. 4. **(a)** Measured ultrasonic multi-bistatic (multi-pitch-catch) time-domain data set for the transmitter at 0°; **(b)** Time-domain SAFT reconstruction; **(c)** Actual profile; **(d)** Frequency-domain CSI reconstruction using data at six different frequencies, no a priori information used; **(e)** Frequency-domain CSI reconstruction using single frequency data and a priori information, the contrast is imaginary valued; **(f)** Frequency-domain CSI reconstruction using data at four different frequencies and a priori information, the contrast is imaginary valued.

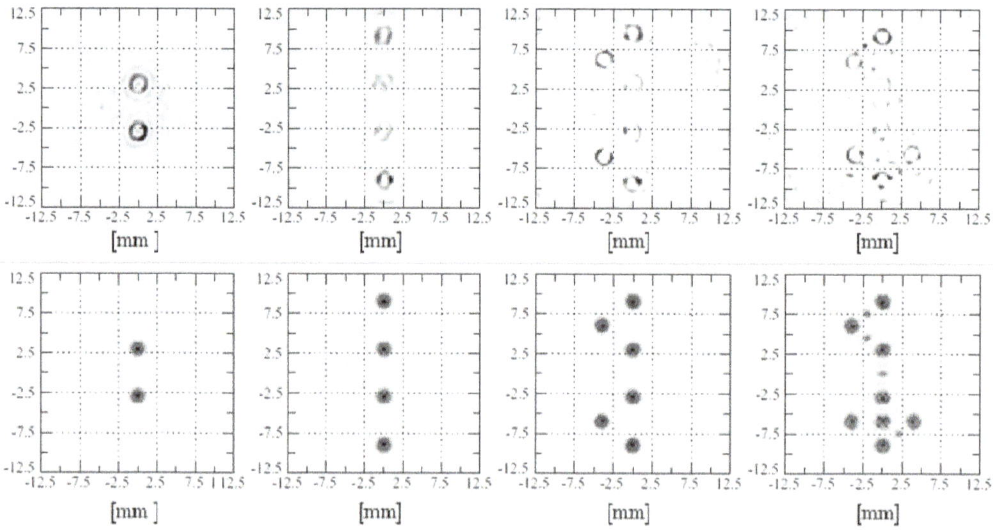

Fig. 5. Top: CSI reconstructions of experimental data and synthetic data; Bottom: CSI reconstructions of synthetic data which are obtained by an Domain Integral Equation (DIE) solver in the frequency domain.

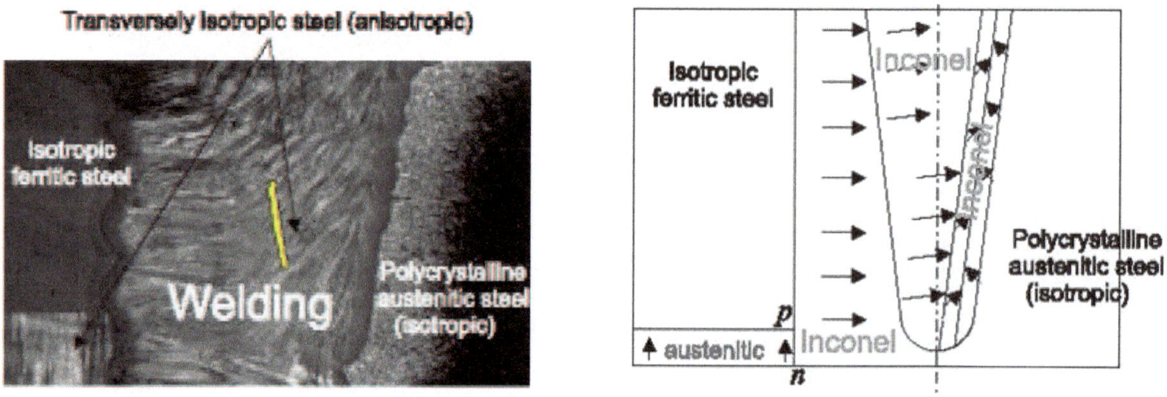

Fig. 6. Left: Polished cut of the dissimilar weld. Note the grain orientation in each of the weld parts; Right: "Blue-print" modeling.

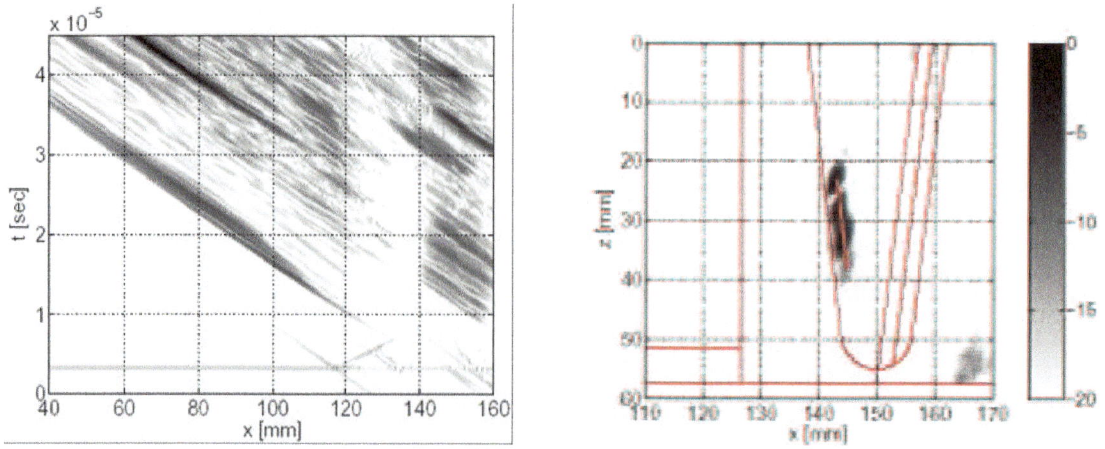

Fig. 7. Left: Synthetic B-scan as a result of the EFIT modeling; Right: InASAFT reconstruction.

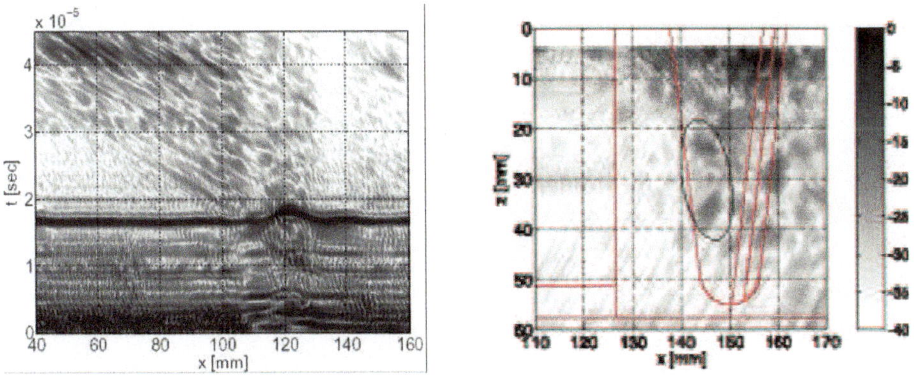

Fig. 8. Left: Experimental B-scan; Right: InASAFT reconstruction.

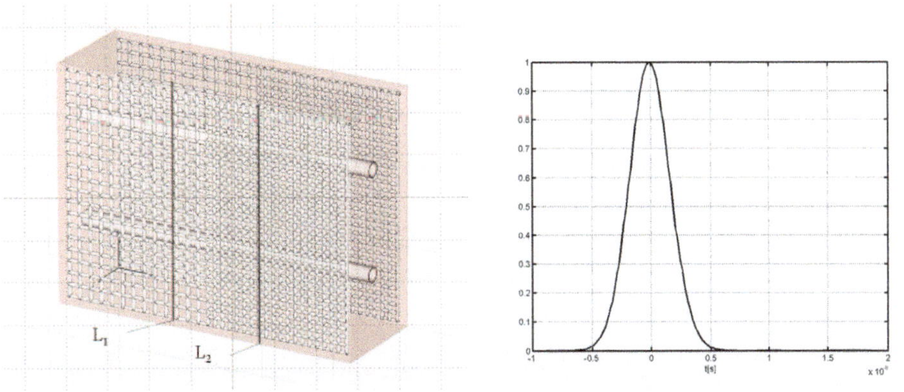

Fig. 9. Left: Geometry of a concrete test specimen as a computer model; Right: Excitation signal.

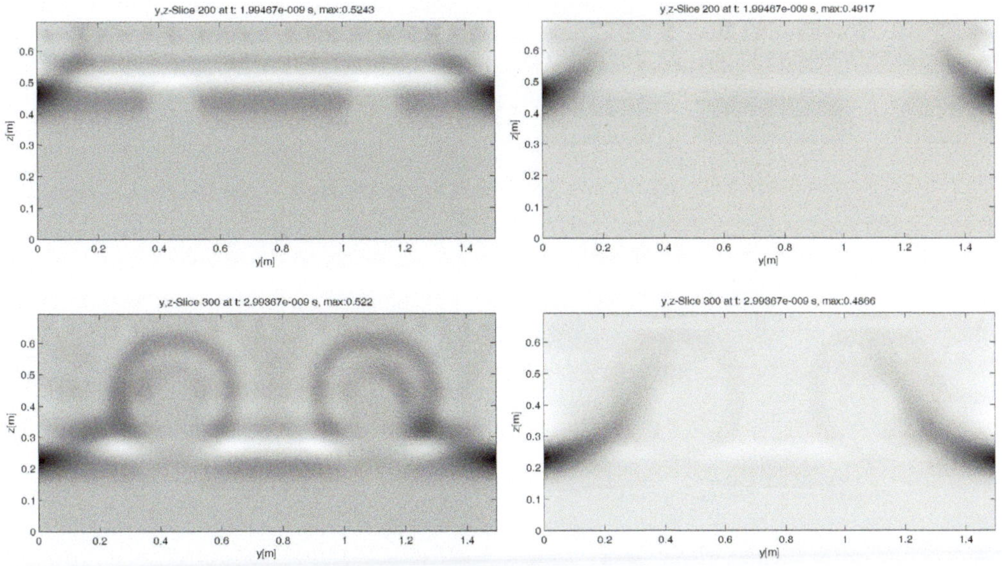

Fig. 10. 2-D slices out of the 3-D wave field along the lines L1 and L2 in Fig. 9a.

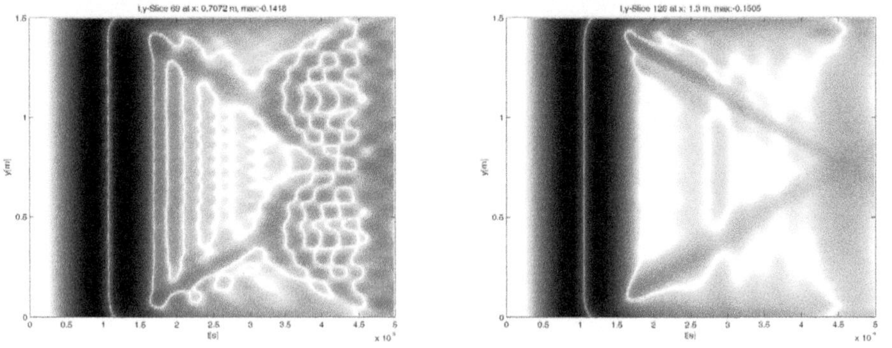

Fig. 11. Received signals along lines L1 (left) and L2 (right) of Fig. 9 (left) (logarithmic scale: −60 dB).

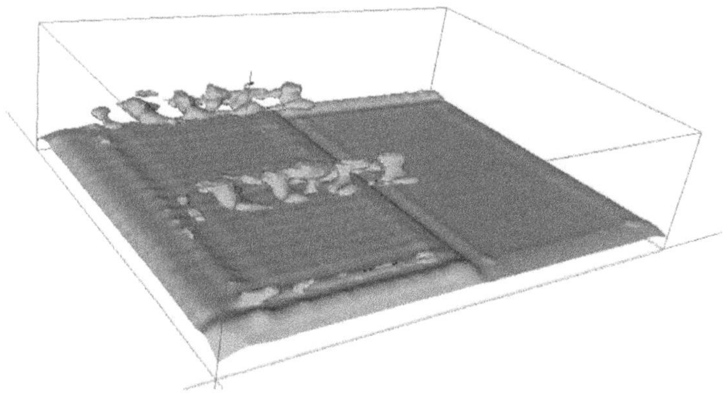

Fig. 12. Linear diffraction tomographic inverse scattering scheme to the 3-D data.

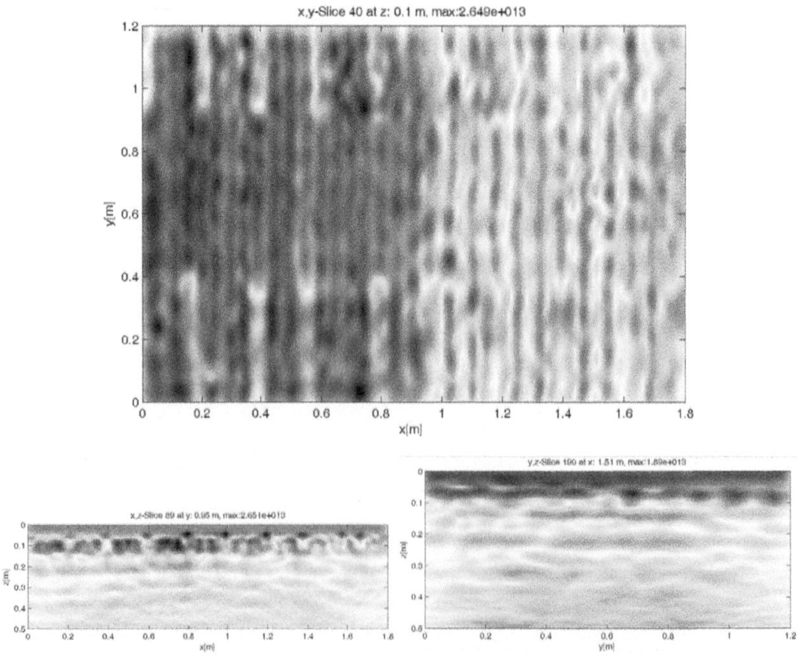

Fig. 13. Three orthogonal slices through the 3-D FT-SAFT image volume of experimental data.

References

Abubakar, A.: Three-Dimensional Nonlinear Inversion of Electrical Conductivity, Delft University Press, Delft, 2000.

CST MICROWAVE STUDIO, http://www.cst.de.

Marklein, R.: The Finite Integration Technique as a General Tool to Compute Acoustic, Electromagnetic, Elastodynamic, and Coupled Wave Fields, in Review of Radio Science: 1999-2002 URSI, edited by: Stone, W. R., IEEE Press, Piscataway, 201-244, 2002.

Marklein, R., Balasubramanian, K., Qing, A., and Langenberg, K. J.: Linear and nonlinear iterative scalar inversion of multi-frequency multi-bistatic experimental electromagnetic scattering data, Inverse Problems, 17, 1597–1610, 2001.

Marklein, R., Balasubramanian, K., Langenberg, K. J., Miao, J., and Sinaga, S. M.: Applied Linear and Nonlinear Scalar Inverse Scattering: A Comparative Study, Proceedings of the XXVIIth General Assembly of the International Union of Radio Science (URSI), CD-ROM, 2002.

Marklein, R., Mayer, K., Hannemann, R., Krylow, T., Balasubramanian, K., Langenberg, K. J., and Schmitz, V.: Linear and nonlinear inversion algorithms applied in nondestructive evaluation, Inverse Problems, 18, 1733–1759, 2002.

Mayer, K., Marklein, R., Langenberg, K. J., and Kreutter, T.: Three-dimensional imaging system based on Fourier transform synthetic aperture focusing technique, Ultrasonics, 28, 241–255, 1990.

Pelekanos, G. and Sevroglou, V.: Journal of Computational and Applied Mathematics, 151, 129–140, 2003.

Shlivinski, A., Langenberg, K. J., and Marklein, R.: Ultrasonic Modelling and Imaging in Dissimilar Welds, Proceedings of the 30th MPA-Seminar in Conjunction with the 9th German-Japanese Seminar, Stuttgart, 2004.

Shlivinski, A., Langenberg, K. J., Marklein, R., Schmitz, V., Müller, W., and Mletzko, U.; Modelling of Elastic Wave Propagation and Imaging of Defects in Dissimilar Welds, Proceedings of the 4th International Conference on NDE in Relation to Structural Integrity for Nuclear and Pressurised Components, London, 2004.

van den Berg, P. M., Abubakar, A., and Fokkema, J. T.: Extended Contrast Source Inversion, Radio Science 38, VIC-1–VIC-10, 2003.

Fast Multipole Acceleration of a MoM Code for the Solution of Composed Metallic/Dielectric Scattering Problems

U. Jakobus and J. van Tonder

EM Software & Systems - S.A. (Pty) Ltd, PO Box 1354, Stellenbosch 7599, South Africa

Abstract. An existing method of moments (MoM) code for the solution of complex scattering bodies has been accelerated by means of a multilevel fast multipole method (MLFMM). We demonstrate the usage of this technique both for metallic structures (wires and surfaces) and for dielectric bodies (volume and surface equivalence principle). Aspects like the effect of the type of integral equation, preconditioning schemes, or iterative solution techniques are discussed. But also limitations are addressed, which are encountered when for instance attempting to model highly lossy dielectric bodies with a high permittivity. Several validation and application examples demonstrate the usefulness of this method, both with regard to the obtained accuracy, but also with respect to the potential saving in memory and run-time as compared to a standard MoM formulation.

1 Introduction to the Multilevel Fast Multiple Method (MLFMM)

1.1 Formulation of the MLFMM

A brief outline of the MLFMM will be presented in this section. The interested reader is referred to Coifman et al. (1993); Song and Chew (1994, 1995); Chew et al. (1997); Song et al. (1997); Gyure and Stalzer (1998); Chew et al. (2001) for more details.

The MLFMM is is based on a hierarchical grid. At the top level (level 0) the whole computational space is enclosed by one large cube. At the next level (level 1) this cube is then subdivided in 3-dimensions into a maximum of 8 child cubes. This process is repeated until at the finest level the cube side length is approximately 0.25λ. Only non-empty cubes are stored at each level forming a tree-like data structure. Figure 1 shows the cubes at the finest level for one automotive example.

In the MoM framework a system of linear equations

$$Z\, I = V \tag{1}$$

needs to be solved. The MLFMM is implemented by writing the impedance matrix Z into the near-field term Z_{near} and the far-field term Z_{far}, i.e. Equation (1) then becomes

$$Z_{near}\, I + Z_{far}\, I = V\,. \tag{2}$$

Z_{near} consists of all matrix elements where basis and weighting functions are within the same box or in adjacent boxes at the finest level. Only this near-field mattrix Z_{near} is computed traditionally and stored in a sparse format. The system of linear equations (1) is solved with an iterative technique where matrix-vector products $Z\, I_k$ are required, with k indicating the iteration counter. The far-field term Z_{far} is never computed explicitly, but the matrix-vector product $Z_{far}\, I_k$ is computed via the MLFMM as

$$Z_{far}\, I_k = D\, T\, A\, I_k \tag{3}$$

with the following three phases:

- The **A**ggregation phase where all the basis functions inside the same source cube (at the finest level) are grouped together,
- The **T**ranslation phase from the source cube to the observation cube,
- The **D**isaggregation phase from the centre of the observation cube to every basis function inside that cube.

All these phases make use of the addition theorem to approximate the free-space Green's function

$$G(\bar{\mathbf{x}}, \bar{\mathbf{x}}') = \frac{e^{-jkR}}{R} \approx$$

$$\frac{-jk}{4\pi} \int d^2\hat{\mathbf{k}}\, e^{-j\bar{\mathbf{k}}\cdot(\bar{\mathbf{x}}-\bar{\mathbf{x}}_m)}\; T_L(\bar{\mathbf{k}}, \bar{\mathbf{X}}_{m'm})\; e^{+j\bar{\mathbf{k}}\cdot(\bar{\mathbf{x}}'-\bar{\mathbf{x}}_{m'})} \tag{4}$$

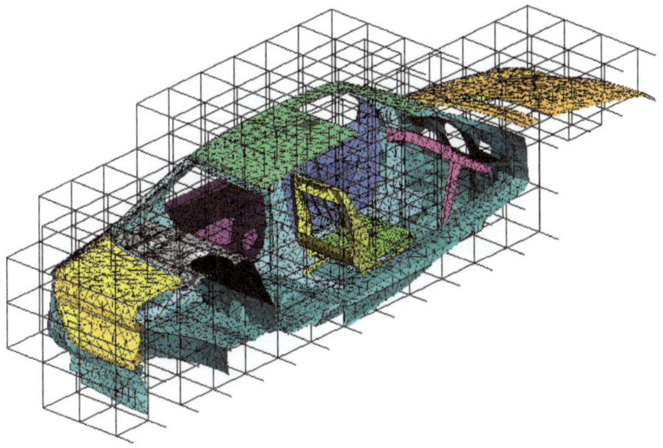

Fig. 1. The MLFMM boxes at the finest level for one automotive example. Only half of the geometry is shown for clarity.

where

$$T_L(\bar{\mathbf{k}}, \bar{\mathbf{X}}_{m'm}) = \sum_{l=0}^{L} (-j)^l (2l+1) h_l^{(2)}(kX_{m'm}) P_l(\hat{\mathbf{k}}\cdot\hat{\mathbf{X}}_{m'm}) \quad (5)$$

and the number of terms L are determined empirically for a given accuracy ε by the formula

$$L = kD + 1.8\ (kD)^{1/3}\ (\log_{10}(1/\varepsilon))^{2/3} \quad (6)$$

with the wavenumber k and the box size D.

For the integration over the sphere a quadrature rule with $2L^2$ points is applied according to Coifman et al. (1993); Song and Chew (1994). Looking at Eq. (4) the aggregation step is given by $e^{+j\bar{\mathbf{k}}\cdot(\bar{\mathbf{x}}'-\bar{\mathbf{x}}_{m'})}$, the translation step by $T_L(\bar{\mathbf{k}}, \bar{\mathbf{X}}_{m'm})$, and the disaggregation step by $e^{-j\bar{\mathbf{k}}\cdot(\bar{\mathbf{x}}-\bar{\mathbf{x}}_m)}$.

Eqation (2) must be solved with iterative techniques (for example CGS, Bi-CGSTAB, etc.) since we only have the sparse Z_{near} and never store Z_{far}. For general open structures the Electric Field Integral Equation (EFIE) is poorly conditioned, causing the iterative technique to converge very slowly (or even diverge). To accelerate the rate of convergence we use a preconditioner that is computed from the near-field matrix Z_{near}. Implemented preconditioners include Incomplete LU (ILU), Block-Jacobi and Block-Jacobi one-level-up.

1.2 Scaling of memory and CPU-time

Let N be the number of unknowns (i.e. number of basis functions). The traditional MoM scales as N^2 in terms of memory (to store the impedance matrix) and as N^3 in terms of CPU-time (to solve the linear set of equations). When N becomes large the MoM will therefore require too much memory and CPU-time. Much more favourable is the MLFMM, which scales as $N\log N$ in memory and as $N\log^2 N$ in terms of CPU-time. Figure 2 shows typical memory and CPU-times for an automotive example (full vehicle of approximate length 4.5 m including seats, windows, etc. at different frequencies). One can see that the actual values for the

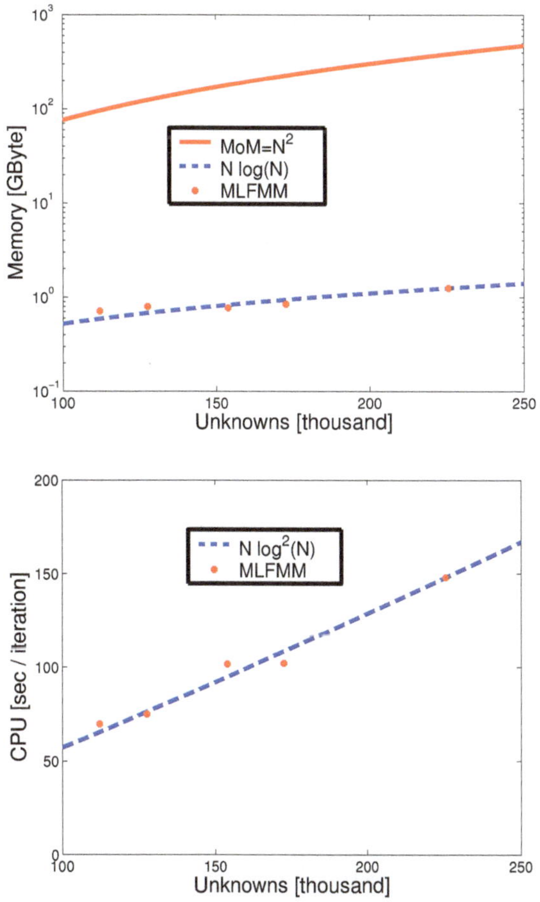

Fig. 2. Memory and CPU-time scaling for the MLFMM for an automotive example.

Table 1. Typical memory requirement and CPU-times for an antenna analysis on an aircraft. All MLFMM runs performed on a 64-bit AMD Opteron 248 (2.2 GHz).

No. of unknowns	MoM [GByte] [GByte]	MLFMM [GByte]	MLFMM [hours]
1 030 891	16 216	10.3	15.8
1 573 620	37 785	13.9	21.8

MLFMM (dots) follow nicely the theoretically expected scaling (dashed lines).

The advantages of the MLFMM become more evident as the geometry becomes larger in terms of the wavelength. The results for an aircraft are given in Table 1. The MLFMM for the 1.57 million unknowns case uses 2718 times less memory than the MoM.

1.3 Implementation details

Our MLFMM implementation in the computer code FEKO (see FEKO, 2004) includes amongst others the following

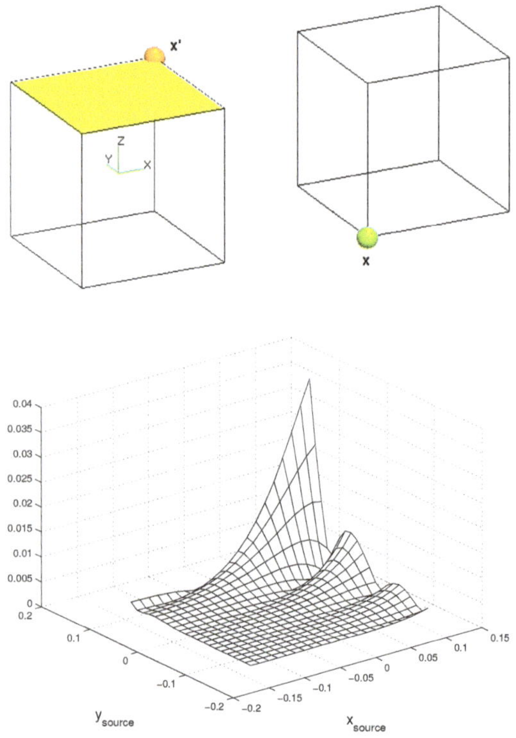

Fig. 3. Relative error in the MLFMM representation of the free-space Green's function as a function of source/observer distance for a fixed number of terms L.

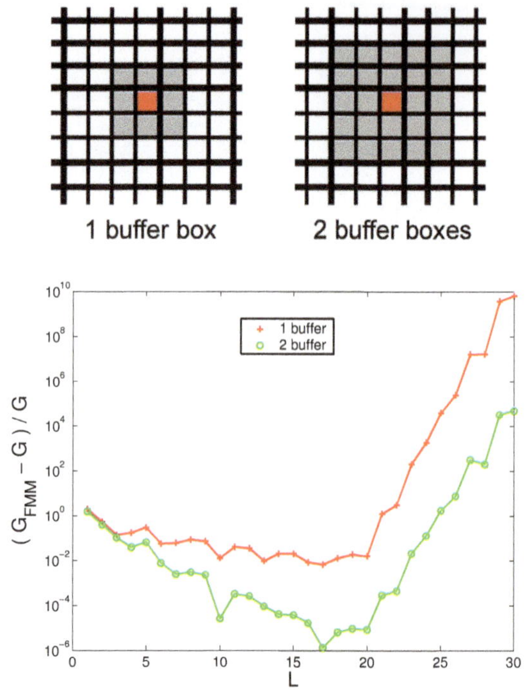

Fig. 4. Using 1 or 2 buffer boxes in the MLFMM gridding and associated error in the Green's function versus the number of terms L.

features, all of which were extensively verified:

- EFIE (valid for general open geometries) for metallic triangular surface patches with basis functions according to Rao et al. (1982),
- Metallic wires,
- Connection basis functions between wires and triangles,
- The combined field integral equation (CFIE) (only valid for closed geometries),
- Dielectric cuboid elements with the volume equivalence principle,
- Dielectric bodies with the surface equivalence principle using the PMCHW formulation,
- The geometry can be located above real ground,
- Thin dielectric sheet approximation to model e.g. thin windows.

It should be mentioned that both the CFIE and the volume equivalence principle result in Fredholm integral equations of the second kind with excellent convergence during the iterative solution. Therefore when we use the CFIE or volume cuboids, then we can use a smaller preconditioner (or even no preconditioner). To obtain a small preconditioner one can use the Block-Jacobi preconditioner (block-diagonal obtained from boxes at finest level), or the Block-Jacobi one-level-up (obtained from the parent boxes of the Block-Jacobi). Reducing the level-of-fill of the ILU preconditioner also reduces the size of the preconditioner. For general open geometries the EFIE is poorly conditioned and requires a good preconditioner (typically an ILU with the level-of-fill=12 is used). The Bi-CGSTAB iterative solver outperformed the other solvers (CGS, RGMRES, etc.) in most of our applications.

2 Considerations regarding the treatment of dielectric bodies

For a fixed value of L in Eq. (5), the error between the exact Green's function and the MLFMM approximation in Eq. (4) is computed in a plane in Fig. 3. It can be seen that the maximum error corresponds to the situation when the source and observation points are located at the corners as indicated by the spheres ((see also Ohnuki and Chew, 2003).

The empirical formula to determine L in Eq. (6) is no longer valid for large dielectric losses. L must then be determined numerically (Geng et al., 2001, Fig. 2) at each level in the MLFMM tree so that the maximum error is below the required threshold. However, if L becomes too large the Hankel function in Eq. (5) will diverge for large order and small argument.

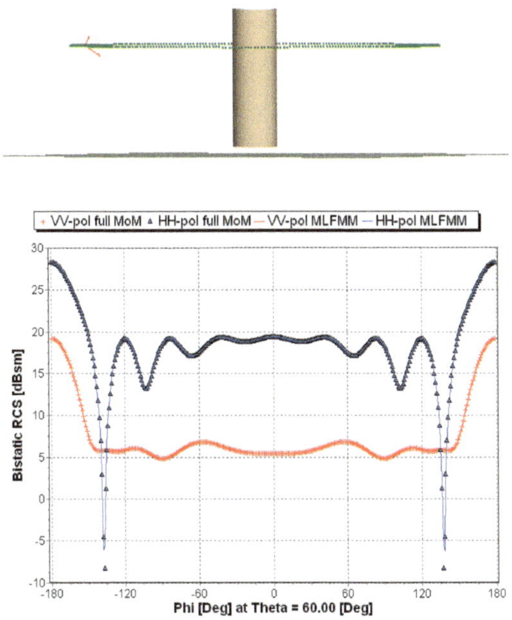

Fig. 5. Bistatic RCS computation of a metallic cylinder above real ground.

One trick which can be used here is to increase the near-field matrix by so-called buffer boxes, so that for the far-field terms the minimum distances where the representation (4) is used are larger.

The lower bound on the argument of the Hankel function is dependent on the number of buffer boxes. By increasing the number of buffer boxes from one to two, it can be seen in Fig. 4 that the error decreases for a fixed L. Therefore, if for a fixed buffer box size the numerically computed error remains above the required threshold, then the number of boxes must be increased. The drawback is that the size of the near field matrix increases dramatically with the number of buffer boxes.

As example, consider human eye tissue with typical permittivity $\varepsilon_r=55-j23$. The required maximum error shall be smaller than 10^{-3}, and for a box size of $0.5\lambda_0$ in the MLFMM tree we determine numerically that we need two buffer boxes and L=70. For comparison, with the same box size and buffer boxes, the free-space case $\varepsilon_r=1$ will need only L=12. Since the integration over the unit Ewald sphere in Eq. (4) uses a quadrature rule with $2L^2$ points, the eye will use 34 more sample points. Therefore, as the dielectric loss increases the MLFMM will become slower and use more memory.

3 Application and validation examples

The MLFMM has been validated extensively with analytical, published, as well as full MoM results. In this section some verification examples shall be presented and discussed.

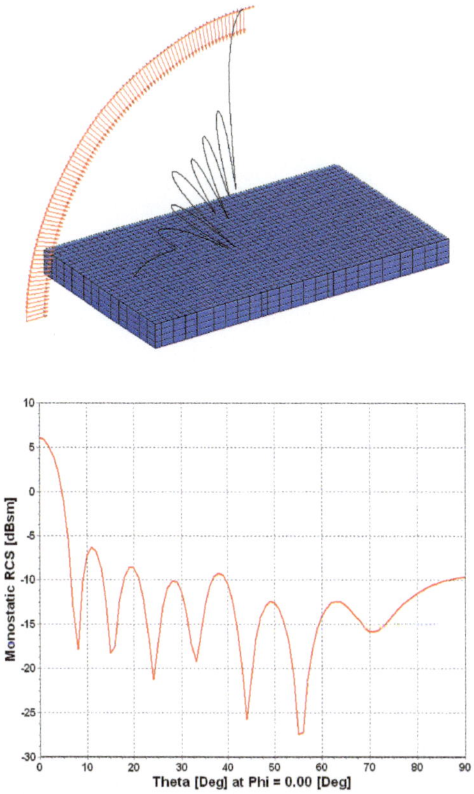

Fig. 6. Monostatic RCS from a dielectric slab computed with the MLFMM using the volume equivalence principle.

To verify the implementation of objects above real ground consider the cylinder located above earth in Fig. 5. The cylinder is of height 3 m and diameter 1 m situated 0.2 m above ground with complex permittivity $\varepsilon_r=6.5-j0.6$. A plane wave is incident from $\vartheta_{inc}=30°$ and $\varphi_{inc}=0°$. The bistatic RCS shall be computed versus the angle φ_{scat} for $\vartheta_{scat}=60°$ and at a frequency of f=600 MHz.

This example is relatively small and the cylinder consists of 6168 metallic triangles resulting in 9252 unknowns. In Fig. 5 the bistatic RCS is depicted for the full MoM (1306 MByte memory) and also for the MLFMM (231 MByte memory). Excellent agreement can be observed between the MoM, MLFMM and the published results in Geng et al. (2000) and Hu and Chew (2001) (not shown in the graph).

Dielectric cuboids treated with the volume equivalence principle have been implemented in the MLFMM and verified to published results. In Fig. 6 the monostatic RCS of a dielectric slab computed with the MLFMM agrees very well with that published in Chew et al. (2001, Fig. 11.16, pp. 522). The dimensions of the slab are $3.5\lambda_0\times2\lambda_0\times0.25\lambda_0$. The frequency is 1 GHz and the permittivity is $\varepsilon_r=3-j0.09$.

To validate the MLFMM for more complex real-life problems we will compare results to those obtained using the full MoM for a mobile phone radiating inside a Lancia car model. The model in Fig. 7 is divided into 20 754 triangles

Fig. 7. Analysis of a mobile phone radiating inside a Lancia car model at 600 MHz.

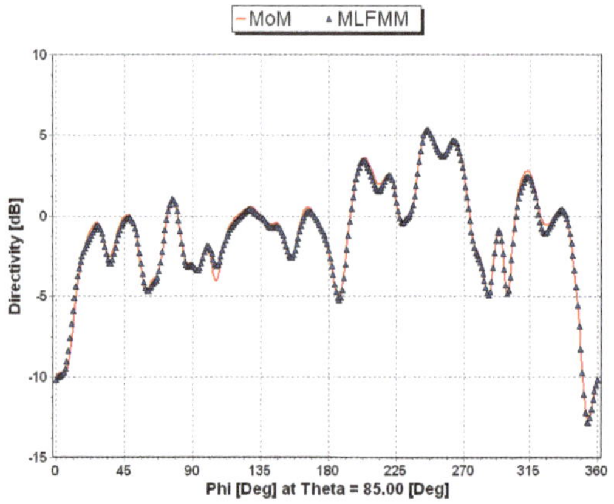

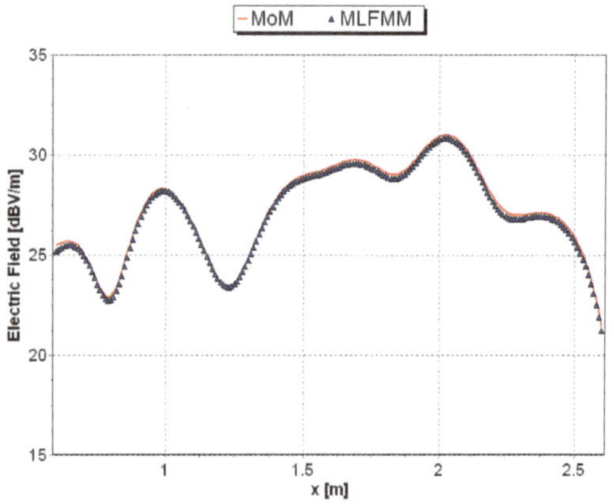

Fig. 8. The far field (top) and near field (bottom) of a mobile phone inside the Lancia car model as shown in Fig. 7.

Table 2. Memory requirements and CPU-times for a dielectrically coated sphere.

Method	Memory [GByte]	Runtime [hours]
MoM	9.43	3.269
MLFMM	1.00	0.734
FEM	3.04	0.388

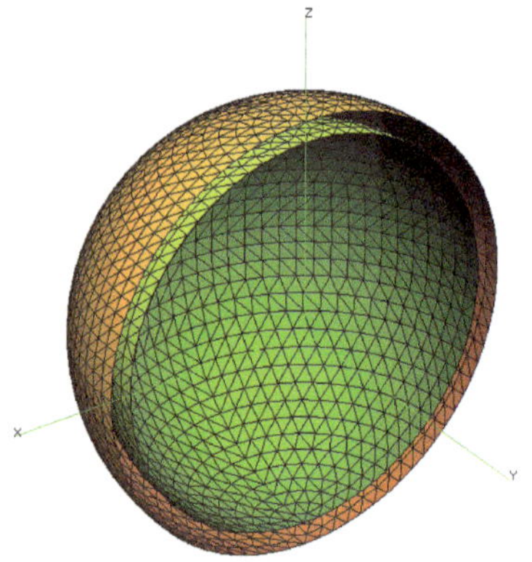

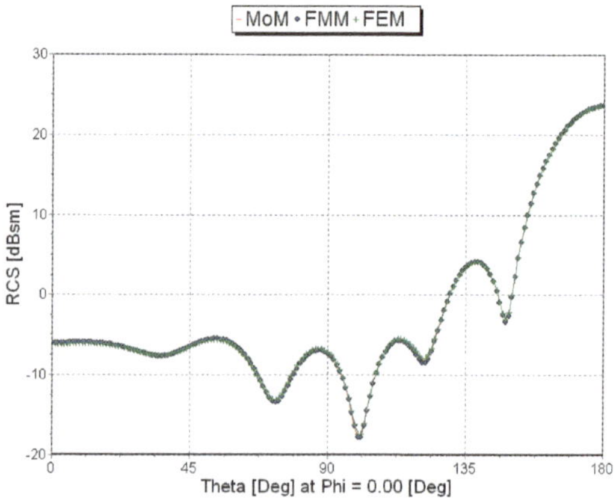

Fig. 9. Bistatic RCS of a dielectrically coated sphere.

and 9 wire segments resulting in 30 915 unknown basis functions. This is a relatively small example, but the full MoM already requires 14 583 MByte of memory. The MoM run was done on a Linux cluster of 16 processors (AMD Athlon 1 GHz) and the sum of the CPU-times of the 16 processes is 33.17 h (2.073 h on average per process). The MLFMM requires only 443 MByte of memory and 8.4 min of CPU-time on a single Intel P4 2.4 GHz processor.

Very good agreement between the full MoM and the MLFMM can be seen in Fig. 8, for both the far field and the near field. The far field is computed versus φ at ϑ=85°(5° above the horizon). The near field is computed along a line inside the Lancia as shown in Fig. 7.

As an example for lossy dielectric structures, we consider the dielectrically coated sphere shown in Fig. 9. This example was also presented by Sertel and Volakis (2004). The inner sphere has a diameter $d_{in}=1.8\lambda_0$ and permittivity $\varepsilon_r^{in}=1.75-j0.3$. The outer sphere has $d_{out}=2.0\lambda_0$ and $\varepsilon_r^{out}=1.25-j1.25$ (thus loss tangent of one). The surface equivalence principle is used in the MoM and MLFMM with the number of unknowns N=25 152. Memory and run-times (no symmetry used in all cases) are given in Table 2. Figure 9 shows the results for the MoM, MLFMM and also a reference solution using FEM (Finite Element Method). It can be seen that the agreement between the three techniques is excellent.

4 Conclusions

We have shown that with the MLFMM large complex electromagnetic problems can be solved using only a fraction of the memory and CPU-time as required by the full MoM. The errors introduced by the MLFMM are fully controllable unlike with other asymptotic techniques as Physical Optics (PO) or Uniform Theory of Diffraction (UTD). This enables the MLFMM to produce very accurate results. Furthermore, the formulation and our implementation have the advantage that the whole gridding and split into near and far field matrices are done automatically. This eliminates any a priori decision by the user to decide what should be in the near field or the far field.

We have demonstrated the usage of this technique both for metallic structures (wires and surfaces) and for dielectric bodies (volume and surface equivalence principle). We have also highlighted the difficulties encountered when attempting to model highly lossy dielectric bodies, and we have presented some solution strategies involving buffer boxes.

References

Chew, W. C., Jin, J. M., Lu, C. C., Michielssen, E., and Song, J. M.: Fast solution methods in electromagnetics, IEEE Transactions on Antennas and Propagation, 45, 533–543, 1997.

Chew, W. C., Jin, J. M., Michielssen, E., and Song, J. M.: Fast and efficient algorithms in computational electromagnetics, Artech House, Boston, 2001.

Coifman, R., Rohklin, V., and Wandzura, S.: The fast multipole method for the wave equation: A pedestrian prescription, IEEE Transactions on Antennas and Propagation, 35, 7–12, 1993.

FEKO: Field Computations Involving Bodies of Arbitrary Shape, EM Software & Systems - S.A. (Pty) Ltd, Stellenbosch, South Africa, http://www.feko.info, 2004.

Geng, N., Sullivan, A., and Carin, L.: Multilevel fast-multipole algorithm for scattering from conducting targets above or embedded in a lossy half space, IEEE Transactions on Geoscience and Remote Sensing, 38, 1561–1573, 2000.

Geng, N., Sullivan, A., and Carin, L.: Fast multipole method for scattering from an arbitrary PEC target above or buried in a lossy half space, IEEE Transactions on Antennas and Propagation, 49, 740–748, 2001.

Gyure, M. F. and Stalzer, M. A.: A prescription for the multilevel Helmholtz FMM, IEEE Computational Science & Engineering, 5, 39–47, 1998.

Hu, B. and Chew, W. C.: Fast inhomogeneous plane wave algorithm for scattering from objects above the multilayered medium, IEEE Transactions on Geoscience and Remote Sensing, 39, 1028–1038, 2001.

Ohnuki, S. and Chew, W. C.: Numerical accuracy of multipole expansion for 2-D MLFMA, IEEE Transactions on Antennas and Propagation, 51, 1883–1890, 2003.

Rao, S., Wilton, D., and Glisson, A.: Electromagnetic scattering by surface of arbitrary shape, IEEE Transactions on Antennas and Propagation, 30, 409–418, 1982.

Sertel, K. and Volakis, J.: Multilevel fast multipole method solution of volume integral equations using parametric geometry modeling, IEEE Transactions on Antennas and Propagation, 52, 1686–1692, 2004.

Song, J. M. and Chew, W. C.: Fast multipole method solution using parametric geometry, Microwave and Optical Technology Letters, 7, 760–765, 1994.

Song, J. M. and Chew, W. C.: Multilevel fast-multipole algorithm for solving combined field integral equations of electromagnetic scattering, Microwave and Optical Technology Letters, 10, 14–19, 1995.

Song, J. M., Lu, C. C., and Chew, W. C.: Multilevel fast multipole algorithm for electromagnetic scattering by large complex objects, IEEE Transactions on Antennas and Propagation, 45, 1488–1493, 1997.

19

A varactor tuned low-cost 24 GHz harmonic VCO

M. O. Olbrich, L. Huang, and E. M. Biebl

Fachgebiet Höchstfrequenztechnik, Technische Universität München, 80290 München, Germany

Abstract. We present a low-cost 24 GHz VCO that is based on a microstrip design combined with discrete packaged devices. The output frequency is generated by a harmonic oscillator. The tunabilty was reached using a varactor diode. Two versions of the VCO were built, one has a wide tuning range of 1.1 GHz and the other one has a high output power of 3.7 dBm.

1 Introduction

A voltage controlled oscillator is the key component in various microwave systems. Tunability of these VCOs is essential in many systems, e.g. in FMCW radars. Moreover fabrication variations or temperature drift can easily be compensated by using VCOs.

For the 24 GHz ISM-band an increasing number of applications like automotive radar systems or short-range communication links are proposed. Since the current ISM bandwidth specification limits these applications, the FCC and the ETSI put significant effort into expanding the specification for usability for UWB systems. Such systems demand for a moderate fixed frequency output power of a few mW and an ultra-wide tuning range to spread the signal. Tuning speed is also a figure of merit as it directly influences the repetition rate of the overall system. Finally, the usual prerequisite of low cost, as well as ease of fabrication is demanded.

Monolithic integrated solutions with a fractional-N-PLL achieve -6 dBm output power and a tuning range of 500 MHz (Stelzer et al., 2003). A harmonic DRO was published (Jeon et al., 1996) with 5.5 dBm output power at 20.5 GHz and a tuning range of 54 MHz. Another common tuning possibility is a YIG tuned oscillator. In Khanna and Hauptman (1991) a YTO is published from 18 GHz to 40 GHz with an output power of 13 dBm including integrated amplifiers.

Our approach to meet the above mentioned requirements is to integrate microstrip structure design with SMD handling compatible discrete packaged active devices. Furthermore, we take advantage of the generally present harmonics in the transistor of the oscillator. This is common practice to gain RF output power, generated at moderate frequencies where low-cost devices are available. The tunabilty was reached using a varactor diode.

2 Principle

Packaged devices are rarely available at 24 GHz due to the parasitics. And accordingly, low parasitics packages yield considerably more expensive devices. To bypass this barrier, a common solution is to generate the oscillation at lower frequencies where adequate devices are available, before mixing the signal to the desired frequency. This can be done by separate elements, or as shown in our design within one element. Harmonics are present in any oscillator due to the nonlinearities of the active element. The principle of a harmonic oscillator is to reflect the power at the fundamental frequency back into the device and to get maximum output power at the desired harmonic.

The oscillator design is based on the one-port approach. The transistor circuit has to fulfill the oscillation conditions

$$|\Gamma_A| \cdot |\Gamma_P| = 1 \tag{1}$$

and

$$\angle\Gamma_A + \angle\Gamma_P = 2\pi n, n = 0, 1, 2, ... \tag{2}$$

at the fundamental frequency, with the reflection coefficients $|\Gamma_A|$, $|\Gamma_P|$ of the active and the passive part. According to Chua et al. (1987) this criterion can be replaced by the Nyquist plot analysis. The circuit is unstable if the Nyquist loop $(\Gamma_A \cdot \Gamma_P)$ encircles the point $1 + j0$ in clockwise direction. The separation plane is set at the source pin of the tran-

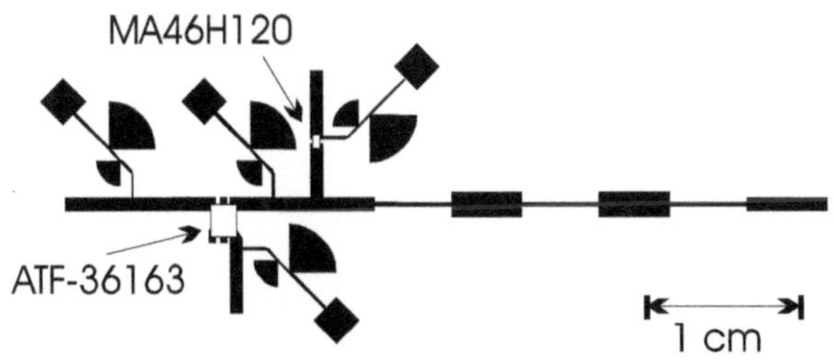

Fig. 1. Layout of the maximum power VCO.

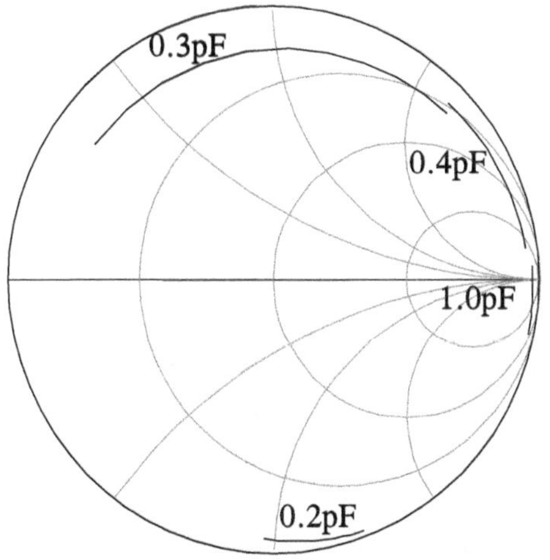

Fig. 2. Reflection coefficient of the passive part for several varactor capacitances (11.9–12.1 GHz).

sistor. The magnitude of Γ_A is an indicator for the achievable output power. The reflection coefficient of the the active part is influenced by the circuit at the gate and drain connectors of the transistor. The transistor is modeled by its s-parameters given by the manufacturer. The given two-port s-parameters were converted into three-port s-parameters according to Khanna (1985) to gain an additional degree of freedom for the design.

In addition to the oscillation condition at the fundamental frequency, the matching condition ($|\Gamma_A| = |\Gamma_P|^*$) has to be fulfilled at the harmonic frequency to achieve high output power.

According to the oscillation criteria there are two ways to tune the oscillator: Varying either the passive or the active part. Our tuning approach is to change the phase by using a varactor diode. The diode is preferable a packaged device that can be handled with standard SMD fabrication technologies. Since its influence at 24 GHz decreases with increasing capacitance, a diode with low capacitance is required. Therefore the M/A-COM MA46H120 GaAs hyperabrupt varactor diode, with a capacitance from 1.1 pF to 0.14 pF is chosen. Taconic TLP-5-0100 substrate is used. The transistor in the design is an Agilent GaAs PHEMT ATF-36163.

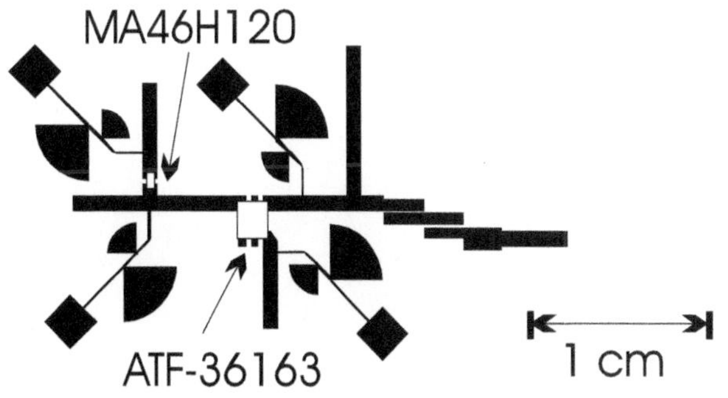

Fig. 3. Layout of the maximum tuning range VCO.

3 Design

We present two different designs. One, referred to as maximum power VCO, was designed for high output power. The other one, referred to as maximum tuning range VCO, showed a high tuning range. The main difference in the design is the location of the varactor diode. In both designs the varactor diode is connected to a virtual ground provided by a $\lambda/4$ transformation in order to keep the biasing effort moderate.

The bias network has to be designed for rejecting the fundamental frequency and the harmonic. This is accomplished by the combination of two radial stubs in each bias network as shown in Fig. 1. The rejection of the fundamental frequency and the matching of the 1st harmonic is done by means of a microstrip filter structure.

3.1 Maximum power VCO

Figure 1 shows the layout of the VCO optimized for maximum power output. A stepped impedance bandpass filter satisfies the oscillation and matching conditions. The varactor diode is attached to this filter, equivalent to the passive part. Thus, the ground bias supplies both, the diode and the source of the transistor. Figure 2 shows a plot of the varactor tuned reflection coefficient of the passive part at the fundamental frequency. Tuning the voltage across the varactor diode results mainly in a displacement of the phase for a frequency point.

3.2 Maximum tuning range VCO

Figure 3 shows the layout of the VCO that exhibits the maximum tuning range. A parallel-coupled, strip-line bandpass filter satisfies the oscillation and matching conditions at the passive part. The varactor diode is attached to the drain stub of the transistor at the active part. In this design, the reflection coefficient of the passive part is fixed, whereas the reflection coefficient of the active part varies with phase and magnitude as shown in Fig. 4.

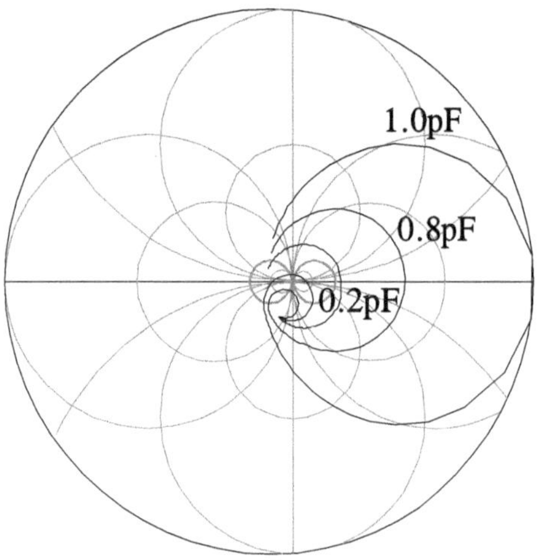

Fig. 4. Reflection coefficient of the active part for several varactor capacitances (11.5-12.5 GHz).

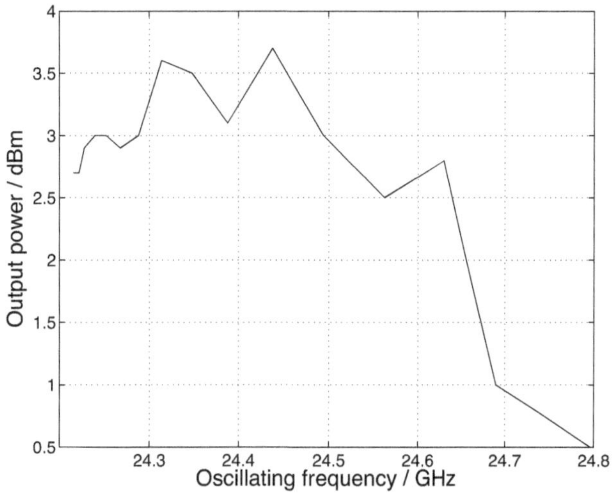

Fig. 5. Output power vs. oscillating frequency of the maximum power VCO.

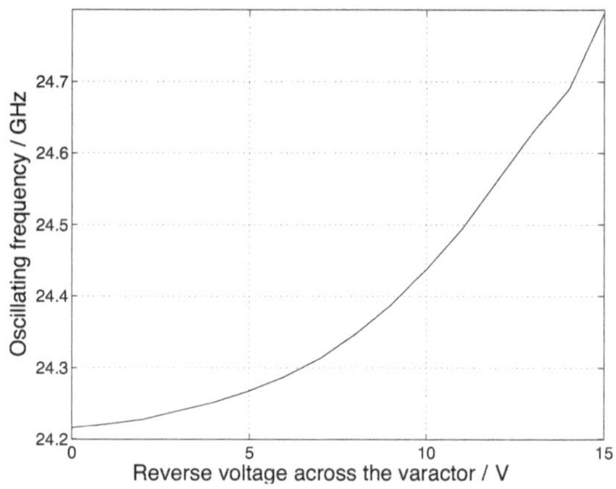

Fig. 6. Oscillating frequency vs. reverse voltage of the maximum power VCO.

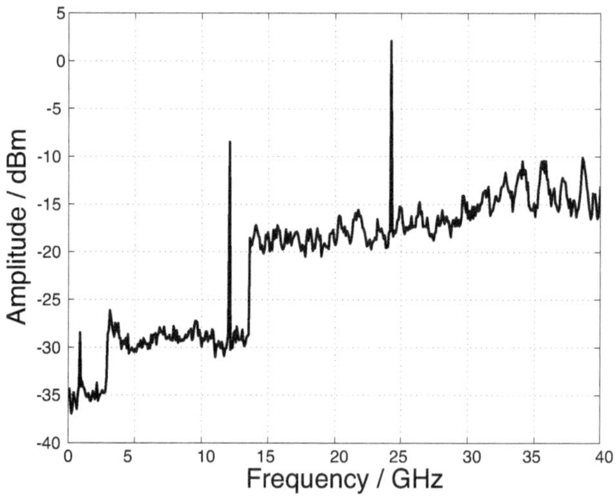

Fig. 7. Output spectrum of the maximum power VCO.

4 Measurements

All measurement data was taken by a spectrum analyzer. The modulation bandwidth was determined by applying a rectangular signal. A waveform generator provides respective voltages for two oscillation frequencies to the varactor diode. The maximum modulation bandwidth is identified at the 3 dB drop of the output power at a certain oscillation frequency when increasing the modulation frequency.

4.1 Maximum power VCO

The design optimized for maximum output power has a peak power of 3.7 dBm and variations of 3.2 dB over its tuning range of 580 MHz, see Fig. 5. A higher frequency shift is reached for low varactor capacitances, see Fig. 6. The phase noise is -90 dBc/Hz for 1 MHz offset. The modulation bandwidth is 2.5 MHz for a frequency shift of 420 MHz, corresponding to 0 V and 13 V at the waveform generator output. The 13 V result in a frequency of 24,63 GHz.

The power variation results from two effects. The capacitance variation of the varactor diode changes not only the phase at the fundamental frequency, but also the matching at the passive part for the harmonic frequency. Additionally, the magnitude of the active part varies over the tuned frequency range, see Fig. 4.

The oscillating frequency is approximately 12 GHz. It is 10 dB below the output power of the 1st harmonic, see Fig. 7.

4.2 Maximum tuning range VCO

The design that showed the maximum tuning range has a tuning range of 1.1 GHz, see Fig. 8. The reverse voltage differs from the other design, since it is superposed with the drain bias of the transistor. The output power varies from -6.2 dBm

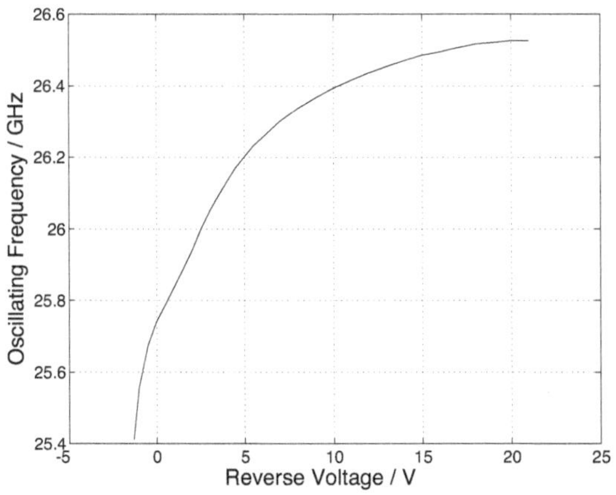

Fig. 8. Oscillating frequency vs. varactor voltage of the maximum tuning range VCO.

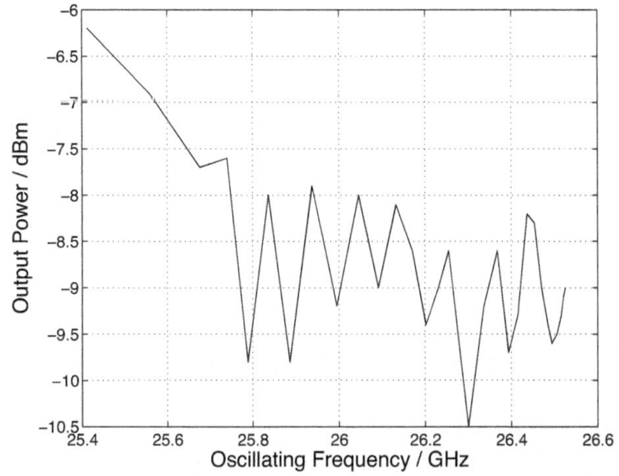

Fig. 9. Output power vs. oscillating frequency of the maximum tuning range VCO.

at 25.4 GHz to -9 dBm at 26.5 GHz, see Fig. 9. Its phase noise is -87 dBc/Hz for 1 MHz offset. The modulation bandwidth is 1.5 MHz for a frequency shift of 900 MHz, corresponding to 3 V and 16 V at the waveform generator output. The 16 V result in a frequency of 26,44 GHz.

The reduction of the output power with increasing frequency corresponds to the simulated reduction of the magnitude as shown in Fig. 4. Additionally, the matching of the harmonic degrades. The spectrum is plotted in Fig. 10. This design results in a 2nd harmonic oscillator with a fundamental frequency at approximately 8 GHz. Though this effect was not designated it was repeatable and stable. The 1st harmonic at about 16 GHz is 9 dB below the output power. Since the frequency shift is converted to the respective harmonics and the influence of the capacitance is more distinct at lower frequencies, the higher tuning range and the lower output power compared to the maximum power VCO design are evident.

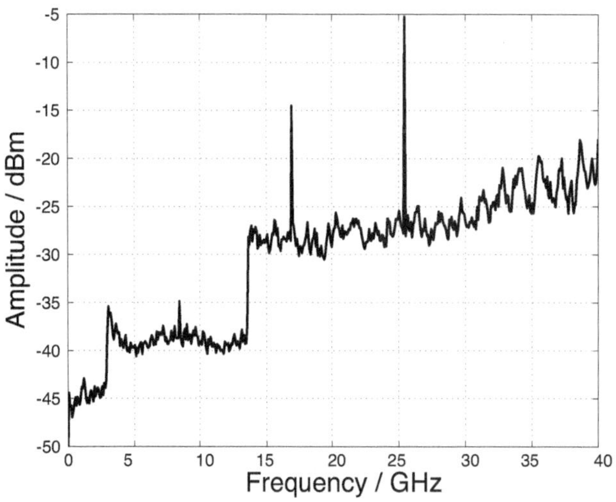

Fig. 10. Output spectrum of the maximum tuning range VCO.

5 Conclusions

We have presented two versions of a 24 GHz varactor tuned harmonic VCO. One exhibits a high tuning range of 1.1 GHz and an output power of -6 dBm. The other one exhibits a lower tuning range of 580 MHz with a higher output power of 3.7 dBm. For the application as a synthesizer in a FMCW radar, the required linearity of the voltage across the varactor diode and the output frequency can be achieved by using a compensation network.

An enhancement of the tuning range can be achieved by tuning the active and the passive part, analogical the matching can be aligned to keep the output power more constant.

References

Stelzer, A., Höftenberger, J., Ettinger, K., and Weigel, R.: Generation of Precise Frequency Ramps for FSCW Applications with a Fractional-N PLL at 24 GHz, European Microwave Conference, EuMC, Munich, Germany, 2003.

Jeon, S., Tchah, K., Yim, C., and Chung, S.: "Voltage Tuned Dielectric Resonance 20.5GHz Harmonic Oscillator with Novel Structure", 2003 IEEE MTT-S Int. Microwave Symp. Dig., 3, 1519–1522, June 1996.

Khanna, A. P. S. and Hauptman, J.: 8-40 GHz 13 dBm low noise GaAs FET YIG tuned oscillator, 1991 IEEE MTT-S Int. Microwave Symp. Dig., 1, 209–212, 1991.

Chua, L., Desoer, C., and Kuh, E.: Linear and Nonlinear Circuits, New York: McGraw-Hill, 1987.

Khanna, A. S.: Three-port S-parameters ease GaAs FET designing, Microwaves and RF, 2, 81–84, 1985.

20

Combining the multilevel fast multipole method with the uniform geometrical theory of diffraction

A. Tzoulis and T. F. Eibert

FGAN-Research Institute for High-Frequency Physics and Radar Techniques (FHR), 53343 Wachtberg-Werthhoven, Neuenahrer Str. 20, Germany

Abstract. The presence of arbitrarily shaped and electrically large objects in the same environment leads to hybridization of the Method of Moments (MoM) with the Uniform Geometrical Theory of Diffraction (UTD). The computation and memory complexity of the MoM solution is improved with the Multilevel Fast Multipole Method (MLFMM). By expanding the $\hat{k}$-space integrals in spherical harmonics, further considerable amount of memory can be saved without compromising accuracy and numerical speed. However, until now MoM-UTD hybrid methods are restricted to conventional MoM formulations only with Electric Field Integral Equation (EFIE). In this contribution, a MLFMM-UTD hybridization for Combined Field Integral Equation (CFIE) is proposed and applied within a hybrid Finite Element - Boundary Integral (FEBI) technique. The MLFMM-UTD hybridization is performed at the translation procedure on the various levels of the MLFMM, using a far-field approximation of the corresponding translation operator. The formulation of this new hybrid technique is presented, as well as numerical results.

1 Introduction

Exact numerical solutions of electromagnetic radiation and scattering problems with composite dielectric/metallic objects are actually obtained with the Finite Element – Boundary Integral (FEBI) method. The involved integral equation (IE) is discretized by the Method of Moments (MoM) using Rao-Wilton-Glisson (RWG) basis functions and a Galerkin's type approach (Rao et al., 1982). The resulting hybrid linear equation system is efficiently solved with iterative solvers in combination with appropriate preconditioning (Eibert, 2003).

Computation complexity and memory requirements of large scale MoM solutions for high frequencies increase rapidly with the dimensions of the objects. This problem is overcome by the Multilevel Fast Multipole Method (MLFMM) resulting in low computation and memory complexity of $O(N \log N)$ (Coifman et al., 1993; Chew et al., 2001). Further saving of a considerable amount of memory without compromising accuracy and numerical speed is achieved by expanding the FMM $\hat{k}$-space integrals into spherical harmonics (Eibert, 2004).

Electrically large conducting objects are handled efficiently with ray-based high-frequency methods. Such methods give accurate asymptotic solutions of the electromagnetic fields for objects with large dimensions as compared to the wavelength and no discretization is needed. The electromagnetic fields are described according to classical ray concepts of Geometrical Optics (GO) (Lo and Lee, 1993) and Uniform Geometrical Theory of Diffraction (UTD) (Kouyoumjian and Pathak, 1974).

Full electromagnetic coupling between dielectric arbitrarily shaped and UTD objects within a common environment is obtained with the hybrid FEBI-UTD method (Alaydrus et al., 2001), which is however restricted to conventional MoM formulations only with EFIE. In this contribution, the hybrid FEBI-UTD technique is extended by combining MLFMM with UTD for CFIE. The MLFMM-UTD hybridization is performed in the translation procedure on the various MLFMM levels. The amplitudes of the UTD rays are determined according to the far-field MLFMM in Chew et al. (2001). Due to different ray directions for outgoing and incoming waves, appropriate interpolation and anterpolation routines must be applied to consider the UTD contributions.

In the following, first the formulation of the FEBI-UTD for CFIE and after that the formulation of the MLFMM-UTD approach with expansion of the $\hat{k}$-space integrals in spherical harmonics is presented. Finally, numerical results for various large scale examples are given.

2 Formulation

2.1 FEBI-UTD hybrid method

Consider the configuration of Fig. 1, where inhomogeneous dielectric FEBI and large conducting UTD objects exist in the same environment in free space. The electromagnetic fields in the interior region of the FEBI objects are expressed according to the Finite Element Method (FEM) (Volakis et al., 1998) and are expanded in terms of edge element basis functions $\boldsymbol{\alpha}_n$ using tetrahedral volume meshes.

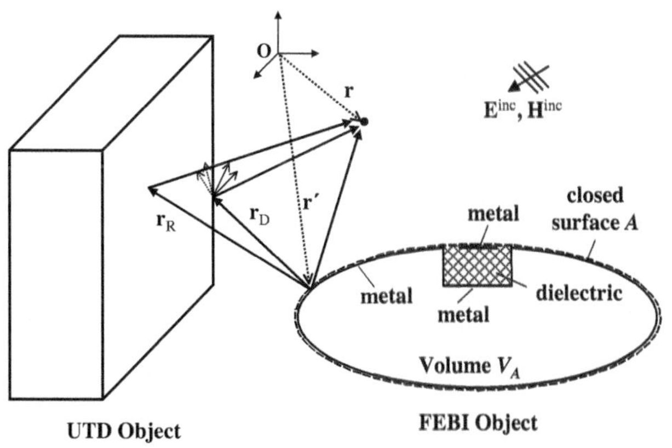

Fig. 1. Hybrid FEBI-UTD concept.

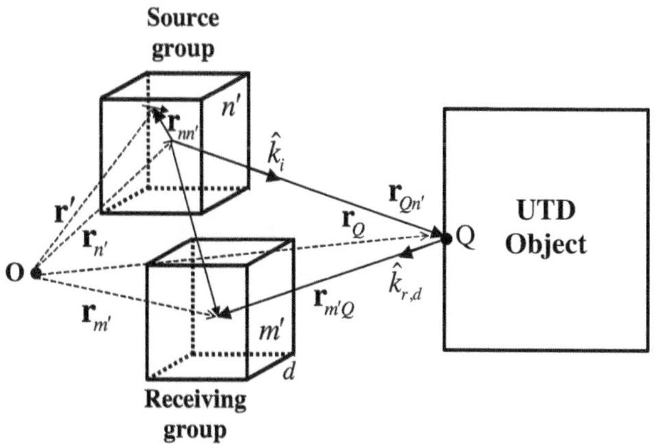

Fig. 2. Hybrid MLFMM-UTD concept.

The fields in the exterior regions are expressed with IE's over the boundaries of the FEBI objects, being solved by a Galerkin-type MoM with RWG basis functions (Rao et al., 1982). The presence of UTD objects is taken into account by modifying the Green's functions, as well as the incident fields of the IE's, with additional high-frequency contributions received at the testing or observation points, respectively (Alaydrus et al., 2001). In this contribution, 1st and 2nd order reflections and diffractions on flat structures are taken into account. The following formulations will be given for clarity only for 1st order mechanisms. The formulation for 2nd order mechanisms can be easily given in the same manner.

The UTD ray contributions are taken into account for the CFIE

$$Z(1-\alpha)MFIE - \alpha EFIE = 0, \tag{1}$$

where α is the combination parameter with values from 0 to 1 and Z the wave impedance of the considered solution space. The EFIE is given by

$$\hat{n} \times \Bigg[\hat{n} \times \Bigg(\iint\limits_A \Big[\bar{\mathbf{G}}^E_{J,tot}(\boldsymbol{r},\boldsymbol{r}')\cdot \boldsymbol{J}_A(\boldsymbol{r}') + \bar{\mathbf{G}}^E_{M,tot}(\boldsymbol{r},\boldsymbol{r}')\cdot \boldsymbol{M}_A(\boldsymbol{r}')\Big]da' + \boldsymbol{E}^{inc}_{tot}(\boldsymbol{r})\Bigg) + \frac{1}{2}\boldsymbol{M}_A(\boldsymbol{r})\Bigg] = 0 \tag{2}$$

and the Magnetic Field Integral Equation (MFIE) by

$$\hat{n} \times \Bigg(\iint\limits_A \Big[\bar{\mathbf{G}}^H_{M,tot}(\boldsymbol{r},\boldsymbol{r}')\cdot \boldsymbol{M}_A(\boldsymbol{r}') + \bar{\mathbf{G}}^H_{J,tot}(\boldsymbol{r},\boldsymbol{r}')\cdot \boldsymbol{J}_A(\boldsymbol{r}')\Big]da' + \boldsymbol{H}^{inc}_{tot}(\boldsymbol{r})\Bigg) - \frac{1}{2}\boldsymbol{J}_A(\boldsymbol{r}) = 0, \tag{3}$$

where $\bar{\mathbf{G}}^{E/H}_{J,tot}(\boldsymbol{r},\boldsymbol{r}')$ and $\bar{\mathbf{G}}^{E/H}_{M,tot}(\boldsymbol{r},\boldsymbol{r}')$ are the total Green's functions of the electric or magnetic field due to electric and magnetic surface currents, respectively. Also, $\boldsymbol{E}^{inc}_{tot}(\boldsymbol{r})$ and $\boldsymbol{H}^{inc}_{tot}(\boldsymbol{r})$ are the total incident electric and magnetic fields at the observation point $\boldsymbol{r}$.

The total Green's functions of the hybrid problem are the superposition of the Green's functions of the direct coupling of the currents $\bar{\mathbf{G}}^{E/H}_{J/M}(\boldsymbol{r},\boldsymbol{r}')$ and of

$$\bar{\mathbf{G}}^{E/H}_{J/M,UTD}(\boldsymbol{r},\boldsymbol{r}') = \sum_s A_{R_s}\bar{\mathbf{R}}^{E/H}_s \cdot \bar{\mathbf{G}}^{E/H}_{J/M}(\boldsymbol{r}_{R_s},\boldsymbol{r}') + \sum_v A_{D_v}\bar{\mathbf{D}}^{E/H}_v \cdot \bar{\mathbf{G}}^{E/H}_{J/M}(\boldsymbol{r}_{D_v},\boldsymbol{r}') + \cdots, \tag{4}$$

which is the Green's functions of the high-frequency ray-based fields due to UTD objects. A_{R_s} and A_{D_v} are the divergence and phase factors for reflection and diffraction, respectively, and $\bar{\mathbf{R}}^{E/H}_s$ and $\bar{\mathbf{D}}^{E/H}_v$ the dyadic reflection and diffraction coefficients for electric or magnetic field defined by basic GO (Lo and Lee, 1993) and UTD concepts (Kouyoumjian and Pathak, 1974). Direct coupling is not taken into account if UTD objects lie between source and testing currents.

Similarly, the high-frequency contributions of the incident electric and magnetic field at the observation point $\boldsymbol{r}$ have the form

$$\boldsymbol{E}^{inc}_{UTD}(\boldsymbol{r}) = \sum_s A_{R_s}\bar{\mathbf{R}}^{E}_s \cdot \boldsymbol{E}^{inc}(\boldsymbol{r}_{R_s}) + \sum_v A_{D_v}\bar{\mathbf{D}}^{E}_v \cdot \boldsymbol{E}^{inc}(\boldsymbol{r}_{D_v}) + \cdots \tag{5}$$

and

$$\boldsymbol{H}^{inc}_{UTD}(\boldsymbol{r}) = \frac{1}{Z}\hat{k}_r \times \sum_s A_{R_s}\bar{\mathbf{R}}^{E}_s \cdot \boldsymbol{E}^{inc}(\boldsymbol{r}_{R_s}) + \frac{1}{Z}\hat{k}_d \times \sum_v A_{D_v}\bar{\mathbf{D}}^{E}_v \cdot \boldsymbol{E}^{inc}(\boldsymbol{r}_{D_v}) + \cdots, \tag{6}$$

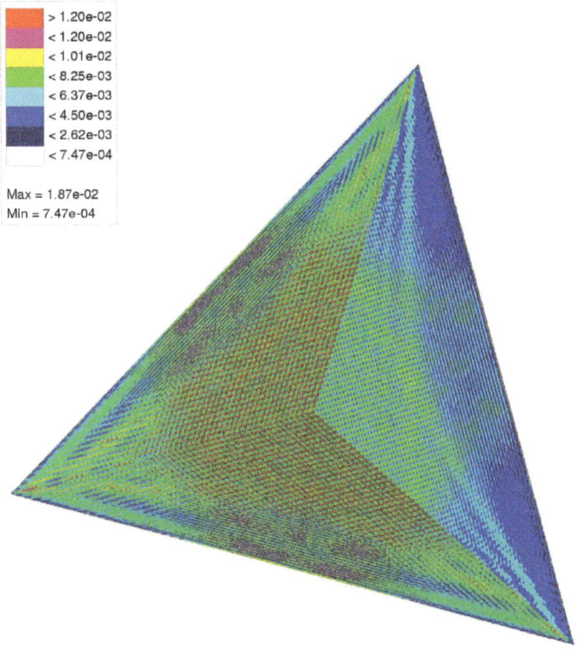

Fig. 3. Surface current density magnitude of trihedral 90° corner reflector for 15 GHz.

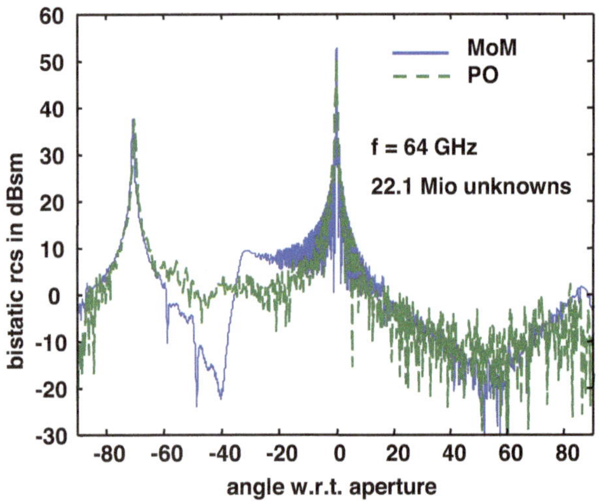

Fig. 4. Bistatic RCS of trihedral 90° corner reflector for 64 GHz.

where $\hat{k}_r$ and $\hat{k}_d$ are the propagation directions of the reflected and diffracted rays, respectively. Again, no direct field contribution is taken into account if the observation point is in the shadow of an UTD object.

The MoM solution of the IE's (1), (2) and (3) is achieved by expanding the surface currents in terms of RWG basis functions $\boldsymbol{\beta}_n$ using triangular surface meshes. The exterior fields are coupled with the interior ones via field continuity conditions at the boundaries of the FEBI objects. The resulting hybrid linear equation system is solved by an iterative solution technique, using a multilevel iterative preconditioning strategy (Eibert, 2003).

The matrix-vector product computations are accelerated by MLFMM. In order to retain low complexity in the hybrid approach, UTD contributions are taken into account in the matrix-vector product MLFMM computations. In the following, the MLFMM formulation for CFIE with expansion of the $\hat{k}$-space integrals in spherical harmonics, as well as the hybrid MLFMM-UTD formulation is presented.

2.2 MLFMM-UTD hybrid method

The hybrid MLFMM-UTD concept is shown in Fig. 2. The source group n' and the receiver group m' belong to the FMM model of the FEBI object at a specific level. The CFIE matrix elements in FMM representation can be written as (Coifman et al., 1993; Chew et al., 2001)

$$Z_{mn,J}^{CFIE} = -c_1 \oiint \tilde{\boldsymbol{\beta}}_m^*(\hat{k}) \cdot T_L(\hat{k}\cdot\hat{r}_{m'n'}) \left(\bar{\mathbf{I}} - \hat{k}\hat{k}\right) \cdot \tilde{\boldsymbol{\beta}}_n(\hat{k})\, d\hat{k}^2 + c_2 \oiint \left(\hat{k} \times \tilde{\boldsymbol{\alpha}}_m^*(\hat{k})\right) \cdot T_L(\hat{k}\cdot\hat{r}_{m'n'}) \tilde{\boldsymbol{\beta}}_n(\hat{k})\, d\hat{k}^2 \tag{7}$$

and

$$Z_{mn,M}^{CFIE} = +c_3 \oiint \left(\hat{k} \times \tilde{\boldsymbol{\beta}}_m^*(\hat{k})\right) \cdot T_L(\hat{k}\cdot\hat{r}_{m'n'}) \tilde{\boldsymbol{\beta}}_n(\hat{k})\, d\hat{k}^2 + c_4 \oiint \tilde{\boldsymbol{\alpha}}_m^*(\hat{k}) \cdot T_L(\hat{k}\cdot\hat{r}_{m'n'}) \left(\bar{\mathbf{I}} - \hat{k}\hat{k}\right) \cdot \tilde{\boldsymbol{\beta}}_n(\hat{k})\, d\hat{k}^2 \tag{8}$$

where $c_1 = j\frac{\omega\mu}{4\pi}\alpha$, $c_2 = jZ_0\frac{k}{4\pi}(1-\alpha)$, $c_3 = j\frac{k}{4\pi}\alpha$ and $c_4 = jZ_0\frac{\omega\varepsilon}{4\pi}(1-\alpha)$. Further,

$$\tilde{\boldsymbol{\beta}}_m(\hat{k}) = \iint_A \boldsymbol{\beta}_m(\boldsymbol{r}) e^{j\boldsymbol{k}\cdot\boldsymbol{r}_{mm'}}\, da \tag{9}$$

and

$$\tilde{\boldsymbol{\alpha}}_m(\hat{k}) = \iint_A \boldsymbol{\alpha}_m(\boldsymbol{r}) e^{j\boldsymbol{k}\cdot\boldsymbol{r}_{mm'}}\, da \tag{10}$$

are the $\hat{k}$-space representations of the basis functions and

$$T_L(\hat{k}\cdot\hat{r}) = \sum_{l=0}^{L} (-j)^l (2l+1) \mathrm{h}_l^{(2)}(kr) \mathrm{P}_l(\hat{k}\cdot\hat{r}) \tag{11}$$

is the translation operator. Also, $\mathrm{h}_l^{(2)}$ is the second kind spherical Hankel function of degree l, P_l is the Legendre polynomial of degree l and $*$ denotes complex conjugation. The MLFMM is implemented with memory efficiency by expanding the representations $\tilde{\boldsymbol{\beta}}_m(\hat{k})$ of the basis functions in spherical harmonics according to

$$\tilde{\boldsymbol{\beta}}_m(\hat{k}) = \sum_{p=0}^{P} \sum_{q=-p}^{p} \boldsymbol{f}_{pq}^n \mathrm{Y}_{pq}(\vartheta, \varphi), \tag{12}$$

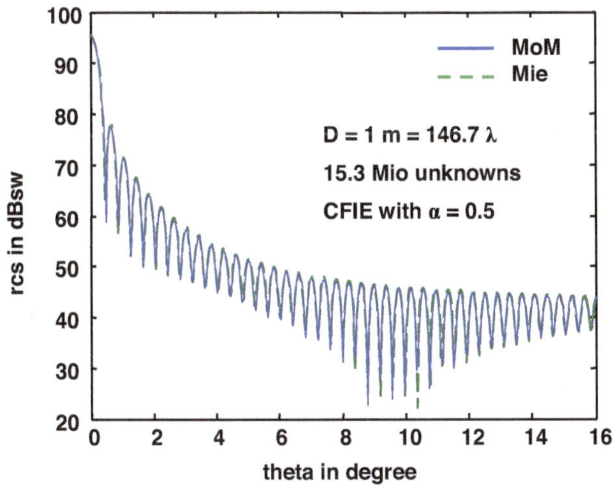

Fig. 5. RCS near forward direction of conducting sphere with diameter $D = 1$ m$= 146.7\lambda$ by CFIE solution.

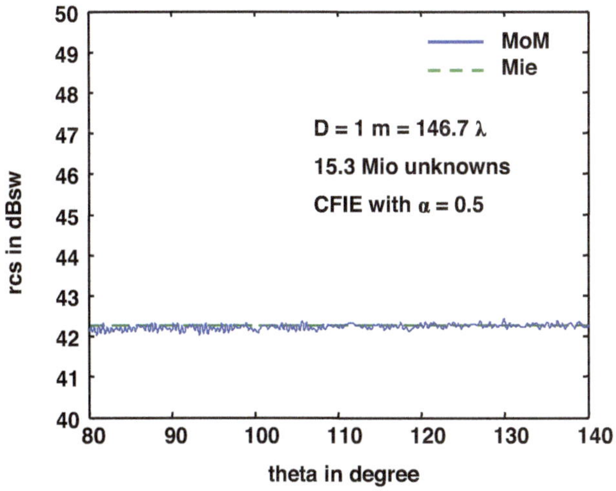

Fig. 6. RCS near backward direction of conducting sphere with diameter $D = 1$ m$= 146.7\lambda$ by CFIE solution.

where

$$Y_{pq}(\vartheta,\varphi) = \sqrt{\frac{(2p+1)(p-q)!}{4\pi(p+q)!}}\, P_p^q(\cos\vartheta) e^{jq\varphi}, \qquad (13)$$

are the orthonormalized spherical harmonics (Harrington, 1961). Also, P_p^q is the associated Legendre polynomial of degree p and order q. According to this approach, the matrix-vector product computations within the iteration loop of the iterative solver are performed by first aggregating all expansion coefficients at the center of each MLFMM group at the finest level. After that, the outgoing waves are computed at the quadrature points for all groups on the finest MLFMM level. The translations of outgoing waves into incoming waves as well as the aggregations and disaggregations between different MLFMM levels including interpolation and anterpolation are performed as in standard MLFMM using the numerical quadrature samples. However, when all incoming waves are collected in a certain group on the finest level, the spherical harmonics expansion

$$T_L(\hat{k}\cdot\hat{r}_{m'n'})\left(\bar{\mathbf{I}} - \hat{k}\hat{k}\right)\cdot\tilde{\boldsymbol{\beta}}_n(\hat{k}) = \sum_{p=0}^{P}\sum_{q=-p}^{p} \boldsymbol{g}_{pq}^n Y_{pq}(\vartheta,\varphi) \qquad (14)$$

is carried out and by utilizing the orthogonality of the spherical harmonics, the closed integral over the Ewald sphere is simplified to the series

$$Z_{mn} = -j\frac{\omega\mu}{4\pi}\sum_{p=0}^{P}\sum_{q=-p}^{p}(\boldsymbol{f}_{pq}^m)^*\cdot\boldsymbol{g}_{pq}^n. \qquad (15)$$

The number of expansion coefficients needed with this approach is in general less than the corresponding number of quadrature samples for numerical integration in the $\hat{k}$-space, resulting in saving of memory requirements.

The ray contributions due to UTD objects are taken into account in the MLFMM matrix-vector product computations by assuming that the ray path from the source group to the UTD object $\boldsymbol{r}_{Qn'}$ is much greater than the dimensions of the group itself. This means, that only one $\hat{k}$ direction is needed to express the radiated field from the source group. Assuming far-field conditions, the scalar Green's function from the source current to the local point on the UTD object is expressed with the far-field MLFMM approximation

$$G(\boldsymbol{r}_Q,\boldsymbol{r}') = \frac{e^{-jk|\boldsymbol{r}_Q-\boldsymbol{r}'|}}{|\boldsymbol{r}_Q-\boldsymbol{r}'|} = e^{j\boldsymbol{k}_i\cdot\boldsymbol{r}_{nn'}}T_L^{FF}, \qquad (16)$$

where $\boldsymbol{k}_i = k\hat{k}_i$, $\hat{k}_i = \hat{r}_{Qn'}$ is the direction of ray incidence and

$$T_L^{FF} = \frac{e^{-jkr_{Qn'}}}{r_{Qn'}} \qquad (17)$$

is the associated far-field translation operator (Chew et al., 2001). Thus, the translation of the high-frequency contributions is performed only for the direction of ray incidence $\hat{k}_i$. The received high-frequency fields are taken into account only for the reflection $\hat{k}_r$ or diffraction $\hat{k}_d$ directions from the local point on the UTD object to the receiver groups. The following MLFMM-UTD formulations are given for simplicity for the case of the EFIE contributions in (7). However, they apply to all contributions in (7) and (8) as well. The ray contributions of the matrix elements of the hybrid MLFMM-UTD approach can be written as

$$\begin{aligned} Z_{mn,J,UTD}^{EFIE} &= \\ &-j\frac{\omega\mu}{4\pi}\tilde{\boldsymbol{\beta}}_m^*(\hat{k}_r)\cdot\sum_s A_{R_s}\bar{\mathbf{R}}_s^E T_L^{FF}\cdot\left(\bar{\mathbf{I}} - \hat{k}_r\hat{k}_i\right)\cdot\tilde{\boldsymbol{\beta}}_n(\hat{k}_i) \\ &-j\frac{\omega\mu}{4\pi}\tilde{\boldsymbol{\beta}}_m^*(\hat{k}_d)\cdot\sum_v A_{D_v}\bar{\mathbf{D}}_v^E T_L^{FF}\cdot\left(\bar{\mathbf{I}} - \hat{k}_d\hat{k}_i\right)\cdot\tilde{\boldsymbol{\beta}}_n(\hat{k}_i) \\ &+\cdots, \end{aligned} \qquad (18)$$

where $\boldsymbol{k}_r = k\hat{k}_r$ and $\boldsymbol{k}_d = k\hat{k}_d$. Again, no direct field contribution is taken into account if the receiver group is in the shadow of an UTD object.

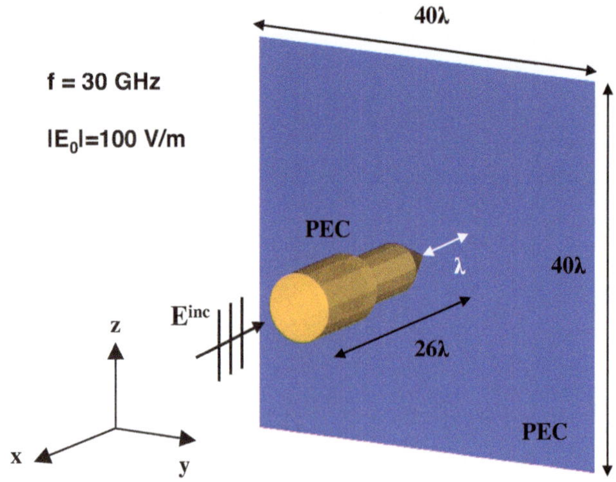

Fig. 7. Conducting cone cylinder in front of conducting plate at distance λ.

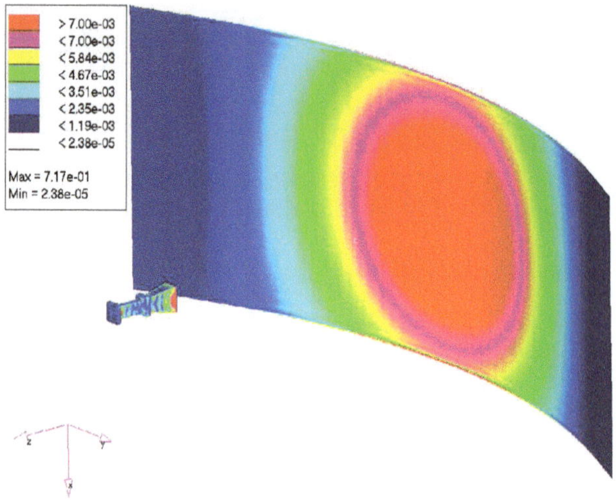

Fig. 9. Magnitude of surface current density on conducting cylindrical parabolic reflector with X-band horn excitation for f=10 GHz.

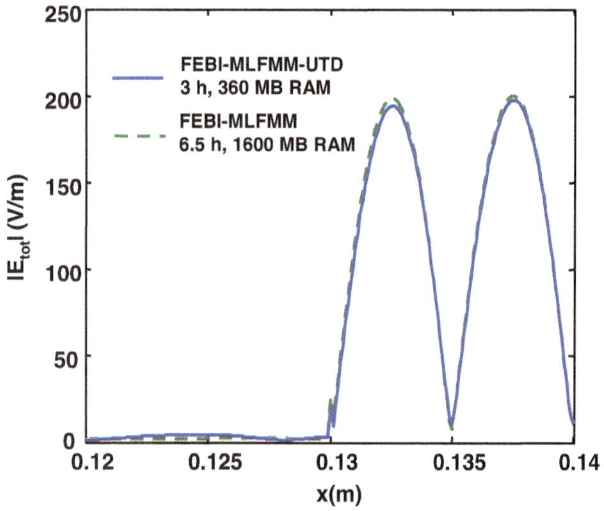

Fig. 8. Electric field in direction of propagation for conducting cone cylinder in front of conducting plate.

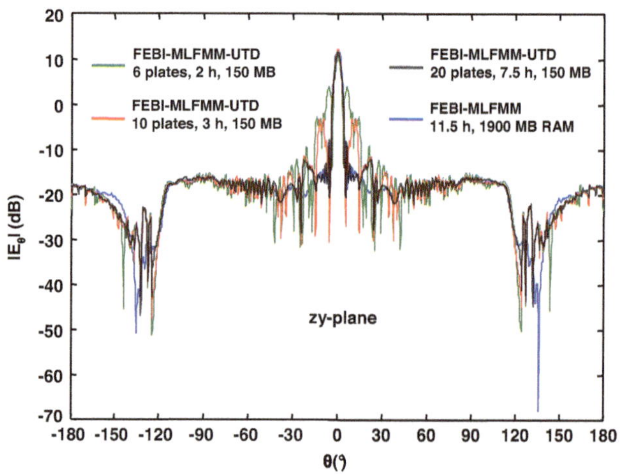

Fig. 10. Co-polar radiation pattern on principal E-plane of conducting cylindrical parabolic reflector with X-band horn excitation for f=10 GHz.

In the numerical implementation of MLFMM, a limited number of sampling points is used to evaluate the $\hat{k}$-space integrals. In general, the ray directions of incidence $\hat{k}_i$, reflection $\hat{k}_r$ or diffraction $\hat{k}_d$ do not match with any of these sampling points. For this reason, the required direction of incidence must be interpolated from the neighboring sampling points. Similarly, after reflection or diffraction, the appropriate ray direction must be anterpolated to the neighboring sampling points.

3 Numerical examples

In this section, numerical results of various large scale examples are presented. First, a trihedral 90° conducting corner reflector with 1.4 m edge length has been computed, as shown in Fig. 3. In the same figure, the magnitude of the surface current density on the corner reflector can be seen, computed for 15 GHz, using a triangular surface mesh. For the same corner reflector, RCS computations have been carried out for 64 GHz, using H-polarized plane wave excitation with normal incidence. In this case, the reflector was discretized with 22.1 Million unknowns and the computation time of the solution was about 122 h on an Opteron 2.2 GHz CPU with about 28 GB memory requirements. In Fig. 4, the co-polar bistatic RCS can be seen, compared to a PO solution. It is noticed, that to our knowledge this computation is currently the world-wide largest MoM solution performed on a single CPU.

Next example is a conducting sphere with diameter D=1 m=146.7λ. The triangular surface mesh on the sphere resulted in 15.3 Million unknowns and the solution was obtained with FEBI for CFIE with α=0.5. In Figs. 5 and 6, the computed RCS is shown, for small regions near forward and backward direction, respectively. The numerical results are compared to the corresponding Mie-series analytical solution. It can be seen, that numerical results have excellent agreement with the exact analytical solution. Especially at backward direction, very small ripple of the numerical solution can be seen.

Next example is the problem of a conducting cone cylinder in front of a conducting plate at distance λ, as shown in Fig. 7. Two computations have been carried out. First, both objects where discretized with triangular surface mesh, resulting in 630 628 unknowns. In the second computation, the same surface mesh was used for the cone cylinder and the plate was treated by UTD. In this case, the number of unknowns was 150 492. In Fig. 8, the calculated electric field in direction of propagation for both computations is shown. In the same figure, computation time and memory requirements on an AMD Athlon 2000+ PC are shown as well. It can be seen, that the hybrid FEBI-UTD results agree very well with the exact numerical FEBI solution, needing much less computation time and memory requirements.

Last example is a conducting cylindrical parabolic reflector with X-band horn excitation for f=10 GHz, as shown in Fig. 9. In the same figure, the magnitude of the surface current density on both objects can be seen, using a triangular surface mesh for both objects with 830 280 unknowns. For the hybrid FEBI-UTD computation, the reflector was approximated by flat plates and computations were done for 6, 10 and 20 plates. In Fig. 10, the co-polar radiation pattern on the principal E-plane for all computations is shown. In the same figure, the computation time and memory requirements on an AMD Athlon 2000+ PC are shown as well. It can be seen, that for 6 and 10 plates the hybrid FEBI-UTD solution presents significant side lobes. For 20 plates, the solution shows very good agreement with the exact numerical FEBI solution, needing less computation time and memory requirements.

4 Conclusions

In this contribution, the hybrid FEBI-MLFMM-UTD method for numerical radiation and scattering computations was presented. Due to memory efficient implementation of the MLFMM, large scale numerical computations can be carried out. The FEBI-UTD hybridization allows optimum treatment of problems including arbitrarily shaped composite dielectric/metallic objects and electrically large conducting objects of simple shape in the same environment. Low computation complexity and memory requirements are retained in the hybrid FEBI-UTD approach, due to MLFMM-UTD combination. The MLFMM-UTD hybridization is performed in the translation procedure on the various MLFMM levels, using a far-field approximation of the appropriate translation operator. The hybrid FEBI-UTD formulation for CFIE, as well as the formulation of the MLFMM-UTD combination with expansion of the $\hat{k}$-space integrals in spherical harmonics was presented. Finally, numerical results for various large scale antenna and scattering problems were given.

References

Alaydrus, M., Hansen, V., and Eibert, T. F.: Hybrid^2: Combining the three-dimensional hybrid finite element – boundary integral technique for planar multilayered media with the uniform geometrical theory of diffraction, IEEE Trans. Antennas Propagat., vol. 50, no. 1, 67–74, Jan. 2002.

Chew, W. C., Jin, J.-M., and Michielssen, E.: Fast and efficient algorithms in computational electromagnetics, Boston: Artech House, 2001.

Chew, W. C., Cui, T. J., and Song, J. M.: A FAFFA-MLFMA algorithm for electromagnetic scattering, IEEE Trans. Antennas Propagat., vol. 50, no. 11, 1641–1649, Nov. 2002.

Coifman, R., Rokhlin, V., and Wandzura, S.: The fast multipole method: A pedestrian prescription, IEEE Antennas Propagat. Mag., vol. 35, no. 3, 7–12, Jun. 1993.

Eibert, T. F.: Advances in hybrid finite element – boundary integral modelling of three-dimensional radiation and scattering problems, Intern. ITG Conf. on Antennas, 157-160, Berlin, Sep. 2003.

Eibert, T. F.: A multilevel fast multipole method with spherical harmonics expansion of the k-space integrals, URSI EM-Theory Symposium, Pisa, Italy, 150–153, May 2004.

Harrington, R. F.: Time harmonic electromagnetic fields, New York: McGraw-Hill, 1961.

Kouyoumjian, R. G. and Pathak, P. R.: A uniform geometrical theory of diffraction for an edge in a perfectly conducting surface, Proc. IEEE, vol. 62, no. 11, 1448–1461, Nov. 1974.

Lo, Y. T. and Lee, S. W.: Antenna handbook, vol.1, Chapman & Hall, New York, 1993.

Rao, S. M., Wilton, D. R., and Glisson, A. W.: Electromagnetic scattering by surfaces of arbitrary shape, IEEE Trans. AP, vol. 30, 409–418, May 1982.

Volakis, J. L., Chatterjee, A., and Kempel, L.-C.: Finite Element Method for Electromagnetics, IEEE, Inc., New York, 1998.

21

Simplified modeling of EM field coupling to complex cable bundles

B. Schetelig[1], J. Keghie[1], R. Kanyou Nana[1], L.-O. Fichte[1], S. Potthast[2], and S. Dickmann[1]

[1]Faculty of Electrical Engineering, Helmut-Schmidt-University/University of the Federal Armed Forces Hamburg, Germany
[2]Bundeswehr Research Institute for Protective Technologies and NBC Protection (WIS), Munster, Germany

Abstract. In this contribution, the procedure "Equivalent Cable Bundle Method" is used for the simplification of large cable bundles, and it is extended to the application on differential signal lines. The main focus is on the reduction of twisted-pair cables. Furthermore, the process presented here allows to take into account cables with wires that are situated quite close to each other. The procedure is based on a new approach to calculate the geometry of the simplified cable and uses the fact that the line parameters do not uniquely correspond to a certain geometry. For this reason, an optimization algorithm is applied.

1 Introduction

To assure the immunity of cable-wired communication systems to external electromagnetic influences, one needs to know the disturbances coupled into the connecting cables to carry out an evaluation at the inputs of the connected devices based on its prescriptive limits. From several research projects, it is known that coupling via the cables is often quite important because of their extended geometry, compared to the direct coupling via the enclosures. Especially on ships, we have to deal with quite extensive cable bundles. Numerical calculations of such cable structures place great demands on computation power. An analytical approach is mostly restricted to quite simple geometries. Within this context, a methodology can be applied that can be found in literature (Andrieu, 2006; Andrieu et al., 2008). This procedure is based on the idea to combine these wires of a cable harness that show similar behaviour while being irradiated by an electromagnetic field. In this way, the number of wires to be considered can be reduced. The behaviour of this simplified cable bundle to electromagnetic irradiation remains unchanged when compared to the original one. Especially when very extensive cable bundles are analyzed, there is the opportunity of a considerably large reduction of the bundle geometry.

After applying this methodology to a cable bundle, its immunity to EMI can be calculated later on by using any well-established tools and procedures. The reduced cable bundle can be treated the same way as any other harness. Any kind of calculation method can be applied, but the application of the reduction technique is particularly appropriate when using numerical methods such as the method of moments. When we apply numerical procedures on the simplified harness we get a significant reduction of the necessary number of mesh cells or segments when compared to the original, much larger cable bundle. This means reduced demands on computation power. Analytical approaches are simplified, too.

The "Equivalent Cable Bundle Method" was formerly used for the reduction of multiconductor transmission lines with a reference conductor that is formed by a common ground plane. The resulting voltages and currents therefore are common mode values referred to the ground plane. In a scenario close to reality, particularly the analysis of differential signal lines matters. Within this context, this paper presents a modification of the reduction technique to allow its application to differential signal lines, too. Nevertheless, the obtained insight can likewise be applied to the calculation of disturbances on cables with reference to ground.

2 Fundamentals of transmission line theory

The methodology of simplified modelling of field coupled disturbances is based on the transmission line analysis of the cable bundles of interest. The use of transmission line theory to describe the coupling of electromagnetic fields is well-known and was described by (Tesche et al., 1997; Paul, 1994)

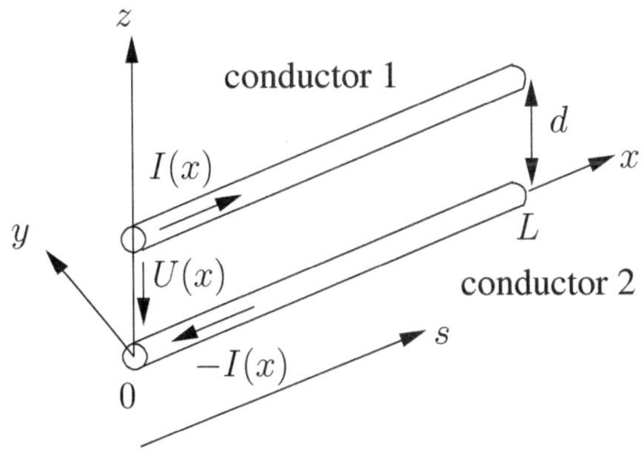

Fig. 1. Geometrical adjustment of the transmission line.

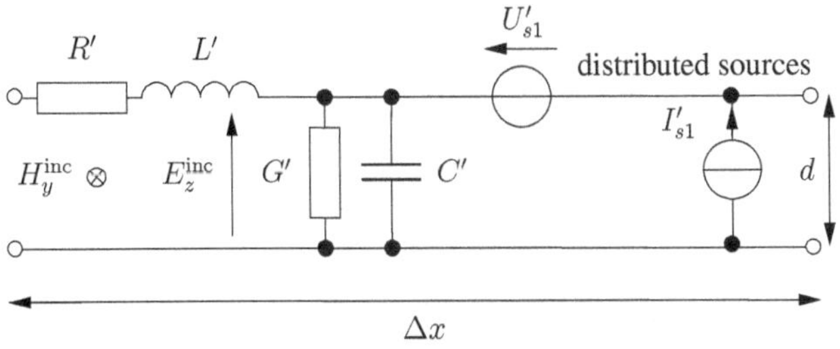

Fig. 2. Equivalent circuit of an irradiated single conductor transmission line by Taylor approach.

among others. In the following, we want to summarize the basic facts, that are used in the context of this study.

The behaviour of voltages and currents on transmission lines can be described by the telegrapher's equations (cf. Figs. 1, 2):

$$\begin{cases} -\dfrac{\mathrm{d}U}{\mathrm{d}x} = (R' + \mathrm{j}\omega L') \cdot I, \\ -\dfrac{\mathrm{d}I}{\mathrm{d}x} = (G' + \mathrm{j}\omega C') \cdot U. \end{cases} \tag{1}$$

R', L', G' and C' are the per-unit-length line parameters.

If we assume a lossless transmission line and therefore neglect R' and G', the characteristic impedance then can be noted as follows:

$$Z_0 = \sqrt{\frac{L'}{C'}}. \tag{2}$$

The coupling of external electromagnetic fields to transmission lines was described by Taylor as well as by Agrawal and Rachidi by using different approaches. As summarized by (Tesche, 1995), the influence of the electromagnetic field can be represented by additional sources in the equivalent circuit diagramm of the transmission line. In the general formulation by Taylor, voltage sources as well as current sources

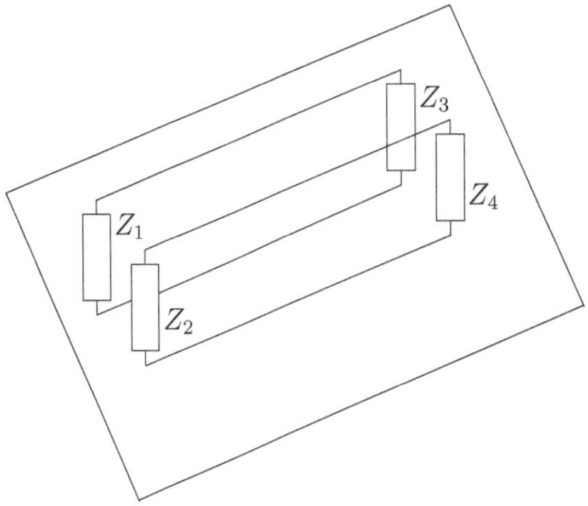

Fig. 3. Transmission lines with differential termination loads.

are used. The telegrapher's equations (Eq.1) are modified as follows:

$$\begin{cases} \dfrac{\mathrm{d}U}{dx} + \mathrm{j}\omega L' \cdot I = U'_{s1}, \\ \dfrac{\mathrm{d}I}{dx} + \mathrm{j}\omega C' \cdot U = I'_{s1}. \end{cases} \tag{3}$$

The distributed sources can be described as

$$U'_{s1} = -\mathrm{j}\omega\mu_0 \int_0^d H_y^{\mathrm{inc}} \mathrm{d}z, \tag{4}$$

$$I'_{s1} = -\mathrm{j}\omega C' \int_0^d E_z^{\mathrm{inc}} \mathrm{d}z, \tag{5}$$

with reference to Fig. 2.

3 Methodological approach

The simplification of cable bundles using differential signal lines is similar to the methodology for signal lines referred to the ground plane. This procedure was presented in detail in (Andrieu et al., 2008).

The scenario of the modified methodology is presented in Figs. 3 and 4: We analyze a cable bundle with a constant distance to the ground. The wires are surrounded by a homogeneous dielectric. The attached devices are represented by their input impedances. As we focus on differential signal transmission, no common mode terminations exist.

The first step of reducing the profile of the bundle is done by assuming a similar coupling of different wires. For these wires we assume a similar current distribution in the lines, so in the following we can categorize all the wires to individual groups with similar attributes. In the sample setup in Fig. 4, there are two differential signal lines consisting of the wires

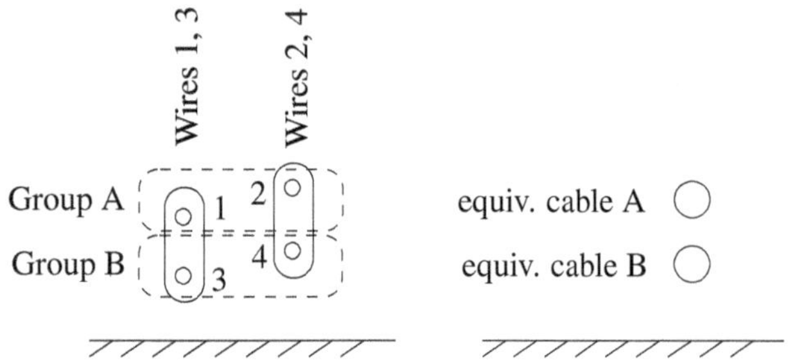

Fig. 4. Cross section of the original (left) and reduced (right) cable of Fig. 3 (principle diagram with two pairs).

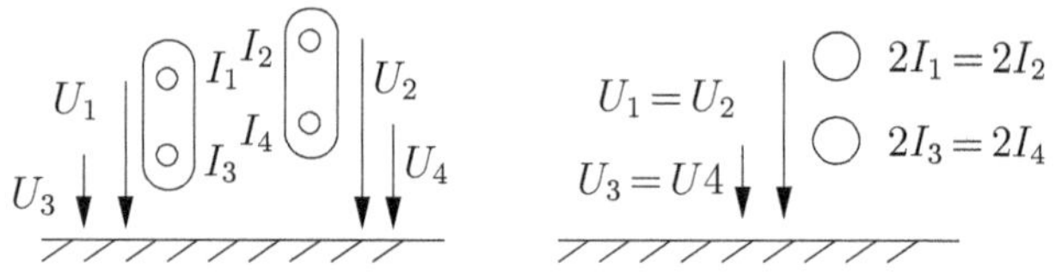

Fig. 5. Voltages and currents of the original (left) and reduced (right) cable.

1, 3 and the wires 2, 4 respectively. We assume that the wires 1, 2 have a similar coupling behaviour and label them group A. Group B consists of the wires 3, 4. If we focus on differential lines, we have to take into account the restriction that it is not useful to allocate both wires of the same wire pair to the same group.

Now, a general rule how to group the wires has to be developed. As already mentioned, the basic arrangement criterion is, that all wires which are placed in one group have to behave the same way (or very similar) in terms of the field to transmission line coupling. The susceptibility strongly depends on the relation of the characteristic impedance and the termination loads. This is why in the referred procedure developed by (Andrieu et al., 2008) the coupling of the wires is characterized by the relation of the common mode terminations to the common mode characteristic impedance Z_{cm}. This rule allows up to four groups: In the first group, both common mode terminations are smaller than the common mode characteristic impedance. In the second group, both terminations are larger than the common mode characteristic impedance. In the third group, e.g. the termination at the left end is smaller and the one at the right end is larger than Z_{cm}. In the fourth group, the size of the terminations is alternated. Following this rule, a differential signal cable with several pairs and without common mode terminations is reduced to a single pair of wires. In applications where identical wires with identical terminations are used (e.g. LAN or any other bus system), we always get a reduction to two groups. As mentioned before, it is not useful to place both wires of a differential pair in one group, even if their susceptibility is very similar.

4 Line parameters of the reduced cable bundle

Now, all the wires in each of these groups have a similar coupling behaviour. In the following we want to reduce each of these groups to single equivalent wires that show the same coupling behaviour as the sum of the single wires of each group. This can be done by making the following assumptions.

1. The induced voltages in each of the wires of one group are the same.
2. All currents in the different wires of one group are equal.

If we compare the original cable bundle with the reduced one (cf. Fig. 4), the following observations can be derived (Fig. 5):

1. The induced voltages in each of the wires of one group of the original cable and the one of the belonging equivalent conductor of the simplified cable are the same.
2. The sum of all currents in all conductors of a group equals the current in the equivalent simplified conductor.

Applying these rules, Eq. (6), which represents the telegrapher's equations for a four-wire cable bundle (cf. Figs. 4, 5), can be simplified:

$$\begin{cases} \dfrac{\mathrm{d}}{\mathrm{d}x}\begin{pmatrix} I_1 \\ I_2 \\ I_3 \\ I_4 \end{pmatrix} = -\mathrm{j}\omega \begin{pmatrix} C_{11} & C_{12} & C_{13} & C_{14} \\ C_{21} & C_{22} & C_{23} & C_{24} \\ C_{31} & C_{32} & C_{33} & C_{34} \\ C_{41} & C_{42} & C_{43} & C_{44} \end{pmatrix}' \cdot \begin{pmatrix} U_1 \\ U_2 \\ U_3 \\ U_4 \end{pmatrix}, \\ \dfrac{\mathrm{d}}{\mathrm{d}x}\begin{pmatrix} U_1 \\ U_2 \\ U_3 \\ U_4 \end{pmatrix} = -\mathrm{j}\omega \begin{pmatrix} L_{11} & L_{12} & L_{13} & L_{14} \\ L_{21} & L_{22} & L_{23} & L_{24} \\ L_{31} & L_{32} & L_{33} & L_{34} \\ L_{41} & L_{42} & L_{43} & L_{44} \end{pmatrix}' \cdot \begin{pmatrix} I_1 \\ I_2 \\ I_3 \\ I_4 \end{pmatrix}. \end{cases} \tag{6}$$

Assuming that group A consists of the wires 1 and 2 and that group B consists of the wires 3 and 4, these equations can be written the following way by applying simple matrix operations:

$$\begin{cases} \dfrac{\mathrm{d}}{\mathrm{d}x}\begin{pmatrix} 2I_1 \\ 2I_3 \end{pmatrix} = -\mathrm{j}\omega \cdot \mathbf{C}'_{\mathrm{red}} \cdot \begin{pmatrix} U_1 \\ U_3 \end{pmatrix}, \\ \dfrac{\mathrm{d}}{\mathrm{d}x}\begin{pmatrix} U_1 \\ U_3 \end{pmatrix} = -\mathrm{j}\omega \cdot \mathbf{L}'_{\mathrm{red}} \cdot \begin{pmatrix} 2I_1 \\ 2I_3 \end{pmatrix}, \end{cases} \tag{7}$$

with:

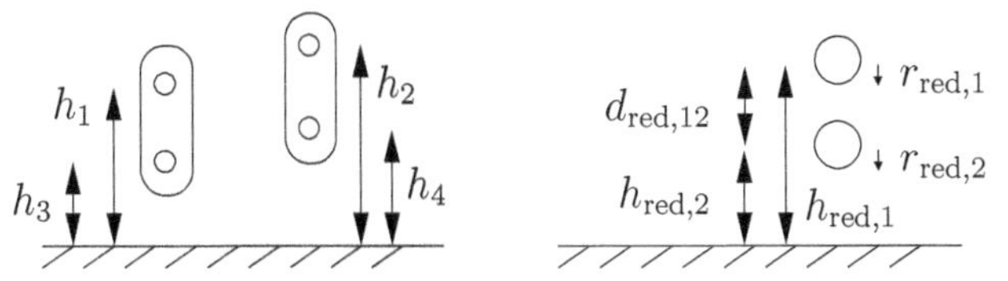

Fig. 6. Geometry of the original (left) and reduced (right) cable.

$$\mathbf{C}'_{\mathrm{red}} = \begin{pmatrix} C_{11}+C_{12}+C_{21}+C_{22} & C_{13}+C_{14}+C_{23}+C_{24} \\ C_{31}+C_{32}+C_{41}+C_{42} & C_{33}+C_{34}+C_{43}+C_{44} \end{pmatrix}',$$

$$\mathbf{L}'_{\mathrm{red}} = \begin{pmatrix} \frac{L_{11}+L_{12}+L_{21}+L_{22}}{4} & \frac{L_{13}+L_{14}+L_{23}+L_{24}}{4} \\ \frac{L_{31}+L_{32}+L_{41}+L_{42}}{4} & \frac{L_{33}+L_{34}+L_{43}+L_{44}}{4} \end{pmatrix}'.$$

This yields a reduced cable bundle with only two equivalent wires with a reduced capacitance matrix $\mathbf{C}'_{\mathrm{red}}$ and a reduced inductance matrix $\mathbf{L}'_{\mathrm{red}}$.

5 Termination loads

If we focus on differential signals, we consider only differential termination loads, too. Arranging the wires of the different pairs to various groups, the differential terminations are set in parallel. So, the resulting terminations of the equivalent pairs equal the parallel connections of the belonging original terminations.

6 Geometry of the reduced cable bundle

Following the procedure presented in the previous chapter, we get the system of differential equations (Eq. 7) that characterizes the reduced cable bundle. If we want to use field simulation tools on the reduced cable bundle we need to extract the resulting geometry of the reduced cable bundle from the determined p.u.l. (per-unit-length) line parameters. This can be done by the following procedure. It was developed for symmetric signal lines, but can also applied to asymmetric ones.

6.1 Determination of estimates

The heights above ground of the equivalent conductors can be estimated by taking the average of the heights of all the conductors of the belonging group (cf. Fig. 6):

$$h_{\mathrm{red},i} = \frac{1}{n}\sum_{k=1}^{n} h_k, \tag{8}$$

with n: number of wires in group i.

The radii of the equivalent wires can be determined by the conversion of the analytical formula of the self inductance $L'_{\mathrm{red},ii}$:

$$L'_{\mathrm{red},ii} = \frac{\mu}{2\pi}\ln\frac{2h_{\mathrm{red},i} - r_{\mathrm{red},i}}{r_{\mathrm{red},i}}. \tag{9}$$

The equation for the mutual inductance (concerning the conductors i, j) can be used to determine the distance $d_{\mathrm{red},ij}$ between two equivalent conductors:

$$L'_{\mathrm{red},ij} = \frac{\mu}{2\pi}\ln\left(\frac{d_{\mathrm{red},ij} - r_{\mathrm{red},j}}{s_2 - r_{\mathrm{red},j}}\right), \tag{10}$$

with:

$$s_2 = \sqrt{(h_{\mathrm{red},i} + h_{\mathrm{red},j})^2 + d^2_{\mathrm{red},ij} - (h_{\mathrm{red},j} - h_{\mathrm{red},i})^2},$$

with $d_{\mathrm{red},ij}$ being the distance between conductor i and conductor j. The distance between the mirror image of conductor i (because of the metallic ground) and conductor j is named s_2.

6.2 Optimization of the geometry

If we use the results of the geometry parameters as described in Sect. 6.1 for further susceptibility calculations of asymmetric signal lines (return conductor: ground), we get quite a good accordance of the sum of the induced currents of the original and the current of the reduced cable bundle. Applying this method to differential cables and particularly to twisted pairs, the accordance is getting very poor. Besides, the results of the first estimation usually are not coherent. Looking at them in detail, we find out that the distance $d_{\mathrm{red},ij}$, calculated from Eq. (10), and the distance which can be derived from the difference of the $h_{\mathrm{red},i}$, $h_{\mathrm{red},j}$ are not equal. This comparison can only be done if we use a geometry where the two wires of a pair are located above each other. This is a restriction if uniform lines are analyzed, but is no problem when stepping over to twisted pairs.

The contradiction of the incoherent geometry parameters can be solved by applying an optimization algorithm on the estimates. The estimated values, calculated above, individually match the corresponding inductance matrix perfectly well (where they were derived from). But the geometry of the simplified cable bundle, associated to the matrix, is not unambiguous and the geometry parameters of the first estimation do not necessarily match to each other. So, the task is, to find a solution where all parameters harmonize and all Eqs. (8, 9, 10) are still fulfilled.

The capacitance matrix $\mathbf{C}'_{\mathrm{red}}$ does not provide further information if a homogene dielectric is assumed, because $\mathbf{L}'_{\mathrm{red}}$ and $\mathbf{C}'_{\mathrm{red}}$ are directly connected via the propagation velocity:

$$\mathbf{C}'_{\mathrm{red}} = \frac{1}{v^2}\mathbf{L}'^{-1}_{\mathrm{red}}. \tag{11}$$

We can state the following conditions, which have to be satisfied by the optimization:

1. The distances of the wires of the reduced cable bundle equal the differences of the first estimates of the heights of the reduced cable bundle, Eq. (8): $d_{\text{red},ij} = |h_{\text{red},i} - h_{\text{red},j}|$. That means that $d_{\text{red},ij}$ remains unchanged during the optimization process. This starting condition guarantees, that the total diameter of the equivalent cable remains the same, compared to the original cable.

2. As $d_{\text{red},ij}$ is fixed, the $h_{\text{red},i}$, $h_{\text{red},j}$ have to be modified to fulfill $d_{\text{red},ij} = |h_{\text{red},i} - h_{\text{red},j}|$.

3. When the optimization process in finished, the final reduced geometry still has to correspond to the the line parameters of the reduced cable bundle (cf. Eqs. 8, 9, 10).

Using these conditions, the optimization procedure is structured as follows:

1. Estimate $h_{\text{red},i}$, $h_{\text{red},j}$ as the average of the original conductors of the corresponding group.

2. Set $d_{\text{red},ij}$ fix as defined in Eq. (1) according to the original geometry.

3. Recalculate the height $h_{\text{red},i}$ of the lower wire i, using $d_{\text{red},ij}$ from Eq. (2) and $h_{\text{red},j}$ by applying Eq. (10).

4. Now, the error can be calculated by comparing the difference of the heights with distance $d_{\text{red},ij}$ Eq. (2).

5. If the error exceeds the predefined tolerance, the height $h_{\text{red},j}$ of the upper wire j is adjusted by adding (subtracting) a portion of the error, depending on the polarity of the calculated error

6. A next iterative loop is started by continuing with Eq. (3).

After some steps and according to the defined tolerance, the curve convergences, as to be seen in Fig. 7.

There are different ways of implementing the conditions mentioned above. They result in a different number of necessary steps to reach the convergence condition. It is very likely to get different results for the geometry of the reduced cable bundle with little differences in coupling behaviour. The variation presented here, is a very simple one, producing quite good results (see Sect. 8).

By applying this code, a second problem to be mentioned is solved, too. During the reduction of the cable bundle, the radii of the simplified cable tend to grow. This depends on the reduction of the L'_{ii} (cf. Eq. 12) when moving from Eq. (6) to Eq. (7). As a consequence it is not possible to apply the cable reduction technique if the original cables have very little distance, since the reduced cable would overlap. When

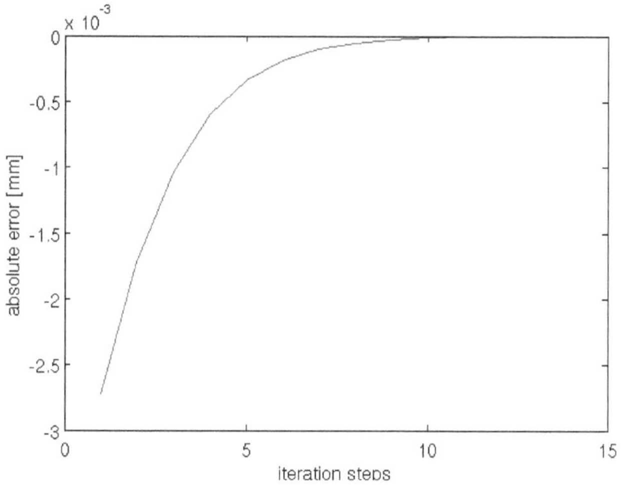

Fig. 7. Error reduction due to recursive iterations.

the geometry is varied recursively in the context of the optimization code, the heights are reduced, compared to the first estimate of the reduced geometry, and as Eq. (12) shows, the result is a reduction of the radii, too:

$$r_{\text{red},i} = \frac{2 \cdot h_{\text{red},i}}{\exp\left(\frac{L'_{\text{red},ii} \cdot 2\pi}{\mu}\right)}. \tag{12}$$

We know from experience that it is even possible to cut away the overlapping radii (without touching the distance) to about 70% causing quite little error.

7 Twisted-pair cables

The application of this procedure on twisted-pair cables seems to be very difficult at the beginning: The procedure described above depends on the line parameters of the original as well as the reduced cable. The determination of these parameters initially requires, by definition, a homogene geometry with a constant distance to the ground. That is not the case for twisted cables. There are several studies that have a close look at the problem how to apply transmission line theory to nonuniform cables. (Nitsch and Gronwald, 1999) uses a generalized telegrapher's equation. (Omid, 1997) adopts a method of an equivalent cascaded network chain. In the context of reduction of twisted-pair cables, a much more simple methodology can be applied and leads to satisfying results.

The approach is based on a chained transmission line analysis too. What we are looking for is the replacement of a large bundle of twisted pairs by a reduced bundle with only one twisted pair (Fig. 8). This is an enormous simplification of the nonuniform structure.

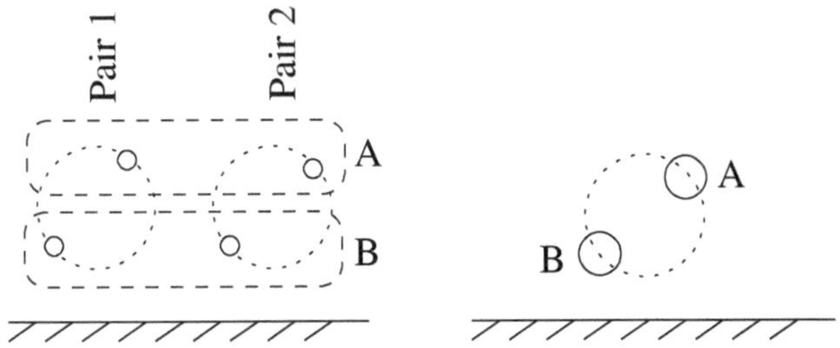

Fig. 8. Principle of reducing twisted-pair cables: Original cable bundle (left) and reduced cable (right).

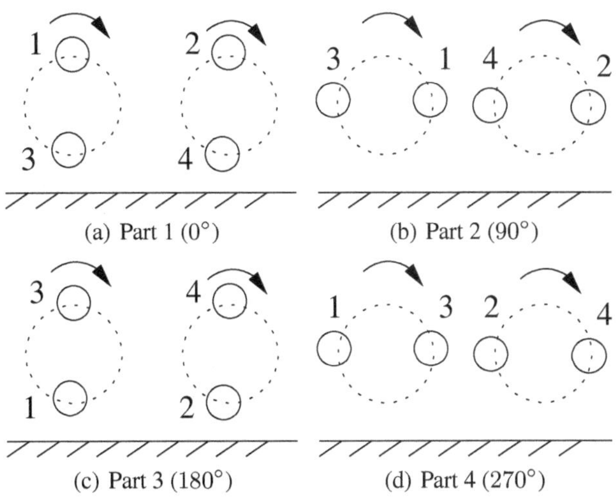

Fig. 9. Stepwise analysis of a 360° twist of a twisted-pair cable.

To demonstrate this simplification approach, we want to analyze one twist of the original cable. The 360° twist is divided exemplarily into four parts (angle of difference $\Delta\alpha = 90°$), as illustrated in Fig. 9. These parts can now be regarded separately. We reduce the twisted pairs in each part of the twist in the way as shown on the left side of Fig. 8. The arrangement of the wires is done in the same way as it is done with uniform cables. When we now have a look at the geometry of the reduced system, we see that the reduction of twisted pairs yields a twisted pair again. This is because the wires we group together always have the same relation to each other in every position of the twist. This can been seen exemplarily in Fig. 9 for four steps. Of course, because of the change of height of each wire, the resulting radii of the reduced cable are not exactly the same. As the difference can be neglected in practical configurations, these radii can be averaged. Hence, it is possible to simplify cable bundles with several twisted pairs to reduced ones with only one pair. Additionally, it is possible to perform the total reduction process and the calculation of the reduced geometry only on the geometry of part 1 of Fig. 9. The results then can be used to model the reduced twisted pair in the CAD environment of the field simulation tool. In this way, it possible to treat twisted-pair cables during the reduction process in the same way as uniform cables, if the pitch length (incl. orientation) as well as the wire distances of the different original pairs are the same. If one of these criteria is violated, the resulting equivalent cable pair is no longer an ideal helix structure and averaging had to be applied.

8 Validation

To verify the presented method, a MATLAB script was implemented to cover the total reduction process starting with the original geometry and ending with the reduced geometry. The total procedure of reduction is structured as follows:

1. Analytic calculation of the line parameters according to the original geometry.
2. Generation of the line parameters of the reduced cable bundle from the original cable bundle.
3. First estimation of the geometry of the reduced cable by applying the same equations as in Eq. (1).
4. Application of the iterative optimization run.
5. Modelling the reduced geometry in a 3D field simulation program (FEKO) and comparing the induced currents to the currents from the simulation run of the original cable.

The basic geometrical parameters of the original cable used for the following validations are:

height of the lower wire	5 mm
wire radius	0.1 mm
distance between twisted wires	0.5 mm
distance between two pairs	0.5 mm
length of twist (for twisted pairs)	20 mm

The excitation of the transmission lines is a vertically polarized plane EM wave with a poynting vector perpendicular to the cable bundle.

First, we want to compare the results for differential pairs of uniform wires. In Fig. 10, the currents in the original and reduced cable bundles are plotted. Note the good accordance between the curves. The amplitudes as well as the resonance peaks match very well.

Figure 11 shows the comparison of the currents in the original and the reduced twisted harness. As we can see, the curves match quite well. There is only a little difference in the resonance peaks. This can be explained by the fact, that by reducing the cable, the propagation velocity is shifted a tiny amount.

$$v^2 = \mathbf{L}'^{-1} \cdot \mathbf{C}'^{-1} \qquad (13)$$

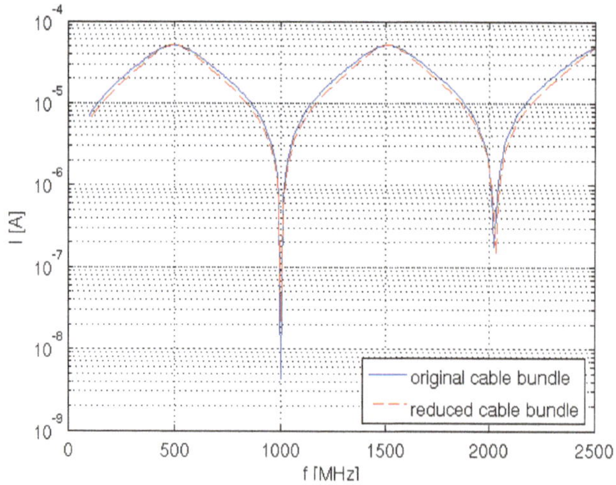

Fig. 10. Comparison of the induced current on a uniform, differential cable.

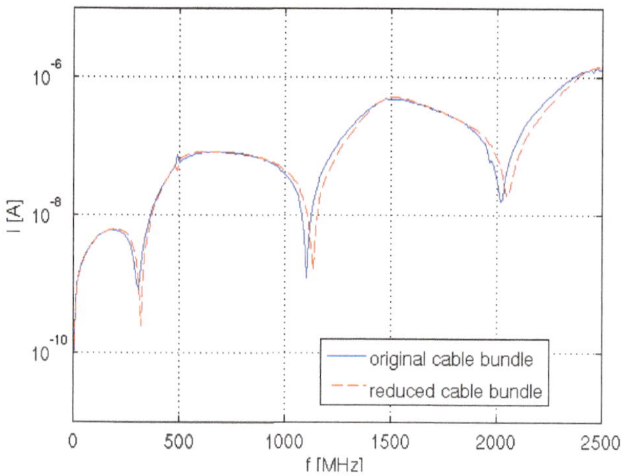

Fig. 11. Comparison of the induced current on a twisted-pair cable.

This is because the product of the p.u.l.-parameters $\mathbf{L}'_{\text{orig}}$ and $\mathbf{C}'_{\text{orig}}$ of the original cable does not equal the product $\mathbf{L}'_{\text{red}}$ and $\mathbf{C}'_{\text{red}}$ of the reduced cable. One way to improve the accordance is to add a dielectric coating to the wires of the reduced cable bundle (Andrieu, 2006). This would affect the capacitance and in that way, the propagation velocity of the reduced cable could be adapted.

Finally, we want to review the improved possibility to apply the reduction method to cables with wires that are situated very close to each other. The advantage can be summarized as a reduced growth of the radii in the context of the simplification. The radii of the simplified cable in this example are $r_{\text{red,est}} = 0.32$ mm if the optimization algorithm is not applied. If the optimized geometry data is used, the radii are reduced to an average of $r_{\text{red,opt}} = 0.21$ mm. This means a reduction to 66% of the original size and the arising space between the wire can be used for a narrower adjustment of the original cable bundle. The agreement between the original and the reduced cable bundles still remains good as shown in Figs. 10 and 11.

9 Conclusions

In this paper an approach was derived to extend the Equivalent Cable Bundle Method to differential cables. We presented an optimization algorithm that enabled the application on twisted-pair cables and the analysis of cable bundles with very little distance between the wires. As shown in the validation chapter, this methodology yields to a good accordance of the simplified cable bundle with the original one. Hence, the methodology allows to perform a simplified modelling of EM fields to differential transmission lines.

References

Andrieu, G.: Elaboration et application d'une méthode de faisceau équivalent pour l'étude des couplages électromagnetiques sur réseaux de câblages automobiles, Ph.D. thesis, Lille University, 2006.

Andrieu, G., Koné, L., Bocquet, F., Démoulin, B., and Parmantier, J.-P.: Multiconductor Reduction Technique for Modeling Common-Mode Currents on Cable Bundles at High Frequency for Automotive Applications, IEEE Transactions on electromagnetic compatibility, 50, 175–184, 2008.

Nitsch, J. and Gronwald, F.: Analytical Solutions in Nonuniform Multiconductor Transmission Line Theory, IEEE Transactions on electromagnetic compatibility, 41, 469–479, 1999.

Omid, M.: Field Coupling to Nonuniform and Uniform Transmission Lines, IEEE Transactions on electromagnetic compatibility, 39, 201–211, 1997.

Paul, C. R.: Analysis of Multiconductor Transmission Lines, Wiley, 1994.

Tesche, F. M.: Principles and applications of EM field coupling to transmission lines, EMC Zurich Symposium, 9, 21–31, 1995.

Tesche, F. M., Ianoz, M. V., and Karlsson, T.: EMC Analysis Methods and Computational Methods, Wiley, 1997.

22

Application of transient pulses on power supply cables by using commercial of the shelf components

J. Hagmann[1], L.-O. Fichte[1], S. Dickmann[1], M. Schaarschmidt[2], and S. Potthast[2]

[1]Helmut-Schmidt-Universität/Universität der Bundeswehr Hamburg, Hamburg, Germany
[2]Wehrwissenschaftliches Institut für Schutztechnologien – ABC-Schutz (WIS), Munster, Germany

Abstract. This paper deals with the development of ultra wide band pulse devices using merchantable hardware. The major goal is to create wide band transients and pulses with simple electronic equipment and to achieve high voltages and low pulse rise times. The propagation of these signals on common power lines are examined to analyse and forecast the effects on attached devices.

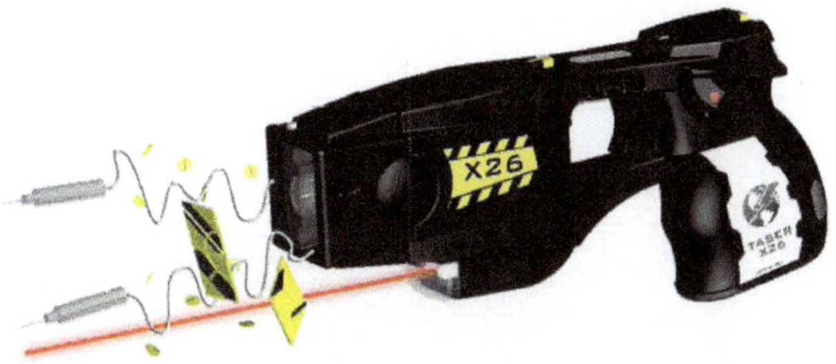

Fig. 1. Taser.

1 Introduction

The basic idea of this research project is the application of pulses on power lines using "commercial of the shelf" components (COTS). Therefore it is necessary to generate wide band transient pulses which can propagate along a power line system and produce cable bounded effects with the attached devices. One of the main elements for this research is a so called Taser (Fig. 1) or stun-gun (Fig. 2), which are used as a high voltage power supply for the forthcoming tests.

The main analysis deals with the disturbing effects of these kinds of power supplies on attached components, e.g. computers and all kinds of electronic devices.

Conducted emission and cable coupling are the main couple mechanisms for this chosen problem. The high voltage transients or disturbances are the source for the IEMI. The power supply cables and data communication cables are used to transfer the transients from the source to the receptor.

2 Stun-gun

This research starts with the analysis of different simple high voltage sources, which are used to generate a high voltage out of simple 9 V batteries. These high voltage sources should be handy to be mobile in a small housing. Therefore, some typical mobile voltage sources were analysed, like stun-guns, tesla-transformers and fence-energizers. The following properties should be complied by the ideal source:

- high output voltage,
- network-independent power supply,
- compact casing box.

Because of the potential high output voltage of ca. 250kV and the compact and lightweight housing a stun-gun, a kind of the electroshock devices, was chosen. The mechanism of action for these devices is based on the well known Taser (Fig. 1). This stun gun fulfils the made demands above. The figure (Fig. 2) shows the components of the examined stun-gun: battery case (9V-Block), dc-ac converter, intermediate circuit with transformer, tesla-stage with spark gap, tesla-transformer, contact electrodes. The stun-gun is easily to obtain on the legal market and they are easily to handle. Older investigations have shown, that these devices have not only been used for the planned application but have also been used for IEMI, e.g. to influence alarm systems and computers by direct contact.

The working mechanism of a stun-gun is shown in the schematic (Fig. 3). The gun is supplied by a

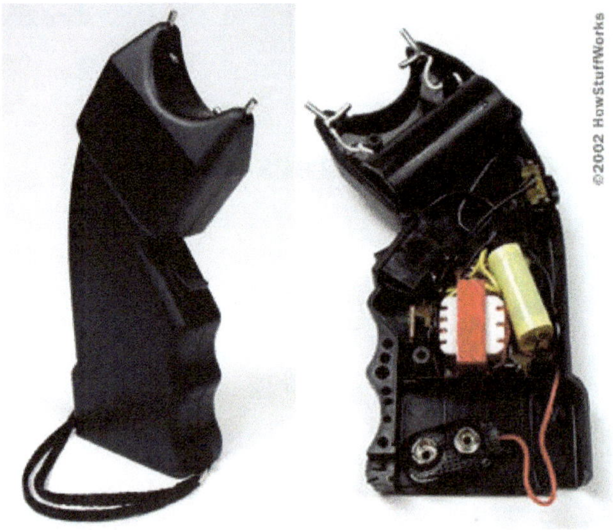

Fig. 2. Stun-Gun.

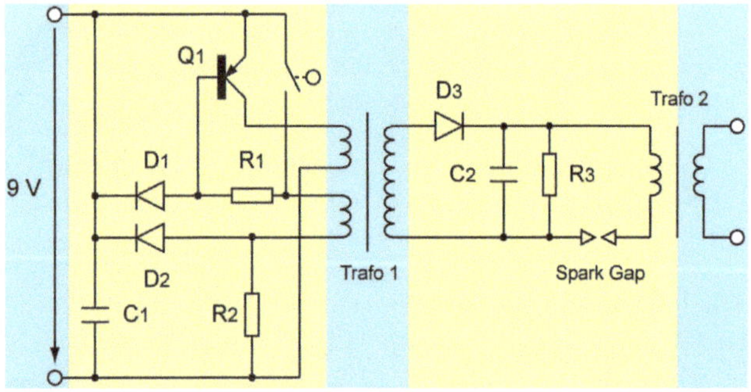

Fig. 3. Schematic of a stun-gun.

9V-block-battery. This dc voltage is converted into an ac voltage by a oscillator circuit. This ac voltage is transformed up to ca. 1.5kV. The second stage is characterized by a tesla transformer. The diode D_3 is used to commutate the high ac voltage and loads the capacitor C_2. The spark gap fires, when the breakthrough voltage of about 1.5kV is reached. While the spark gap fires, the voltage at the electrodes is theoretically transformed up to the range of 100–250 kV, before inner sparks could damage the transformer or internal electronics in the stun gun.

3 Using the stun-gun as pulse source

For the actual application of transient pulses on the power supply cables the stun-gun can be used directly. Therefore the contact electrodes are connected to the power lines and the stun gun pulse is applied on the lines. The effects of disturbances has to be analysed.

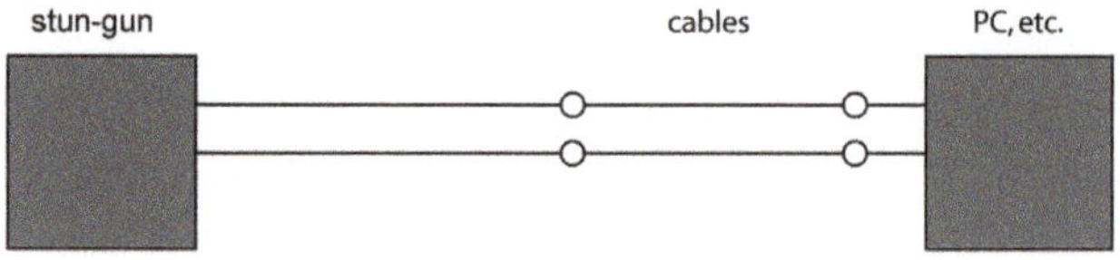

Fig. 4. Stun-gun, directly attached.

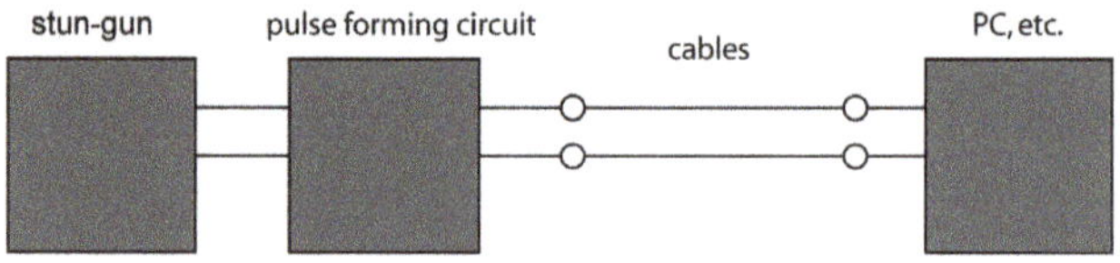

Fig. 5. With pulse forming circuit.

This arrangement (Fig. 4) has some obvious advantages:

- hardly any additional hardware is used,
- the transients are easily to apply,
- high voltage transients can be generated,
- the system is compact and mobile.

The stun gun is commonly designed to interact with biological tissue. Therefore a current (...) is driven through the human body to immobilize a possible attacker. The shape of the pulse is formed to be optimized for the nervous system. The pulses are bipolar and have only a narrow bandwidth in the lower frequency range. For the planned applications as a UWB-pulse-source in the upper frequency range the stun-gun is not useful directly. As an alternative for the direct use of the stun gun a pulse forming network was considered (Fig. 5. This pulse forming stage is loaded by the stun-gun. The stun-gun is used as a current source, which can provide a high voltage. The pulse forming stage is designed to generate a double exponential transient or a damped sinus. Therefore a high amplitude for the transient and a short rise time of the pulse and high frequency range are needed and taken into account.

4 Pulse forming network

The schematic of the used pulse forming circuit is shown in (Fig. 6). A capacitor C is used to store the electric charge. It is charged by a high voltage source via the diode D, with a high breakdown voltage 15kV. The spark gap should fire at a voltage of some 10kV and can be interpreted as a bypass, while firing. The designated voltage for the spark gap can be set by the distance of the electrodes. The pulse is applied on a line, while the capacitor is discharged via a parasitic inductance to the applied line.

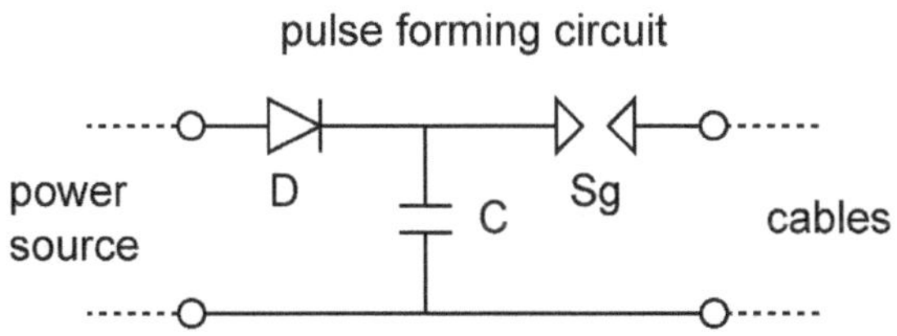

Fig. 6. Pulse forming network.

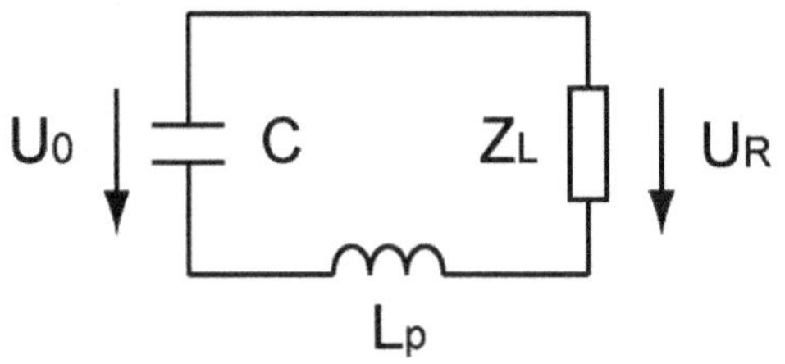

Fig. 7. Parasitic elements.

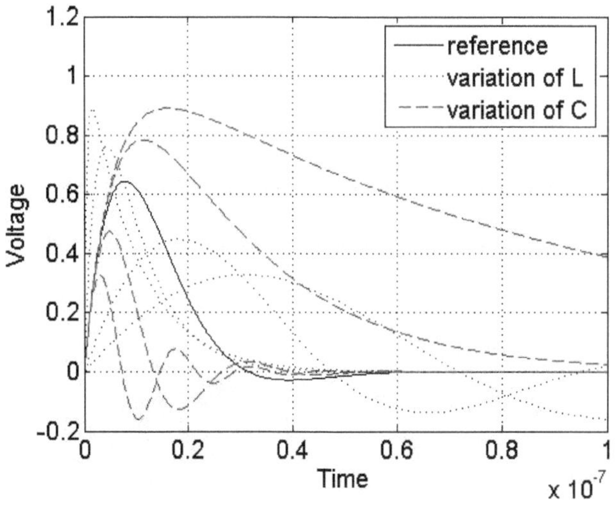

Fig. 8. Double-exponential pulse.

The pulse forming network should be useful for wide frequency band. Therefore it is useful to match the pulse forming network to the line impedance of the attached cables. The circuit has to be compact, due to the parasitic effects of the circuit. Especially the current loop and it's spanned area of the capacitor-sparkgap-circuit form an inductance (Fig. 6), which increases the rise time of the pulse and decreases the frequency range. In the following analysis the line impedance of the lines, which are assumed to be ohmic, are expressed by the resistance R (Fig. 7). The parasitic inductance, which is described by the area spanned by the flowing current, is described by L_p. Which voltage U_R can be applied on the line and how can the voltage be calculated, according to the charge voltage at the capacitor U_0. Using the Laplace's transformation, the solution for this problem can be derieved. The solution starts with:

$$U_R(s) = U_0 \cdot \frac{R}{L} \cdot \frac{1}{s^2 + s\frac{R}{L} + \frac{1}{CL}}. \tag{1}$$

For the transformation of this equation (Eq. 1) from the frequency domain back into the time domain the correspondence tables are used via a substitution:

$$U_R(s) = U_0 \cdot \frac{R}{L} \cdot \frac{1}{(s-s_1)(s-s_2)}. \tag{2}$$

With the according Values:

$$s_{1,2} = -\frac{R}{2L} \pm \sqrt{\frac{R^2}{4L^2} - \frac{1}{CL}}. \tag{3}$$

As a solution in the time domain can be dereived by the transformation:

$$U_R = U_0 \cdot \frac{R}{L} \cdot \frac{1}{\sqrt{\frac{R^2}{4L^2} - \frac{1}{CL}}} \cdot e^{-\frac{R}{2L}t} \cdot \left[\sinh\left(\sqrt{\frac{R^2}{4L^2} - \frac{1}{CL}} \cdot t\right)\right]. \tag{4}$$

In Eq. (4) the common solution for the voltage U_R on the start point of the lines is given. In this equation an exponentiell damping factor and a sinh-function are given with a prefactor according to the voltage amplitude. These combined terms generate a double-exponential function, which can be transformed into a damped sinus according to the circuit elements L, C, R (Fig. 8). It can be seen, that the increase of the inductance L and the decrease of the capacitor C produce a damped sinus.

The expected voltage shape for U_R is a double-exponential discharge and the case for the critical damping. Therefore the squareroot term has to be

$$\sqrt{\frac{R^2}{4L^2} - \frac{1}{CL}} \to 0. \tag{5}$$

Therefore the following inequality is used to determine the case:

$$\frac{R^2}{4} \geq \frac{L}{C}. \tag{6}$$

To achieve high frequency spectral parts, it is necessary to increase the values for s_1 and s_2. Therefore the parasitic inductance of the circuit has to be very small. The other elements have to be balanced to to fulfil the Eqs. (5) and (6). Especially the parasitic inductance L_p and the capacitor C have to be balanced.

5 Design

The main efforts at the design of the pulsforming circuit are: compact design and housing, low parasitic inductance, low peeking-capacitor, high voltage capability of all elements

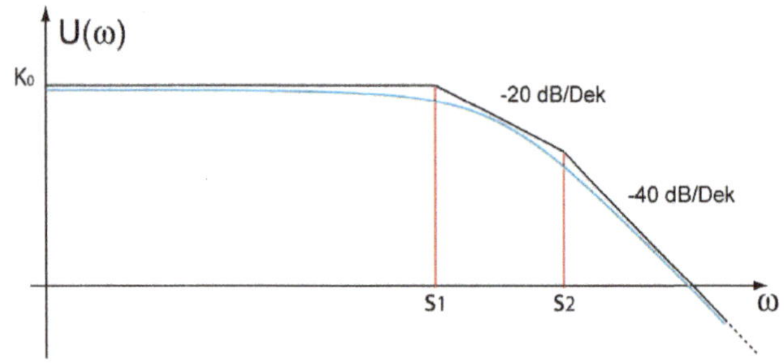

Fig. 9. Pulse in frequency domain.

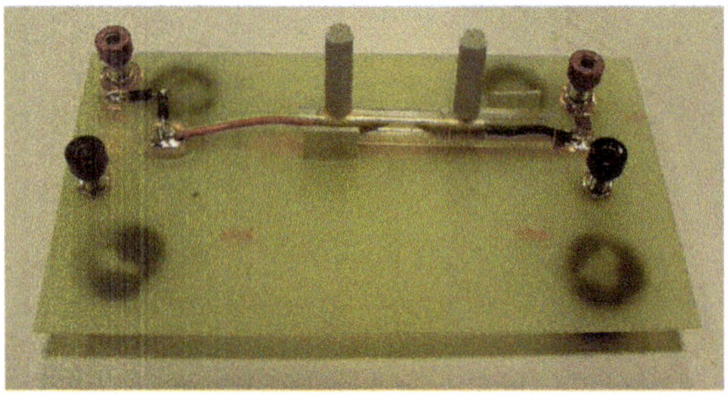

Fig. 10. Simle pulse forming circuit.

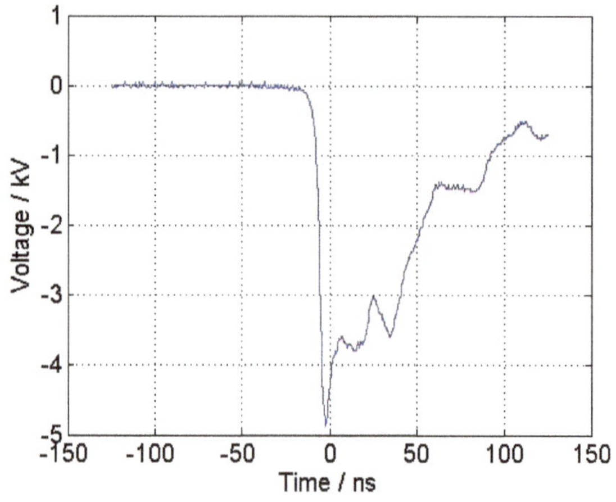

Fig. 11. Measured transient.

Fig. 12. Pulse on the line.

and the PCB layers, matching the line impedance of the cables to be attached. The Fig. 10 shows a simple pulse forming network of the first gerneration. The Elements used for this circuit can be seen. These elements include a diode, capacitor and a spark gap together with some contact clamps for voltage supply and for the cables to be attached.

It is possible to generate double-exponential transient pulses, which can propagate on the lines to a destinated receptor. The pulsforming stage is charged onto a voltage of ca. 5kV and the spark gap fires. The Fig. 11 shows the measured transient on the line. This measured pulse shows a voltage amplitude of 4.9kV and a rise time of 4.5ns. During the measurements the capacitive probe P6015 was used. This probe can only resolute a minimum rise time of 4.5ns. Therefore the transient's rise time can be lower than the measured.

6 Transient on lines

The next analysis deals with the influence of the cables on the pulses and according to the quality of the lines with the attenuation and dispersion along the lines.

In principle, the following tasks have to be analysed:

- breakthrough voltage
- attenuation
- dispersion
- line impedance

To determine the influence of the cables on the pulses, NYM (1.5mm^2) power supply cables were used, which are commonly used as power supply cables in the civil households. The line impedance of these lines are ca. $Z_L = 82\Omega$ for flexible cable and $Z_L = 75\Omega$ for semi rigid cables. To determine the attenuation for the cables measurements were done (Fig. 13). The measured values depend on the measured scatter-parameters of the line and show the reflection parameter S_{11} and the transmission parameter S_{12}, which represents the attenuation of the CW-signal.

The analysis shows, that the used test cable of 1m length provides a linear damping factor depending on the frequency. At the frequency of 1GHz a damping factor of 8dB can be measured, while at a frequency of 200MHz an attenuation of 2dB can be determined. This lines are principally applicable up to a frequency of about 500MHz, because the most relevant cable conducted effects in the attached devices are determined in the frequency range below $f_0 = 300\text{MHz}$. Assuming linear behaviour of the line system, we can determine dispersion on the line system due to the propagation velocity, which is frequency dependent. This dispersion effect influences the shape of the transient pulse along the line. The rise time t_A of the pulse increases. To determine the nonlinear behaviour of the lines, pulses with different frequency ranges and amplitudes were applied on the power lines. The pulses at the front and back side of the 10m-line have the same shape, but differences in running time of ca. 50ns (Fig. 14).

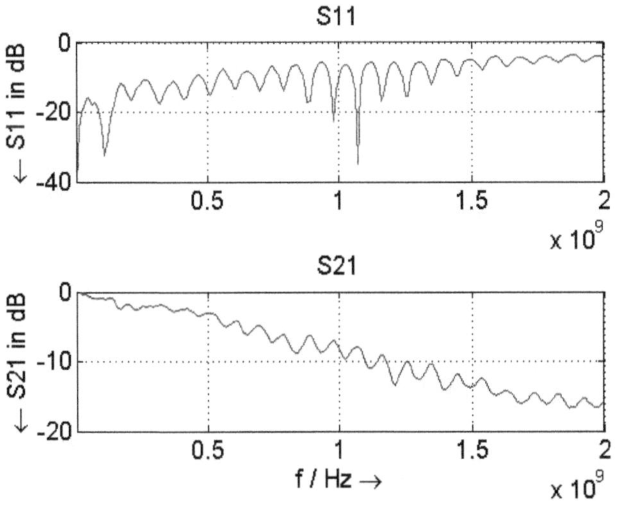

Fig. 13. S-Parameters of a NYM-cable (1 m).

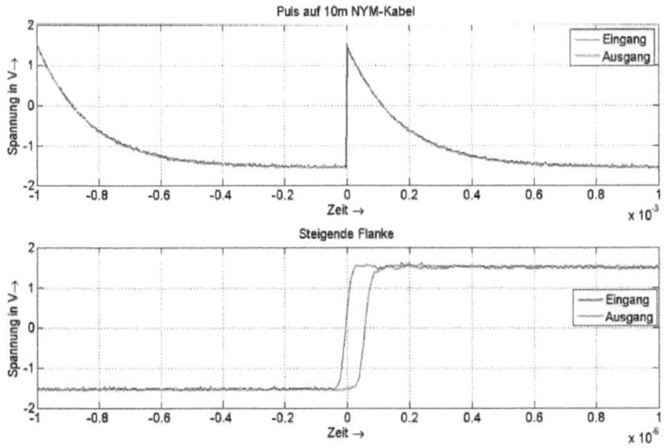

Fig. 14. Pulse propagation on the lines.

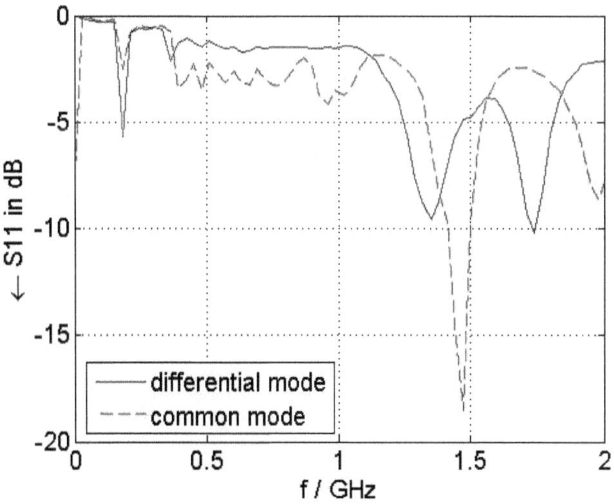

Fig. 15. Filter quality, measured between phase and ground.

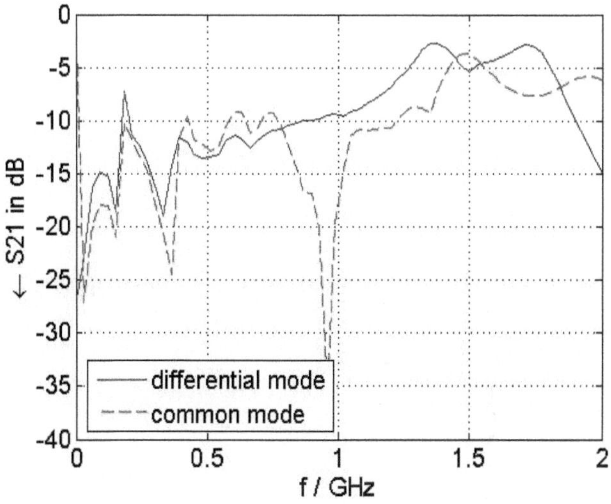

Fig. 16. Filter quality, measured between phase and ground.

7 Filtering

Common netfilters in personal computers and electronic devices are designed to filter surges and harmonics near the net frequency $f = 0.1 - 2\text{kHz}$ from the supply voltage. Another goal of this research is to generate pulses with spectral parts in frequnecy ranges, which are not filtered by the net filters to pass over to the sensitive stages of a device. The next picture shows the filter quality S_{12} of a common personal computer.

The filter shows different behaviour under a common mode and a differential mode source. Some gaps in the filtering ability of the filter can be seen (Figs. 15, 16). Using a differential mode signal a gap in the filtering ability is shown at about 200 MHz. The attenuation effect of the filter is low in the frequency range of $1 - 2\text{GHz}$ (Fig. 15). Using a common mode some other frequency ranges below 1 GHz with a low attenuation below 10dB can be found (Fig. 16). These analysis of the filter attenuation depends on CW signals and steady state vibrations of the filters. To include the transient oscillation of the capacities of the filters, it is necessary to analyse the filters in the time domain. The transient pulses can be that fast, that the filters are not able to react. This is a field for further research.

8 Conclusions

In this paper some steps to produce a simple pulse generator were shown. The generated pulses have a high voltage and low puse rise times. Therefore the pulses have a wide frequency bandwidth. For the application in the case of IEMI, some power supply cables were attached and filters were analysed to determine the propagation behaviour of the transients on these lines.

References

Benford, J., Swegle, J. A., and Schamiloglu, W.: High Power Microwaves, Second Edition, Taylor & Friends, New York, 2007.

Hilberg, W.: Impulse auf Leitungen, Oldenbourg, Mnchen, 1981.

Hölzler, E. and Holzwarth, H.: Pulstechnik, Band 1, Springer, Berlin, 1982.

Hölzler, E. and Holzwarth, H.: Pulstechnik, Band 2, Springer, Berlin, 1984.

Mesyats, G. A.: Pulsed Power, Springer, Netherlands, 2004.

Paul, C. R.: Analysis of Multiconductor Transmission Lines, Wiley Series in Microwave and Optical Engineering, John Wiley & Sons, 1994.

Paul, C. R.: Introduction to Electromagnetic Compatibility, Wiley Series in Microwave and Optical Engineering, John Wiley & Sons, 2006.

Schildt, G. H.: Impulstechnik: Grundlagen und Anwendungen, Lyk Informationstechnik, 2008.

Zinke, O. and Brunswig, H.: Hochfrequenztechnik I: Hochfrequenzfilter, Leitungen, Antennen, Springer, Berlin, 1999.

23

Computation of antenna pattern correlation and MIMO performance by means of surface current distribution and spherical wave theory

O. Klemp, G. Armbrecht, and H. Eul

Universität Hannover, Institut für Hochfrequenztechnik und Funksysteme, Hannover, Germany

Abstract. In order to satisfy the stringent demand for an accurate prediction of MIMO channel capacity and diversity performance in wireless communications, more effective and suitable models that account for real antenna radiation behavior have to be taken into account. One of the main challenges is the accurate modeling of antenna correlation that is directly related to the amount of channel capacity or diversity gain which might be achieved in multi element antenna configurations. Therefore spherical wave theory in electromagnetics is a well known technique to express antenna far fields by means of a compact field expansion with a reduced number of unknowns that was recently applied to derive an analytical approach in the computation of antenna pattern correlation. In this paper we present a novel and efficient computational technique to determine antenna pattern correlation based on the evaluation of the surface current distribution by means of a spherical mode expansion.

1 Introduction

In this article the analysis of the multiple-input multiple-output (MIMO) transmission behavior of antenna modules will be evaluated based upon an expansion of antenna surface current distribution represented by a series expansion involving spherical eigenmodes as in Stratton (1941) and Werner and Mittra (2000). The latter mathematical description yields a representation of antenna radiation patterns with a reduced order and therefore provides the advantage of an analytical approach in the description of multielement antenna transmission parameters. Therefore we present a modified technique of spherical mode expansion as in Chen et al. (1992) based upon a computation of vector expansion coefficients of the field approach that were directly derived by the surface-current distribution on the antenna elements taken into consideration. Based upon a suitable statistical model to include for the spatial properties of a multipath propagation scenario as given in Jakes (1974), a closed form representation for antenna correlation may be derived as shown in Leifer (2002) and Klemp and Eul (2005). This representation of antenna correlation may be directly related to the computation of MIMO channel capacity as for the description of MIMO communication systems employing a certain number of antenna elements at the transmitter and at the receiver.

The computational models will be evaluated considering symmetrical planar, logarithmically-periodic four-arm antennas of different geometrical shapes that were designated for broadband operations in the frequency range between 1.0 GHz and 6.0 GHz as in Klemp et al. (2005). The antenna elements taken into consideration provide four symmetrical shaped and metalized antenna arms. Each adjacent pair of antenna arms represents one transmit- or receive branch of a MIMO-antenna system providing dual-linear polarization if fed with an opposite phase relation. Surface currents will be determined from appropriate computer models applying a conventional field simulator based on the finite-element-method (FEM). Applying simplified stochastic models of the radio channel that account for the spatio-temporal properties of multipath fading, we will extend our studies to the computation of MIMO channel capacities furthermore. The results of the applied current-based approach as well as the spherical mode expansion of the surface current distribution is accomplished according to Armbrecht et al. (2006) and compared to the standard methods given in Taga (1990) using a numerical integration in the far field domain of the considered antenna elements. Based on the results of convergence studies we will show the agreement between the different methods for the computation of antenna pattern correlation and MIMO channel capacity. The presented analysis establishes an explicit separation between antenna related properties and channel dependent quantities and could be therefore favorably applied to an optimization of antenna correlation

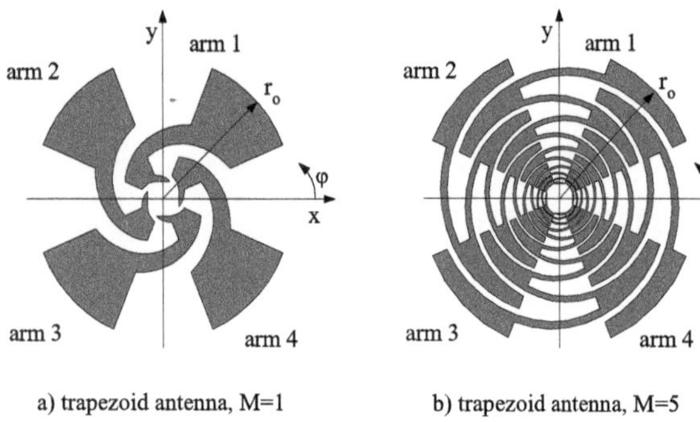

Fig. 1. Trapezoidal four-arm antenna elements with $M = 1$ and $M = 5$ periods of the trapezoidal unit cell.

or MIMO capacity. Based on the field expansion in terms of spherical eigenmodes, an investigation of the modal contributions to the overall antenna correlation can be performed additionally. In particular the analysis can be used to predict the MIMO performance of antenna elements in distinct channel scenarios by simply specifying their spectra of spherical eigenmodes.

The article is organized as follows, Sect. 2 summarizes the properties of self-complementary, four-arm, log.-per. planar antennas for applications with dual polarization. In Sect. 2.1 we review the fundamental properties of the spherical mode expansion technique with respect to the electric- and magnetic vector potential of a given elementary source distribution and apply this approach to the surface current distribution of the antenna elements taken into consideration. Section 3 is used to introduce the basic principles of MIMO transmission involving a channel structure based on the Kronecker model. Computing the modal representation of antenna radiation patterns as in Sect. 3.1 we derive an analytical approach for antenna power correlation based on the surface currents on the interactive antenna elements. The presented analysis is adopted to the computation of MIMO channel capacity in Sect. 4 for dual-polarized log.-per. planar four-arm antennas on each side of the communication link. Section 5 concludes this article.

2 Antennas for broadband polarization diverse transmission

Logarithmically-periodic antenna elements may theoretically provide frequency independent transmission characteristics (DuHamel and Isbell, 1957; Rumsey, 1957) if they adhere to the principle of self-complement. Designed as a planar logarithmically-periodic four-arm antenna element it offers dual linear polarization as shown in e.g. Klemp et al. (2005). Dual linear operation results from the odd-phase excitation of each of the two pairs of adjacent antenna arms. Antenna elements of that kind can be applied in a broadband antenna system for MIMO operations using dual orthogonal polarizations. In comparison to antenna configurations exploiting spatial diversity concepts, polarization diverse air interfaces provide independent branches for signal reception with a much smaller spatial extent.

In this paper, the planar, log.-per. four-arm trapezoid antenna as in Klemp et al. (2004) is used in order to demonstrate its capabilities as a compact polarization-diverse antenna element for MIMO communications based on an eigenmode characterization of its surface current distribution. Two different geometrical types of the trapezoid antenna will be investigated according to Fig. 1. Referring to Klemp et al. (2005) the trapezoid antennas taken into consideration are given by $M = 1$ and $M = 5$ periods of the trapezoidal unit cell that provides a trapezoidal slew rate of 50%.

The outer radii of the antenna elements amount to $r_o = 50$ mm and the antenna radiation behavior will be evaluated in free-space.

In order to derive a compact representation of antenna radiation behavior, the field quantities can be expanded by a compact set of spherical eigenmodes whose weights are computed from the appropriate current distribution on the antenna surface (e.g. Chen et al. (1992)). This field expansion yields a distinct set of vector expansion coefficients that is solely dependent on the excitation and the geometrical shape of the antennas taken into consideration. Therefore the spherical mode expansion may yield a significant order reduction in the description of antenna radiation behavior that speeds up computations of field-related quantities as e.g. the computation of antenna correlation and MIMO channel capacities. In Sect. 2.1 the results of the field expansion based on the surface current distribution are summarized.

2.1 Spherical mode spectra of log.-per. antenna elements

Following the procedure of spherical mode expansion for planar, log.-per. antenna elements based on its surface current distribution as given by Armbrecht et al. (2006), this section summarizes the spectra of spherical eigenmodes for the log.-per. trapezoid antennas with $M = 1$ and $M = 5$ periods of the trapezoidal unit cell as shown in Fig. 1.

As an example, Fig. 2 shows the normalized SME coefficients at a simulation frequency of 3 GHz. The normalization is performed with respect to the absolute value of the expansion coefficient of degree $m = 0$ and order $n = 0$, $\underline{a}_{00}$, which yields normalized weights $\|\underline{a}_{nm}\| \,/\, \|\underline{a}_{00}\|$ in terms of expansion coefficients related to even spherical harmonics $Y^e_{nm}(\vartheta, \varphi)$ and $\|\underline{b}_{nm}\| \,/\, \|\underline{a}_{00}\|$ for expansion coefficients of odd spherical harmonics $Y^o_{nm}(\vartheta, \varphi)$ (see Stratton, 1941).

As can be seen from the decaying behavior of modal contributions with increasing degree and order of the harmonics, it becomes feasible to limit the number of modes in the field expansion with respect to a finite value N. Therefore it is advantageous to conduct a radiation pattern expansion by spherical eigenmodes because the derived set of expansion coefficients yields an approximative low-order characterization of the radiation profile of the related antenna structure.

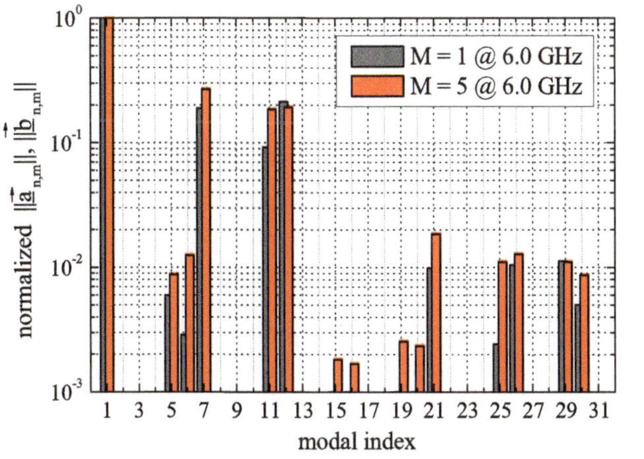

Fig. 2. Normalized coefficients of the spherical mode expansion for trapezoid antennas $M = 1$ and $M = 5$ at 3 GHz. See Armbrecht et al. (2006) for the appropriate enumeration of the abscissa.

3 Multiple-input multiple-output antenna systems

Multiple-input multiple-output antenna systems may efficiently exploit the spatial domain by parallel subchanneling employing decoupled transmit- and receive paths as shown in Foschini (1996), Yu and Ottersten (2002). Due to the ability of transmitting simultaneous data streams incorporating antenna arrays at both sides of the radio link, a great increase of channel capacities in wireless communications may be achieved. The establishment of parallel subchannels is a result of uncorrelated fading processes between individual transmission paths of the communication link and can be a result of spatial-, pattern- or polarization diverse properties of the applied antenna elements. Depending on the properties of the regarded multipath channel, the application of dual-polarized antenna modules at the receiver and at the transmitter is possible. Compared to MIMO techniques exploiting spatial diversity concepts, using polarization-diverse air interfaces leads to the design of compact transmission terminals with enhanced transmission capacity.

Figure 3 shows the model of the considered wireless MIMO system incorporating log.-per. antenna elements on both sides of the radio link. The four-arm antenna structures provide orthogonal-polarized operational modes by feeding one of the two adjacent antenna arm pairs with an odd phase relation. Using one antenna element with dual linear polarization at the transmit (TX) side and one antenna element with dual linear polarization at the receive (RX) side yields a (2×2)-MIMO communication system.

Given frequency flat signal vectors of the TX- and the RX antenna array, $\mathbf{x}$ and $\mathbf{y}$, the channel matrix equation of the corresponding frequency flat channel following Yu and Ottersten (2002) is given as follows:

$$\mathbf{y} = \mathbf{H}\,\mathbf{x} + \mathbf{n}\,. \tag{1}$$

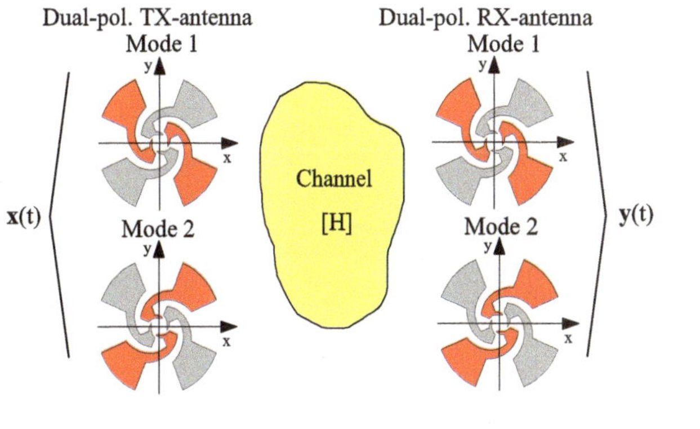

Fig. 3. Model of the MIMO radio channel incorporating n^{TX} antennas at the TX-side and n^{RX} antennas at the RX-side.

$\mathbf{H}$ is the narrowband transmission matrix of the MIMO channel and $\mathbf{n}$ describes zero-mean additive white Gaussian noise with a variance of σ_n^2. The complex correlation coefficient between independent operational modes i and j at the transmitter- or at the receiver side is given in accordance with Taga (1990) for incident phase centers as:

$$\rho_{\mathrm{i,j}}^{\mathrm{TX,RX}} = \frac{R_{\mathrm{i,j}}^{\mathrm{TX,RX}}}{\sqrt{\sigma_{\mathrm{i}}^{\mathrm{TX,RX}^2}\,\sigma_{\mathrm{j}}^{\mathrm{TX,RX}^2}}}\,. \tag{2}$$

From Eq. (2) antenna correlation $\rho_{\mathrm{i,j}}^{\mathrm{TX,RX}}$ depends on antenna covariance $R_{\mathrm{i,j}}^{\mathrm{TX,RX}}$ and the variances $\sigma_{\mathrm{i}}^{\mathrm{TX,RX}^2}$ and $\sigma_{\mathrm{j}}^{\mathrm{TX,RX}^2}$. $R_{\mathrm{i,j}}$ can be computed from the radiation patterns $\boldsymbol{C}(\vartheta,\varphi)$ of operational mode i and operational mode j, as follows:

$$\begin{aligned} R_{i,j} = \mathrm{K} \int_0^{2\pi}\!\!\int_0^{\pi} & \Big[C_{\vartheta,\mathrm{i}}(\vartheta,\varphi)\, C^*_{\vartheta,\mathrm{j}}(\vartheta,\varphi) \\ & + \mathrm{XPD}\, C_{\varphi,\mathrm{i}}(\vartheta,\varphi)\, C^*_{\varphi,\mathrm{j}}(\vartheta,\varphi)\Big]\, p_{\vartheta,\varphi}(\vartheta,\varphi) \sin\vartheta\, d\vartheta\, d\varphi\,. \end{aligned} \tag{3}$$

In Eq. (3), $p_{\vartheta,\varphi}(\vartheta,\varphi)$ denotes a two-dimensional probability density function that accounts for the spatial property of outgoing waves at the transmitter and impinging waves at the receiver. The ratio of the mean incident power in vertical direction to the mean incident power in horizontal direction at the locations of receiver and transmitter is given by the cross-polarization discrimination XPD. '*' means conjugate operation and K is a proportionality constant. Replacing $i = j$ in Eq. (3) yields the variances of the operational modes i and j, $\sigma_{\mathrm{i}}^{\mathrm{TX,RX}^2}$, $\sigma_{\mathrm{j}}^{\mathrm{TX,RX}^2}$, respectively.

The correlation values between any possible combination of adjacent port-related antenna patterns at the TX- and at the RX-side will be combined to correlation matrices $\mathbf{R}^{\mathrm{TX}}$ and $\mathbf{R}^{\mathrm{RX}}$ (Kermoal et al., 2002). Performing the Kronecker-product between the correlation matrices, $\mathbf{R}^{\mathrm{TX}} \otimes \mathbf{R}^{\mathrm{RX}}$ yields the correlation properties of the channel matrix $\mathbf{H}$ of the communication link. In order to determine the theoretical channel capacity of the proposed MIMO radio link, Monte-Carlo simulations are performed, involving $N_{\mathbf{H}} = 5000$ realizations of **H**. Initially, the elements of **H** are assumed to be uncorrelated zero-mean complex Gaussian variables with a unit variance. The appropriate correlation values between the channel coefficients are computed from matrix multiplication with the correlation matrix $\mathbf{R}^{\mathrm{TX}} \otimes \mathbf{R}^{\mathrm{RX}}$ as shown in Kermoal et al. (2002).

Provided a uniform power allocation between the independent operational modes at the transmit antenna, the ergodic MIMO channel capacity s given as a sum over K subchannels:

$$C = \sum_{k=1}^{K} \log_2 \left(1 + \lambda_{\mathrm{k}} \frac{P_{\mathrm{k}}}{\sigma_n^2}\right) . \tag{4}$$

In Eq. (4), P_{k} denotes the amount of power assigned to the kth subchannel and σ_n^2 is the total noise power at the receiver. λ_{k} gives the eigenvalue of the k-th subchannel.

3.1 SME-based computation of antenna pattern correlation

In order to simplify the computation of antenna pattern correlation and MIMO channel capacity which are directly related to the radiation patterns of the involved antenna elements as shown in Sect. 3, a spherical mode expansion of the surface current distribution will be accomplished. Due to the fact, that the radiation patterns of each antenna element may be described by a distinct set of modal expansion coefficients this analysis is predestined to provide a deeper insight into the radiation mechanisms that cause high values of pattern correlation and low channel capacities. Furthermore it will become possible to optimize delimiting antenna geometries with respect to certain cost functions of related transmission parameters that take into account the real antenna radiation behavior.

As shown in Armbrecht et al. (2006), antenna radiation patterns $\boldsymbol{C}(\vartheta, \varphi)$ of an arbitrary antenna geometry can be expressed by an expansion of the current distribution on the respective antenna geometry by means of an orthogonal set of spherical basis functions as shown in Eq. (5). The subsequent analysis is based on planar antenna geometries, whose metalization surfaces are located in parallel to the xy-plane of a cartesian coordinate system. Surface current distributions $\boldsymbol{J}(x', y') = \left[J_x(x', y'), J_y(x', y')\right]$ (where x' and y' indicate the cartesian coordinates of the source region) are derived using a conventional full-wave simulator based on the finite-element-method (FEM). Assuming far-field conditions of the electromagnetic field, the spherical mode expansion of the surface current distribution yields:

$$\begin{bmatrix} C_{\vartheta,i} \\ C_{\varphi,i} \end{bmatrix} = -jk\eta \sum_{n=0}^{\infty} \sum_{m=0}^{n} j^{n} \begin{bmatrix} \cos\vartheta \cos\varphi & \cos\vartheta \sin\varphi \\ -\sin\varphi & \cos\varphi \end{bmatrix} \begin{bmatrix} \underline{a}^{x,i}_{n_i m_i} & \underline{b}^{x,i}_{n_i m_i} \\ \underline{a}^{y,i}_{n_i m_i} & \underline{b}^{y,i}_{n_i m_i} \end{bmatrix} \begin{bmatrix} Y^{e}_{n_i m_i}(\vartheta, \varphi) \\ Y^{o}_{n_i m_i}(\vartheta, \varphi) \end{bmatrix} . \tag{5}$$

In Eq. (5), spherical harmonics $Y^e_{nm}(\vartheta, \varphi)$ and $Y^e_{nm}(\vartheta, \varphi)$ are mapped to the spherical components $C_{\vartheta,i}$, $C_{\varphi,i}$ of antenna radiation pattern i incorporating a field expansion matrix including vector expansion coefficients $\underline{\boldsymbol{a}}^{i}_{n_i m_i}$, $\underline{\boldsymbol{b}}^{i}_{n_i m_i}$ as derived from the surface current distribution of operational mode i and an additional matrix that accounts for coordinate transformation. The coordinate transform of the cartesian components of the surface currents $J_x(x', y')$ and $J_y(x', y')$ into spherical components is mandatory in order to apply the Fraunhofer approximation of the electromagnetic field (Lo and Lee, 1988). For convenience, in Eq. (5), the upper infinite summation index of the spherical eigenfield decomposition has to be replaced by a finite value N leading to impairments in the synthesized far-field representation. Those impairments can be reduced by adhering to the convergence criteria for spherical wave expansions as in Narasimhan et al. (1985). Therefore a sufficient degree of field approximation is derived for $N \geq ka$, a denoting the radius of a sphere that is completely enclosing the considered antenna and k is the scalar wave number.

Splitting the integral for covariance computation $R_{i,j}$ as given in Eq. (3) into two parts, $R^{\vartheta}_{i,j}$ and $R^{\varphi}_{i,j}$ that account for the contributions of antenna covariance in the directions of ϑ and φ and replacing $\boldsymbol{C}(\vartheta, \varphi)$ by the eigenfield decomposition as given in Eq. (5), it follows:

$$R^{\vartheta,\varphi}_{i,j} = \mathrm{K} \sum_{n_i=0}^{\infty} \sum_{m_i=0}^{n_i} \sum_{n_j=0}^{\infty} \sum_{m_j=0}^{n_j} j^{n_i} j^{n_j^*} \left\{ \mathbf{T}^{ee} \begin{bmatrix} I^{ee,M_1^{\vartheta,\varphi}}_{n_i m_i, n_j m_j} \\ I^{ee,M_2^{\vartheta,\varphi}}_{n_i m_i, n_j m_j} \\ I^{ee,M_3^{\vartheta,\varphi}}_{n_i m_i, n_j m_j} \end{bmatrix} + \mathbf{T}^{eo} \begin{bmatrix} I^{eo,M_1^{\vartheta,\varphi}}_{n_i m_i, n_j m_j} \\ I^{eo,M_2^{\vartheta,\varphi}}_{n_i m_i, n_j m_j} \\ I^{eo,M_3^{\vartheta,\varphi}}_{n_i m_i, n_j m_j} \end{bmatrix} + \mathbf{T}^{oe} \begin{bmatrix} I^{oe,M_1^{\vartheta,\varphi}}_{n_i m_i, n_j m_j} \\ I^{oe,M_2^{\vartheta,\varphi}}_{n_i m_i, n_j m_j} \\ I^{oe,M_3^{\vartheta,\varphi}}_{n_i m_i, n_j m_j} \end{bmatrix} + \mathbf{T}^{oo} \begin{bmatrix} I^{oo,M_1^{\vartheta,\varphi}}_{n_i m_i, n_j m_j} \\ I^{oo,M_2^{\vartheta,\varphi}}_{n_i m_i, n_j m_j} \\ I^{oo,M_3^{\vartheta,\varphi}}_{n_i m_i, n_j m_j} \end{bmatrix} \right\} . \tag{6}$$

In Eq. (6), $I^{e,oe,o,M_\xi^{\vartheta}}_{n_\nu m_\nu, n_\mu m_\mu}$ (with $\nu = i, j$, $\mu = i, j$ and $\xi = 1, \ldots, 3$) denote coupling integrals between two spherical harmonics $Y^{e,o}_{n_j m_j}(\vartheta, \varphi)$ in a distinct channel scenario $p_{\vartheta,\varphi}(\vartheta, \varphi)$, weighed by a field-point dependent factor $M_\xi^{\vartheta}(\vartheta, \varphi)$ as follows:

$$\begin{Bmatrix} I^{e,oe,o,M_\xi^{\vartheta}}_{n_\nu m_\nu, n_\mu m_\mu} \\ I^{e,oe,o,M_\xi^{\varphi}}_{n_\nu m_\nu, n_\mu m_\mu} \end{Bmatrix} = \int_{\varphi=0}^{2\pi} \int_{\vartheta=0}^{\pi} \begin{Bmatrix} M_\xi^{\vartheta}(\vartheta, \varphi) \\ M_\xi^{\varphi}(\vartheta, \varphi) \end{Bmatrix} Y^{e,o}_{n_i m_i}(\vartheta, \varphi)\, Y^{e,o}_{n_j m_j}(\vartheta, \varphi)\, p_{\vartheta,\varphi}(\vartheta, \varphi) \sin\vartheta \, d\vartheta \, d\varphi . \tag{7}$$

The antenna-dependent scaling factors $\mathbf{T}^{ee}$, $\mathbf{T}^{eo}$, $\mathbf{T}^{oe}$ and $\mathbf{T}^{oo}$ in Eq. (6) are given entirely with respect to the field expansion coefficients:

$$\mathbf{T}^{ee} = \begin{bmatrix} \underline{a}^{x,i}_{n_i m_i} \underline{a}^{x,j*}_{n_j m_j} \\ \underline{a}^{x}_{n_i m_i} \underline{a}^{y*}_{n_j m_j} + \underline{a}^{y,i}_{n_i m_i} \underline{a}^{x,j*}_{n_j m_j} \\ \underline{a}^{y,i}_{n_i m_i} \underline{a}^{y,j*}_{n_j m_j} \end{bmatrix} , \tag{8}$$

$$\mathbf{T}^{eo}=\begin{bmatrix}\underline{a}^{x,i}_{n_im_i}\underline{b}^{x,j^*}_{n_jm_j}\\ \underline{a}^{x}_{n_im_i}\underline{b}^{y^*}_{n_jm_j}+\underline{b}^{y,i}_{n_im_i}\underline{a}^{x,j^*}_{n_jm_j}\\ \underline{a}^{y,i}_{n_im_i}\underline{a}^{y,j^*}_{n_jm_j}\end{bmatrix}, \tag{9}$$

$$\mathbf{T}^{oe}=\begin{bmatrix}\underline{b}^{x,i}_{n_im_i}\underline{a}^{x,j^*}_{n_jm_j}\\ \underline{b}^{x}_{n_im_i}\underline{a}^{y^*}_{n_jm_j}+\underline{b}^{y,i}_{n_im_i}\underline{a}^{x,j^*}_{n_jm_j}\\ \underline{b}^{y,i}_{n_im_i}\underline{a}^{y,j^*}_{n_jm_j}\end{bmatrix} \tag{10}$$

and

$$\mathbf{T}^{oo}=\begin{bmatrix}\underline{b}^{x,i}_{n_im_i}\underline{b}^{x,j^*}_{n_jm_j}\\ \underline{b}^{x}_{n_im_i}\underline{b}^{y^*}_{n_jm_j}+\underline{b}^{y,i}_{n_im_i}\underline{b}^{x,j^*}_{n_jm_j}\\ \underline{b}^{y,i}_{n_im_i}\underline{b}^{y,j^*}_{n_jm_j}\end{bmatrix}. \tag{11}$$

The field-point dependent factors $M_\xi^\vartheta(\vartheta,\varphi)$ related to the ϑ-component of the antenna radiation pattern can be computed from:

$$\begin{Bmatrix}M_1^\vartheta(\vartheta,\varphi)\\ M_2^\vartheta(\vartheta,\varphi)\\ M_3^\vartheta(\vartheta,\varphi)\end{Bmatrix}=\begin{Bmatrix}\cos^2\vartheta\cos^2\varphi)\\ \cos^2\vartheta\cos\varphi\sin\varphi\\ \cos^2\vartheta\sin^2\varphi\end{Bmatrix}. \tag{12}$$

The scaling factors $M_\xi^\varphi(\vartheta,\varphi)$ that are related to the φ-component of $\boldsymbol{C}(\vartheta,\varphi)$ yield:

$$\begin{Bmatrix}M_1^\varphi(\vartheta,\varphi)\\ M_2^\varphi(\vartheta,\varphi)\\ M_3^\varphi(\vartheta,\varphi)\end{Bmatrix}=\begin{Bmatrix}\sin^2\varphi\\ -\cos\varphi\sin\varphi\\ \cos^2\varphi\end{Bmatrix}. \tag{13}$$

As can be seen from Eq. (6), a complete separation between antenna- and channel dependent quantities is achieved. It is also an important advantage that the values of the coupling integrals in Eq. (7) remain fixed within a distinct channel scenario $p_{\vartheta,\varphi}(\vartheta,\varphi)$. Due to the fact that antenna radiation behavior is characterized in terms of an SME by its vector field expansion coefficients $\underline{a}^i_{n_im_i}$ and $\underline{b}^i_{n_im_i}$, the effect of changing the antenna topology or geometry of single antenna elements within the considered channel scenario can be easily obtained from Eq. (6).

3.2 Correlation properties of log.-per. four-arm antennas

In this section the correlation properties of log.-per. antenna elements at the transmitter or at the receiver are investigated. The analysis is accomplished using the novel approach for the computation of power correlation as derived in Sect. 3.1 based on the SME expansion of the surface current distributions of the related antenna elements. Results were computed in terms of a two-dimensional probability density $p_{\vartheta,\varphi}(\vartheta,\varphi)$ applying a Laplacian dependence in azimuth φ and a Gaussian dependence in elevation ϑ in accordance to Taga (1990) featuring mean values $m_\varphi=90°$, $m_\vartheta=90°$ and angular spreads $\sigma_\varphi=10°$, $\sigma_\vartheta=10°$.

Antenna power correlation is computed between the vector antenna radiation patterns $\boldsymbol{C}_{i,j}(\vartheta,\varphi)$ related to the operational modes i and j of the antenna element with dual linear polarization. Figure 4 shows the results for antenna power correlation ρ_P evaluated by the standard approach involving electric far-fields derived by methods of FEM (see Eq. (3)) and a spherical mode expansion of the surface-current distribution. Results are shown in the predestined frequency range of the antenna elements between 1 GHz and 6 GHz. Increasing the order of the field approach ameliorates the fit between the two computational approaches (FEM and SME) as can be seen from Fig. 4. Therefore including a maximum order $N=4$ seems adequate for the trapezoid antenna $M=1$ that is depicted in Fig. 1a. Due to the imperfect polarization decoupling between the two operational modes of this antenna element, ρ_P yields a peak at the frequency of 4.3 GHz and a local maximum at the upper frequency of 6 GHz of the considered frequency range. According to the trapezoid antenna including $M=5$ periods of the trapezoidal unit cell as depicted in Fig. 1b, the results of power correlation ρ_P are shown in Fig. 5. Due to excellent polarization decoupling between the two pairs of adjacent antenna arms for this antenna element, power correlation remains below 0.015 in the entire frequency range. Despite those low numerical values for ρ_P the results derived by spherical eigenfield decomposition of the surface current distribution including a maximum expansion degree $N=3$ are in an excellent agreement.

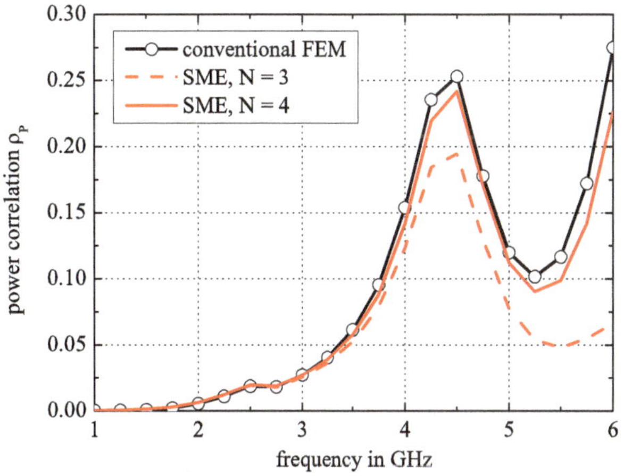

Fig. 4. Power correlation between orthogonal operational modes for trapezoid antenna $M=1$.

4 Analysis of MIMO antenna performance by means of SME

Following the stochastic model given in Sect. 3 for a (2×2)-MIMO antenna configuration using identical trapezoid antenna elements with dual-linear polarization at the transmitter and at the receiver, the ergodic channel capacities were computed from Eq. (4). In order to evaluate the applicability of the spherical wave approach as given in Sect. 3.1 in terms of MIMO transmission performance, the results for transmission capacity were plotted for the spherical wave approach

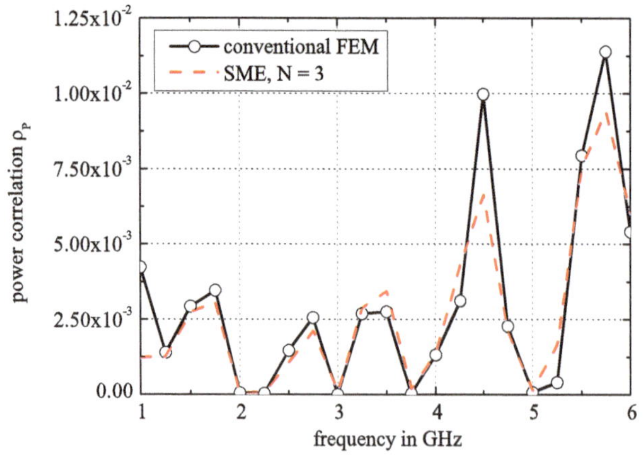

Fig. 5. Power correlation between orthogonal operational modes for trapezoid antenna $M = 5$.

in comparison to the conventional, FEM-based results. The results in this section are limited to a maximum degree of $N = 3$ for the spherical wave expansion.

Figures 6 and 7 show the results for MIMO channel capacity in terms of cumulative distribution functions (CDFs) of the outage capacity in bit/s/Hz for the trapezoid antenna $M = 1$. At 3 GHz, the results of conventional FEM-based computation and spherical mode expansion are in a very good agreement, as can be seen from Fig. 6. Regarding the sum capacity for the (2×2)-MIMO transmission link, both computational techniques consistently yield a 10%-outage capacity of 10.45 bit/s/Hz. At a simulation frequency of 6 GHz, where slight deviations between the two computational techniques were reported in terms of antenna power correlation ρ_P (see Fig. 4), the results for conventional computed channel capacity and channel capacity as derived by means of SME ($N = 3$) are not in an excellent agreement. Following the values for the 10%-outages of the channel capacity, spherical mode expansion exhibits a slight overestimation of the channel capacity resulting in $C_{10\%,6\text{GHz}} = 10.4$ bit/s/Hz. Conventional, FEM-based computations yield a value of 10.2 bit/s/Hz. Especially the CDF for the weaker channel of the transmission link can not be approximated very well by means of SME. This result is due to the small contributions of this spatial subchannel to the overall capacity of the MIMO link, where impairments in the field-representations due to an inadequate degree of modal field expansion become visible first.

This result is in a close agreement to the values for antenna power correlation at 6 GHz as shown in Fig. 4. In this case, SME yields a power correlation of 0.07 whereas conventional, FEM-based computations lead to $\rho_P = 0.28$. As can be seen from these results, frequency variation of the ergodic channel capacity is extremely low.

The results for using trapezoid antennas with $M = 5$ periods of the trapezoidal unit cell (see Fig. 1b) in a (2×2)-MIMO communication link are depicted in Fig. 8 at 3 GHz

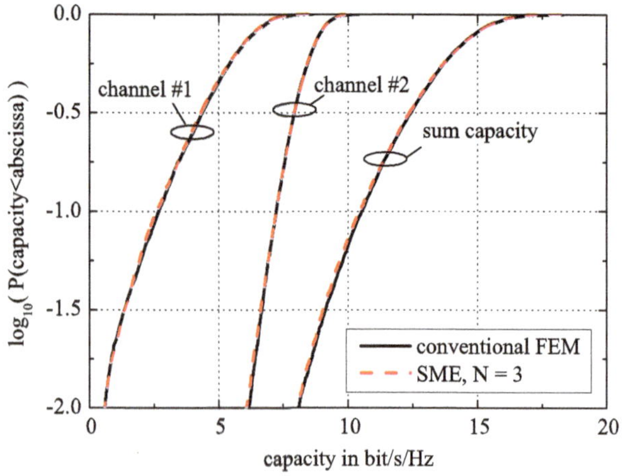

Fig. 6. CDF of MIMO channel capacity for $M = 1$ antenna at 3 GHz derived from FEM computations and SME, $N = 3$.

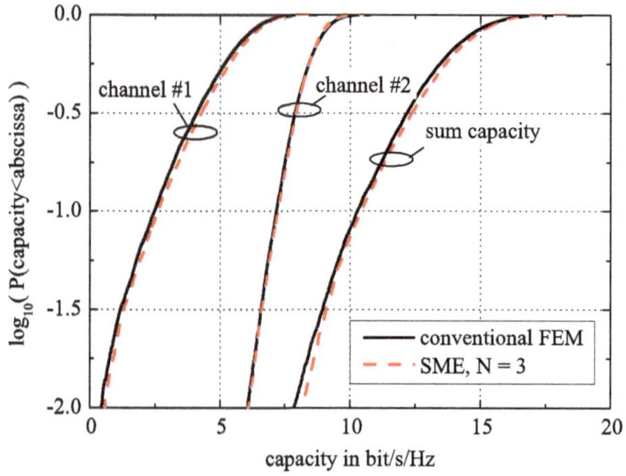

Fig. 7. CDF of MIMO channel capacity for $M = 1$ antenna at 6 GHz derived from FEM computations and SME, $N = 3$.

and in Fig. 9 for a simulation frequency of 6 GHz, respectively. As already shown for the trapezoid antenna $M = 1$, frequency variation of the ergodic channel capacity is found to be very low. The approximations of the CDFs for the channel capacities based on the SME of the surface current distributions are in a very close agreement to the results as derived from the conventional FEM-based computations. As shown in Fig. 8, at 3 GHz, the 10%-outage capacity based on FEM-results indicates a value of 10.5 bit/s/Hz that is in exact agreement with the results based on the spherical mode expansion ($N = 3$). At a simulation frequency of 6 GHz FEM-based computations yield a value of 10.5 bit/s/Hz for the 10%-outage capacity that is in a good agreement to the result for SME-based computations that yields a value of 10.4 bit/s/Hz. Due to the approximation of Rayleigh distributed signal fading by a finite occurrence of 5000 channel realizations as shown in Sect. 3 this result remains within computational exactness of the applied model of the radio channel.

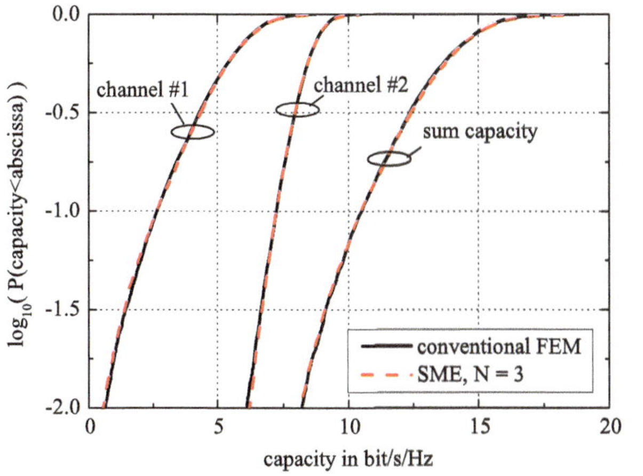

Fig. 8. CDF of MIMO channel capacity for $M = 5$ antenna at 3 GHz derived from FEM computations and SME, $N = 3$.

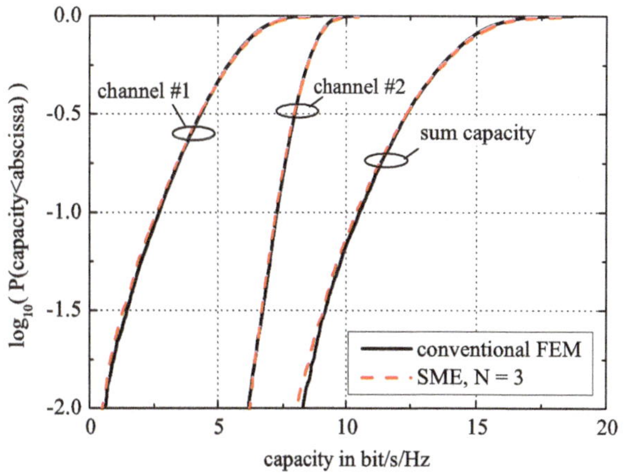

Fig. 9. CDF of MIMO channel capacity for $M = 5$ antenna at 6 GHz derived from FEM computations and SME, $N = 3$.

5 Conclusions

This paper presented a generalized approach for the computation of antenna power correlation and MIMO channel capacity in multielement antenna arrangements based on a spherical mode expansion of the respective surface current distributions. The analysis was applied in order to characterize the power correlation and channel capacity for planar, dual-polarized log.-per. four-arm antenna elements. The characterization of the MIMO radio channel was based on a simple stochastic model that accounted for the spatio-temporal behavior of signal fading at the transmitter and at the receiver. Results based on the introduced spherical mode expansion were found to be in a very close agreement to the results as derived by conventional FEM-based computations. The major advantages of this work is the complete analytical description of the link-related transmission parameters like antenna power correlation and channel capacity by means of spherical harmonics. Due to the fact that the coefficients of field expansion that are directly related to the surface current distribution on the related antenna elements are separated mathematically from the quantities that characterize the channel model, this approach allows a clearer insight into the field-related properties of MIMO communications systems. It may be additionally used in order to optimize or fit certain antenna geometries to distinct channel scenarios.

References

Armbrecht, G., Klemp, O., and Eul, H.: Spherical Mode Analysis of Planar Frequency-Independent Multi-Arm Antennas Based on Its Surface Current Distribution, Accepepted for publ. in Adv. Radio Sci., 2006.

Chen, Y., Simpson, T. L., and Ho, T. Q.: Highly efficient technique for solving radiation and scattering problems, IEE Proc.-H, 139, 7–10, 1992.

DuHamel, R. H. and Isbell, D. E.: Broadband Logarithmically Periodic Antenna Structures, IRE Intern. Conv. Rec., 5, 119–128, 1957.

Foschini, G. J.: Layered space-time architecture for wireless communication in fading environment when using multiple antennas, Bell Labs Tech. J., 41–59, 1996.

Jakes, W. C.: Microwave Mobile Communications, IEEE Press, Inc., New York, 1974.

Kermoal, J. P., Schumacher, L., Pedersen, K. I., Morgensen, P. E., and Frederiksen, F.: A Stochastic MIMO Radio Channel Model With Experimental Validation, 20, 1211–1226, 2002.

Klemp, O. and Eul, H.: Radiation Pattern Analysis of Antenna Systems for MIMO and Diversity Configurations, Adv. Radio Sci., 3, 157–165, 2005.

Klemp, O., Schultz, M., and Eul, H.: A Planar Broadband Antenna for Applications in WLAN and 3G, Proc. of the 2004 Intern. Sympos. Signals, Systems and Electronics ISSSE, Austria, 2004.

Klemp, O., Schultz, M., and Eul, H.: Novel Logarithmically Periodic Antenna Structures for Broadband Polarization Diversity Reception, AEUE, Intern. J. on Electronics and Commun., 59, 268–277, 2005.

Leifer, M. C.: Signal Correlations in Coupled Cell and MIMO antennas, Antennas and Propagation Soc. Intern. Symp., 16-22 June 2002, 3, 194–197, 2002.

Lo, Y. T. and Lee, S. W.: Antenna Handbook - Theory, Applications and Design, Van Nostrand Reinhold Company, New York, 1988.

Narasimhan, M. S., Christopher, S., and Varadarangan, K.: Modal Behavior of Spherical Waves from a Source of EM Radiation with Application to Spherical Scanning, 33, 350–354, 1985.

Rumsey, V. H.: Frequency Independent Antennas, IRE Intern. Conv. Rec., 5, 114–118, 1957.

Stratton, J. A.: Electromagnetic Theory, McGraw Hill, New York, 1941.

Taga, T.: Analysis for Mean Effective Gain of Mobile Antennas in Land Mobile Radio Environments, 39, 117–131, 1990.

Werner, D. H. and Mittra, R.: Frontiers in Electromagnetics, IEEE Press Series on Microwave Technology and RF, New York, 2000.

Yu, K. and Ottersten, B.: Models for MIMO Propagation Channels, Wiley Journal on Wireless Communications and Mobile Computing, Spec. Issue on Adapt. Antennas and MIMO Syst., 2002.

24

Spherical mode analysis of planar frequency-independent multi-arm antennas based on its surface current distribution

G. Armbrecht, O. Klemp, and H. Eul

Universität Hannover, Institut für Hochfrequenztechnik und Funksysteme, Hannover, Germany

Abstract. Deployment in the design of mobile radio terminals focuses on the implementation of multiradio transmission systems, using a multiplicity of different radio standards combined with high-speed data communication over multiple-input multiple-output (MIMO) and multimode diversity techniques. Hence, planar log.-per. four-arm antennas are predistined to meet the requirements of future mobile multiradio RF-frontends and will be introduced and analysed in terms of an efficient spherical mode analysis by means of surface current distribution in order to derive an analytic access to MIMO- and polarisation-diversity performance computation. A remarkable parameter reduction and a faster numerical analysis with respect to conventional techniques may be achieved. The sources in the near-field antenna region are based on the numerical computation of surface currents involving the finite element method (FEM). Relations between the variations of the geometrical antenna parameters and the excitation of discrete spherical modes are presented and will be analysed in detail.

1 Introduction

This article focuses on the in-depth analysis of planar multi-arm antennas for the application in future mobile broadband communication environments. For the practical realisation of multiradio terminals which combine the reception of a multiplicity of different communication standards like cellular phone (e.g. GSM, WCDMA, UMTS), wireless networks (e.g. WLAN, HIPERLAN, WiMAX), proximity (e.g. Bluetooth, RFID, UWB) and broadcasting services (e.g. FM, DVB-H) in one box, antennas with a compact footprint and almost frequency-independent electromagnetic antenna transmission behaviour are required. Furthermore the characteristics of multipath propagation by independently fading broadband and orthogonal transmission channels are exploited for the increased demand of high-speed data transmission. Following Jensen and Wallace (2004) this may lead to the application of MIMO and multimode diversity techniques in the future standardisation of novel communication systems. In this case especially planar log.-per. four-arm antennas in a frequency range from 1.0 to 6.0 GHz with self-complementary antenna shapes may contribute to an excellent RF-frontend performance, due to the orthogonality between the two linear excitation modes of the antenna that results in a high degree of cross-polarisation suppression and marginal envelope correlation behaviour (see Klemp et al., 2005). Such a multi-arm antenna performance is mandatory for the use in future mobile MIMO and polarisation diversity applications.

Hence, the efficient analysis based on a spherical mode expansion (SME) is applied to planar log.-per. multi-arm antennas and especially evaluated in case of four-arm antennas with a trapezoidal geometry, as proposed in Klemp and Eul (2005). The theory of spherical mode expansion first investigated by Stratton (1941) may be effectively applied to solve the scalar, homogeneous Helmholtz-PDE, as shown extensively e.g. in Werner and Mittra (2000) into a series of well-known spherical eigenfunctions. In the past, many authors reported about the successful application to a multiplicity of wave propagation problems, e.g. Cassegrainian-fed paraboloids (see Potter, 1967) and the investigation of multiple scattering of electromagnetic waves (see Bruning and Lo, 1971). Even for todays techniques of spherical antenna near field measurement this series expansion is a fundamental basis, as described by Ludwig (1971). In Chen and Simpson (1991) and Chen et al. (1992) this kind of spatial field approach is expanded to a straight-forward spherical series representation by means of the antenna surface current distribution and was applied to an infinitely thin dipole structure. The main advantage of this field approach is the opportu-

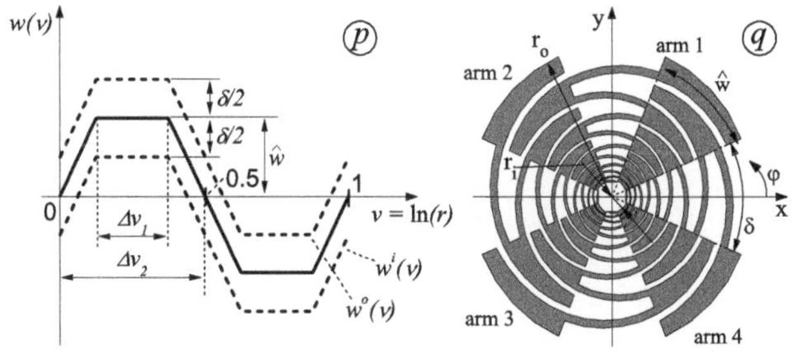

a) original plane b) image plane

Fig. 1. Trapezoidal antenna geometry in the original (p) and image plane (q).

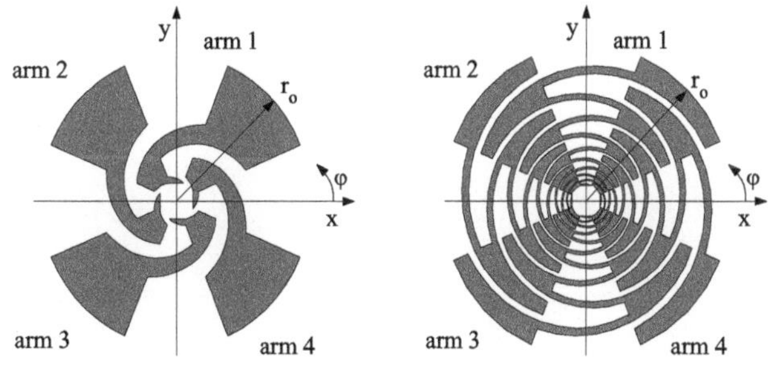

a) trapezoid antenna, M=1 b) trapezoid antenna, M=5

Fig. 2. Trapezoidal antenna geometry in the image plane q with $M=1$ and $M=5$ basis function periods.

nity of separating the given source-region from the requested far-field region which leads to an efficient parameter reduction and a faster numerical analysis with respect to conventional computational techniques. The article is organized as follows: A short introduction into the generatation methology of planar log.-per. antennas will be given in Sect. 2. In Sect. 3.1 and 3.2 the details of the applied SME field approach are presented, in which the sources in the antenna near-field region are obtained from the surface current distribution derived from a commercial field-simulator based on the finite element method (FEM).

In consequence of the coordinate-transform between the source- and the far-field region a new, modulated eigenmode spectrum will be introduced in Sect. 3.3. Furthermore, we will highlight the spatial- and frequency-based dependence of the spherical eigenmode excitation for this special kind of frequency-independent log.-per. planar antennas in Sect. 3.4. A verification of the spherical field approach is accomplished in Sect. 4 involving a conventional, moment-based field calculation method in the special case of planar antennas. We will investigate the convergence properties of the series expansion with respect to the underlying coordinate system. Finally, Sect. 5 derives a relationship between the geometrical properties of log.-per. four-arm antennas of different geometrical shapes and the distinct spectra of spherical eigenmodes.

2 Log.-per. multi-arm antenna design

In this section the planar logarithmically-periodic self-complementary trapezoidal antenna designed for a use in the frequency range of 1.0 to 6.0 GHz is introduced that serves as an initial point for the spherical eigenmode analysis. In Klemp et al. (2005) this kind of antenna structure was identified as an optimum in terms of cross-polarization decoupling, cut-off frequency and footprint. Such characteristics in combination with an almost frequency-independent input-impedance and radiation pattern behaviour above the cut-off frequency are favorable for the application in multiradio transmission systems. One of the specialized designs is a

Table 1. Typical geometry settings of planar trapezoidal log.-per. antenna structures

parameter	r_i/mm	r_o/mm	N_A	M	$\hat{w}/°$	$s_{fl}/\%$	$\delta/°$
value	5.0	50.0	4	5	45	50	45

four-arm ($N_A=4$) trapezoidal antenna, which provides two operational modes with dual-linear orthogonal polarisation fed by odd-phase port excitation of each pair of two opposite arms. The properties of dual-polarisation may be used in order to enhance channel capacity in MIMO transmission system or reduce signal fading due to polarisation diversity reception.

Referring to Fig. 1, the log.-per. antenna geometry is generated using Eq. (1) for the inner r^i and outer r^o limitation radii in the image plane q, as follows:

$$\begin{Bmatrix} r^i(\varphi) \\ r^o(\varphi) \end{Bmatrix} = e^v \cdot \exp \begin{Bmatrix} jw^i(v) \\ jw^o(v) \end{Bmatrix} . \tag{1}$$

In Eq. (1), w^i and w^o denote the inner and outer basis function in the original plane p, as depicted in Fig. 1a. A trapezoidal basis geometry is defined in terms of multiple parameter settings. A typical set of these parameters is shown in Table 1, where N_A denotes the number of antenna arms and M the number of periods of the trapezoidal basis function in the p-plane. Additional parameters are the angular spread $\hat{w}$, the slope of the trapezoidal flanks s_{fl} and the arm width δ, where $\delta=45°$ is used for self-complementary geometries. As given in Klemp et al. (2005), the inner and outer radii determine the upper and the lower antenna cut-off frequencies, respectively.

In this paper the distinct trapezoidal antenna geometries with $M=1$ and $M=5$ periods and cut-off frequencies at approximately $f_c^{M_1} \approx 4.2$ GHz and $f_c^{M_5} \approx 1.6$ GHz are extensively evaluated. To get a profound insight of these structures, they are depicted in Fig. 2.

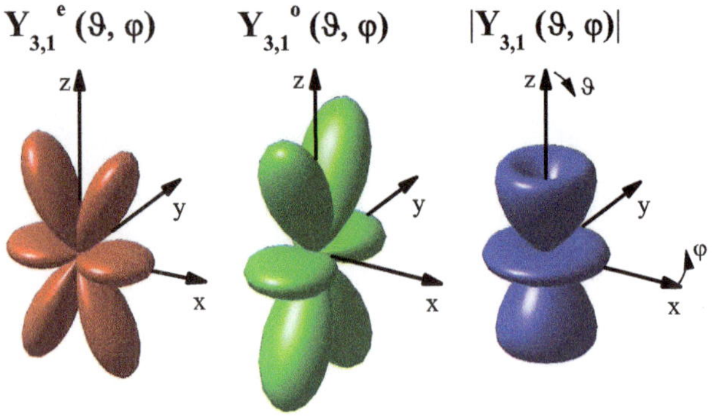

Fig. 3. Spherical harmonics $Y_{3,1}(\vartheta, \varphi)$ subdivided into even, odd and magnitude spectrum.

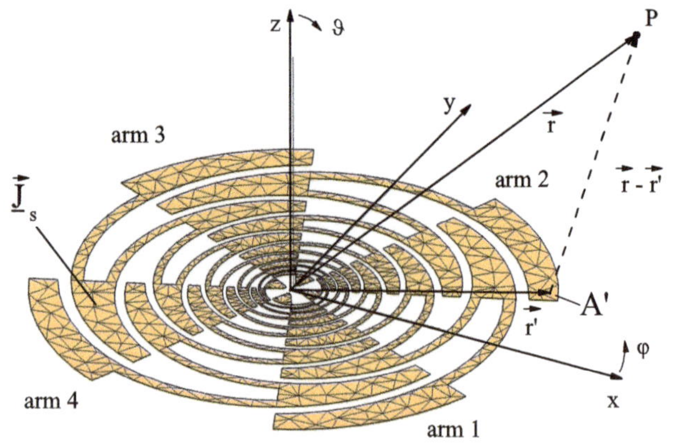

Fig. 4. Trapezoidal antenna ($M = 5$) in spherical coordinates with a given electric current distribution $\underline{\boldsymbol{J}}_s(\boldsymbol{r}')$ affecting the magnetic vector potential $\underline{\boldsymbol{A}}(\boldsymbol{r})$ in the requested far-field point P.

3 Spherical mode expansion (SME)

According to Werner and Mittra (2000) the spherical mode expansion is a systematic field approximation technique for arbitrary source-free regions in linear and partial homogeneous materials. The results possess a rather mathematical than physical character and need to be interpreted in terms of coordinate plane dependency. In this article the spherical multipole expression is utilized to characterize the radiation behaviour of planar log.-per. antennas introducing a novel modal description. In this section a brief introduction of the chosen, straight-forward field approach will be given.

3.1 Modal solutions of scalar Helmholtz equation

Any radiation and field propagation problem in ordinary spherical coordinates exhibits a solution to the scalar, homogeneous Helmholtz partial differential equation (PDE) for any functional dependency $\Psi(\boldsymbol{r})$, given by:

$$\Delta\underline{\Psi}(\boldsymbol{r}) + k^2\underline{\Psi}(\boldsymbol{r}) = 0\,. \tag{2}$$

Solving the Eq. (2) in terms of coordinate separation techniques into a radial and a transversal functional dependency leads to a double sum representation of orthogonal spherical eigenfunctions $Y_{\mathrm{n,m}}(\vartheta, \varphi)$ with integer eigenvalues n and m describing the degree and the order of the spherical harmonics (see Stratton, 1941), as follows:

$$\underline{\Psi}(\boldsymbol{r}) = \sum_n \sum_m \underline{c}_{\mathrm{n}}^{\mathrm{m}} \cdot z_{\mathrm{n}}^{\mathrm{m}}(kr) \cdot Y_{\mathrm{n,m}}(\vartheta, \varphi)\,. \tag{3}$$

In Eq. (3) $\underline{c}_{\mathrm{n}}^{\mathrm{m}}$ is a complex weighting factor and $z_{\mathrm{n}}^{\mathrm{m}}$ denotes a spherical cylinder function.

In detail the Eq. (4) expresses the composition of the spherical eigenfunctions subdivided into even (e) and odd (o) modes which are given by orthogonal order-dependent trigonometric dependency (periodic Lamé products) of the azimuth angle φ.

$$\left\{ \begin{array}{c} Y_{\mathrm{n,m}}^{\mathrm{e}}(\vartheta, \varphi) \\ Y_{\mathrm{n,m}}^{\mathrm{o}}(\vartheta, \varphi) \end{array} \right\} = N_{\mathrm{n}}^{\mathrm{m}} \cdot P_{\mathrm{n}}^{\mathrm{m}}(\cos\vartheta) \cdot \left\{ \begin{array}{c} \cos(m\varphi) \\ \sin(m\varphi) \end{array} \right\} \tag{4}$$

Furthermore $P_{\mathrm{n}}^{\mathrm{m}}$ denotes the elevation dependency in terms of associated Legendre polynomials as non-periodic Lamé products. $N_{\mathrm{n}}^{\mathrm{m}}$ represents a normalisation coefficient in accordance to Werner and Mittra (2000). Both quantities depend on the degree n and order m of the spherical harmonics, respectively.

Figure 3 depicts the spatial magnitude distribution in terms of the even- and odd-parts as well as the magnitude plot of the spherical eigenfunction $Y_{3,1}(\vartheta, \varphi)$ of the degree $n = 3$ and the order $m = 1$.

3.2 Current-based field approach

In order to determine the magnetic vector potential $\underline{\boldsymbol{A}}(\boldsymbol{r})$ in any requested field point, we will refer to a planar region A' that exhibits a source distribution of electric surface currents $\underline{\boldsymbol{J}}_s(\boldsymbol{r}')$, as shown in Fig. 4. According to Eq. (5), the magnetic vector potential $\underline{\boldsymbol{A}}(\boldsymbol{r})$ may be computed involving the Kirchhoff integral of a given surface current distribution $\underline{\boldsymbol{J}}_s(\boldsymbol{r}')$ that causes free-space radiation represented by the Green's free-space function $G(\boldsymbol{r}, \boldsymbol{r}') = e^{-jk|\boldsymbol{r}-\boldsymbol{r}'|} / |\boldsymbol{r} - \boldsymbol{r}'|$.

$$\underline{\boldsymbol{A}}(\boldsymbol{r}) = \frac{1}{4\pi} \iint_{(A')} \underline{\boldsymbol{J}}_s(\boldsymbol{r}') \cdot \frac{e^{-jk|\boldsymbol{r}-\boldsymbol{r}'|}}{|\boldsymbol{r} - \boldsymbol{r}'|}\, dA'\,. \tag{5}$$

Therefore, the Green's free space function for homogeneous, source-free regions is a solution of the Helmholtz-PDE (2) and may be expanded into a spherical eigenmode series, as given in Sect. (3). The spherical wave expansion corresponds directly to Klemp and Eul (2005) and will therefore not be demonstrated in detail.

Magnetic current distributions $\underline{\boldsymbol{J}}_m$ may also be analysed in the same manner leading to the electric vector potential $\underline{\boldsymbol{F}}$. An enhancement to magnetic currents is not required, because planar log.-per. antenna structures may be completely

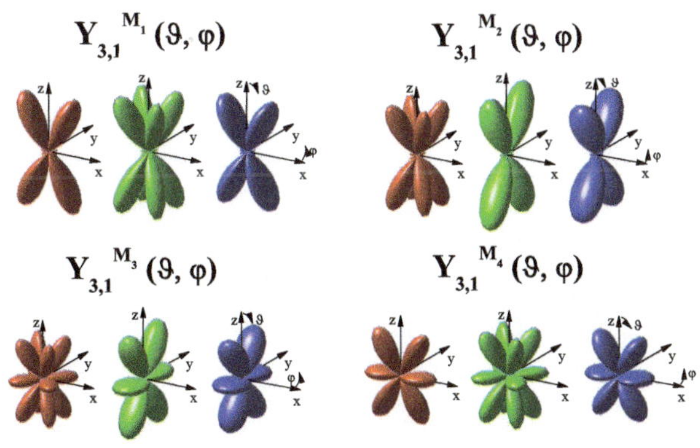

Fig. 5. Modulated spherical harmonics spectrum $Y_{3,1}(\vartheta, \varphi)$.

described in terms of real electric surface currents $\underline{\boldsymbol{J}}_s$.

$$\underline{\boldsymbol{A}}(\boldsymbol{r}) = \frac{e^{-jkr}}{r} \sum_{n=0}^{\infty} \sum_{m=0}^{n} j^{\mathrm{n}} \left[\underline{\boldsymbol{a}}_{\mathrm{n}}^{\mathrm{m}} \cdot Y_{\mathrm{n,m}}^{\mathrm{e}}(\vartheta, \varphi) + \underline{\boldsymbol{b}}_{\mathrm{n}}^{\mathrm{m}} \cdot Y_{\mathrm{n,m}}^{\mathrm{o}}(\vartheta, \varphi) \right]. \quad (6)$$

For that reason Eq. (6) provides a compact spherical eigenfield approach for $\underline{\boldsymbol{A}}(\boldsymbol{r})$ that is characterized by a complete coordinate separation between near- and far-field quantities adhering to the Fraunhofer approximation. The current sources in the near-field region given by the surface current distribution $\underline{\boldsymbol{J}}_s$ of the planar antenna described by the complex vector coefficients $\underline{\boldsymbol{a}}_{\mathrm{n}}^{\mathrm{m}}$ and $\underline{\boldsymbol{b}}_{\mathrm{n}}^{\mathrm{m}}$ of the field expansion. The Green's function accounts for the spherical wave propagation of the magnetic vector potential, whereas the remaining terms incorporate its spatial profile.

For a numerical evaluation of $\underline{\boldsymbol{A}}(\boldsymbol{r})$ given by Eq. (6), the limitation of the series expansion to finite SME degrees N will be essential leading to impairments in the approximated far-field representation. For that purpose Ludwig (1971), Narasimhan et al. (1985) and Chen and Simpson (1991) introduced a SME convergence criteria to reduce the field impairments significantly, if the SME degree N is derived for $N \geq ka$, whereas k denotes the scalar wavenumber of the free-space and a represents the radius of the sphere including all current sources in the near-field. In our special case of planar trapezoidal antennas the radius a correlates with the maximum radius $r_{\mathrm{o}} = 50$ mm of the antenna footprint.

3.3 Modulated eigenmode spectra

To accomplish the numerical antenna analysis based on the antenna's surface currents, we applied a conventional FEM field-simulator in order to obtain the correct near-field source distribution on the antenna surface. For further analysis the far-field representation of the antenna's radiation characteristics is required, which leads to the application of the far-field approximation by Fraunhofer. By this means, a simple relationship between the SME expanded magnetic vector potential $\underline{\boldsymbol{A}}(\boldsymbol{r})$ according to Eq. (6) and the resulting electric field components E_ϑ and E_φ in spherical coordinates is exhibited, given by,

$$\boldsymbol{E}(\vartheta, \varphi) = \begin{bmatrix} E_\vartheta \\ E_\varphi \end{bmatrix} = -jk\eta \begin{bmatrix} A_\vartheta \\ A_\varphi \end{bmatrix}, \quad (7)$$

where $\eta = 120\pi\,\Omega$ denotes the value of the characteristic free-space impedance. The field approximation given by Eq. (7) is only applicable in spherical coordinates (r, ϑ, φ) and neglects the radial field component dependency. A certain coordinate transform between cartesian and spherical coordinates has to be taken into account, which may be expressed by four metric coefficients $M_i(\vartheta, \varphi)$, as follows:

$$M_1(\vartheta, \varphi) = \cos(\vartheta)\cos(\varphi); \; M_2(\vartheta, \varphi) = \cos(\vartheta)\sin(\varphi)$$
$$M_3(\varphi) = -\sin(\varphi); \qquad M_4(\varphi) = \cos(\varphi). \quad (8)$$

The metric coefficients $M_i(\vartheta, \varphi)$, with $i = 1, ..., 4$, represent the transform of cartesian components $\{A_x, A_y\}$ of the magnetic vector potential to the spherical components $\{A_\vartheta, A_\varphi\}$, which leads to a new set of orthogonal spatial eigenfunctions. The new set of eigenfunctions $Y_{\mathrm{n,m}}^{\mathrm{M_i}}$ may be represented in terms of the original set of spherical harmonics, as in Eq. (4), using the coefficients $M_i(\vartheta, \varphi)$, as given in Eq. (8). This leads to a quadruple of modified spherical harmonics $Y_{\mathrm{n,m}}^{\mathrm{M_i}}$ that exhibits a distinct modulation with respect to the spatial coordinates ϑ and φ, as shown in Eq. (9). Using the metric coefficients given in Eq. (8) the original spherical wave function of the example $Y_{3,1}(\vartheta, \varphi)$, as depicted in Fig. 3, yields the set of modulated harmonics $Y_{3,1}^{\mathrm{M_i}}(\vartheta, \varphi)$, as shown in Fig. 5. Equation (9) shows the new, functional dependence between Eq. (4) and the metric coefficients according to Eq. (8):

$$\begin{Bmatrix} Y_{\mathrm{n,m}}^{\mathrm{e,M_i}} \\ Y_{\mathrm{n,m}}^{\mathrm{o,M_i}} \end{Bmatrix} = M_i(\vartheta, \varphi) \cdot \begin{Bmatrix} Y_{\mathrm{n,m}}^{\mathrm{e}}(\vartheta, \varphi) \\ Y_{\mathrm{n,m}}^{\mathrm{o}}(\vartheta, \varphi) \end{Bmatrix}, \quad (9)$$

with $i = 1, ..., 4$.

In case of comparing different SME field approxmations by complex vector coefficients $(\underline{a}_{\mathrm{n}}^{\mathrm{m}}, \underline{b}_{\mathrm{n}}^{\mathrm{m}})$, the equality of the underlying spherical harmonics between different field approaches is fundamental. Obviously, this is only realisable when the field approaches are based on the same spectra of spherical harmonics $Y_{\mathrm{n,m}}(\vartheta, \varphi)$. Applying Eqs. (8) and (9) in order to map the surface current distribution to spherical coordinates and replacing the magnetic vector potential $\underline{\boldsymbol{A}}(\boldsymbol{r})$ in Eq. (7) by the expression given in Eq. (6) yields:

$$\begin{bmatrix} \underline{E}_\vartheta \\ \underline{E}_\varphi \end{bmatrix} = K(r) \sum_{n=0}^{\infty} \sum_{m=0}^{n} j^{\mathrm{n}} \cdot \left\{ \begin{bmatrix} \underline{a}_{\mathrm{x}}^{\mathrm{n,m}} \cdot Y_{\mathrm{n,m}}^{\mathrm{e,M_1}}(\vartheta, \varphi) + \underline{a}_{\mathrm{y}}^{\mathrm{n,m}} \cdot Y_{\mathrm{n,m}}^{\mathrm{e,M_2}}(\vartheta, \varphi) \\ \underline{a}_{\mathrm{x}}^{\mathrm{n,m}} \cdot Y_{\mathrm{n,m}}^{\mathrm{e,M_3}}(\vartheta, \varphi) + \underline{a}_{\mathrm{y}}^{\mathrm{n,m}} \cdot Y_{\mathrm{n,m}}^{\mathrm{e,M_4}}(\vartheta, \varphi) \end{bmatrix} + \begin{bmatrix} \underline{b}_{\mathrm{x}}^{\mathrm{n,m}} \cdot Y_{\mathrm{n,m}}^{\mathrm{o,M_1}}(\vartheta, \varphi) + \underline{b}_{\mathrm{y}}^{\mathrm{n,m}} \cdot Y_{\mathrm{n,m}}^{\mathrm{o,M_2}}(\vartheta, \varphi) \\ \underline{b}_{\mathrm{x}}^{\mathrm{n,m}} \cdot Y_{\mathrm{n,m}}^{\mathrm{o,M_3}}(\vartheta, \varphi) + \underline{b}_{\mathrm{y}}^{\mathrm{n,m}} \cdot Y_{\mathrm{n,m}}^{\mathrm{o,M_4}}(\vartheta, \varphi) \end{bmatrix} \right\}, \quad (10)$$

Table 2. Coordinate plane dependency of the SME vector coefficient excitation

	n = 0		n = 1				n = 2					
	$\underline{a}_0^0$	$\underline{b}_0^0$	$\underline{a}_1^0$	$\underline{b}_1^0$	$\underline{a}_1^1$	$\underline{b}_1^1$	$\underline{a}_2^0$	$\underline{b}_2^0$	$\underline{a}_2^1$	$\underline{b}_2^1$	$\underline{a}_2^2$	$\underline{b}_2^2$
xy	X	-	-	-	X	X	X	-	-	-	X	X
xz	X	-	X	-	X	-	X	-	X	-	X	-
yz	X	-	X	-	-	X	X	-	-	X	X	X
index *i*	1	2	3	4	5	6	7	8	9	10	11	12

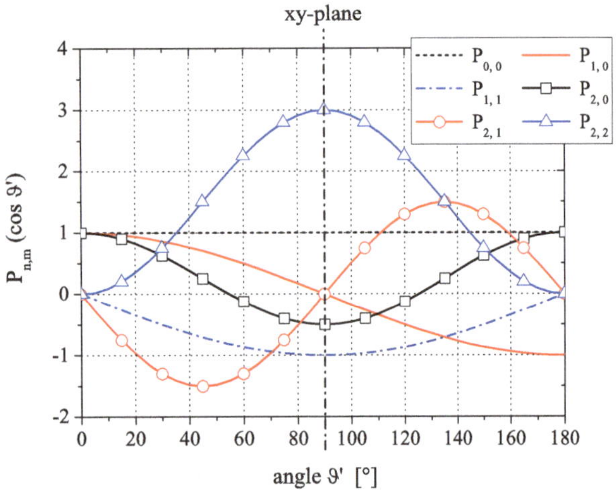

Fig. 6. Associated Legendre polynomials P_n^m depending on the elevation angle ϑ' of the underlying coordinate system.

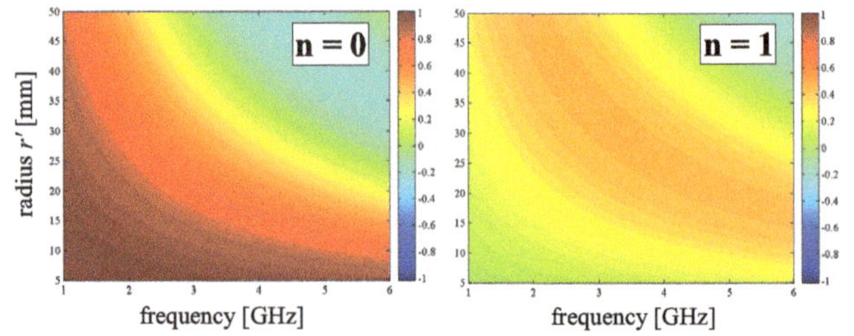

Fig. 7. Vector coefficient weighting characterized by spherical Bessel functions $j_n(kr')$.

where $K(r)$ merges the total radial functional dependency. In Eq. (6), $\underline{a}_{x,y}^{n,m}$ and $\underline{b}_{x,y}^{n,m}$ correspond to the vector expansion coefficients of the spherical field approach that can be directly computed from Eq. (11) using the surface current distribution on the antenna, as derived from the FEM solution.

Applying Fraunhofer relation, given by Eq. (7), to the electric far-fields, the obtained SME field approximation according to Eq. (10) is complete but its compactness is reduced due to the adaption of the spherical harmonics, which was inevitable in order to perform the requested coordinate transform. In detail, the next section focuses on the characterisation of antenna's surface currents in terms of complex modal vector expansion coefficients $\underline{\boldsymbol{a}}_n^m, \underline{\boldsymbol{b}}_n^m$.

3.4 Modal characteristics of planar, log.-per. antennas

In this section the representation of the antenna's surface current distribution $\underline{\boldsymbol{J}}_s$ in terms of SME vector coefficients $\underline{\boldsymbol{a}}_n^m$ and $\underline{\boldsymbol{b}}_n^m$ will be extensively analysed by evaluating the following integral equation:

$$\begin{Bmatrix} \underline{\boldsymbol{a}}_n^m \\ \underline{\boldsymbol{b}}_n^m \end{Bmatrix} = \iint_{(A')} N_n^m \cdot \underline{\boldsymbol{J}}_s(\boldsymbol{r}') \cdot j_n(kr') \cdot P_n^m(\cos\vartheta') \cdot \begin{Bmatrix} \cos(m\varphi') \\ \sin(m\varphi') \end{Bmatrix} dA'. \quad (11)$$

Besides the already established expressions in Eq. (11), j_n denotes a spherical Bessel function that has to be integrated across the entire planar antenna footprint. j_n performs a scaling of the surface current $\underline{\boldsymbol{J}}_s(\boldsymbol{r}')$ in dependence of its operational frequency and the position $\boldsymbol{r}'$. Subsequently, the computation of the vector expansion coefficients $\underline{\boldsymbol{a}}_n^m$ and $\underline{\boldsymbol{b}}_n^m$ will be carried out for antenna elements located in parallel to the xy-plane. In this context the argument of the associated Legendre polynomials P_n^m remains unchanged due to a constant elevation angle $\vartheta' = 0$ in the whole xy−plane.

In Fig. 6 the associated Legendre polynomials P_n^m for different field expansion degrees and orders are depicted for the complete range of the elevation angle ϑ'. Nulls of discrete Legendre polynomials lead to a distinct fade-out of certain modal vector coefficients in the xy-plane in combination with the nulls of the trigonometric functions. Table 2 shows a complete set of vector expansion coefficients up to a maximum degree $n = 2$ for surface current distributions limited to one of the three main planes of the cartesian coordinate system. Considering the symmetry relationship of the spherical eigenfunctions, the xy−plane might be the optimum choice with respect to an optimum far-field approximation.

Yet another modal characteristic is expressed by the spherical Bessel function $j_n(kr')$, resulting in a radius- and frequency-dependent weighting factors. In Fig. 7 the amplitude scaling of the vector expansion coefficients $\underline{\boldsymbol{a}}_n^m$ and $\underline{\boldsymbol{b}}_n^m$ is analyzed with respect to a variable radius r' in an interval from 5 to 50 mm that covers the entire footprint of the antenna and the operational frequency range of the antennas from 1.0 to 6.0 GHz.

In Fig. 7 the spherical Bessel function $j_n(kr')$ is evaluated

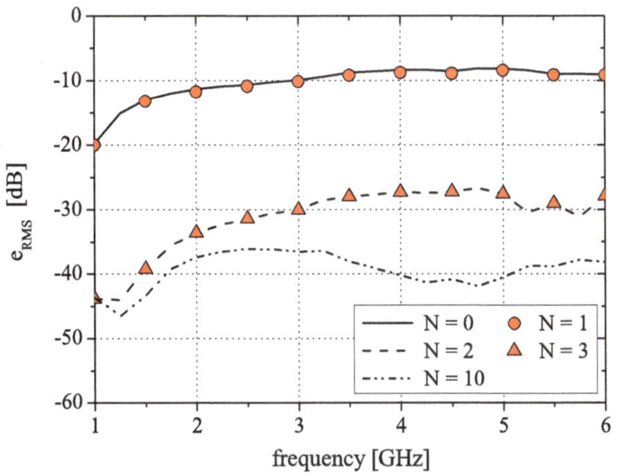

Fig. 8. RMS error in terms of different SME degrees N for a trapezoidal antenna ($M = 5$).

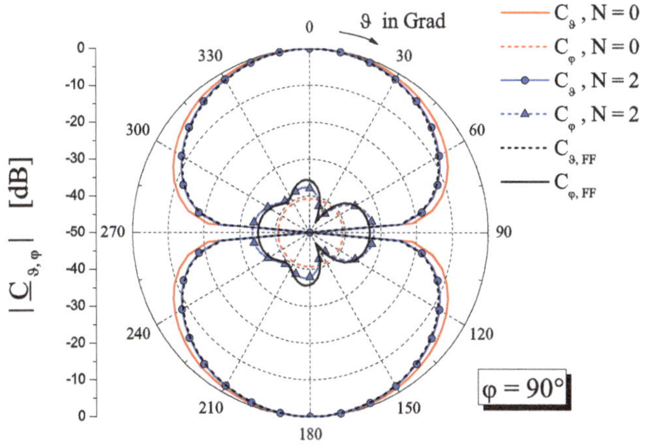

Fig. 9. Normalized radiation pattern $C_{\vartheta,\varphi}\left(\vartheta, \varphi = 90°\right)$ of a trapezoidal antenna ($M = 5$, $f = 2.5$ GHz) in free space.

for the first two consecutive SME degrees $n = 0, 1$ that illustrate the wavelike amplitude scaling of vector expansion coefficients with increasing radius of the antenna element and increasing frequency. Thus, the vector coefficient $\underline{\boldsymbol{a}}_0^0$ accounts for a maximum power allocation to the near-field sources close to the antenna center, whereas subsequent coefficients of the series expansion (as shown for SME degree $n = 1$) emphasize regions that are aligned offset the center. As can be seen from Fig. 7, additional regions on the antenna surface exist, where the spherical Bessel function vanishes for a given frequency point. Therefore current sources for this given combination of location on the antenna surface and operational frequency do not contribute to the overall far-field radiation due to the fact that they are suppressed by the properties of the spherical Bessel function $j_n(kr')$.

4 Verification and convergence analysis

In this section the verification of the spherical eigenfield approach is aspired and will be combined with an analysis of the convergence behaviour of the series expansion in terms of limiting the upper expansion degree to a finite value N. Only in this case a distinct numerical evaluation of the given SME approximation can be accomplished. The results derived by the SME expansion of the surface current distribution will be compared to the far-field (FF) results that were computed using a conventional moment-based calculation of Eq. (5).

According to Chen and Simpson (1991) the root-mean-square (RMS) error e_{RMS} is used for the analysis in order to evaluate the exactness of the far-field quantities as derived from SME with the reference results. The RMS error computes the deviation between the far-field quantities for both electric field components $\boldsymbol{E}_\vartheta$ and $\boldsymbol{E}_\varphi$ in every requested far-field point, as follows:

$$e_{RMS} = \left[\frac{1}{KL}\sum_{k=1}^{K}\sum_{l=1}^{L}\left\{\left(C_\vartheta^{\text{ref}}(\vartheta_k,\varphi_l) - C_\vartheta^{\text{comp}}(\vartheta_k,\varphi_l)\right)^2 + \left(C_\varphi^{\text{ref}}(\vartheta_k,\varphi_l) - C_\varphi^{\text{comp}}(\vartheta_k,\varphi_l)\right)^2\right\}\right]^{1/2} \tag{12}$$

with $$C_{\vartheta,\varphi}^{\text{ref,comp}}(\vartheta_k,\varphi_l) = \frac{\left|\underline{E}_{\vartheta,\varphi}^{\text{ref,comp}}(\vartheta_k,\varphi_l)\right|}{\left|\max\left\{\underline{E}_{\vartheta,\varphi}^{\text{ref,comp}}\right\}\right|},$$

where $C_{\vartheta,\varphi}^{\text{ref,comp}}$ denotes the normalized radiation characteristics in terms of the co- (CP) and cross-polarized (XP) components of the far-field of the related antenna element. In this context, the RMS error gives the total amount of field error in terms of one singular value regardless in which component or direction the field deviation occurs. In Fig. 8 the RMS error in dB is evaluated and depicted for a log.-per. four-arm antenna with $M = 5$ periods of trapezoidal unit cell in the considered frequency range from 1.0 to 6.0 GHz for different SME expansion degrees N.

As shown in Fig. 8, the spherical field approximation exhibits an erratic error behaviour with an increasing degree N of the field expansion. Whereas the RMS error ranges above -20 dB for expansion degrees $N = 0, 1$ the error is significantly reduced by applying expansion degrees $N = 2, 3$ and provides an already excellent far-field approximation with a marginal value of the RMS error ranging below $e_{RMS} < -25$ dB over the entire frequency range of operation. In addition, increasing the maximum SME degree up to $N = 10$ just offers a slight error improvement. Obviously, the discontinuities in the approximation depend on the discrete, modal excitation behaviour due to the vector coefficient's functional nulls, as exposed in Sect. 3.4.

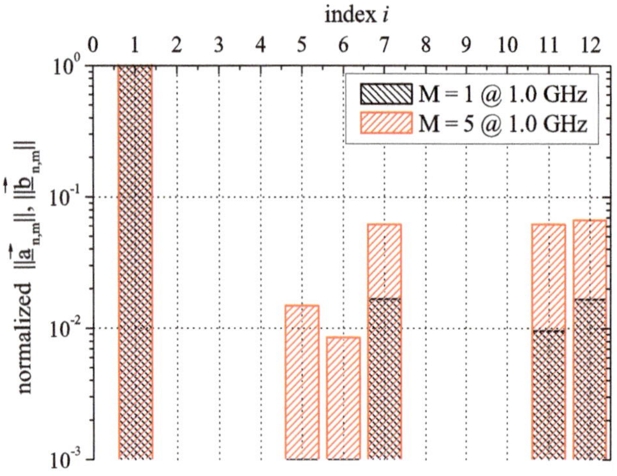

Fig. 10. Normalized vector coefficients at $f = 1.0\,\mathrm{GHz}$ for the antenna structures with $M = 1.5$

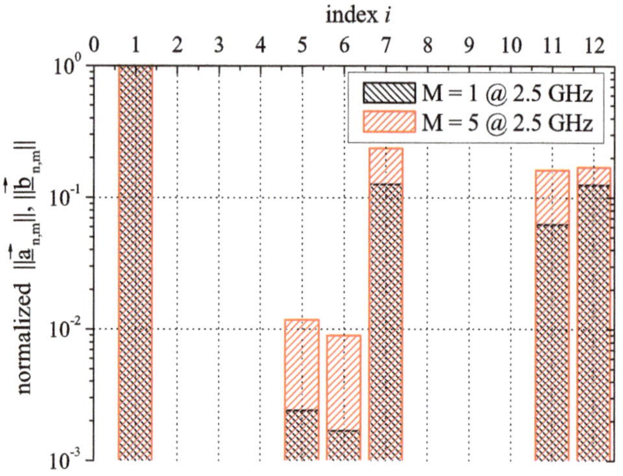

Fig. 11. Normalized vector coefficients at $f = 2.5\,\mathrm{GHz}$ for the antenna structures with $M = 1, 5$

Furthermore, in Fig. 9 the various radiation characteristics of the considered trapezoidal antenna are depicted, as utilized for the RMS error analysis that was shown in Fig. 8. The normalized antenna radiation patterns are depicted in the distinct far-field cutting plane of $\varphi = 90°$ at an operational frequency of $f = 2.5\,\mathrm{GHz}$ and separated with respect to its co- (ϑ-) and cross-polarized (φ-) far-field components. This diagram directly highlights angular regions with a very high degree of field approximation and marks other regions that mainly contribute to an enhanced RMS error.

Figure 9 demonstrates the excellent copolar SME field approximation that is already obtained for very small degrees $N \geq 0$ in antenna main beam direction at $\vartheta = 0°$. Even the radiation pattern nulls at $\vartheta = 90°$ and $\vartheta = 270°$ are approximated almost perfectly. This is a remarkable property of the field expansion with modulated spherical harmonics as shown in Sect. 3.3 due to the fact that the original spherical harmonic $Y^{e}_{0,0}(\vartheta, \varphi)$ represents a sphere in spherical coordinates ϑ and φ and therefore provides no nulls at the angular positions $\vartheta = 90°$ and $\vartheta = 270°$. In contrast, the cross-polar field component C_{φ} exhibits an attenuation by about 35 dB in antenna main beam direction and requires an SME field approximation of a higher degree. Therefore the cross polar component of the far-field exhibits a lower convergence behaviour than the co-polar component C_{ϑ}. An improvement of the far-field approximation is obtained in terms of an increasing degree of the field expansion.

Recapitulating this Section, the chosen spherical current-based eigenfield approach was verified and analyzed in terms of its field convergence behaviour, providing a basis for a significant parameter reduction with respect to the computation of antenna radiation fields and therefore may yield a much faster numerical evaluation.

5 Modal four-arm trapezoidal antenna analysis

As demonstrated in Sect. 3.4, the entire antenna's source region can be completely specified by vector coefficients of a spherical mode expansion. E.g. applying an approximation with a maximum degree $N = 2$ of the field expansion incorporating 12 SME vector coefficients yields an acceptable field approximation for both components of the far-field. Hence, the SME-based antenna analysis may be reduced to an in-depth consideration of the related modal SME vector coefficients as an equivalent source for far-field radiation.

For this purpose the extraction of dominant modal vector coefficients $\underline{\boldsymbol{a}}^{\,\mathrm{m}}_{\mathrm{n}}$ and $\underline{\boldsymbol{b}}^{\,\mathrm{m}}_{\mathrm{n}}$ will be identified and compared by means of the complex vector norm $\|\underline{\boldsymbol{a}}^{\,\mathrm{m}}_{\mathrm{n}}\|$ and $\|\underline{\boldsymbol{b}}^{\,\mathrm{m}}_{\mathrm{n}}\|$, as follows:

$$\left\{ \begin{matrix} \|\underline{\boldsymbol{a}}^{\,\mathrm{m}}_{\mathrm{n}}\| \\ \|\underline{\boldsymbol{b}}^{\,\mathrm{m}}_{\mathrm{n}}\| \end{matrix} \right\} = \sqrt{ \left\{ \begin{matrix} |\underline{a}^{\,\mathrm{nm}}_{\mathrm{x}}|^2 \\ |\underline{b}^{\,\mathrm{nm}}_{\mathrm{x}}|^2 \end{matrix} \right\} + \left\{ \begin{matrix} |\underline{a}^{\,\mathrm{nm}}_{\mathrm{y}}|^2 \\ |\underline{b}^{\,\mathrm{nm}}_{\mathrm{y}}|^2 \end{matrix} \right\} } . \tag{13}$$

Furthermore, the complex vector norm given by Eq. (13) is normalized to the dominant coefficient, which is given by $\underline{\boldsymbol{a}}^{\,0}_{0}$. The real values are logarithmically scaled and depicted in Figs. 10 to 12 for three different exclusive frequencies of 1.0, 2.5 and 6.0 GHz, respectively. The analysis is conducted for two trapezoidal antennas with a variable number of basis function periods $M = 1$ and $M = 5$, as given in Sect. 2. The abscissa scaling is given in terms of an index value i according to Table 2 with its respective vector coefficients.

Besides the magnitude of the vector coefficient $\underline{\boldsymbol{a}}^{\,0}_{0}$ (index 1) the evaluation of Figs. 10 to 12 points out three dominant coefficients, namely $\underline{\boldsymbol{a}}^{\,0}_{2}$ (index 7), $\underline{\boldsymbol{a}}^{\,2}_{2}$ (index 11) and $\underline{\boldsymbol{b}}^{\,2}_{2}$ (index 12). The magnitude of the other two excited coefficients $\underline{\boldsymbol{a}}^{\,1}_{1}$ (index 5) and $\underline{\boldsymbol{b}}^{\,1}_{1}$ (index 6) is significantly weaker with a level of about $-20\,\mathrm{dB}$ compared to the reference mode $\underline{\boldsymbol{a}}^{\,0}_{0}$.

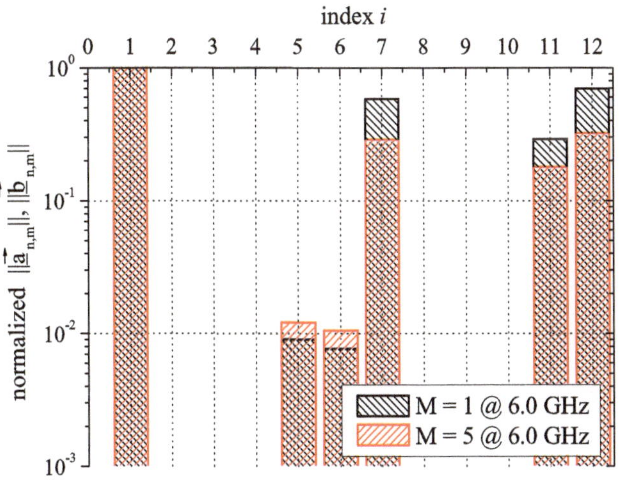

Fig. 12. Normalized vector coefficients at $f = 6.0$ GHz for the antenna structures with $M = 1, 5$

From $f = 1.0$ GHz depicted in Fig. 10 to $f = 6.0$ GHz shown in Fig. 12 the modal vector differences between both trapezoidal antennas are noticeable decreasing. The variation especially at $f = 1.0$ GHz between the two vector magnitudes is related to the reflections at the end of the arms. Otherwise, at $f = 6.0$ GHz both antenna geometries operate in their frequency-independent radiation behaviour which can be seen from the almost identical magnitudes of the normalized vector coefficients (see Fig. 12). In Fig. 11 at $f = 2.5$ GHz the structure with $M = 5$ periods radiates in a frequency-independent travelling wave mode, whereas the second structure with $M = 1$ periods of the trapezoidal unit cell is still predominated by resonant radiation characteristics.

As shown in this section, the consideration and the analysis of four dominant current-based vector quantities allows a deeper insight into the radiation behaviour of planar, trapezoidal antennas, combined with a faster numerical far-field antenna analysis due to a significant parameter reduction compared to conventional methods.

6 Conclusions

In this paper, planar frequency-independent log.-per. trapezoidal antenna structures are analysed by means of surface current distribution and a spherical eigenfield approach. Antennas of that kind are characterized by a high impedance bandwidth combined with a dual linear-polarized radiation behaviour which can be favorably used in future MIMO- and multimode diversity communication systems. The generalized spherical mode expansion was adapted and verified by different convergence criteria with respect to field-quantities as derived from conventional full-wave analysis applying the FEM. Leading to a significant parameter reduction and a complete coordinate separation between the sources in the near-field and the requested far-field region, a modified spherical wave expansion was accomplished that yielded a description of antenna radiation fields by means of a distinct and solely antenna-dependent set of complex expansion coefficients. The radiation behaviour analysis of two different trapezoidal antenna structures was accomplished by comparing the two distinct sets of modal expansion coefficients that were directly related to the operational mode of the antenna, in order to highlight the intermediate relationship between the set of vector expansion coefficients and the shape of the antenna, respectively. Furthermore, in Klemp et al. (2006) this kind of current-based SME technique is utilized to deduce equations for a modal antenna correlation analysis in terms of MIMO transmission schemes using trapezoidal planar antennas with dual linear excitation.

References

Bruning, J. H. and Lo, Y. T.: Multiple Scattering of EM Waves by Spheres Part I-Multipole Expansion and Ray-Optical Solutions, IEEE Transactions on Antennas and Propagation, 19, 378–390, 1971.

Chen, Y. and Simpson, T.: Radiation Pattern Analysis of Arbitrary Wire Antennas Using Spherical Mode Expansions with Vector Coefficients, IEEE Transactions on Antennas and Propagation, 39, 1716–1721, 1991.

Chen, Y., Simpson, T. L., and Ho, T. Q.: Highly efficient technique for solving radiation and scattering problems, IEE Proc.-H, 139, 7–10, 1992.

Jensen, M. A. and Wallace, J. W.: A Review of Antennas and Propagation in MIMO Wireless Communications, 52, 2810–2824, 2004.

Klemp, O. and Eul, H.: Radiation Pattern Analysis of Antenna Systems for MIMO and Diversity Configurations, Adv. Radio Scie., 3, 157–165, 2005.

Klemp, O., Schultz, M., and Eul, H.: Novel logarithmically periodic planar antennas for broadband polarization diversity reception, Int. J. Electron. Commun. (AEUE), 59, 268–277, 2005.

Klemp, O., Armbrecht, G., and Eul, H.: Computation of Antenna Pattern Correlation and MIMO Performance by means of Surface Current Distribution and Spherical Wave Theory, Accepted for publication in Adv. Radio Scie., 2006.

Ludwig, A. C.: Near-Field Far-Field Transformations Using Spherical-Wave Expansions, IEEE Transactions on Antennas and Propagation, 19, 214–220, 1971.

Narasimhan, M. S., Christopher, S., and Varadarangan, K.: Modal Behavior of Spherical Waves from a Source of EM Radiation with Application to Spherical Scanning, 33, 350–354, 1985.

Potter, P. D.: Application of Spherical Wave Therory to Cassegrainian-Fed Paraboloids, IEEE Transactions on Antennas and Propagation, 15, 727–736, 1967.

Stratton, J. A.: Electromagnetic Theory, McGraw Hill, New York, 1941.

Werner, D. H. and Mittra, R.: Frontiers in Electromagnetics, IEEE Press Series on Microwave Technology and RF, New York, 2000.

25

Fault tree analysis for system modeling in case of intentional EMI

E. Genender[1], M. Mleczko[1], O. Döring[1], H. Garbe[1], and S. Potthast[2]

[1]Institute of Electrical Engineering and Measurement Technology, Leibniz Universität Hannover, Hannover, Germany
[2]Bundeswehr Research Institute for Protective Technologies and NBC-Protection Munster, Germany

Abstract. The complexity of modern systems on the one hand and the rising threat of intentional electromagnetic interference (IEMI) on the other hand increase the necessity for systematical risk analysis. Most of the problems can not be treated deterministically since slight changes in the configuration (source, position, polarization, ...) can dramatically change the outcome of an event. For that purpose, methods known from probabilistic risk analysis can be applied. One of the most common approaches is the fault tree analysis (FTA). The FTA is used to determine the system failure probability and also the main contributors to its failure. In this paper the fault tree analysis is introduced and a possible application of that method is shown using a small computer network as an example. The constraints of this methods are explained and conclusions for further research are drawn.

1 Introduction

The complexity of modern electronic systems such as IT networks does not allow to examine the susceptibility of that systems against intentional EMI (IEMI) solely deterministically (Garcia, 2008; Mansson et al., 2009; Holland and John, 1999). The reason for this are the uncertainties associated with the analysis:

- Sources: There are a lot of different types of sources with different characteristics (Giri and Tesche, 2004). To each source the system might respond differently.
- Coupling paths: Depending on the position of the source or its orientation different coupling paths are dominant.
- System behavior: For the same source, same position and same orientation the system might fail or not, which is a random event regarding the huge number of internal states of the system (Camp et al., 2004).

The aspects described above are typically out of analysts control. It is not possible to determine which type of source the attacker is going use or how he is going to position it. Hence, these uncertainties need to be treated statistically. Furthermore, there are aspects which can be determined exactly however because of the complexity of the system have to be estimated. The electromagnetic field inside a shielded room depends on so many factors that the deterministic calculation becomes impossible. The application of worst case scenarios and safety factors will make the system overprotected and thus too expensive. That is why simple hardening against all sources and all possible situations cannot be done. Furthermore, the question about the risk that the systems is exposed to remains still unanswered. Hence, the system needs to be analyzed statistically taking the uncertainties into consideration.

Methods for analyzing the system reliability and risk have been developed and used for many years. Especially the nuclear and the aerospace industries have made a lot of progress in this area (Stamatelatos, 2002). When analyzing the risk of a complex system it is important to proceed systematically. For that, the fault tree analysis (FTA) is an important tool. In Sect. 2 a short introduction to fault tree analysis is given. Then in Sect. 3 a computer network is introduced for which the reliability analysis is carried out. Section 4 is the main section and discusses how the fault tree for the computer network can be developed. First the functional and then IEMI specific dependencies are modeled. The restrictions of FTA associated with that are discussed.

2 Fault tree analysis

It is difficult to predict the failure probability of the whole system in a certain electromagnetic environment at once. By decomposing its failure behavior in smaller scenarios the description becomes possible. This principle is also used in

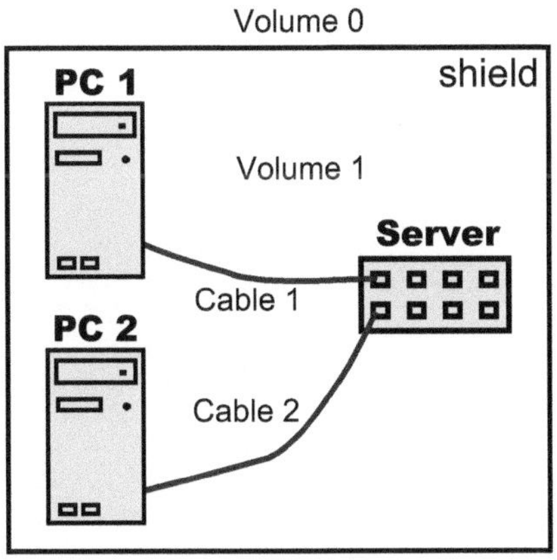

Fig. 1. Network which reliability is to be determined.

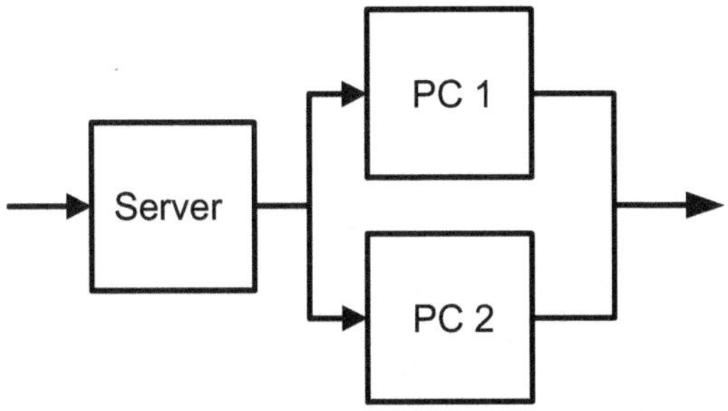

Fig. 2. Reliability block diagram for the network from Fig. 1.

the electromagnetic topology concept. In order to analyze the system reliability the fault tree analysis (FTA) can be applied (Vesely, 2002). Using the FTA for analyzing a system exposed to an EMP was first mentioned by National Research Council (1984). However, as far as the authors know, FTA has never been applied for that purpose. The FTA is a deductive method: beginning with an undesired event (also called the top event) the FTA is used to find the causes for this top event. When determining the causes, a fault tree is constructed from top to bottom. For its construction several symbols are used that indicate the relation between different events. The main symbols are presented in Table 1.

3 Computer network under test

In order to present the application of FTA, a computer network is introduced in Fig. 1. The network is placed in a (shielded) room and consists of one server and two computers (PCs). The two PCs are connected to the server. Both PCs fulfill the same function and are thus redundant. Now the task is to analyze the system reliability when it is exposed to an electromagnetic interference. It is assumed that the interference source is outside of the shielded room.

The first step of the analysis is to determine the functional dependencies of the system. A reliability block diagram is

Table 1. Fault tree symbols.

Symbol	Name	Meaning
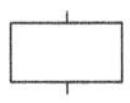	Intermediate Event	An event is caused by combination of other events
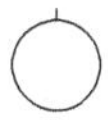	Basic Event	A basic event requiring no further development
	AND Gate	Output occurs if all of the input events occur
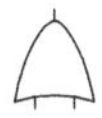	OR Gate	Output occurs if at least one of the input events occurs

used to visualize the dependency of the overall system on the functionality of its elements. Each element of the system is represented by a block. The blocks are connected depending on their function in the system. A serial connection of two blocks means that both blocks are needed for the successful operation of the system. A parallel connection of blocks means that at least one of the blocks is needed. The system operates successfully if a path between the left and the right end of the reliability block diagram exists in which all blocks are functioning. The reliability block diagram for the introduced computer network is shown in Fig. 2.

4 Fault tree analysis of the computer network

4.1 Fault tree for system failure

At the beginning of each FTA the top event is defined. For this event the causes are then determined successively. For the network under test different events might be defined. The determination of all possible top events cannot be done with FTA. For that other tools need to be used. Possible top events could be permanent damage of the network, degradation of the performance, temporal uncontrollability of the network and so on. In this paper the top event is simply defined as the failure of the whole system which is denoted as F_{Sys}.

From the reliability block diagram in Fig. 2 it is concluded that the system fails, if the server fails or the redundant (parallel) system consisting of the two PCs fails. That relation is depicted in Fig. 3 by the OR connection of the events *Server fails* F_{Ser} and *Redundant system (PC1&2) fails* $F_{\mathrm{PC1\&2}}$. The event $F_{\mathrm{PC1\&2}}$ is further developed as the AND connection of the events F_{PC1} and F_{PC2}. Using Boolean algebra the system failure event can be described as:

$$\begin{aligned} F_{\mathrm{Sys}} &= F_{\mathrm{Ser}} \cup F_{\mathrm{PC1\&2}} \\ &= F_{\mathrm{Ser}} \cup (F_{\mathrm{PC1}} \cap F_{\mathrm{PC2}}) \end{aligned} \tag{1}$$

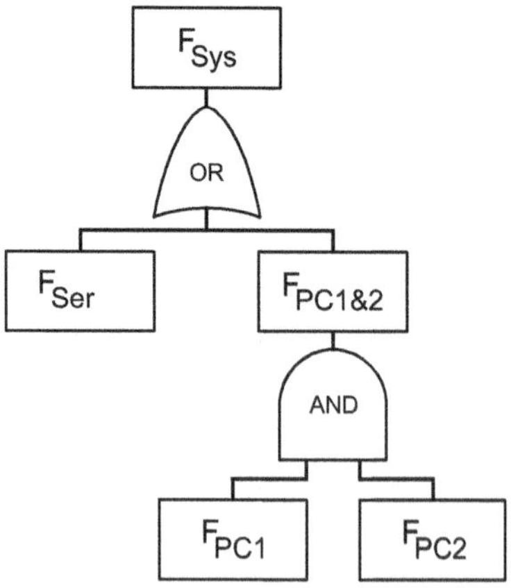

Fig. 3. Fault tree for the top event *System fails* F_{Sys}.

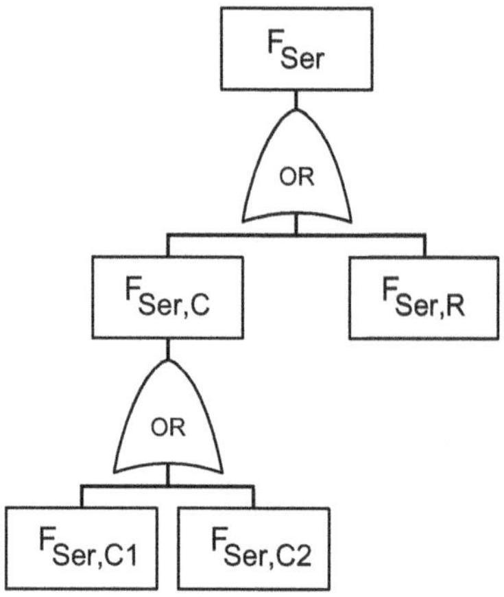

Fig. 4. Fault tree for the top event *Server failed.*

4.2 Fault tree for the server failure event

The fault tree which is developed until here is not directly related to IEMI problems and the main focus is on functional dependencies. The causes of the bottom events in Fig. 3 however are related to IEMI. For each of the bottom events a fault tree needs to be developed. In this paper we take a closer look at the failure behavior of the server F_{Ser}. Hence, the new top event is defined as *Server failed* (F_{Ser}). The fault tree which is going to be developed for this event can later just be attached at the corresponding position of the fault tree in Fig. 3.

Typically, electric equipment can be disturbed by radiated or conducted interference. Hence, the next step is to model the failure of the server as an OR connection of the events *Server failed through conducted disturbance* ($F_{\mathrm{Ser,C}}$) and *Server failed through radiated disturbance* ($F_{\mathrm{Ser,R}}$), see Fig. 4.

Those two events require further analysis. In Fig. 1 it can be seen that two cables are connected to the server. Hence, the event *Server failed through conducted disturbance* ($F_{\mathrm{Ser,C}}$) can be developed as the combination of the events *Server failed through conducted disturbance on cable 1* ($F_{\mathrm{Ser,C1}}$) and *Server failed through conducted disturbance on cable 2* ($F_{\mathrm{Ser,C2}}$). The top event has been decomposed into lower level events that are easier to analyze for themselves. Using Boolean algebra the server failure event can be described as:

$$\begin{aligned} F_{\mathrm{Ser}} &= F_{\mathrm{Ser,C}} \cup F_{\mathrm{Ser,R}} \\ &= F_{\mathrm{Ser,C1}} \cup F_{\mathrm{Ser,C2}} \cup F_{\mathrm{Ser,R}} \end{aligned} \qquad (2)$$

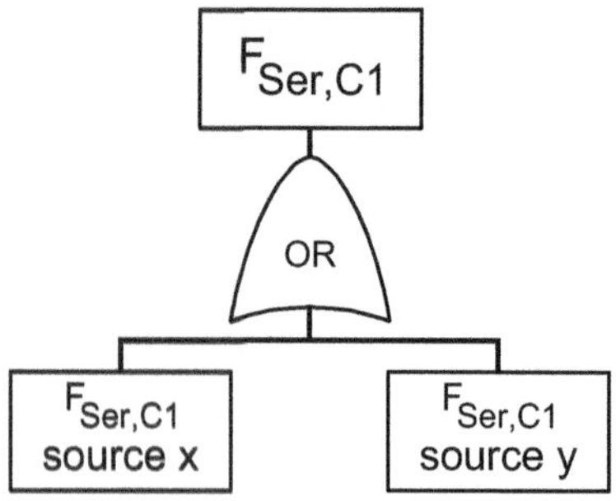

Fig. 5. Fault tree for the event *Server failed through conducted disturbance on cable 1*.

4.3 Fault tree for the server failure through disturbance on cable 1

At this point there is a need to know how susceptible the server at different ports is. The susceptibility of each port depends on the type of the source. In Brauer et al. (2009) it is shown that different types of cables are susceptible to different types of pulses. For example in Brauer et al. (2009) it is shown that, an ultra wideband (UWB) source couples well to network cables, damped sinusoid (DS) source to power supply and high power microwave (HPM) into the casing of the analyzed equipment. Hence, at this point of the fault tree failures through different sources are distinguished. For each cable only those sources need to be taken into consideration, which are important for this type of cable. In this paper the event *Server failed through conducted disturbance on cable 1* ($F_{\mathrm{Ser,C1}}$) is further developed. It is assumed that the server is susceptible on cable 1 to source x and sources y, which is depicted in the fault tree in Fig. 5.

In order for the server to fail through source x this source needs to occur and given the presence of the source x the server has to fail. Let's assume that the source is a high-altitude EMP (HEMP). This event has a very low probability, let's say 1%. However, given the occurrence of the source the probability that the server fails is very high, let's say 80%. Then the probability of the server to fail through a HEMP is $0.01 \cdot 0.8 = 0.008$. This relation can be depicted as an AND connection as can be seen in Fig. 6.

As both events at the bottom of the fault tree are basic events the fault tree development for this branch is finished

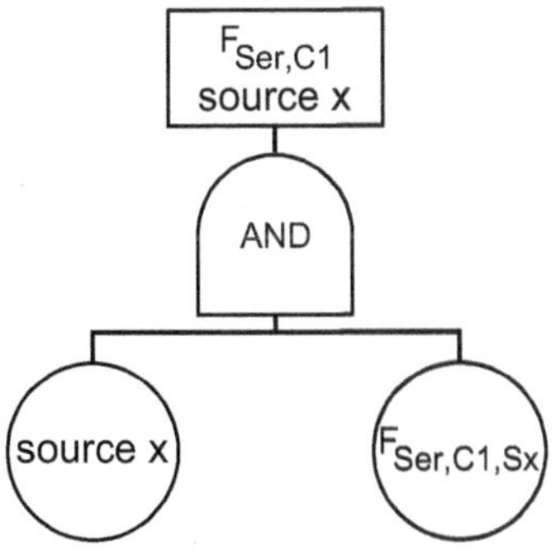

Fig. 6. Fault tree for the event *Server failed through conducted disturbance on cable 1 through source X*.

and has to be continued for other branches. When all branches are finished the fault tree can be solved.

5 Solution and restrictions of the fault tree

The fault tree is developed top-down and solved bottom-up. This means that the probabilities of the basic events have to be defined and then propagated through the fault tree. In the fault tree in Fig. 6 the basic events are the probability of the source and the failure probabilities of the elements given a certain source. These basic events are analyzed in the following.

5.1 Source likelihoods

In order to determine the source likelihoods (Sabath and Garbe, 2009) suggest to evaluate the mobility and the technological challenge of a source. The smaller and lighter the source is the higher is its mobility. The technological challenge combines quantities such as the level of knowledge needed to design or operate the system, availability of components and costs. The lower the technological challenge and the higher the mobility of a source is, the higher is its likelihood.

5.2 Failure probability of components

The failure probability of single elements can be estimated from the knowledge of strength (failure threshold) of the equipment and the stress which the equipment is exposed to.

5.2.1 Estimation of the strength of the components

The strength can be measured or estimated in a separate arrangement by inducing different voltages on the cable and then collecting the failure statistics of the device. Another option is to estimate the strength from the experience of the EMC engineer or from the known strengths of similar equipment.

In previous research, Camp et al. (2004) have shown that the strength (breakdown failure rate (BFR)) can be described by the Weibull distribution. The parameters of that distribution are however dependent on the shape of the radiated electromagnetic pulse. For different pulse shapes different breakdown behavior is expected. Using the window norms in frequency domain the energy- and amplitude efficiency factors (Nitsch et al., 2004) can be calculated. With the help of the energy- and amplitude efficiency factors the breakdown behavior for one pulse shape can be estimated from the knowledge of the breakdown behavior for another pulse shape.

5.2.2 Estimation of the stress

Besides the strength of the components the knowledge of the stress that the elements are exposed to is required. For that the principles of the electromagnetic (EM) topology can be applied (Baum, 1980). The main idea of EM topology is the same as in fault tree analysis: a complex system is decomposed into smaller subsystems. That way, the subsystems can be analyzed independently. In order to analyze the EM topology the whole system is divided into subvolumes which are separated by surfaces. A surface can be represented by a transfer function from one volume to another. Having the absolute knowledge of the situation including information like the exact positions of the objects, material parameters, angle of incident of electromagnetic wave, there would be no need to analyze the system statistically. In reality, we don't posses the absolute knowledge of the situation but a certain degree of belief or expectation about the true situation and thus we have uncertainties. Statistical analysis of coupling on cables in shielded environments is well described by Holland and John (1999). In the classical EM topology concept however, the uncertainties are not taken into account.

In order to take the uncertainties into consideration, the classical topological concept needs to be extended. Not only the transfer functions need to be considered but also the uncertainties of these transfer functions.

In order to analyze the system statistically there are two conceivable ways. The first way is to extend the topology concept in order to include the uncertainties and then to use the results as input data for the fault tree. The second way is to further extend the fault tree and to include the effects of shielding and propagation in this fault tree. By choosing the fault tree analysis, the whole system can be described by one model. There are however restrictions of the fault tree. The events in the tree are binary. Continuous values, such as field strength or shielding effectiveness, can be included if the symbols of the fault tree are extended. However the more serious restriction is that a disturbance in the electromagnetic topology graph can take several paths to each system element. Trying to implement this parallel paths in the fault tree is difficult and for some cases even impossible. Because of that the classical topology should be used in order

Fig. 7. Redundant two component system.

to describe the stress which the components are exposed to. For that the classical topology needs to be extended in order to include the uncertainties. This can be achieved by assigning the transfer function between two volumes uncertainties. This uncertainty can be described by single values such as the standard deviations or by whole probability density function. For example, instead of using just a shielding effectiveness value of 40 dB, a pair consisting of the expected shielding value and its standard deviation can be used and propagated through the topology tree such as (40 dB, $\pm$10 dB).

6 Common cause failures

In our previous research we have measured the failure probability of a redundant two component system depicted in Fig. 7. It was observed that depending on the wiring of the system there was a strong dependency between the two components. This dependency has a very strong influence on the redundancy and thus on the failure probability of the whole system. In classical risk analysis this type of failure is called Common Cause Failures (CCF). There are several models that try to include the CCF into the overall model (Bedford and Cooke, 2001, pp. 140–152). In order to include the common cause failures in our model the fault tree can be extended. However, before that the common cause have to be analyzed. It is part of our future work to predict the dependencies of the failures in interconnected systems.

7 Summary and conclusions

In this paper an example is shown of how a fault tree for a complex system exposed to IEMI can be developed. For that, the first step is the analysis of functional dependencies and scenarios that can lead to system failure. In the next step a differentiation is made between the different ways the EM interference can disturb the system. For that, radiated and conducted interference are modeled separately in the fault tree. Furthermore a distinction is made between the different entrance points of conducted interference. Hereafter, a distinction between different IEMI sources is made. It is discussed how the source likelihoods can be estimated and how the breakdown strength can be analyzed and determined in a separate setup. In order to model the disturbance two approaches are possible. The first approach is the classical electromagnetic topology approach. However, the uncertainties associated with modeling and the randomness of events must be included into the topology concept. The second approach is the further development of the fault tree. Because of the restrictions of the fault tree to model parallel propagation of the disturbance a sugestion is made to extend the classical topology concept in order to include statistical aspects.

The calculation of the probability of the top event and of the uncertainties associated with it is one part of the fault tree analysis. The other and not less important parts are the sensitivity and importance analysis. With the help of sensitivity analysis it is possible to determine how sensitive the model (result) reacts regarding changes of the model parameters. That way, it can be determined which model parameters should be modeled more precisely and which model parameters are less important and can thus be estimated roughly. In so doing, the efficiency of modeling time and effort can be improved. The importance analysis (Genender et al., 2010) is closely related to sensitivity analysis. With the help of importance analysis the significance of the system elements can be analyzed in regard to different aspects of the system behavior. Using the quantities like critical importance, risk reduction worth importance or risk achievement worth importance it is possible to determine the elements from which improvement the system will benefit most or the elements from which degradation the system will suffer least. Taking the importance measures into consideration, the system reliability can be improved systematically and cost-efficiently.

References

Baum, C.: Electromagnetic topology: A formal approach to the analysis and design of complex electronic systems, Interaction Notes, 400, 1980.

Bedford, T. and Cooke, R.: Probabilistic risk analysis: foundations and methods, Cambridge University Press, 2001.

Brauer, F., Sabath, F., and ter Haseborg, J. L.: Susceptibility of IT network systems to interferences by HPEM, 2009 IEEE International Symposium on Electromagnetic Compatibility, pp. 237–242, 2009.

Camp, M., Gerth, H., Garbe, H., and Haase, H.: Predicting the breakdown behavior of microcontrollers under EMP/UWB impact using a statistical analysis, IEEE Transactions on Electromagnetic Compatibility, 46, 368–379, 2004.

Garcia, E.: On the Use of Statistics in EMC for Industrial Complex Systems, in: CEM08, 2008.

Genender, E., Mleczko, M., and Garbe, H.: Importnace Analysis to Describe Reliablity of Systems, EMC Europe 2010, International Symposium on Electromagnetic Compatibility, 2010.

Giri, D. V. and Tesche, F. M.: Classification of Intentional Electromagnetic Environments (IEME), IEEE Transactions on Electromagnetic Compatibility, 46, 322–328, 2004.

Holland, R. and John, R. S.: Statistical Electromagnetics, Taylor & Francis Inc, 1999.

Mansson, D., Thottappillil, R., and Backstrom, M.: Methodology for Classifying Facilities With Respect to Intentional EMI,

IEEE Transactions on Electromagnetic Compatibility, 51, 46–52, 2009.

National Research Council: Evaluation of Methodologies for Estimating Vulnerability to Electromagnetic Pulse Effects, NATIONAL ACADEMY PRESS, 1984.

Nitsch, D., Camp, M., Sabath, F., ter Haseborg, J., and Garbe, H.: Susceptibility of some electronic equipment to HPEM threats, IEEE Transactions on Electromagnetic Compatibility, 46, 380–389, 2004.

Sabath, F. and Garbe, H.: Risk potential of radiated HPEM environments, 2009 IEEE International Symposium on Electromagnetic Compatibility, pp. 226–231, 2009.

Stamatelatos, M.: Probabilistic Risk Assessment, Procedures Guide for NASA Managers and Practitioners, NASA, version 1.1 edn., 2002.

Vesely, W.: Fault Tree Handbook with Aerospace Applications, NASA, 2002.

26

Development of a multiport test bench and systematic measurements of cable bundles for automotive EMC tests

M. Gonser[1], C. Keller[1], J. Hansen[1], and R. Weigel[2]

[1]Corporate Sector Research and Advance Engineering, Robert Bosch GmbH, Stuttgart, Germany
[2]Institute for Electronics Engineering, University of Erlangen-Nürnberg, Germany

Abstract. In this paper the development of a cable bundle test bench is described and exemplary results of the systematic measurement of cable bundles for automotive EMC tests are presented. The test bench consists of particularly developed adapter boxes and switch matrices, which allow together with a network analyzer to perform a network analysis with up to 32 ports and up to 1 GHz. Calibration and deembedding procedures are described and validated. Cable bundles that are characteristic to automotive EMC tests are investigated with respect to the number of wires within the cable bundle, the class of the cable bundle and the type of wires.

1 Introduction

In recent years, the rising complexity of automotive electronics along with shorter development periods challenges many design aspects of automotive products. One major issue is electromagnetic compatibility (EMC) and simulation of associated measurement setups even at early design stages become desireable.

Practical experience with nominally equal setups as well as round robin tests reveal considerable measurement uncertainty. Besides other sources for this uncertainty, the flexible wires within the cable bundle contribute largely to deviations in measurement results. A stochastic cable bundle model for respective simulations therefore becomes inevitable. In order to validate such stochastic models, statistical measurement data of cable bundles are needed.

While the literature shows a broad variety of stochastic cable bundle models (e.g., Ciccolella and Canavero, 1995; Salio et al., 1999; Sun et al., 2007; Steinmetz, 2006; Castanié, 2002), only few contain measurement data. Within most of these (Ciccolella and Canavero, 1995; Sun et al., 2007; Steinmetz, 2006) the reasonable approach is, to measure one resulting quantity (e.g. common mode current) and compare it to respective simulations. Influence of some parameters by measurement are investigated in Castanié (2002).

This work aims at a more extended investigation of cable bundle parameters and at founding a sound data base for the validation of a stochastic cable bundle model. While the model is presented in Gonser et al. (2010a,b) this paper describes the particularly developed test bench required for measuring and presents measurements of real cable bundles.

2 Cable bundle test bench

2.1 Overview

Figure 1 shows an overview of the complete test bench. The *cable bundle*, which rests on styrofoam at height h above the table, is connected to *adapter boxes* at each end. Each of the adapter boxes is connected by *secondary measurement cables* to a *switch matrix*. Unused connectors of the adapter box are terminated with 50 Ω SMA terminations. Finally, each of the switch matrices is connected through the *primary measurement cables* to a *vector network analyzer* (NWA). The adapter boxes together with the actual cable bundle form the *device under test* (DUT) and the respective reference planes are located as indicated in Fig. 1. The setup is controlled by a PC, which is connected to the switch matrices via USB and to the NWA via GPIB (Fig. 3). Automated calibration, measurement and deembedding procedures are implemented in MATLAB. A measurement with all 32 ports and 1000 frequency points takes approximately 8 min.

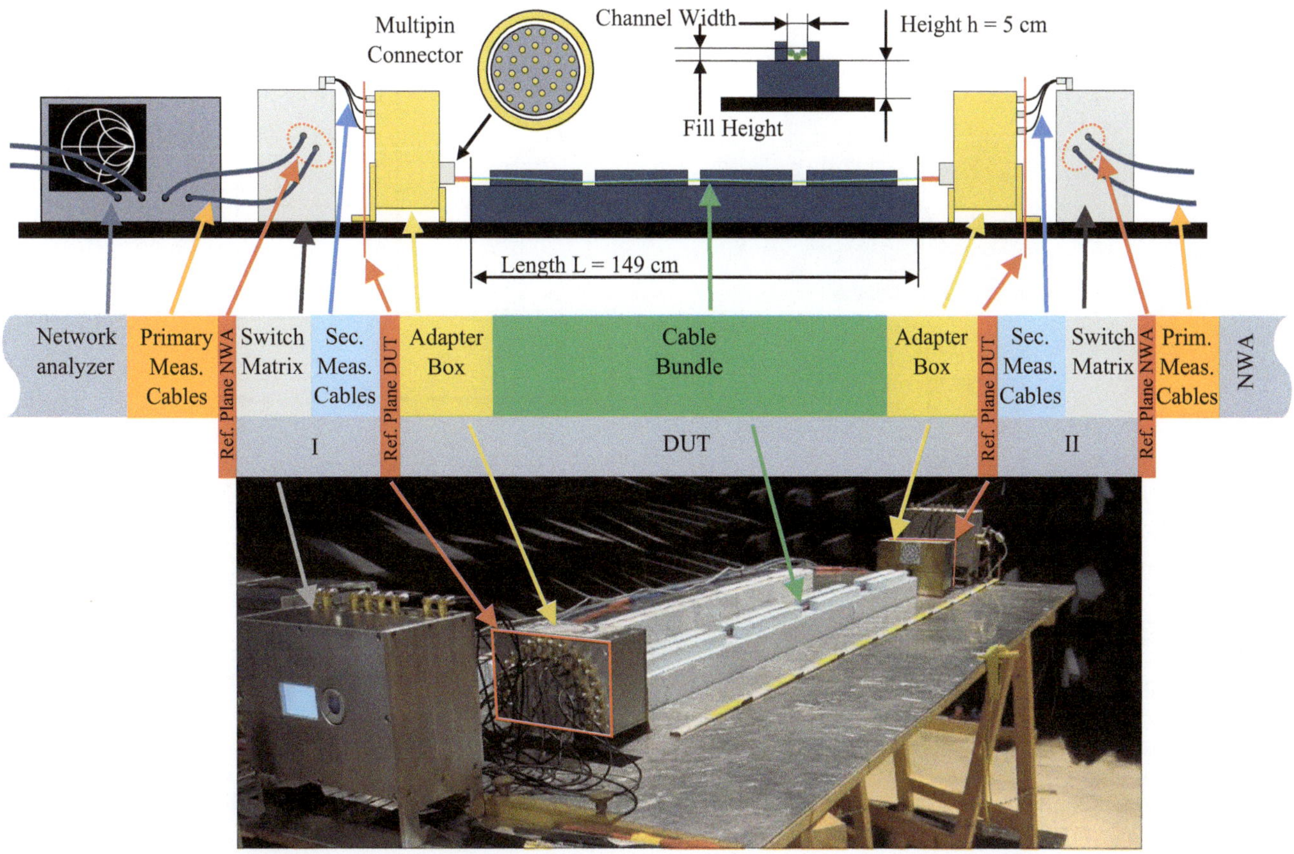

Fig. 1. Overview of the complete test bench.

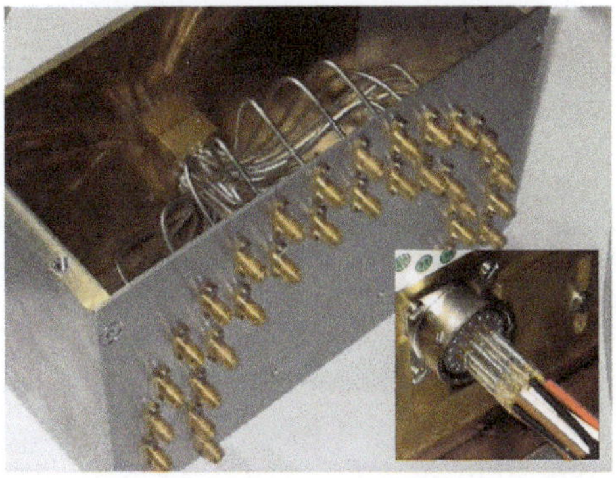

Fig. 2. Adapter box and multipin connector.

2.2 Adapter box

Figure 2 shows the inside of the adapter box and close-up view of the multipin connector. The use of the latter provides a quick and repeatable connection of the wires. Using coaxial semi rigid cables inside the box assures 50 Ω line impedance up to the back of the multipin connector, and is favourable in order to perform network analysis. Each adapter box can accommodate up to 32 wires.

2.3 Switch matrix

As shown in Fig. 3 the switch matrix consists of three main printed circuit boards (PCB) in a housing made of alloy:

- *High Frequency PCB*: The four layer PCB is made of high frequency substrate. The two inner layers are ground layers and a total of 64 relays (2 per housing) are hosted on both outer layers of the PCB. The electromechanical high frequency relays are connected via grounded coplanar microstrip lines with a line impedance of $Z_W = 50\ \Omega$ to each other in a topology, which allows the routing of each of the two NWA-sided ports arbitrarily to any of the 16 DUT-sided ports. All other DUT-sided ports are terminated with 50 Ω.
- *Driver PCB*: The driver PCB serves two purposes: providing the relays with current when they are selected and shielding the high frequency part of the box from the digital part of the box.
- *Microcontroller PCB*: A microcontroller, USB to UART converter, LCD and control buttons are hosted on a third board, mounted on the back of the front panel.

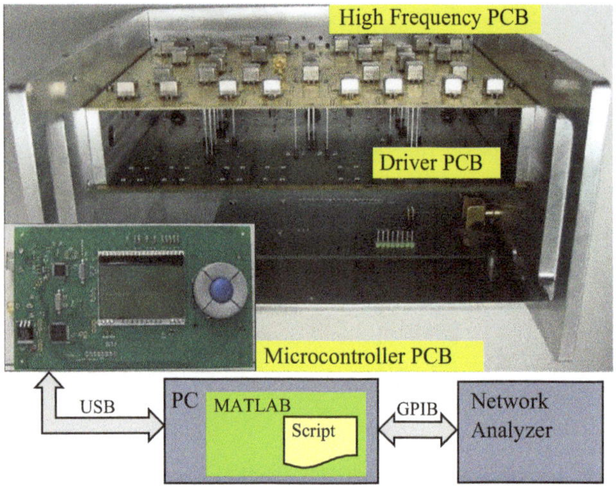

Fig. 3. Components and logical setup of the switch matrix

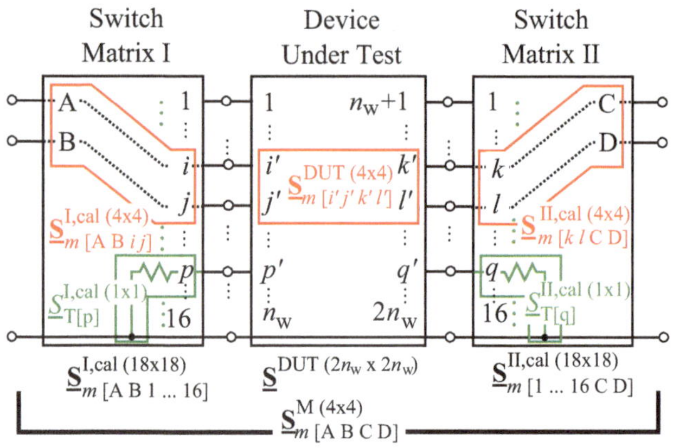

Fig. 4. Schematic of the setup.

2.4 Multiport network analysis

2.4.1 Working principle

In order to measure cable bundles with many wires a system for network analysis with many ports is required. This can be achieved by a switching matrix, which automatically reroutes the ports available on the NWA to the different ports on the DUT. Two switch matrices were built. Each of the two switch matrices I and II has two ports on the side of the NWA, denoted A and B for switch matrix I and denoted C and D for switch matrix II. On the side of the DUT, each switch matrix has 16 ports, denoted 1 through 16.

While the network analyzer together with the switch matrices allow for many setups, only the case of using a four port NWA and both switch matrices is considered in this paper. Further, the total (even) number of used ports shall be split equally between both switch matrices. Without loss of generality, the secondary measurement cables are not mentioned separately, as they can be treated as part of their switch matrix. The primary measurement cables are accounted for by the calibration of the NWA itself.

Figure 4 shows a schematic of this setup. In order to measure the scattering parameter $\underline{\mathbf{S}}^{\mathrm{DUT}\,(2n_\mathrm{w}\times 2n_\mathrm{w})}$ of the DUT, many measurements $\underline{\mathbf{S}}_{m\,[\mathrm{A\,B\,C\,D}]}^{\mathrm{M}\,(4\times4)}$ with the NWA have to be performed. For the m-th measurement, ports A and B of switch matrix I are routed to ports i and j and ports C and D of switch matrix II are routed to the ports k and l respectively. This shall be expressed by the *path setting* $\boldsymbol{P}_m = \begin{bmatrix} i & j & k & l \end{bmatrix}$. Ports i, j, k, and l of the switch matrices are connected to ports $i' = i$, $j' = j$, $k' = k + n_\mathrm{w}$, and $l' = l + n_\mathrm{w}$ of the DUT. One possible set of path settings, which covers all ports of the DUT is:

$$\boldsymbol{P}_m \in \left\{ \begin{bmatrix} i & j & k & l \end{bmatrix} \mid (1 \le i \le n_\mathrm{W}) \wedge (i < j \le n_\mathrm{W}) \wedge (k = i) \wedge (l = j) \right\} \tag{1}$$

Which results in a number of

$$N_\mathrm{M} = \binom{n_\mathrm{W}}{2} \tag{2}$$

total measurements, which have to be performed.

2.4.2 Calibration

Theoretically, for each of the N_M path settings $\boldsymbol{P}_m$ the full 18 port scattering parameters $\underline{\mathbf{S}}_{m\,[\mathrm{A\,B\,1}\ldots16]}^{\mathrm{I,cal}\,(18\times18)}$ and $\underline{\mathbf{S}}_{m\,[1\ldots16\,\mathrm{C\,D}]}^{\mathrm{II,cal}\,(18\times18)}$ had to be measured, which would result in impractical amount of calibration data. However, the design of the switch matrix provides high isolation between all ports that are not involved in the current measurement, as well as high isolation of these ports to the ports involved in the current measurement. Therefore, only the subsets $\underline{\mathbf{S}}_{m\,[\mathrm{A\,B}\,i\,j]}^{\mathrm{I,cal}\,(4\times4)}$ and $\underline{\mathbf{S}}_{m\,[k\,l\,\mathrm{C\,D}]}^{\mathrm{II,cal}\,(4\times4)}$ are determined. Additionally, for all ports that are connected to the DUT, their non-ideal terminations $\underline{S}_{\mathrm{T}\,[p]}^{\mathrm{I,cal}\,(1\times1)}$ and $\underline{S}_{\mathrm{T}\,[q]}^{\mathrm{II,cal}\,(1\times1)}$ seen by the DUT, need to be determined.

2.4.3 Measurement

In order to perform the actual measurement, for every path setting $\boldsymbol{P}_m$ of Eq. (1) a measurement $\underline{\mathbf{S}}_{m\,[\mathrm{A\,B\,C\,D}]}^{\mathrm{M}\,(4\times4)}$ with the NWA is performed. The measurement is then transformed to transfer parameters $\underline{\mathbf{T}}_{m\,[\mathrm{A\,B\,C\,D}]}^{\mathrm{M}\,(4\times4)}$ as described in Appendix B. Similarly, the calibration data $\underline{\mathbf{S}}_{m\,[\mathrm{A\,B}\,i\,j]}^{\mathrm{I,cal}\,(4\times4)}$ and $\underline{\mathbf{S}}_{m\,[k\,l\,\mathrm{C\,D}]}^{\mathrm{II,cal}\,(4\times4)}$ are transformed to the respective transfer parameters. As switch matrices and DUT form a chain of network elements, the transfer parameters of the DUT are then:

$$\underline{\mathbf{T}}_{m\,[i'\,j'\,k'\,l']}^{\mathrm{DUT}\,(4\times4)} = \underline{\mathbf{T}}_{m\,[\mathrm{A\,B}\,i\,j]}^{\mathrm{I,cal}\,(4\times4)\;-1}\,\underline{\mathbf{T}}_{m\,[\mathrm{A\,B\,C\,D}]}^{\mathrm{M}\,(4\times4)}\,\underline{\mathbf{T}}_{m\,[k\,l\,\mathrm{C\,D}]}^{\mathrm{II,cal}\,(4\times4)\;-1} \tag{3}$$

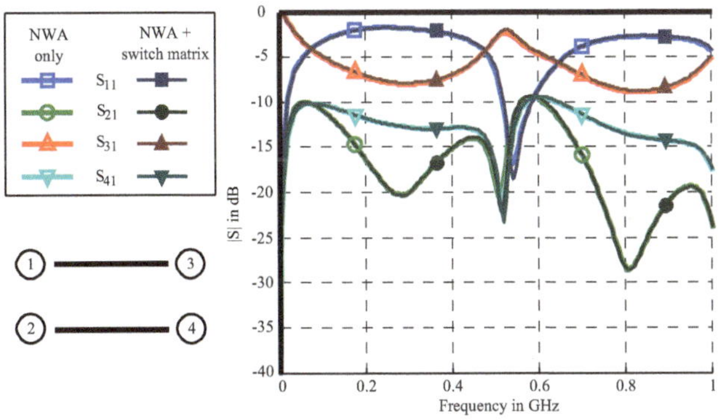

Fig. 5. Validation of the test bench.

These transfer parameters are under the premise, that all other ports of the DUT are terminated by ideal 50 Ω. In reality this is not the case and to account for the non-ideal termination at the DUT the method in Rautio (1983) is applied. It requires the calculation of the Gamma-R parameters (see Appendix C) of this measurement:

$$\underline{\mathbf{T}}_{m\,[i'\ j'\ k'\ l']}^{\mathrm{DUT}\,(4\times4)} \rightarrow \underline{\mathbf{S}}_{m\,[i'\ j'\ k'\ l']}^{\mathrm{DUT}\,(4\times4)} \tag{4}$$

$$\underline{\mathbf{S}}_{m\,[i'\ j'\ k'\ l']}^{\mathrm{DUT}\,(4\times4)} \xrightarrow{\underline{\Gamma}_{[i\ j\ k\ l]}=\underline{\mathbf{S}}_{\mathrm{T}[i\ j\ k\ l]}} \underline{\mathbf{R}}_{m\,[i'\ j'\ k'\ l']}^{\mathrm{DUT}\,(4\times4)} = \underline{\mathbf{R}}_m \tag{5}$$

These Gamma-R parameters $\underline{\mathbf{R}}_m$ are inserted in the matrix

$$\underline{\mathbf{R}}_{[1\ldots 2n_\mathrm{w}]}^{\mathrm{DUT}\,(2n_\mathrm{w}\times 2n_\mathrm{w})} = \begin{array}{c} \\ \\ i'\rightarrow \\ \\ j'\rightarrow \\ \\ k'\rightarrow \\ \\ l'\rightarrow \\ \\ \end{array} \begin{array}{c} \begin{array}{cccc} i' & j' & k' & l' \\ \downarrow & \downarrow & \downarrow & \downarrow \end{array} \\ \begin{bmatrix} \underline{R}_{m\,11} & \cdots & \underline{R}_{m\,12} & \cdots & \underline{R}_{m\,13} & \cdots & \underline{R}_{m\,14} & \cdots \\ \vdots & \ddots & \vdots & \ddots & \vdots & \ddots & \ddots & \vdots \\ \underline{R}_{m\,21} & \cdots & \underline{R}_{m\,22} & \cdots & \underline{R}_{m\,23} & \cdots & \underline{R}_{m\,24} & \cdots \\ \vdots & \ddots & \vdots & \ddots & \vdots & \ddots & \ddots & \vdots \\ \underline{R}_{m\,31} & \cdots & \underline{R}_{m\,32} & \cdots & \underline{R}_{m\,33} & \cdots & \underline{R}_{m\,34} & \cdots \\ \vdots & \ddots & \vdots & \ddots & \vdots & \ddots & \ddots & \vdots \\ \underline{R}_{m\,41} & \cdots & \underline{R}_{m\,42} & \cdots & \underline{R}_{m\,43} & \cdots & \underline{R}_{m\,44} & \cdots \\ \vdots & \ddots & \vdots & \ddots & \vdots & \ddots & \ddots & \vdots \end{bmatrix} \end{array} \tag{6}$$

Diagonal entries will be covered by more than one measurement, however all measurements for one diagonal entry provide the same data. Once all N_M measurements are pasted in $\underline{\mathbf{R}}^{\mathrm{DUT}}$ it can be transformed back to the appropriate scattering parameter

$$\underline{\mathbf{R}}_{[1\ \ldots\ 2n_\mathrm{w}]}^{\mathrm{DUT}\,(2n_\mathrm{w}\times 2n_\mathrm{w})} \xrightarrow{\underline{\Gamma}=\underline{\mathbf{S}}_{\mathrm{T}[1\ \ldots\ 2n_\mathrm{w}]}^{(2n_\mathrm{w}\times 2n_\mathrm{w})}} \underline{\mathbf{S}}_{[1\ \ldots\ 2n_\mathrm{w}]}^{\mathrm{DUT}\,(2n_\mathrm{w}\times 2n_\mathrm{w})} \tag{7}$$

which is the final deembedded measurement $\underline{\mathbf{S}}_{[1\ \ldots\ 2n_\mathrm{w}]}^{\mathrm{DUT}\,(2n_\mathrm{w}\times 2n_\mathrm{w})}$.

Fig. 6. Cable bundle classes (from left to right: channel of styrofoam, braided sleeve, taped).

2.5 Validation

The test bench is validated by a short two wire structure ($n_\mathrm{w} = 2$), which is inserted into the adapter boxes. The resulting 4 ports were measured directly with the NWA, as well as with the two switch matrices in place. The comparison is shown in Fig. 5 and both measurements agree very well with each other, confirming the presented multiport measurement method up to 1 GHz.

3 Cable bundle measurements

3.1 Overview and analysis

Table 1 shows an overview of the measured cable bundle configurations. The number of wires n_w, the size of the wire cross section A_w and the cable bundle class are varied. As shown in Fig. 6, *classes* of cable bundles are wires in a channel made of *styrofoam*, wires in a *braided sleeve* and wires wrapped by *tape*. While not all possible configurations are covered, parameters were chosen in such a way, that every parameter can be studied.

For every cable bundle configuration many specimen are generated and measured. In the case of styrofoam, the loose wires just have to be put newly in place. For the braided sleeve and the taped cable bundle, every time the cable bundle has to be first disassembled and then assembled again. It has been taken care that enough specimen were measured for a sound set of statistical data – i.e. the convergence of the statistical analyzed data are assured. Up to $N_\mathrm{C} = 100$ specimen are necessary, wires with a higher number of wires requiring less measurements N_C.

Finally, the measured scattering parameters $\underline{\mathbf{S}}_i^{\mathrm{DUT}}$ are categorized into reflection $\underline{S}_{j,\mathrm{REFL}}$, transmission $\underline{S}_{j,\mathrm{TRAN}}$, near end crosstalk $\underline{S}_{j,\mathrm{NEXT}}$, and far end crosstalk $\underline{S}_{j,\mathrm{FEXT}}$. Each of the groups is analyzed by calculating the statistical mean value

Table 1. Overview of the measured cable bundle configurations.

#	n_W	A_W	Class	N_C
1	4	$0.25\,\mathrm{mm}^2$	Styrofoam	100
2	4	$0.50\,\mathrm{mm}^2$	Styrofoam	100
3	4	$1.00\,\mathrm{mm}^2$	Styrofoam	100
4	9	$1.00\,\mathrm{mm}^2$	Styrofoam	60
5	15	$0.25\,\mathrm{mm}^2$	Styrofoam	30
6	15	$0.50\,\mathrm{mm}^2$	Styrofoam	50
7	15	$1.00\,\mathrm{mm}^2$	Styrofoam	60
8	32	$1.00\,\mathrm{mm}^2$	Styrofoam	60
9	15	$1.00\,\mathrm{mm}^2$	Braided sleeve	50
10	15	$1.00\,\mathrm{mm}^2$	Taped	50

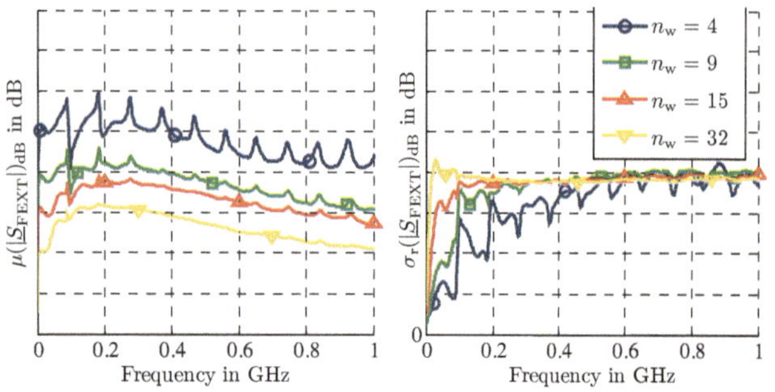

Fig. 7. Comparison of measurement results for $A_W = 1\,\mathrm{mm}^2$ of class styrofoam for the variation of n_W (left: 5 dB/div, right: 1 dB/div).

$$\mu\left(\left|\underline{S}_{\mathrm{cat}}\right|\right) = \frac{1}{N_{\mathrm{cat}}} \sum_{j=1}^{N_{\mathrm{cat}}} \left|\underline{S}_{j,cat}\right| \tag{8}$$

and empirical standard deviation

$$\sigma\left(\left|\underline{S}_{\mathrm{cat}}\right|\right) = \frac{1}{(N_{\mathrm{cat}}-1)} \sum_{j=1}^{N_{\mathrm{cat}}} \left(\underline{S}_{j,cat} - \mu(\left|\underline{S}_{\mathrm{cat}}\right|)\right)^2 \tag{9}$$

where

$$\mathrm{cat} \in \{\mathrm{REFL};\ \mathrm{TRAN};\ \mathrm{NEXT};\ \mathrm{FEXT}\}. \tag{10}$$

N_{cat} is the respective amount of measurements. For displaying purposes the logarithm is applied:

$$\mu(\left|\underline{S}_{\mathrm{cat}}\right|)_{\mathrm{dB}} = 20\log\left(\mu\left(\left|\underline{S}_{\mathrm{cat}}\right|\right)\right)\mathrm{dB} \tag{11}$$

The empirical standard deviation $\sigma\left(\left|\underline{S}_{\mathrm{cat}}\right|\right)$ is normalized by the mean value $\mu\left(\left|\underline{S}_{\mathrm{cat}}\right|\right)$ before applying the logarithm, which results in the relative empirical standard deviation:

$$\sigma_{\mathrm{r}}(\left|\underline{S}_{\mathrm{cat}}\right|)_{\mathrm{dB}} = 20\log\left(\frac{\sigma\left(\left|\underline{S}_{\mathrm{cat}}\right|\right)}{\mu\left(\left|\underline{S}_{\mathrm{cat}}\right|\right)} + 1\right)\mathrm{dB} \tag{12}$$

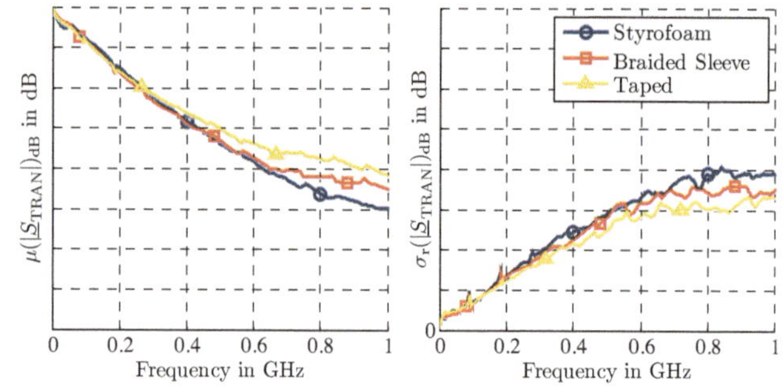

Fig. 8. Comparison of measurement results for $A_W = 1\,\mathrm{mm}^2$ and $n_W = 15$ and for the variation of the class (left: 5 dB/div, right: 1 dB/div).

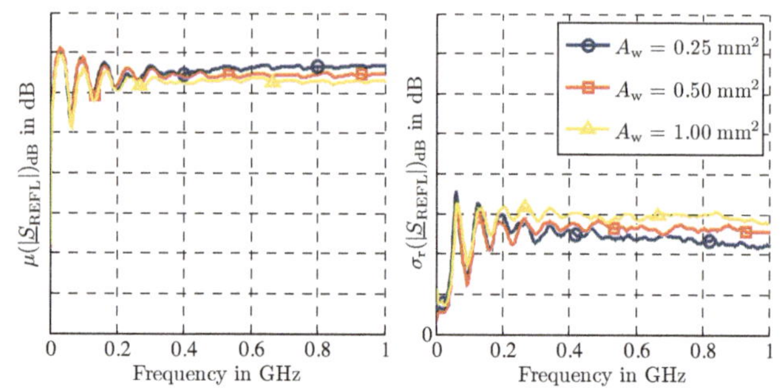

Fig. 9. Comparison of measurement results for $n_W = 15$ of class styrofoam for the variation of A_W (left: 5 dB/div, right: 1 dB/div).

3.2 Results

In the following, exemplary results are shown and discussed. If not otherwise mentioned, similar behaviour is observed also for the other categories *cat*.

Figure 7 shows the parameter study with respect to the number of wires $n_W \in \{4; 9; 15; 32\}$ by the example of far end crosstalk FEXT. Increasing the number of wires n_W leads to a decrease in the mean value $\mu(\left|\underline{S}_{\mathrm{FEXT}}\right|)_{\mathrm{dB}}$. The reason is, that the power of the source is coupled to an increasing number of adjacent wires $n_W - 1$. Further, resonances deviate more in frequency for the single specimen resulting in a mean value $\mu(\left|\underline{S}_{\mathrm{FEXT}}\right|)_{\mathrm{dB}}$ with smaller resonance peaks. The relative standard deviation $\sigma_{\mathrm{r}}(\left|\underline{S}_{\mathrm{FEXT}}\right|)_{\mathrm{dB}}$ is approximately the same for all configurations for frequencies above 500 MHz. Cable bundle configurations with many wires reach this asymptotical value at lower frequencies than configurations with less wires.

Variation in the cable bundle class generally shows only small deviations. Figure 8 shows this for the mean value $\mu(\left|\underline{S}_{\mathrm{TRAN}}\right|)_{\mathrm{dB}}$ and relative standard deviation $\sigma_{\mathrm{r}}(\left|\underline{S}_{\mathrm{TRAN}}\right|)_{\mathrm{dB}}$ for the category transmission TRAN. Examining the bundles, it can easily be seen that cable bundles of the class taped wires are much closer together than cable bundles of the class styrofoam. Braided sleeve is in between. As the same order is shown in Fig. 8 it is concluded that this change is due to the cable bundle density. Further, experience with a stochastic cable bundle model (Gonser et al., 2010a,b) show the same behavior in simulations, when the parameter density ρ of the model is varied.

Similar behavior can be observed from variation of the size of the wire cross section A_W as shown in Fig. 9, which displays the mean value $\mu(|\underline{S}_{REFL}|)_{dB}$ and relative standard deviation $\sigma_r(|\underline{S}_{REFL}|)_{dB}$ for the category reflection REFL. By experience during actual measurement it is assumed that the small variation of mean value $\mu(|\underline{S}_{REFL}|)_{dB}$ and relative standard deviation $\sigma_r(|\underline{S}_{REFL}|)_{dB}$ is also caused by variations of the cable bundle density. Thicker wires usually result in a more densely packed channel of styrofoam, than thinner wire types.

4 Conclusions

A test bench for measuring cable bundles is presented. It consists of a network analyzer, two adapter boxes and two switch matrices, which were particularly developed for this task. The test bench allows to perform a fully automated network analysis with 32 ports up to 1 GHz. Control and deembedding are implemented in MATLAB. The mathematical background of the calibration and measurement tasks are presented and validated against direct measurement with the network analyzer.

The test bench is used to systematically measure cable bundles of different configurations. The size of the wire cross section A_W, number of wires n_W and cable bundle class are varied and measurements for many of the resulting cable bundle configurations are performed. For each configuration up to 100 specimen were measured and analyzed. The influence of the parameters n_W, A_W and class are presented. As expected, the number of wires n_W has the greatest influence on the scattering parameter. Remarkably, at higher frequencies the relative standard deviation $\sigma_r(|\underline{S}_{cat}|)_{dB}$ is unaffected by a change of the number of wires n_W. The variation of the cable bundle class and the variation of the wire type A_W – both resulting in a change of cable bundle density – only show small effect on the scattering parameter.

The measurements presented are the basis for the validation of a stochastic cable bundle model as presented in Gonser et al. (2010a,b). The test bench will be used in future investigations of more complex cable bundles.

Appendix A

Scattering parameters

All networks in this paper have an even number of ports, which share the same ground node. Hence, a network with $2n$ ports has $(2n+1)$ nodes as shown in Fig. A1, the ports being equally split between both sides. As stated in Kurokawa

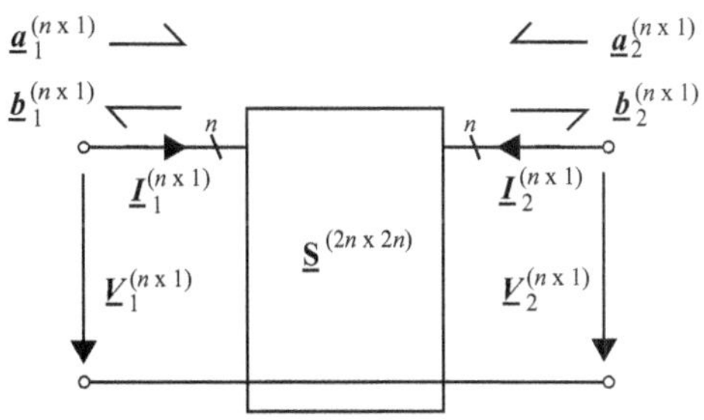

Fig. A1. Scattering parameter.

(1965), by using the nodal voltages $\underline{V}$ and the currents $\underline{I}$

$$\underline{V}^{(2n\times1)}=\begin{bmatrix}\underline{V}_1^{(n\times1)}\\ \underline{V}_2^{(n\times1)}\end{bmatrix}\qquad \underline{I}^{(2n\times1)}=\begin{bmatrix}\underline{I}_1^{(n\times1)}\\ \underline{I}_2^{(n\times1)}\end{bmatrix}\tag{A1}$$

the following waves can be defined:

$$\underline{a}^{(2n\times1)}=\underline{\mathbf{F}}^{(2n\times2n)}\left(\underline{V}^{(2n\times1)}+\underline{\mathbf{G}}^{(2n\times2n)}\underline{I}^{(2n\times1)}\right)\tag{A2}$$

$$\underline{b}^{(2n\times1)}=\underline{\mathbf{F}}^{(2n\times2n)}\left(\underline{V}^{(2n\times1)}-\underline{\mathbf{G}}^{+\,(2n\times2n)}\underline{I}^{(2n\times1)}\right)\tag{A3}$$

$\underline{\mathbf{G}}$ and $\underline{\mathbf{F}}$ are diagonal matrices where the i-th entry is the impedance of the i-th port $\underline{Z}_i$ and $1/|\mathrm{Re}(\underline{Z}_i)|$. The "+" denotes the complex conjugate transpose of the matrix. The scattering parameters $\underline{\mathbf{S}}$ are defined as:

$$\underline{b}^{(2n\times1)}=\underline{\mathbf{S}}^{(2n\times2n)}\underline{a}^{(2n\times1)}\tag{A4}$$

Or divided by sides as in Fig. A1:

$$\begin{bmatrix}\underline{b}_1^{(n\times1)}\\ \underline{b}_2^{(n\times1)}\end{bmatrix}=\begin{bmatrix}\underline{\mathbf{S}}_{11}^{(n\times n)} & \underline{\mathbf{S}}_{12}^{(n\times n)}\\ \underline{\mathbf{S}}_{21}^{(n\times n)} & \underline{\mathbf{S}}_{22}^{(n\times n)}\end{bmatrix}\begin{bmatrix}\underline{a}_1^{(n\times1)}\\ \underline{a}_2^{(n\times1)}\end{bmatrix}\tag{A5}$$

Appendix B

Transfer parameters

According to Klein (1976) the transfer parameters are

$$\begin{bmatrix}\underline{b}_1^{(n\times1)}\\ \underline{a}_1^{(n\times1)}\end{bmatrix}=\begin{bmatrix}\underline{\mathbf{T}}_{11}^{(n\times n)} & \underline{\mathbf{T}}_{12}^{(n\times n)}\\ \underline{\mathbf{T}}_{21}^{(n\times n)} & \underline{\mathbf{T}}_{22}^{(n\times n)}\end{bmatrix}\begin{bmatrix}\underline{a}_2^{(n\times1)}\\ \underline{b}_2^{(n\times1)}\end{bmatrix}\tag{B1}$$

Transformation between $\underline{\mathbf{S}}$ and $\underline{\mathbf{T}}$ is

$$\underline{\mathbf{T}}=\begin{bmatrix}\underline{\mathbf{S}}_{12}-\underline{\mathbf{S}}_{11}\underline{\mathbf{S}}_{21}^{-1}\underline{\mathbf{S}}_{22} & \underline{\mathbf{S}}_{11}\underline{\mathbf{S}}_{21}^{-1}\\ -\underline{\mathbf{S}}_{21}^{-1}\underline{\mathbf{S}}_{22} & \underline{\mathbf{S}}_{21}^{-1}\end{bmatrix}\tag{B2}$$

$$\underline{\mathbf{S}}=\begin{bmatrix}\underline{\mathbf{T}}_{12}\underline{\mathbf{T}}_{22}^{-1} & \underline{\mathbf{T}}_{11}-\underline{\mathbf{T}}_{12}\underline{\mathbf{T}}_{22}^{-1}\underline{\mathbf{T}}_{21}\\ \underline{\mathbf{T}}_{22}^{-1} & -\underline{\mathbf{T}}_{22}^{-1}\underline{\mathbf{T}}_{21}\end{bmatrix}\tag{B3}$$

Appendix C

Gamma-R parameters

In order to correct non-ideal terminations, Gamma-R parameters $\underline{\mathbf{R}}$ are used as described in Rautio (1983). A new set of waves is defined as

$$\underline{\alpha}^{(2n\times 1)} = \underline{a}^{(2n\times 1)} - \underline{\boldsymbol{\Gamma}}^{(2n\times 2n)}\underline{b}^{(2n\times 1)} \tag{C1}$$

$$\underline{\beta}^{(2n\times 1)} = \underline{\boldsymbol{\Gamma}}^{+(2n\times 2n)}\underline{a}^{(2n\times 1)} + \underline{b}^{(2n\times 1)} \tag{C2}$$

$\underline{\boldsymbol{\Gamma}}$ is a diagonal matrix, where the i-th entry is the non-ideal termination $\underline{\Gamma}_i$ at the respective port. The important property is, that $\underline{\alpha}_i$ becomes zero when the i-th port is terminated by the non-ideal load $\underline{\Gamma}_i$. The definition of the Gamma-R parameters is

$$\underline{\beta}^{(2n\times 1)} = \underline{\mathbf{R}}^{(2n\times 2n)}\underline{\alpha}^{(2n\times 1)}. \tag{C3}$$

Transformation from scattering parameters $\underline{\mathbf{S}}$ to Gamma-R parameters $\underline{\mathbf{R}}$ and vice versa is:

$$\underline{\mathbf{R}} = \left(\underline{\boldsymbol{\Gamma}}^{+} + \underline{\mathbf{S}}\right)\left(\mathbf{1} - \underline{\boldsymbol{\Gamma}}\,\underline{\mathbf{S}}\right)^{-1} \tag{C4}$$

$$\underline{\mathbf{S}} = \left(\mathbf{1} + \underline{\mathbf{R}}\,\underline{\boldsymbol{\Gamma}}\right)^{-1}\left(\underline{\mathbf{R}} - \underline{\boldsymbol{\Gamma}}^{+}\right) \tag{C5}$$

References

Castanié, C.: Modele de lignes de transmission non uniformes pour l'étude de couplages électromagnétiques sur des réseaux de câbles, Ph.D. thesis, L'Université Paul Sabatier, Paris, 2002.

Ciccolella, A. and Canavero, F.: Stochastic Prediction of Wire Coupling Interference, in: IEEE Int. Symp. on Electrom. Compat., 51–56, 1995.

Gonser, M., Keller, C., Hansen, J., and Weigel, R.: Advanced Simulations of Automotive EMC Measurement Setups Using Stochastic Cable Bundle Models, in: Asia Pacific EMC 2010, Bejing, China, 2010a.

Gonser, M., Keller, C., Hansen, J., and Weigel, R.: Stochastisches Kabelbaummodell und dessen messtechnische Validierung für Kabelbäume in EMV-Komponentenmessverfahren der Automobilindustrie, in: EMV Düsseldorf, Düsseldorf, Germany, 2010b.

Klein, W.: Mehrtortheorie, Akademie-Verlag Berlin, Berlin, Germany, 1976.

Kurokawa, K.: Power Waves and the Scattering Matrix, IEEE Trans. on Microwave Theory and Technique, 13, 194–202, 1965.

Rautio, J. C.: Techniques for Correcting Scattering Parameter Data of an Imperfectly Terminated Multiport When Measured with a Two-Port Network Analyzer, IEEE Trans. on Microwave Theory and Technique, 3, 407–412, 1983.

Salio, S., Canavero, F., Lefebvre, J., and Tabbara, W.: Statistical Description Of Signal Propagation On Random Bundles Of Wires, in: 13th Int. Zurich Symp. on EMC, 499–504, 1999.

Steinmetz, T.: Ungleichförmige und zufällig geführte Mehrfachleitungen in komplexen, technischen Systemen, Ph.D. thesis, Otto-von-Guericke-Universität Magdeburg, Magdeburg, 2006.

Sun, S., Liu, G., Drewniak, J., and Pommerenke, D.: Hand-Assembled Cable Bundle Modeling for Crosstalk and Common-Mode Radiation Prediction, IEEE Trans. on Electromagn. Compat., 49, 708–718, 2007.

27

Photonic crystals as host material for a new generation of microwave components

F. Daschner[1]**, R. Knöchel**[1]**, E. Foca**[2]**, J. Carstensen**[2]**, V. V. Sergentu**[3]**, H. Föll**[2]**, and I. M. Tiginyanu**[3]

[1]Microwave Laboratory, Christian-Albrechts-University of Kiel, Kaiserstr. 2, 24143 Kiel, Germany
[2]Chair for General Materials Science, Christian-Albrechts-University of Kiel, Kaiserstr. 2, 24143 Kiel, Germany
[3]Institute of Applied Physics, Academy of Sciences of Moldova, 5 Academy str., 2028 Chisinau, Moldova

Abstract. In order to verify simulations of a concave lens based upon a photonic crystal, it was scaled and built for the application in the microwave range. Its field distribution was measured between 5.5 and 12 GHz. Due to the effective refractive index smaller than 1, focusing points were found in spite of the concave shape. The survey of the field distributions at the investigated frequencies leads to other possible applications of photonic crystals in the microwave and millimetre wave region, like beam-splitter and prism.

1 Introduction

Photonic Crystals (PC) are artificial materials that give the possibility to control the propagation of electromagnetic waves (Joannopoulos et al., 1999). Generally they will form periodic dielectric structures with a certain lattice constant *a* which is one of the parameters allowing the control of propagation properties of the PC. One can design a PC in such a way that for particular frequencies no propagation is possible or the propagation paths are modified strongly. The main focus of this paper, however, is the description of experiments with some primary considerations for the possible applications of the presented structures.

In the present work we treat new microwave components based on those PC that have an effective index of refraction $n_{eff} < 1$. That is, under certain circumstances, PC can be viewed as homogenous materials described by an effective material parameter, although being highly anisotropic and dispersive. However, this is only possible for the case that the dielectric contrast in the dielectric constant is high enough (Notomi, 2000). More precisely we focus our attention on possible lenses based on such PC. In the virtue of their refractive index these lenses would have a concave shape in order to be able to focus the radiation.

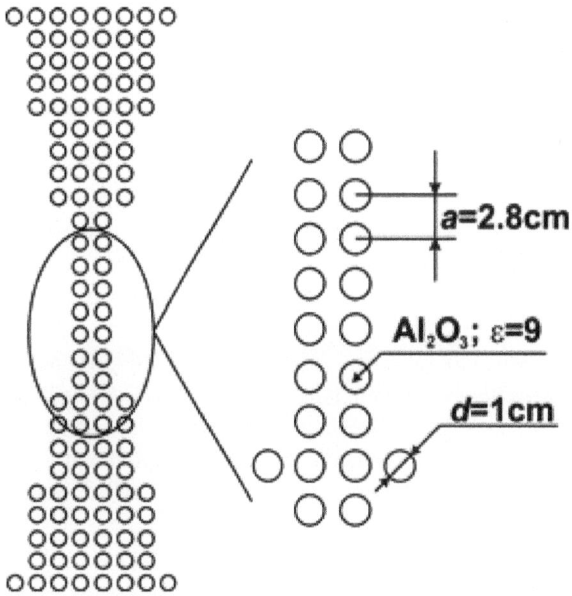

Fig. 1. Top view of the manufactured concave lens; structural dimensions are indicated.

2 Experiments

The PC used as "bulk" for the lens presented here consisted of an periodic arrangement of alumina cylinders, i.e. formed a 2-D structure. For the frequency space involved in the presented measurements alumina has a dielectric constant $\varepsilon = 9$. The lattice constant of the PC is $a = 2.8$ cm and the radius of the rods is $r = 0.5$ cm. Generally the theoretical calculations of the properties of a PC are completely correct only for a PC having infinite dimensions. In practical applications a limitation of the size is needed. For the experiments described here a lens was built using only 112 dielectric rods, arranged in a square lattice as shown in Fig. 1. Due to the small size, the set-up is not any longer called "photonic crystal" in this text,

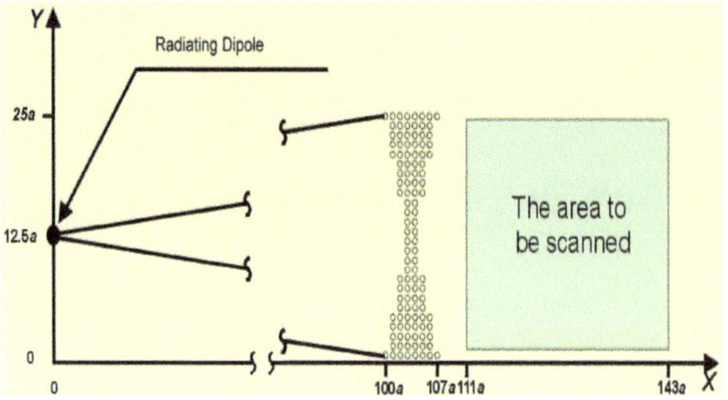

Fig. 2. Drawing of the Measurement setup. The field distribution was measured in a constant-z-plane.

but "lens". It has a concave shape and for frequencies where $n_{\mathrm{eff}} < 1$ a focusing point should be observed.

In order to approximate an incident homogenous, plane wave at the frontal surface of the lens, a dipole was placed at a distance of $100\,a = 2.8$ m. in front of it. The sending dipole was connected to an HP83752B synthesized sweeper with a power of 16 dBm. In order to measure the electro-magnetic field distribution behind the lens, a receiving antenna was mounted on a xyz-table (see Figs. 2 and 3). It was moved over an area of 700 mm by 800 mm and every 5 mm the field was measured. Hence, the total number of points measured was 22 701.

The received signal was amplified by an HP83051A microwave system amplifier. This amplified signal was fed into an ACSP2506-SN0898 detector diode. Since the source signal was amplitude modulated, the detected signal had a frequency of 1 kHz. This signal was processed by a Stanford Research Systems SR530 lock-in amplifier. The time constant used for the filter was selected to 300 ms, hence the sensitivity of the field scanner was relatively high. The whole system was controlled automatically by a microcomputer.

Because the measurement of the field distributions may be distorted by external signals and interferences caused by reflections, the setp-up was placed into an anechoic chamber. The field distributions were measured from $\lambda = 2.5$ to 5.5 cm (12 GHz down to 5.45 GHz). The polarizations measured were TE (the magnetic field components of the incident wave are in parallel to the dielectric rods) and TM (electric field parallel to rods). Doing this, the direction of the sending and receiving dipole was turned and the measurement procedure was done again.

In spite of the high effort using an anechoic chamber, one cannot completely avoid reflections: the distortion of the electromagnetic field distribution has to be considered. Moreover the dipoles (35 mm length) exhibit a certain frequency characteristic. Therefore it was decided to relate the measurements ($E(x, y)$) to free space measurements without the lens ($E_0(x, y)$). The performance of the lens is expressed using a ration called intensity gain defined as

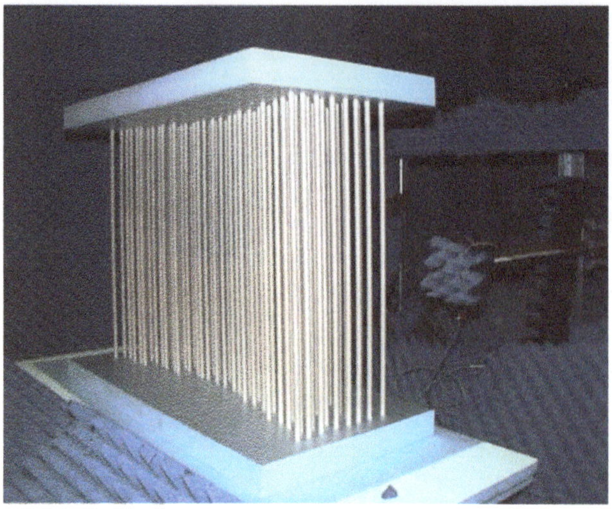

Fig. 3. Photograph of the investigated lens. The lens was placed in an anechoic chamber in front of a field scanner.

$$IG = E^2_{(x,y)} / E^2_{0(x,y)}.$$

3 Results

Figure 4 shows some selected field distributions. Thereby the simulations are confronted with the measurements. A good qualitative and quantitative identicalness can be seen. This implicates firstly that the simulations descried in Foca et al. (2005a,b) are successfully verified with real measurements. This agreement between simulations and measurements can be observed for all measured wavelengths. In case (a) ($a/\lambda = 0.96$, TM) and at many other wavelength a fuzzy field distribution can be observed. In case (b) ($a/\lambda = 0.78$, TE) the lens works as a beam-splitter. A lens should have a focusing point and it can bee found in (c) ($a/\lambda = 0.78$, TM). More features of the same structure could be found elsewhere (Foca et al., 2005a,b).

The frequency characteristics of the intensity gain of both investigated polarizations are shown in Fig. 5. The highest values found for the IG are 5.5 (TE) and 6.5 (TM). This is not even high, but one has to bear in mind that the lens is only one dimensional. For the two dimensional case (the lens would be built here using a 3 dimensional PC), this values has to be squared. Furthermore the lens has only the size of a few wavelengths (here 13..29, depending on the measured wavelength). Hence in comparison to optical lenses the area illuminated is very small.

The quality of focusing is not only addressed by the IG value, but also by the focus spot size. In order to assess this point, the measured intensity distribution in the focal spot was approximated by a bell shape function (i.e. a 2-D Gaussian function in this case). The focal spot area was defined as the full width at half maximum (FWHM). According to

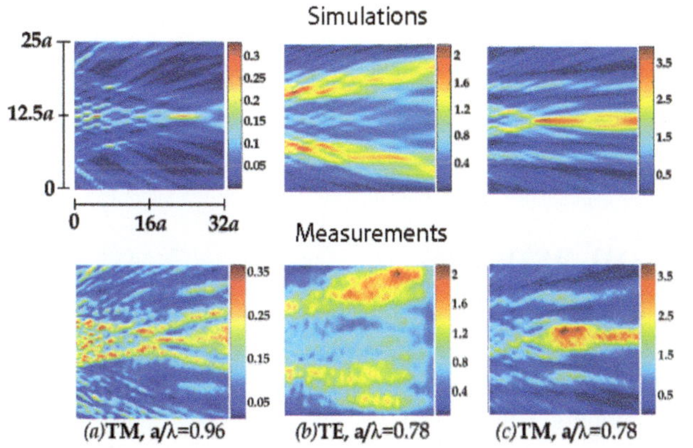

Fig. 4. **(a)** Simulations agree well with experiments; even for the case of a "fuzzy" distribution of the field; **(b)** For a certain wavelengths bands the beam will be split for the TE polarization; **(c)** Good focusing is found for the TM polarization.

Pendry (2000), "flat" lenses made form a negative index material could surpass this limit, but as the experiment seems to indicate, the concave lens investigated here can also overcome this limit at least for the TE polarization; while in the TM polarization it reaches this limit. The smallest ratio of the measured focusing area to the λ^2 is 0.24 for the TE polarization, occurring at $a/\lambda = 0.58$, and 1.02 for the TM polarization, occurring at $a/\lambda = 0.7$. The origin of the subwavelength focusing phenomenon is still questionable. We think that in our case this is rather related to the deterioration of the beam front at the exit of the lens and participation of the evanescent waves in the focus formation. This can be partially explained by the so called Zernike effect. However, due to the lack in space we do not go here in too many details since the generic Zernike effect will have its limitations in this case and more sophisticated discussions are needed in order to address all related issues. More results could be seen in Foca et al. (2006).

For the lack of space not all field distributions measured can be shown here. But there is another interesting behavior observed in the range between $\lambda = 3.1$ cm and $\lambda = 3.6$ cm (TM): the focusing point moves along the symmetry axis ($y = 12.5\,a$). Therefore an application of the lens as a prism is thinkable. However, for this purpose the lens was not designed and an the arrangement may be still considerably improved if a prism design is intended.

4 Conclusions

The primary objective of the measurements described was to verify the simulations done with a concave lens made from dielectric rods. The measurements are in well agreement with the simulation and focusing points were found for the TE-polarization as well for the TM-polarization. Beside this, a qualitative view on the measured electromagnetic field distributions leads to other possible applications. For some frequencies, the lens works as a beam splitter. The band gap can be used for filter applications. Furthermore, in a specific range, the set-up works like a prism.

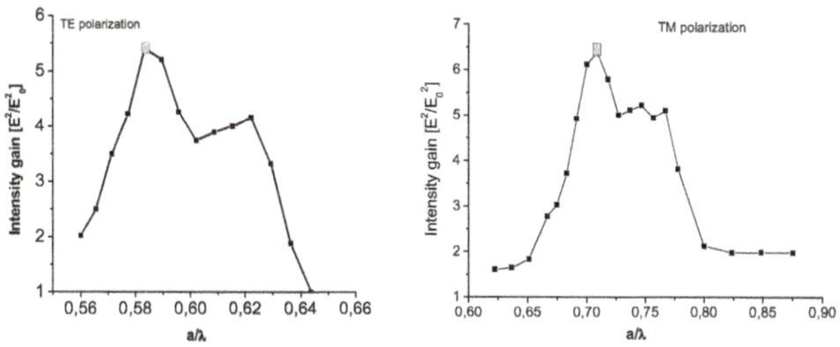

Fig. 5. Frequency Characteristic of the intensity gain of the TE- and TM-polarization.

However, due to the huge degrees of freedom designing such arrangements (shape, lattice constant, dielectric contrast), the challenge is to find an optimal photonic crystal for a given task. For the use in optical electronic circuits, many applications are imaginable. Components based upon photonic crystals may help to close the terahertz-gap in the future, when nano-technologies will make it possible to manufacture such small PC. Because PC are scaleable, theoretically all frequency ranges are applicable.

In the microwave region their behavior can be studied in detail due to the manageable size. But many interesting components in the micro- and mm-wave range could also be developed using photonic crystals.

References

Joannopoulos, J. D., Meade, R. D., and Winn, J. N.: Photonic Crystals: Modeling the Flow of Light, Princenton Univ. Press, Princenton, NJ, 1995.

Notomi, M.: Theory of light propagation in strongly modulated photonic crystals: Refractionlike behavior in the vicinity of the photonic band gap, Phys. Rev. 62, 10 696–10 705, 2000.

Felbacq, D., Tayeb, G., and Maystre, D.: Scattering by a random set of parallel cylinders, J. Opt. Soc. Am., 11, 2526, 1994.

Foca, E., Föll, H., Daschner, F., Sergentu, V. V., Carstensen, J., Knöchel, R., and Tiginyanu, I. M.: Efficient focusing with an Ultra-Low Effective-Index Lens based on Photonic Crystals, Proc. of the MRS Spring Meeting 2005, Symposium D, 2005a.

Foca, E., Föll, H., Daschner, F., Sergentu, V. V., Carstensen, J., Frey, S., Knöchel, R., and Tiginyanu, I. M.: Efficient focusing with a concave lens based on a photonic crystal with an unusual effective index of refraction, Phys. Stat. Sol., 202, 35, 2005b.

Foca, E., Föll, H., Carstensen, J., Sergentu, V. V., Tiginyanu, I. M., Knöchel, R., and Daschner, F.: Strongly frequency dependent focusing efficiency of a concave lens based on two-dimensional photonic crystals, Appl. Phys. Lett., 88, 011102, 2006.

Pendry, J. B.: Negative refraction makes a perfect lens, Phys. Rev. Lett., 85, 3966–3968, 2000.

28

Common mode chokes and optimisation aspects

T. Kut[1], A. Lücken[2], S. Dickmann[1], and D. Schulz[2]

[1]Institute of Fundamentals of Electrical Engineering, Helmut Schmidt University, Hamburg, Germany
[2]Institute of Electrical Power Systems, Helmut Schmidt University, Hamburg, Germany

Correspondence to: T. Kut (tobias.kut@hsu-hh.de)

Abstract. Due to the increasing electrification of modern aircraft, as a result of the More Electric Aircraft concept, new strategies and approaches are required to fulfil the strict EMC aircraft standards (DO-160/ED-14–Sec. 20). Common mode chokes are a key component of electromagnetic filters and often oversized because of the unknown impedance of the surrounding power electronic system. This oversizing results in an increase of weight and volume. It has to be avoided as far as possible for mobile applications. In this context, an advanced method is presented to measure these impedances under operating conditions. Furthermore, the different parameters of the inductance design is explained and an optimisation for weight and volume is introduced.

1 Introduction

Power electronic converters, for example an interleaved boost converter as shown in Fig. 1, consist of switching devices, which are turned on or off, based on a modulation scheme to adjust the output voltage. The main drawback is the generated level of electromagnetic interferences on power lines and other communication wires due to the non-ideal behaviour of the used semiconductors (Farhadi and Jalilian, 2006; Mainali and Oruganti, 2007). The spectrum is usually spread from 10 kHz to 100 MHz. Filtering the transmission wires is a challenge, which often leads to redesigned filters, tested in a "trial and error" procedure and optimised only rarely in terms of weight, size or costs (Mayer, 1998). These optimisations are required, especially for applications in modern aircraft (Derouineua, 2009; Kut et al., 2013a; Chand et al., 2013). Therefore, an advanced method is presented to measure the impedances of the power converter to improve the performance of the filter (Tarateeraseth et al., 2011; See and Deng, 2004; Kut et al., 2013b). Furthermore, the different design parameters such as frequency, saturation flux density and operating current are compared for various new ferromagnetic materials (Vacuumschmelze®, 2013; Magnetics®, 2013). These approaches may reduce or avoid the need for construction of several, often oversized, filter prototypes. They provide an additional insight into the behaviour of the choke by giving an understanding, how variations of parameters influence the final filter performance.

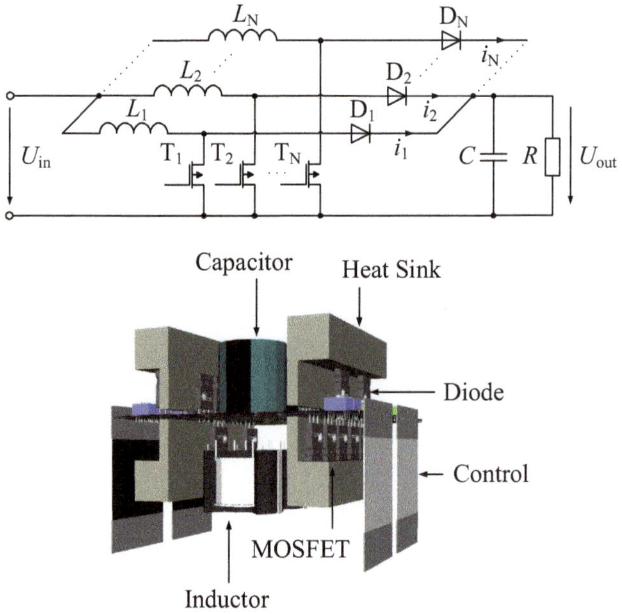

Figure 1. Schematic of a boost converter in interleaved structure (above) with prototypical realization (3-D-model) as four-step interleaved converter (below) as an exemplary application for modern aircraft.

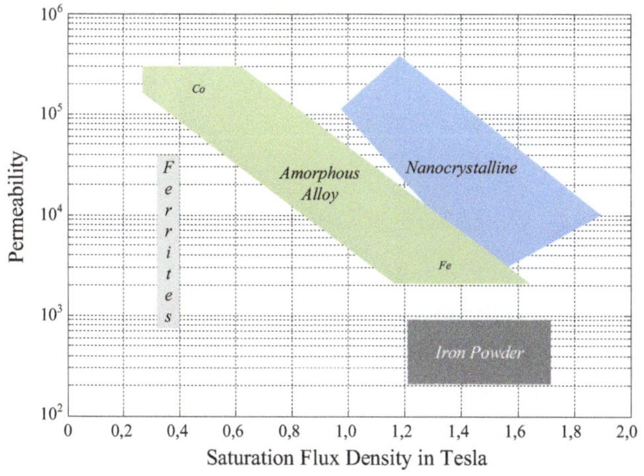

Figure 2. Magnetic properties for ferrites, iron powder and metal alloys: permeability vs. saturation flux density (Roc'h and Leferink, 2013).

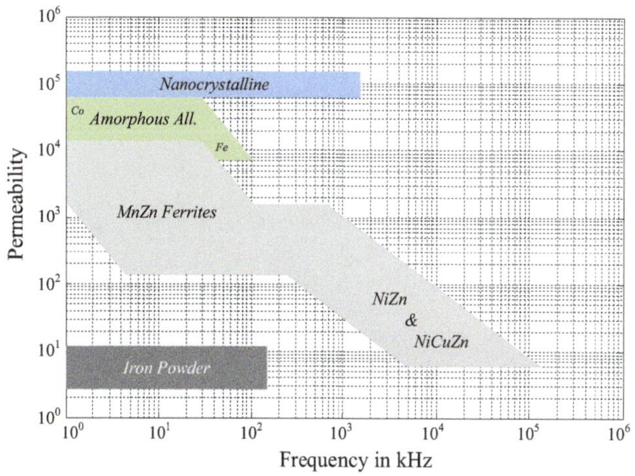

Figure 3. Magnetic properties for ferrites, iron powder and metal alloys: permeability vs. frequency (Roc'h and Leferink, 2013).

2 Designable parameters and impedance modelling

This section focuses on the designable parameters of a common mode choke or a storage inductor such as core material, operating frequency, saturation and impedance behaviour.

2.1 Core material

The main purpose of common mode chokes is to transform the electromagnetic interferences to heat rather than for energy storage as a storage inductor. The application of the core material is quite similar. The material can be used to increase the inductance, to absorb energy and to transform it to heat. The permeability μ as the ratio of the magnetic flux density B to H is the most important magnetic material parameter. Typical requirements of an optimised choke are a high saturation level and a wide frequency range. Figure 2 shows the dependence of the permeability as a function of the saturation flux density and Fig. 3 of the frequency, respectively. Three basic materials, i.e. ferrites, iron powder and metal alloys (nanocrystalline and amorphous structure) are used in the design of traditional common mode chokes. The highest permeabilities are found in nanocrystalline materials. Amorphous alloys have somewhat smaller values whereas iron powder cores have relatively low permeabilities. Ferrite cores can be used over a wide frequency range but exhibit a significantly lower level of saturation. Higher flux densities can be found in nanocrystalline and amorphous materials as well as in iron powder cores but are limited in the used frequency range. Nanocrystalline materials are a good alternative to the traditional chokes made of ferrite or iron powder. So, the main advantage is the high level of saturation and the resulting smaller size. The use of this material as a filter element or the use as a storage inductor due to the distributed air gap, in contrast to a conventional ferrite core is a step towards volume reduction. Reaching 50–80 % lower compared to a ferrite volume and more than 80 % lower compared to an iron core. The resulting weight reduction is an important property in aircraft applications or any other areas where weight is a design constraint (Wen, 2011).

2.2 Impedance modelling

A main objective of a common mode choke is to block the electromagnetic interferences. The study of the behaviour involves five types of impedances. Figure 4 presents the general equivalent circuit with the coupling factor K:

- the common mode (L_{cm})
- the differential mode (L_{dm})
- the inter-winding (C_{p})
- the intra-winding (C_{w})
- the line impedance (R_{wire})

Equation (1), the value of the common mode impedance is the most important value and is strongly related to the permeability (assuming linearity) of the core. In addition to the operating current, the inductance and the capacitance – the designable parameters in these context are the core material $\left(\mu = \mu_0\left(\mu'_{\text{RL}} - \text{j}\mu''_{\text{RR}}\right)\right)$, the dimension and configuration (A_{mat}), the wire material (frequency, diameter, isolation), the number of turns (N) and the effective length (l_{mat}). This can be modified by the designer and the impedance can be represented by a series equivalent circuit consisting of an ideal inductor L_{ser} and a resistor R_{ser}.

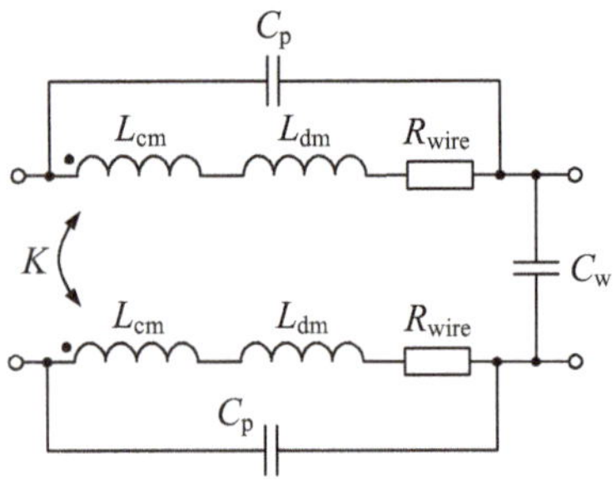

Figure 4. Equivalent circuit of the common mode choke.

$$\underline{Z}_{cm} = j\omega L_{cm} =$$

$$= j\omega \underbrace{\frac{\mu_0 \cdot \mu'_{RL} \cdot A_{mat} \cdot N^2}{l_{mat}}}_{L_{ser}} + \omega \underbrace{\frac{\mu_0 \cdot \mu''_{RR} \cdot A_{mat} \cdot N^2}{l_{mat}}}_{R_{ser}} \tag{1}$$

$$L_{cm} = \mu_0 \left(\mu'_{RL} - j\mu''_{RR}\right) \cdot \frac{A_{mat} \cdot N^2}{l_{mat}}$$

2.3 Measurements

In order to evaluate the influence of each parameter, measurements with a precision LCR meter with dc bias unit are performed for various core materials, which can be assigned to the respective classes of materials. The influence of the frequency (Fig. 5 for $f = 100\,\text{kHz}$, Fig. 6 for $f = 300\,\text{kHz}$) and the flux density is presented. Only the suitable materials have been measured for the higher frequency in Fig. 6. The maximum saturation flux density B_{max} ranges from 0.5 T (PM 50/39) to 2 T (HFlux) and can be described by Eq. (2) with the maximum operating current $\hat{I}_{max}$, the minimum A_{min} and effective cross-sectional area A_{eff} for ferrite cores and with the material-dependent inductance factor A_L for toroidal cores.

$$B_{max} = \underbrace{\frac{N \cdot \hat{I}_{max}}{\frac{l_{mat}}{\mu_0 \cdot \mu_{mat}} + \frac{l_{air}}{\mu_0}} \cdot \frac{A_{eff}}{A_{min}}}_{\text{Ferrites}} = \underbrace{\frac{N \cdot \hat{I}_{max} \cdot A_L}{A_{mat}}}_{\text{Toroid}} \tag{2}$$

As already mentioned, the common mode chokes, inductors or storage inductors constitute a high influence in power electronic systems. The design of these elements has been a challenging task with the increasing complexity of the power electronic. In order to optimize this in the context of weight savings, the area product method is used. This is defined as the product of the core area A_{core} and the window area A_{window} for a core – according to Eq. (3). The various parameters involved in the design are divided into three categories: design requirements, material and manufacturing constraints. Here, the value AP_{min} can be determined and should be minimised from the output power P and operating frequency

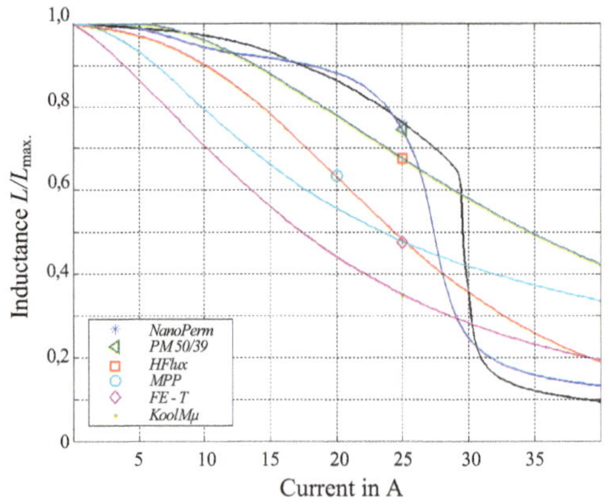

Figure 5. Core materials at $f = 100\,\text{kHz}$.

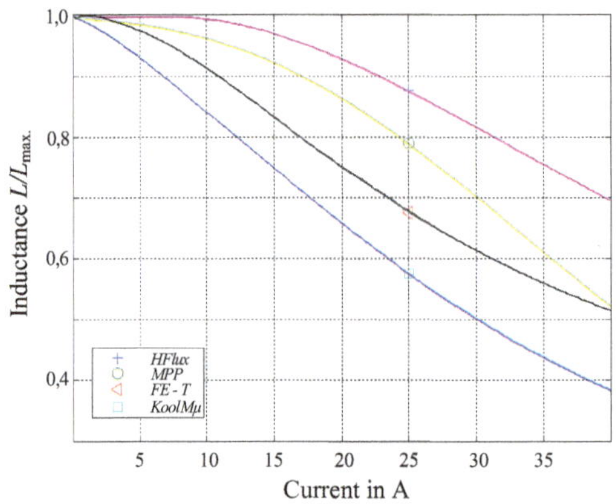

Figure 6. Core materials at $f = 300\,\text{kHz}$.

f as design requirements, the maximum flux density B_{max} and maximum current density J_{max} as material constraints and the window utilisation factor C as manufacturing constraint. The area product method is a very useful expression involving mechanical and electrical parameters. The increase of the operating frequency and the material-dependent saturation flux density contribute to the weight saving effect.

$$AP_{min} = A_{core} \cdot A_{window} = \frac{2 \cdot P}{C \cdot B_{max} \cdot f \cdot J_{max}} \tag{3}$$

In addition to the relationship of the inductance as a function of the flux density and the frequency, the impedance of the common mode choke is another important parameter for the attenuation of electromagnetic interferences. For the common mode impedance measurements the two input and the two output ports with respect to Fig. 4 are connected to each other. Therefore, the impedances have been measured for various cores (Fig. 8) with an impedance analyser between 9 kHz and 100 MHz. The result is presented in Fig. 7. For comparison, two conventional filter elements (VitroVac®, NanoPerm®) are compared with the new materials. It is clear that, for example, AmoFlux can also be used as a filter element, which has a significantly lower weight as

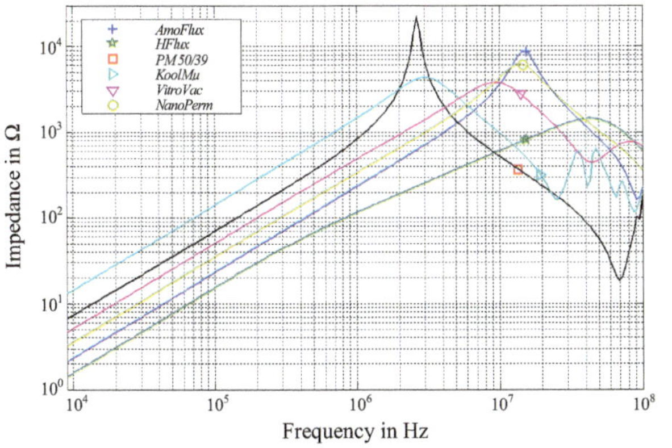

Figure 7. Measured common mode impedance for various cores.

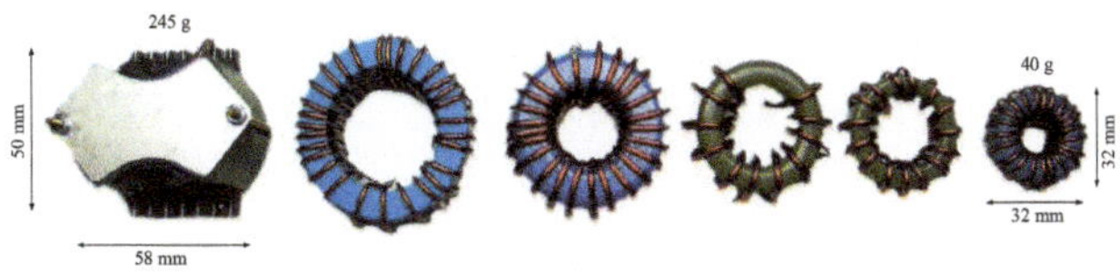

Figure 8. Overview about the core materials.

indicated in Fig. 8. In summary, it can be noted that weight can be saved by the choice of newer materials for common mode chokes as well as for storage inductors. A further advantage is that they have a higher saturation flux density and can also be used for a much higher frequency range.

3 System impedance measurement

An accurate measurement method to extract the common and differential mode impedance of a power electronic system under operating conditions is presented in this section. For this purpose a pre-measurement calibration process and the extraction of common and differential mode is described to systematically design an electromagnetic filter with minimum effort. The system impedance is very important for filter design because only with this information it can be best adapted to the system, without oversizing the filter. In contrast to other systems due to converter topology, component parasitics, printed circuit board layout etc. the impedance is not 50 Ω and must be taken into account. For example, the differential mode is influenced by the reverse recovery effect of the diode or the equivalent series resistance and inductance of a capacitor. As for the common mode, the deciding components are parasitic effects between the semiconductor, heat-sink, chassis and printed circuit board. To illustrate, why the system impedances are important parameters for filter design, the simple schematic as shown in Fig. 9 is analysed by Eq. (4), which represents the transfer function of the entire system. The parameters a_{xy} are the elements of a cascaded filter structure. This clearly illustrates that for optimum EMI filter design, the impedances of source and load should be

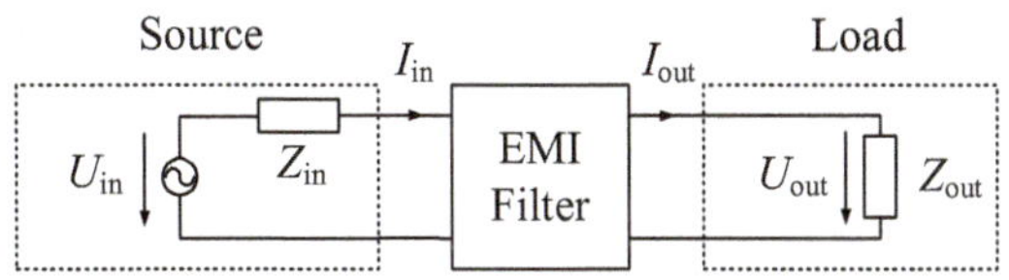

Figure 9. Consideration of the system impedances in filter design.

taken into account in power electronic applications. A detailed description is given by (Kut et al., 2013b).

$$\frac{U_{out}}{U_{in}} = \frac{1}{a_{11} + \frac{a_{12}}{Z_{out}} + a_{21} \cdot Z_{in} + a_{22} \cdot \frac{Z_{in}}{Z_{out}}} \tag{4}$$

Therefore, a two-probe measurement procedure for common and differential mode impedances is advantageous for power electronic systems. It was originally developed for built-in power line electromagnetic filters as small-signal behaviour (Tarateeraseth et al., 2011). It has been extended as follows:

- Measurements under operating conditions in each operating point of the power electronic system
- Independent of each used measurement tool (calibrated out of cables, current probes etc.)
- Minimal frequency range from 9 kHz to 30 MHz
- Separation between common and differential mode
- Operating parameters of several hundred volts and several tens of amperes

The basic concept of the method to measure the unknown impedance Z_X is shown in Fig. 10 (above). It consists of an injection probe, a detection current probe, a vector network analyser (VNA) and the unknown impedance. Port 1 generates a signal into closed-loop through the injection probe and the resulting current is measured at port 2. Figure 10 (below) shows the complete equivalent circuit of the measurement setup with port 1 as source U_1 as well as the internal resistance Z_i, the source voltage U_{C1} as well as the resulting voltage U_{C2}, the primary L_1/L_2 and the mutual inductances M_1/M_2, the formed wiring elements L_w and R_w. These variables are summarised in addition to Z_{setup} (Kut et al., 2013b). By solving the Eq. (5) as equation of the system as indicated in Fig. 10 (below), the unknown impedance Z_X can be expressed by Eq. (6), where F is a frequency-independent factor (Z_T is the ratio of U_{C2} and I_{out}). The ratio of the two-probe voltages is given by Eq. (7).

$$\begin{pmatrix} U_1 \\ 0 \\ -Z_X \cdot I_{out} \end{pmatrix} = \begin{pmatrix} Z_i + j\omega L_1 & 0 & -j\omega M_1 \\ 0 & Z_i + j\omega L_2 & j\omega M_2 \\ -j\omega M_1 & j\omega M_2 & R_w + j\omega L_w \end{pmatrix} \cdot \begin{pmatrix} I_1 \\ I_2 \\ I_{out} \end{pmatrix} \tag{5}$$

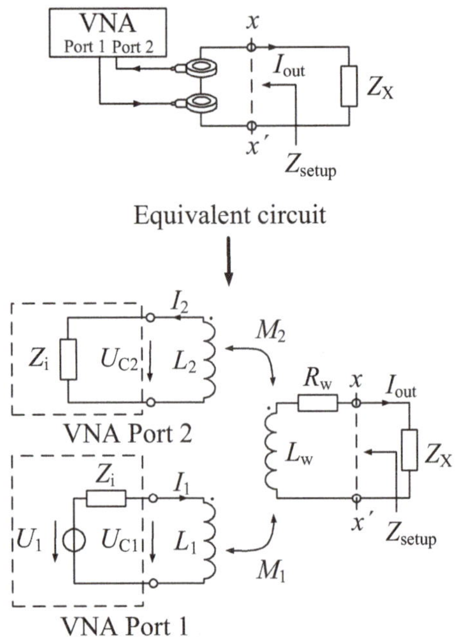

Figure 10. Basic concept and equivalent circuit of the measurement setup.

$$Z_X = F \cdot \left(\frac{U_{C1}}{U_{C2}} \right) - Z_{setup} \quad ; \quad F = \frac{M_1 \cdot Z_T}{L_1} \tag{6}$$

$$\frac{U_{C1}}{U_{C2}} = \frac{S_{11} + 1}{S_{21}} \tag{7}$$

The coefficient F and the setup impedance Z_{setup} can be obtained by two additional measurements: First, by replacing the unknown impedance Z_X in Fig. 10 with a known precision resistor R. The resistor should be chosen in the middle range of the unknown impedance to be measured. The next measurement should be done without the resistor R. Therefore, R can be short-circuited in Fig. 10. With these additional measurements and Eqs. (5) and (6), F and the setup impedance Z_{setup} can be obtained by solving this system of equations. All parameters, which are influenced by the measurement setup are determined and can be eliminated for the determination of Z_X. The measurement setup to extract the "unknown" common mode impedance is shown in Fig. 11. In order to avoid any unwanted electromagnetic interferences from the source supplying the power electronic, these connections are terminated by line impedance stabilisation networks (LISN). Basically, the setup and the connecting wires should correspond to those of the calibration measurements. It is the same procedure with the electronic load to be supplied in Fig. 11. To measure the differential mode impedance one of the two wires must be rotated by 180° and routed back through the two current probes. Figure 12 shows exemplary the magnitude of the extracted common mode impedance. The intermediate steps for determining the Z_{setup} and differential mode impedance are omitted for clarity, at this point.

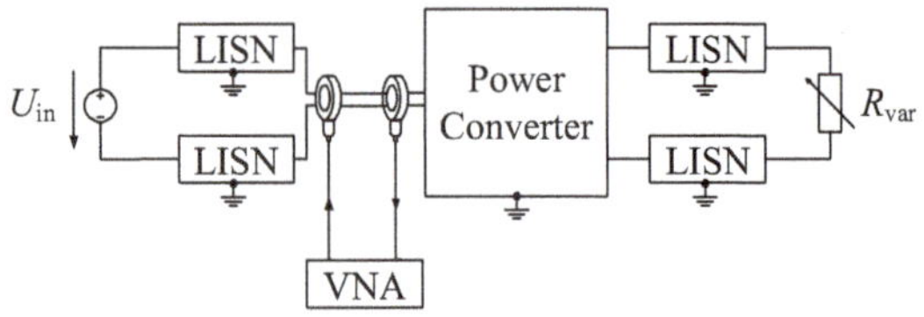

Figure 11. Setup for common mode impedance measurement.

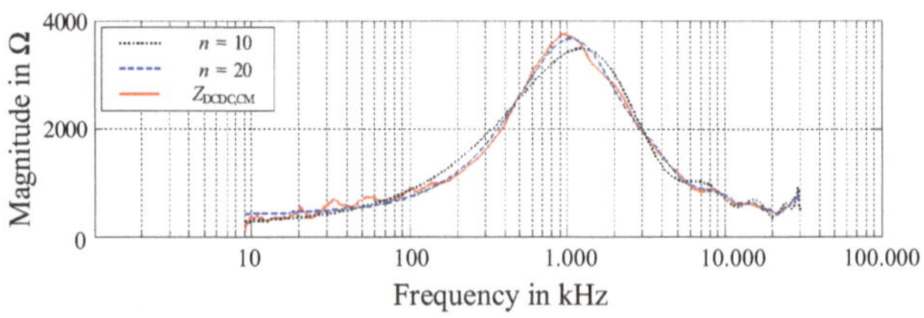

Figure 12. Measured common mode impedance with interpolated values of polynomial $n = 10$ and $n = 20$.

It can also be seen in Fig. 12 that for calculation of the filter elements, the measured curve is represented by a polynomial function, in order to use them in Eq. (4). In summary, with accurate measurements of common and differential mode impedances over the frequency range of interest, an appropriate filter can be chosen or can be designed with optimal component values and is no longer overdimensioned.

4 Conclusions

The EMC requirements increased as a result of the required DC/DC converters due to the installation of electric system components instead of heavy mechanical, pneumatic and hydraulic systems in modern aircraft. In order that these filters do not take up a large part of the overall system weight, they must be optimally adjusted to the converter system. Two aspects should be considered: on the one hand there are weight savings possible by the optimised selection of the inductor material. For this purpose, the dependencies with respect to frequency and saturation flux density for various modern materials have been identified and the potential savings were clearly pointed out. In addition, an accurate measurement method to extract common mode and differential mode impedance of power electronic converters under operation conditions was presented. With these measurements and with the knowledge of the correct material an appropriate EMI filter can be chosen or can be designed with optimal component values and is no longer overdimensioned.

Edited by: F. Sabath

References

Chand, B., Kut, T., and Dickmann, S.: Optimal design of active EMC filters, Adv. Radio Sci., 11, 243–249, doi:10.5194/ars-11-243-2013, 2013.

Derouineua, J.-L.: Power Optimized More Electrical Aircraft, in: Proceedings of the European Conference towards eEnvironment, 293–295, Prag, 2009.

Farhadi, A. and Jalilian, A.: Modeling and Simulation of Electromagnetic Conducted Emission Due to Power Electronics Converters, in: Power Electronics, Drives and Energy Systems, International Conference on, 1–6, New Delhi, 2006.

Kut, T., Chand, B., Lücken, A., Dickmann, S., and Schulz, D.: EMC studies on systems with hybrid filter circuits for modern aircraft applications, Adv. Radio Sci., 11, 237–242, doi:10.5194/ars-11-237-2013, 2013a.

Kut, T., Lücken, A., Dickmann, S., and Schulz, D.: Design von Gleich- und Gegentaktfiltern unter Berücksichtigung der Ein- und Ausgangsimpedanzen für Gleichspannungswandler im HVDC-Netz, in: 62. Deutscher Luft- und Raumfahrtkongress, urn:nbn:de:101:1-2013112910658, Stuttgart, September, 2013b.

Magnetics: A Critical Comparison of Ferrites with other Magnetic Materials, in: Technical Information, 2010.

Mainali, K. and Oruganti, R.: Simple Analytical Models to Predict Conducted EMI Noise in a Power Electronic Converter, in: Industrial Electronics Society, 33rd Annual Conference of the IEEE, 1930–1936, Taipei, 2007.

Mayer, F.: Electrical Power and Signal Distribution in Modern Aircrafts, combines Weight Advantages and EMC Compatibility, in: Electromagnetic Compatibility, IEEE International Symposium on, 1, 281–283, Denver, 1998.

Roc'h, A. and Leferink, F. B. J.: Analysis of Common Mode Inductors and Optimization Aspects, in: Power Electronics and Power Systems, Electromagnetic Interference Issues, Bentham Science Publishers, Australien, 3–37, ISBN 978-1-60805-240-0, 2012.

See, K. Y. and Deng, J.: Measurement of Noise Source Impedance of SMPS Using a Two Probes Approach, in: Power Electronics, IEEE Transactions on, 19, 862–868, Mai, 2004.

Tarateeraseth, V., See, K. Y., Wang, L. B., and Canavero, F. G.: Systematic Power Line EMI Filter Design for SMPS, in: EMC Europe, 586–591, York, September, 2011.

Vacuumschmelze GmbH & Co. KG: Nanocrystalline Material in Common-Mode Chokes, in: Technical Information, 2013.

Wen, B.: Weight Estimation of Electronic Power Conversion Systems, in: Master's Thesis, Virginia Polytechnic Institute and State University, 2011.

Industrial interference and radio astronomy

A. Jessner

Max-Planck-Institut fuer Radioastronomie, Auf dem Huegel 69, 53121 Bonn, Germany

Correspondence to: A. Jessner (jessner@mpifr-bonn.mpg.de)

Abstract. The interferer – victim scenario is described for the case of industrial interference affecting radio astronomical observatories. The sensitivity of radio astronomical receivers and their interference limits are outlined. EMC above 30 MHz is a serious problem for Radio Astronomy. Interferer (CISPR) and victim (ITU-R RA 769) standards are not harmonised. The emissions from the interferer and their spectral characteristics are not defined sufficiently well by CISPR standards. The required minimum coupling losses (MCL) between an industrial device and radio astronomical antenna depends on device properties but is shown to exceed 140 dB in most cases. Spatial separation of a few km is insufficient on its own, the terrain must shield > 30–40 dB, additional mitigations such as extra shielding or suppression of high frequency emissions may be necessary. A case by case compatibility analysis and tailored EMC measures are required for individual installations. Aggregation of many weak rfi emitters can become serious problem. If deployment densities are high enough, the emission constraints can even exceed those for a single interferer at a short distance from the radio observatory. Compatibility studies must account not only for the single interferer but also for many widely distributed interference sources.

1 The victim

Radio astronomy installations (observatories) are a large-scale and cost intensive part of the scientific infrastructure that is fully financed by the tax payer. They have to be large and expensive, because the radio signals we receive from cosmic sources are extremely attenuated and one needs large antennas with large collecting areas and sophisticated cryogenic receivers for their detection. Radio astronomy belongs to the realm of fundamental science. But it has also led to quite a few technical applications (e.g. feed horns for satellite dishes), and its techniques are used in vital measurements such as the establishment of a very precise global coordinate system which even accounts for the motion of the continents but these have been serendipitous side effects, which are nevertheless welcome by all. The aim of radio astronomical research is however to understand the structure of the cosmos we find ourselves in, the development of the universe, its galaxies, stellar systems and planets as well as the very structure of space-time itself. Radio astronomy also plays a significant role in the education of new scientists and engineers in astrophysics as well as in the fields of radio engineering, image processing, data acquisition and many more. As such it is a long-term investment of society into its own future, something outside the scope of commercial interests that demand an identifiable return within a predictable timescale. It is only logical that governments try to protect their investments like anyone else and perhaps for that reason, radio astronomy became a recognised radio service within the radio regulations (RR) of the ITU[1] (Radio Regulations, 2012), a global set of rules governing the allocation and use of radio spectrum which is updated regularly by the world radio conferences. Radio astronomy is a passive (listen only) radio service detecting natural radio emissions in contrast to most other usage of the radio spectrum that involves technical radio transmitters (active services).

One of the reasons for the unequalled sensitivity of radio astronomical measurements lies in the principle of comparing the received radio flux density from an unknown source to that of the know radio background (Dicke, 1946), whereby sensitivities of more than 50dB below the background noise ($S/N < -50$ dB) can be achieved. Such sensitive *radiometric* measurements are only possible, if the reference sky (background) is free from interference itself, and that to a degree exceeding the most sensitive measurements. As a result, radio astronomy has been allocated specific protected bands, mostly around frequencies where molecules of the tenuous cosmic gases emit their spectral lines (Fig. 1).

[1] ITU: International Telecommunication Union (Geneva)

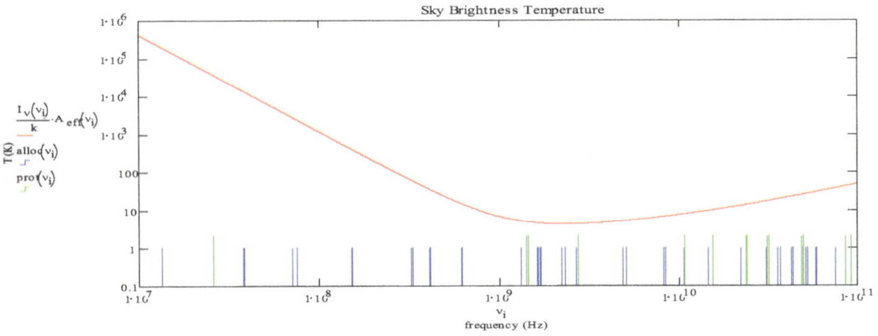

Fig. 1. The Radio sky temperature (Cane, 1979; Ellington, 2005) and radio astronomical allocations between 10 MHz and 100 GHz. Red: Sky temperature for the Zenith, green: passive bands allocated to radio astronomy, blue: bands shared with other services.

Radio astronomy has an exclusive use of only 0.7 % of the radio spectrum below 30 GHz, the majority of the allocated bands are shared with other radio services. The choice of these bands is not free and neither is that of the interference thresholds. Both are determined by the physics of the distant emitter and the physical limits of a radio receiver and its use in sensitive radio astronomical measurements. Exclusive bands for radio astronomy are covered by the RR Footnote 5.340 (Radio Regulations, 2012) which says that "no emissions are permitted in a list of bands" with emissions being termed as radiation coming from a radio station (of any kind) as the result of its active operation. Radiation from other parts of the station, e.g. the local oscillator or the IF of a receiver is not an emission in that sense, an important distinction as we will see later on.

The ITU Recommendation RA 769 (ITU-R, 2003) lists the specific threshold levels of radio interference for which the interference becomes detrimental to the operation of the radio observatory. These levels are specified as 10 dB below the *radiometric receiver sensitivity*, typically an I/N of −60 dB, implying that measurement results should not be affected by an additional error of more than 10 %.

Radio astronomical antennas have forward gains depending on frequency, but typically of the order of 50–60 dBi within a narrow beam of a width of typically a few minutes of arc. Interference enters the receiver mainly via sidelobes with varying degrees of gain or attenuation. For interference estimates the radio astronomical receiver is assumed to have an average gain of 0 dBi for reception from random directions. Measurement errors of radiometric noise power measurements decrease with the square root of the product of band width and averaging time (Dicke, 1946) requiring an appropriate scaling of interference thresholds. As a result of these considerations, the interference thresholds are given by Eq. (1)

$$S_H = 0.1 \cdot \frac{4\pi\, k\, f^2 \cdot (T_{\mathrm{ant}} + T_{\mathrm{rec}})}{c^2 \sqrt{\Delta f\, \tau}} \tag{1}$$

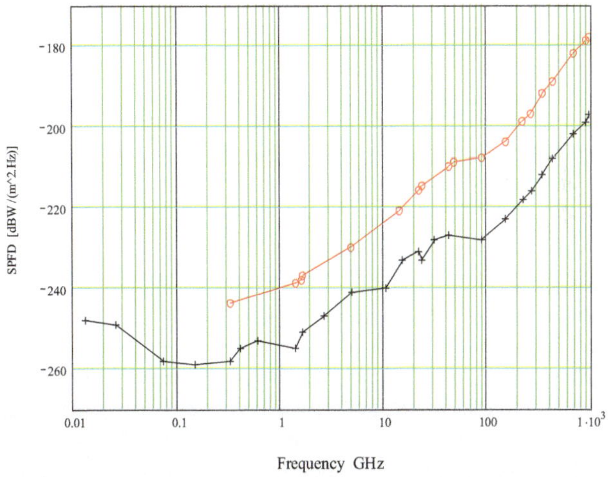

Fig. 2. Detrimental spectral flux density values for radio astronomy according to ITU-R RA 769. Black: continuum limit, red: spectroscopic limit.

taken from ITU-R RA 769. As usual, k is the Boltzmann constant and c the speed of light. The result S_H is the spectral flux density threshold (W m^{-2} Hz^{-1}), with f being the centre frequency of the band, T_{ant} the typical antenna temperature and T_{rec} the typical receiver temperature, Δf is the measurement bandwidth and τ the duration of the signal integration. ITU-R RA 769 lists these quantities in its Tables 1 and 2 for continuum (broad band) and spectroscopy (narrow band) measurements. Figure 2 shows the thresholds as a function of frequency for an integration time of 2000 s.

The interference limits are defined for the victim in the case of radio astronomy as averages over time and bandwidth.

Astronomical sources are weak and therefore the astronomical unit of spectral flux density is 1 Jansky (Jy) = 10^{-26} W m^{-2} Hz^{-1}, named after Karl Jansky (1905–1950), the discoverer of cosmic radio waves. The interference thresholds are of the order of 1 Jy in UHF and L-band and rising roughly proportional to the observation frequency.

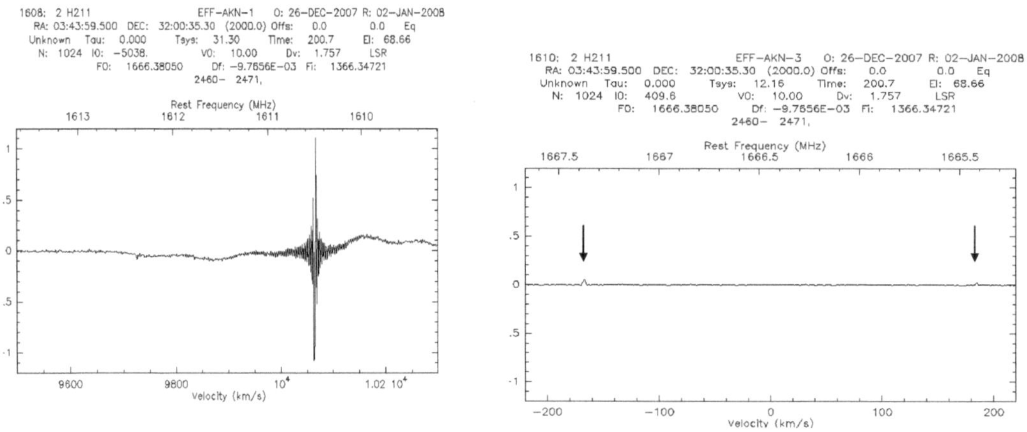

Fig. 3. Interference received from a low earth orbit mobile phone satellite (right) and a typical cosmic radio source (two small bumps in the left panel) at 1610 GHz (diagrams courtesy of A. Kraus).

A few numbers may serve to illustrate the magnitude of the quantities and of the achieved sensitivities: one Jy is the spectral flux density that one would receive on the ground from a 1 W e.i.r.p IMT-2000 mobile phone operating at 2.1 GHz and orbiting the earth at a distance of 8.867×10^5 km, more than twice the distance to the moon! The main beam sensitivity of a large radio telescope at that frequency can exceed 50 dBi. If we assume a 10 MHz detection bandwidth, the 100 m Effelsberg radio telescope is capable of detecting such a 1 W e.i.r.p. source at a distance of 6.9×10^7 km after 2000 s. In this case we receive a spectral flux density of 0.17 mJy, which is a typical strength of the signals received from distant cosmic sources. Figure 3 illustrates the difference between the signals strengths of interference and a cosmic source.

The interference source was one of the IRIDIUM global mobile phone satellites at a distance of 1500 km. They employ a spot beam with a peak e.i.r.p. of 12.5 W and our measurements have shown that their seventh-order intermodulation products with an e.i.r.p. of 1 mW (ECC Rep. 171, 2011; ECC Doc. SE40(09)092, 2009) are causing the interference even when the satellite is not crossing the telescope beam.

Radio observatories are also vulnerable to regular and irregular impulsive interference. When they observe Pulsars, a class of radio sources that emits short radio pulses with periods ranging from 1.5 ms to more than eight seconds, they employ sensitive high time resolution detectors. Simple electric cattle fences in about two km distance from the telescope have frequently been detected as regular pulsed interference.

It becomes clear, that local radio emissions, even those that are considered weak by normal standards can become the source of severe interference for a radio observatory. This is particularly true, when the emission occurs within the radio astronomical bands. Frequency regulators have recognised the problem and implemented a complex set of protection measures for radio astronomy, based on extensive studies that investigated the compatibility of radio astronomy with active radio services.

2 Industrial interference

The ITU came into being because there was a need to coordinate the frequencies and locations of radio transmitters world-wide, for radio waves can travel far and by themselves do not respect conventions such as national boundaries. Hence the focus on ITU regulations is on transmissions intended for communication and information transfer in the widest possible sense. That includes not only broadcasting, radio amateurs, location tracking (radar), but also remote sensing and radio astronomy.

ITU regulations however do not deal extensively with radiation emanating from technical equipment that is not intended for communication. Even in the times before the ubiquitous use of digital data processing and fast power switching gear, the emissions from industrial and consumer electrical installations were already present and came under the headings of "Electromagnetic Compatibilty (EMC)", and "Electromagnetic Interference (EMI)", in contrast to "radio interference" from actual radio transmitters. In the past EMI affected mainly the spectrum below 100 MHz. Figure 4, taken from an ITU source (ITU-R P. 372-9, 2007) reflects this. The earth's ionosphere effectively blocks the reception of radio waves from non-terrestrial sources below 30 MHz. One may note, that trace E of Fig. 4 indicates that the contribution of EMI to the radio background in typical cities exceeds that of "galactic radio noise" by about 20 dB, converted in radio astronomical units we are talking of an ambient noise floor of several hundred Jy. This may serve to illustrate the reason why radio observatories are built in remote areas where man-made noise is expected to be less than the strength of cosmic radio signals (Fig. 4, trace D). The difference was mainly caused by the attenuation of interference emissions coming

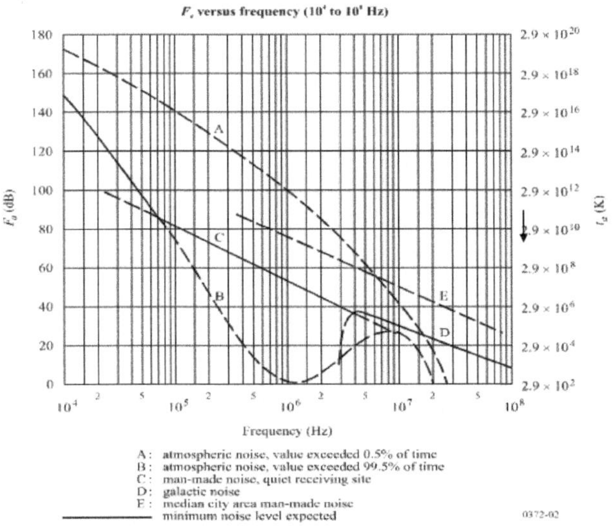

Fig. 4. Contributions to the radio noise environment (from ITU-R P. 372-9, 2007).

from localised concentrations of fixed installations in cities and industrial areas by their distance to the radio observatory and the terrain.

However with the increased use of high speed digital devices contained in license exempt and uncontrolled and often mobile consumer electronics as well as the spread of efficient high speed switching gear in consumer and industrial power conversion equipment, ranging from simple switch mode power supplies for a laptop to MW-size generators and converters in wind power plant, brings the sources of industrial interference closer to the radio observatories. Not only that, their interference spectrum will also be broader as it is naturally related to the rise and fall times of the digital switching used and these have become much faster with the recent advances in semiconductor electronics.

Standards are needed to limit the emission from such devices in order to suppress interference to communication equipment and also to other sensitive electrical equipment. These have been devised by CISPR[2] for a range of equipment separated into different categories and classes according to their use and construction. A list is given in the CISPR guide of the IEC (IEC, 2007). There two main classes:

1. equipment intended for use in a commercial and/or light industrial environment,

2. equipment intended for domestic use.

Class B equipment has to fulfil more stringent requirements than that from class A.

CISPR limits are defined for the interferer. Time-averages and spectral properties are no always specified.

3 Compatibility

Many different interference scenarios that we did not have in the past have started to appear. The domestic use of computers, micro wave ovens, energy saving electric lights, has become common not only in the city but also in the rural environment close to radio observatories. The new scenario requires a detailed study of the impact of new industrial devices on radio astronomical measurements. In this paper we will look at the example of industrial switch-mode power converters in the kW to MW range which are often found in association with wind power generators or solar power stations. Like radio observatories, these are also erected in remote areas in order to avoid densely populated areas[3].

The radio disturbance characteristics and limits are given as field strengths at a distance of 30 m in table 17 of the European standard CISPR-11 (= EN DIN 55011 in Germany) for such devices belonging to group 1 class A:

> 30 dB μV m^{-1} for $f < 230$ MHz and 120 kHz measurement bandwidth
>
> 37 dB μV m^{-1} for 230 MHz $< f < 1$ GHz and 1 MHz measurement bandwidth
>
> No limits are prescribed for $f > 1$ GHz and we will therefore continue to use the limit of
>
> 37 dBμV m^{-1} for all $f > 1$ GHz

A quasi peak detector is to be used for the measurements and no further specification of the spectrum apart from measurement bandwidth has been made. Many devices may emit narrow band radiation in the form of harmonics of their principal operating frequency, with a spacing that corresponds to the fundamental clock frequency. Radio astronomical bands are several MHz to several tens of MHz wide and powerful switches with switching rates of the order of a few ten kHz will create many harmonics within a radio astronomical band, creating an interference that often approximates a broad band signal.

However the increasing complexity and sophistication of the control and electronics of such devices, may also result in broad band interference, e.g. when switching frequencies and duty cycles vary on short timescales. Specifying peak or quasi-peak values only, makes the duration of the interference uncertain, it could last only for a few ms or for many

[2] Comité International Spécial des Perturbations Radioélectriques, the "special international committee on radio interference" is part of the international electro-technical commission (IEC).

[3] This document should however not be interpreted as one of fundamental opposition to modern electronics or to the measures to save energy and energy generation from renewable sources. On the contrary, scientists see that as an important contribution to averting a climatic catastrophe and to secure a sustainable level of human civilisation.

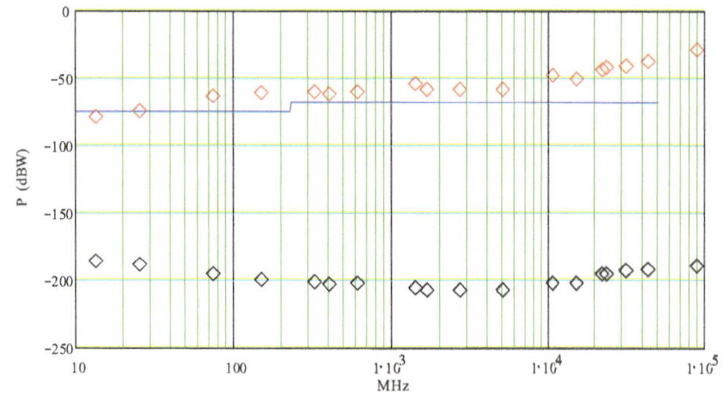

Fig. 5. Emission and reception power limits as specified by CISPR-11 and ITU-R RA 769. The blue trace is the e.i.r.p. per measurement channel as given by CISPR-11, the red diamonds indicate the maximum broad band emission within the allocated radio astronomy band and the black diamonds show the reception limits from ITU R-RA 769.

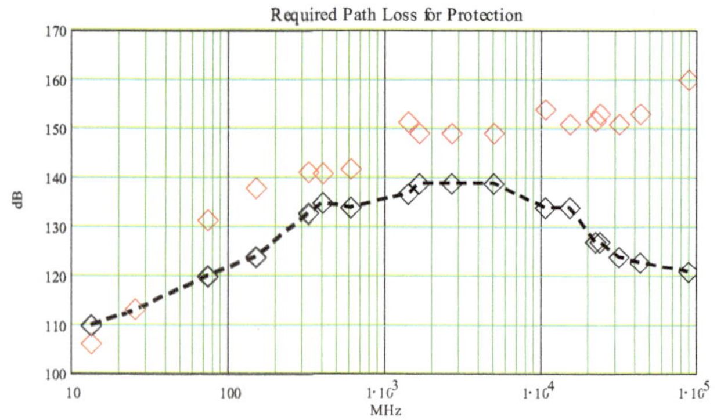

Fig. 6. MCL values for the protection of radio astronomy from devices fulfilling the CISPR-11 standard. Black line: Narrow band emission spectrum with one line per radio astronomical band, red diamonds: MCL for broad band emission still within the CISPR-11 limits.

days. This will affect radiometric measurements differently (see Eq. 1). The unspecific nature of the spectral and temporal characteristics of potential interferers makes generic compatibility assessments uncertain. In compatibility studies for passive services one is forced to err on the safe side and that may impose a greater constraint on emission levels than actually needed if a more informative specification of the interferer emissions were available.

It is useful to calculate the equivalent e.i.r.p. of the interference source for the given field strengths E and distances d:

$$P_{\text{eirp}} = E + 20\log(d) - 134.8 \tag{2}$$

where the power is given in dB(W), the electrical field strength E in dB(μV m^{-1}) and the distance d from the source in m. For the listed values of 30 and 37 dB(μV m^{-1}) we obtain emission e.i.r.p of -75 and -68 dB(W), emission powers that are only fractions of a μW. This has to be contrasted with the 60 dBW power level that is going to be processed in the switching converter. Although one knows that e.g. the harmonics generated by fast switching decay rapidly with frequency, it is also conceivable that meeting a target of 135 dB for the suppression of high frequency harmonics can be difficult.

ITU RA-769 gives in its Table 1 the threshold power ΔP_{H} in dB(W) for a radio astronomical receiver that would be considered as detrimental interference. The received power is meant to be averaged over the total allocated bandwidth for a specified integration time. Threshold values scale according to Eq. (1) for different receiver bandwidths and averaging times. Figure 5 shows the emission and reception thresholds in one graph: the difference between the emission and reception limits constitutes the minimum coupling loss (MCL) that is required to shield the radio receiver from the interferer. It is of the order of 140–160 dB and Fig. 6 shows it for narrow and broad band emissions as a function of frequency.

The effects of uncertainty about spectral properties result in significant variation of the MCL at frequencies above 1 GHz and even more so above 10 GHz. For large power plants high frequency emissions in the 10 GHz range are not expected to be a problem, but they can be important for other types of devices that have emission standards that are similarly defined but where there is a potential of significant broad band emissions in the GHz range.

Assuming unobstructed line of sight propagation without atmospheric absorption one can find the minimum distance required from

$$d = 10^{\frac{\text{MCL}-20\log(f)-92.5-A_{\text{ht}}}{20}} \tag{3}$$

given in km, with the minimum coupling loss in dB, the frequency in MHz and A_{ht} a correction factor for the absorption by ground clutter that depends on terrain properties and height of the transmitter. It ranges from 0 dB at heights exceeding 20 m to roughly 20 dB for ground level emissions (ITU-R P.452, 2005). Figure 7 shows the estimated separation distances as a function of frequency.

It is evident that electronic equipment complying with the industrial standard CISPR-11 has the potential to interfere on VHF and UHF bands over distances significantly greater than the line of sight horizon (typically 25 km) for the radio telescope (nominal height 50 m) and any ground level site of electronic equipment. However, the curvature of the earth and local topography will often provide additional shielding. A case by case study is required when electronic devices are to be operated within the line of sight of a radio telescope. As a result one may have to specify stricter limits than CISPR-11, in some locations but not in others.

An example of how the terrain affects the local attenuation is given in Fig. 7 for a solar power station close to a projected radio astronomical observatory in South Africa. The terrain is flat and there are large areas over distances of more than 30–40 km for which emissions from a power plant may

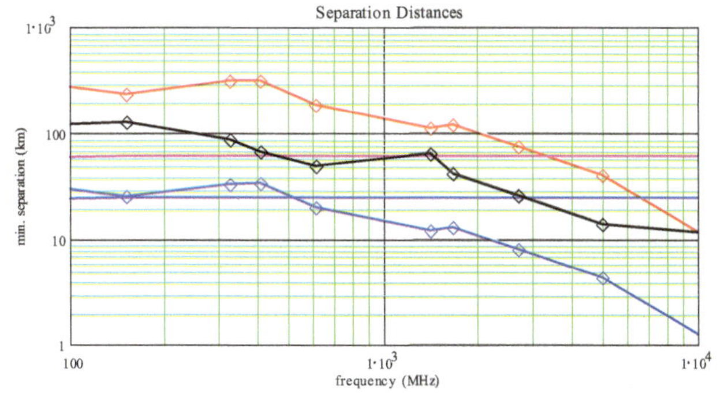

Fig. 7. Separation distances between CISPR-11 certified devices and radio astronomical stations. The red line shows the unobstructed free-space separation distance for equipment complying with CISPR-11 (EN55011) having only one narrow emission line within the radio astronomical band. The blue line with diamonds shows the separation distance for narrow band emitters on the ground (clutter absorption) and the black line that for broad band emitters. The 25 km line of sight horizon of the radio telescope and the 61 km visibility horizon for the hub of a windmill at a height of 100 m are shown by horizontal lines (blue and mauve).

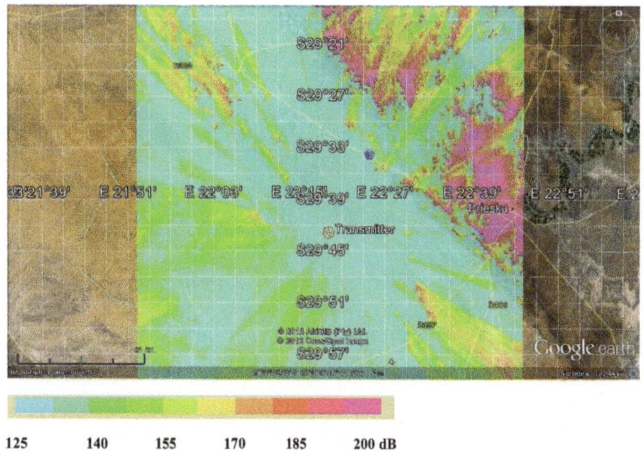

Fig. 8. Path loss map calculated for a South African site (indicated by "Transmitter") for emissions at 610 MHz and an emission height of 5 m using "Pathprofile" (CRAF, 2012). The MCL for narrow band emissions at that frequency is 134 dB and for broad band emissions it is 141 dB. All green areas are unsuitable as sites for a power station that will have emissions around the CISPR-11 limits.

degrade the measurements if their emission levels actually reach the limits specified in the CISPR-11 standard.

4 Specific mitigations and their assessment

Operational factors, such as intermittent activity, improved shielding or design that has inherently lower emissions (steep spectrum with only a few lines) will mitigate the impact the emitted interference. However these factors need to be carefully assessed and verified before they can enter a compatibility study. Standard measurement techniques, such as those prescribed in the CISPR standards may not be sensitive enough to verify the emission levels of 20–40 dB below the norm which may be needed for compatibility in some situations. An ordinary spectrometer, even combined with a low-noise frontend will usually not show any interference on such low levels as its intrinsic noise level is too high, especially when the normally short effective integration times are considered. But radiometric measurements are always possible if the location is radio-quiet. For that one would need to subtract the integrated *linear scaled power spectrum* of the background (device turned off) from the one taken with the device in operation. Equation (1) may be used to calculate the required averaging times for a desired sensitivity.

Separation distances can change significantly and in a non-linear way with mitigation, but terrain effects clearly dominate as can be seen in Fig. 8. The graph illustrates the example of interference on the radio astronomical band 1400–1427 MHz where according to RR footnote 5.340 "no emissions are permitted". For devices on the ground, we find that narrow band rfi can be tolerated for distances greater than 12.3 km and broad band rfi would require a separation of 34.8 km. A 20 dB mitigation through improved shielding, intrinsically lower emissions or a 1 % percent duty cycle will reduce the distance to 1.2 km for narrow band and 6.2 km for broad band rfi. In order to have an unrestricted deployment as close as 100 m from the antenna, one would require a reduction of more than 46 dB for broad band emissions down to a level of −114 dB(W/MHz). Note that within the EU, a limit of −120dB(W/MHz) is specified for license-free UWB devices for frequencies below 1.6 GHz (ECC Decision (06)04, 2007), which is 52 dB more stringent than CISPR-11! For installations in the hub of a wind power generator at a height of 100 m, and a distance of 2 km the requirements are similar: 36 dB reductions are needed in the case of narrow-band emission and a 50 dB reduction is required for broad band emissions (MCL of 150 dB!). It is conceivable that an AC-generator combined with filtered supply cables down to a shielded power converter on the ground may even achieve such a low level of emissions, however the verification will become quite difficult.

5 Aggregation of interference

Depending on the type of equipment, we may see a large deployment of devices over the area around the observatory and even if one can exclude or mitigate the effects of potential interferers close to the radio astronomy site, their aggregate effect may raise the ambient noise levels as seen for the general noise background in Fig. 4. The study of aggregation will provide another set of threshold levels, usually derived for an idealised situation where a large number of trials are made with devices placed randomly over an area centred

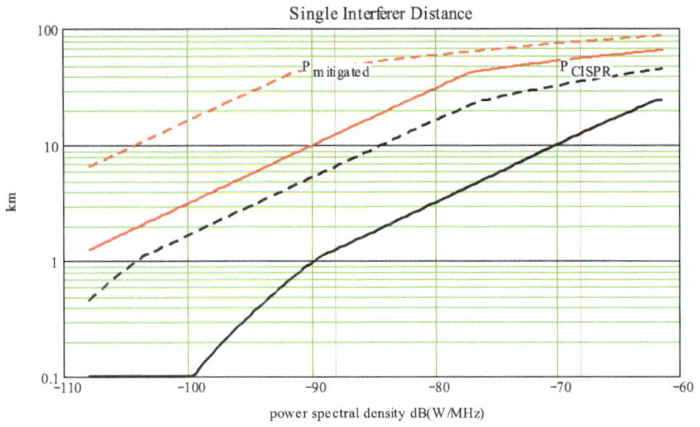

Fig. 9. Single interferer separation distances calculated for $f = 1.41$ GHz using the propagation model of ITU-R P.452 for flat terrain, but including the effects of ground clutter and curvature of the earth. Black solid line: narrow band emission at a height of 1 m, black dashed line: broad band emission. Red solid line: narrow band emission at a height of 100 m. Red dashed line: broad band emission at a height of 100 m.

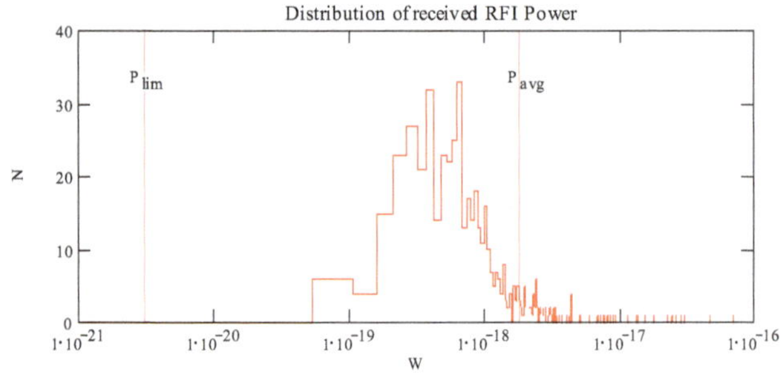

Fig. 10. Histogram of received power levels for 500 trials with random placements of narrow band CISPR-11 certified devices at 1.4 GHz within a range of 92 km from the radio telescope. The contributions from two operational heights (1 m and 100 m) were averaged for each of the 84 placements per trial for a deployment density of 1 km^{-2}. Average signal values are 27 dB above the ITU-R RA 769 limits designated by P_{lim}.

around the receiver and at different heights if so required for a realistic modelling of the deployment scenario. Their average distribution is set to match the average deployment values and emission levels. One then calculates the sum of their signal powers using known propagation mechanisms to obtain the received interference power for each trial. The results statistically evaluated and plotted in histograms like the one shown in Fig. 9 for the 1.4 GHz scenario outlined above. For an assumed deployment density of 1 km^{-2} and two possible operational heights we find that the average received power level amounts to -177 dB(W), exceeding the threshold ITU-R 769 threshold of -204 dB(W) by 27 dB, the median is at -181 dB(W) and in 2 % of all cases, the threshold was exceeded by more than 36 dB.

Although transmitter emissions are not permitted in the band we have just studied, radiation by electronic devices cannot be totally avoided and, an aggregation study may indicate the probability of interference when the deployment density of devices is known and their emissions are already reduced by technical or operational means. Different interference probabilities are needed depending on the allocation and protection status of the band. For shared bands (Radio Regulations FN 5.149) a 2 % interference probability is accepted as a practical measure, however purely passive bands require much lower interference probabilities derivable from the threshold values and the noise statistics. In addition one has to be aware of and take precautions for the fact, that a prediction of deployment characteristics for any new device is always conjectural and the actual level of interference from widely distributed license-free devices can only be known when deployment numbers have stabilised.

6 Conclusions

In the past, radio observatories and other passive services have had the opportunity of choosing remote, radio quiet rural locations which had a significantly lower level of man-made interference. The proliferation of fast switching power equipment and other digital technologies is a necessary step in the direction of greater energy conservation and use of renewable energy sources. This can however result in severe interference problems for passive services such as radio astronomy if compatibility questions not addressed at an early stage. Regulatory administrations have recognised the problem for unlicensed and uncontrolled consumer communications equipment, but this is not the case for industrial devices.

If a radio astronomical observation is obliterated by interference, the information that could be obtained is lost. It is irrelevant if the origin of rfi is from industrial or communication equipment, the distinction becomes more and more artificial with the progress of digital technology and it is now time to harmonise the emission standards for both the communications and the industrial realm.

Acknowledgements. The author likes to thank Mike Willis for the use of his pathprofile program and Alex Kraus, Benjamin Winkel, Harry Smith and Reinhard Keller for support and discussions.

References

Cane, H. V.: MNRAS, 189, 465, 1979.

CISPR Guide, IEC, Geneva, available at: http://www.iec.ch/emc/iec_emc/iec_emc_players_cispr.htm, 2010.

CRAF Newsletter 25, 7, available at: http://www.craf.eu/nwsl25.pdf, 2012.

Dicke, R. H.: The measurement of thermal radiation at Microwave frequencies, Rev. of Sci. Instr., 268, 17, 1946.

ECC Decision (06)04 on the harmonised conditions for devices using Ultra-Wideband (UWB) technology in bands below 10.6 GHz, Constanta, 2007.

ECC Doc. SE40(09)092: Assessment of Intermodulation products for the satellite transmitters and the victim receivers at the radio astronomy station, Copenhagen, 2009.

ECC Rep. 171: Impact of unwanted emissions of IRIDIUM satellites on radio astrononomy operations in the band 1610.6–1613.8 MHz, Tallinn, 2011.

Ellington IEEE Trans on Antennas and Propagation, 53, No. 8, 2480, 2005.

ITU-R RA. 769-2: Protection criteria used for radio astronomical measurements, ITU, Geneva, 2003.

ITU-R P. 372-2: Radio noise, ITU, Geneva, 2007.

ITU-R P. 452 Prediction procedure for the evaluation of microwave interference between stations on the surface of the earth at frequencies above 0.7 GHz, ITU, Geneva, 2005.

Radio Regulations, ITU, Geneva, 2012.

30

Optimal design of active EMC filters

B. Chand, T. Kut, and S. Dickmann

Fundamentals of Electrical Engineering, Helmut Schmidt University / University of the Federal Armed Forces Hamburg, Germany

Correspondence to: B. Chand (benian.chand@hsu-hh.de)

Abstract. A recent trend in automotive industry is adding electrical drive systems to conventional drives. The electrification allows an expansion of energy sources and provides great opportunities for environmental friendly mobility. The electrical powertrain and its components can also cause disturbances which couple into nearby electronic control units and communication cables. Therefore the communication can be degraded or even permanently disrupted. To minimize these interferences, different approaches are possible. One possibility is to use EMC filters. However, the diversity of filters is very large and the determination of an appropriate filter for each application is time-consuming. Therefore, the filter design is determined by using a simulation tool including an effective optimization algorithm. This method leads to improvements in terms of weight, volume and cost.

1 Introduction

The EMC of electronic systems is often achieved by filter networks. The used components take an increasing part of the total system weight and volume. Therefore, the optimal filter for different requirements is determined with previous simulation estimations. The filters are characterized in terms of the number and type of filter stages, component values or attenuation. Therefore passive, active and hybrid filter structures are used. The active filters are investigated by using a feedforward structure (Fig. 1). Furthermore, the frequency characteristic of the filter is described by the use of $\underline{\mathbf{A}}$-matrices. The existing common- and differential-mode interferences are attenuated by an interconnection of common- and differential-mode filters. The input and output impedances are important design criteria. Taking into account these criteria, the manual design is very time-consuming due to the large number of possible filter structures. Therefore, a algorithm was implemented which determines the optimal filter network depending on the required parameters. To determine the optimal filter for given boundary conditions the Particle Swarm Optimization (PSO) process is used. The output of the software includes not only the filter network topology and the component values, but also a comparison of the cost, weight and the simulation time. These results are verified by measurements. Finally, the determined filter is studied in a real, practical construction of an electrical powertrain.

2 Concepts of reducing the interferences by using EMC filters

We consider filter circuits which attenuate conducted interference. Three groups of filter circuits are explained in this chapter: Passive, active and hybrid filters for common-mode and differential-mode. Parasitic effects are not taken into account. Another topic of this section is the description of a two-port network by an $\underline{\mathbf{A}}$-matrix. With these methods it is possible to determine the frequency characteristics of an electric filter circuit in a short time.

Passive filters

are the simplest form of electric filters (Hagmann, 2008). Since the components require no power supply, the combination of these components are called passive filters. In this case, the filter should attenuate harmonics. For this reason, a low-pass topology will be used.

Active filters

reduce the emitted interference from the source by means of controllable active components (Schinkel, 2009). The conducted interference is minimized by means of a positive feedback structure, see Fig. 1. In that case, the signal including

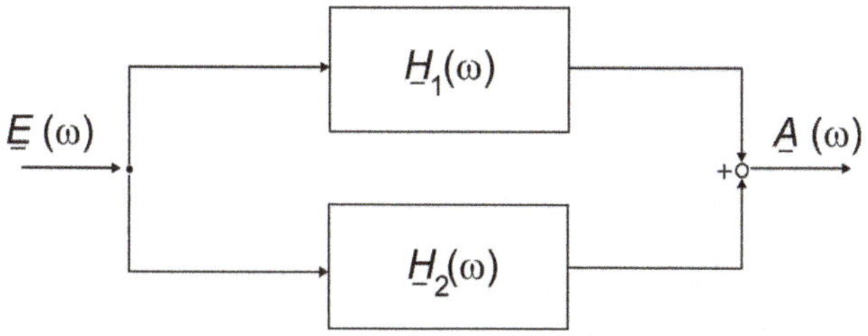

Fig. 1. Positive feedback structure for active filtering.

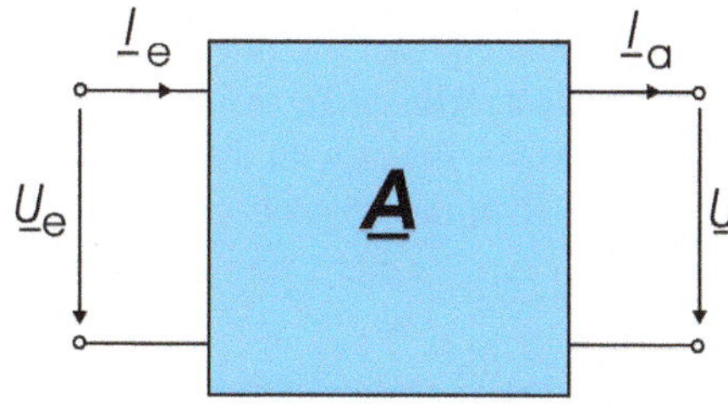

Fig. 2. Schematic description of a four-terminal network.

the interference $\underline{E}(\omega)$ is conducted in parallel over the functions $\underline{H}_1(\omega)$ and $\underline{H}_2(\omega)$, so the unwanted interference can be canceled and the required output signal $\underline{A}(\omega)$ is produced.

Hybrid filters

circuits consist of active and passive filter circuits. The advantage with such a configuration is the attenuation of the interference to a large amount by means of an active filter. Therefore, the passive components can be designed smaller, which can add to weight and cost savings.

Description of the filter using A-Matrices

To determine the frequency characteristic of a filter circuit, A-matrices can be used. In Fig. 2 a four-terminal network with its connections, the output and input current and the output and input voltage is shown.

The system can be represented using a matrix notation:

$$\begin{pmatrix} \underline{U}_e \\ \underline{I}_e \end{pmatrix} = \underline{\mathbf{A}} \cdot \begin{pmatrix} \underline{U}_a \\ \underline{I}_a \end{pmatrix} \text{ with } \underline{\mathbf{A}} = \begin{pmatrix} \underline{A}_{11} & \underline{A}_{12} \\ \underline{A}_{21} & \underline{A}_{22} \end{pmatrix} \tag{1}$$

The individual elements of the matrix $\underline{\mathbf{A}}$ are:

$$\begin{aligned} \underline{A}_{11} &= \left.\frac{\partial \underline{U}_e}{\partial \underline{U}_a}\right|_{\underline{I}_a=0} \quad \underline{A}_{12} = \left.\frac{\partial \underline{U}_e}{\partial \underline{I}_a}\right|_{\underline{U}_a=0} \\ \underline{A}_{21} &= \left.\frac{\partial \underline{I}_e}{\partial \underline{U}_a}\right|_{\underline{I}_a=0} \quad \underline{A}_{22} = \left.\frac{\partial \underline{I}_e}{\partial \underline{I}_a}\right|_{\underline{U}_a=0} \end{aligned} \tag{2}$$

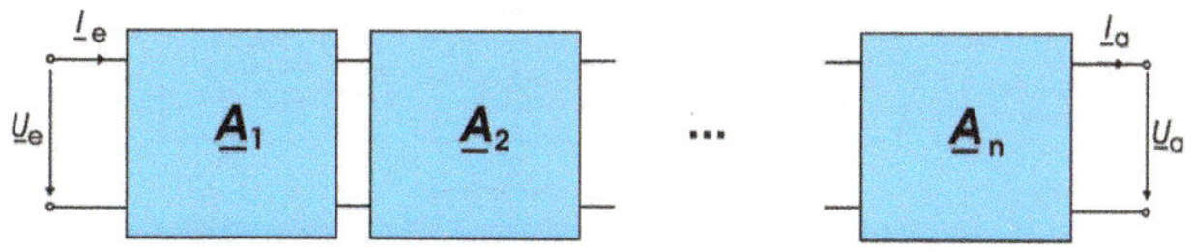

Fig. 3. Series connection of A-matrices.

The advantage for a series connection of A-matrices, as described in Fig. 3, is the possibility to multiply the individual matrices easily. The product results in an overall matrix. For the given overall matrix $\underline{A}_{\mathbf{total}}$, it is now possible to determine the ratio of the input to output voltage. It follows from the inverse of the matrix element $\underline{A}_{11}$ (Marti, 2007).

$$\begin{pmatrix} \underline{U}_e \\ \underline{I}_e \end{pmatrix} = \underline{\mathbf{A}}_\mathbf{1} \cdot \underline{\mathbf{A}}_\mathbf{2} \cdot \underline{\mathbf{A}}_\mathbf{3} \cdot \ldots \cdot \underline{\mathbf{A}}_\mathbf{n-1} \cdot \underline{\mathbf{A}}_\mathbf{n} \cdot \begin{pmatrix} \underline{U}_a \\ \underline{I}_a \end{pmatrix} \tag{3}$$

$$\begin{pmatrix} \underline{U}_e \\ \underline{I}_e \end{pmatrix} = \underline{\mathbf{A}}_\mathbf{tot} \cdot \begin{pmatrix} \underline{U}_a \\ \underline{I}_a \end{pmatrix} \tag{4}$$

For later considerations of active and hybrid filters the A-matrix of the active filter results from Fig. 5 (Rehfeld, 2012):

$$\underline{\mathbf{A}} = \begin{pmatrix} \frac{s^2 \cdot k_1 \cdot L_1 \cdot C_1 + s \cdot C_1 \cdot R_C + 1}{s^2 \underline{k}_2 \cdot L_1 \cdot C_1 + s \cdot C_1 \cdot R_C + 1} & \frac{s \cdot L_1 \cdot (s \cdot C_1 \cdot R_C 1)}{s^2 \cdot L_1 \cdot C_1 (\underline{\kappa} \cdot (1+v) + \frac{v}{\ddot{u}}) + s \cdot C_1 \cdot R_C + 1} \\ \frac{s \cdot C_1 (1 - \underline{\kappa}(1+v))}{s^2 \cdot \underline{k}_2 \cdot L_1 \cdot C_1 + s \cdot C_1 \cdot R_C + 1} & 1 \end{pmatrix}$$

with $k_1 = 1 + \frac{v}{\ddot{u}}$, $\underline{k}_2 = \underline{\kappa} \cdot (v+1) + \frac{v}{\ddot{u}}$, $\underline{\kappa} = \frac{\underline{U}_{OP}}{\underline{U}_e}$, $v = \frac{R_2}{R_1}$, $\ddot{u} = \sqrt{\frac{L_1}{L_2}}$, $s = j \cdot 2 \cdot \pi \cdot f$

Common-mode and differential-mode filters

Since interferences are often composed of common-mode and differential-mode interferences, a seperate consideration of common-mode and differential-mode filters is recommended. This allows a more effective attenuation of the interferences. X-capacitors or differential-mode chokes are used as classical elements for reduction of differential-mode interferences. In DC and single phase systems, X-capacitors are located between the supply and the return conductor. Common-mode filters attenuate common-mode interferences and represent at the same time short circuit for differential-mode interferences. The main components of this filter group are common-mode inductors and Y-capacitors. The leakage currents through the Y-capacitor should be kept small, because otherwise the capacitor bypasses the insulation. According to (DIN-EN, 50178), depending on the model, leakage currents of 0.75 mA to 3.5 mA are allowed. Therefore for the filter design, two different simulations are studied. One for the differential-mode filter and one for the common-mode filter. Thereby the common-mode filter is simulated for

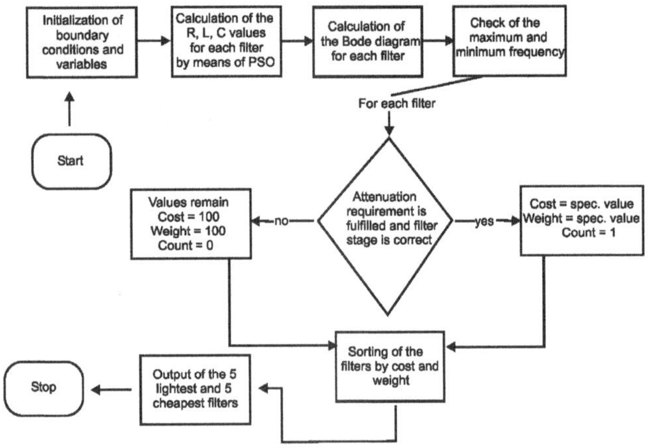

Fig. 4. Program Flowchart.

one wire and the ground and the same common-mode filter is used for the second wire as well. So the approach of 2 × 2 $\underline{A}$-matrices can be used here.

3 Design algorithm for EMC filters

3.1 Program flow of the simulation script

A program for designing electrical filters has been created. After entering the required attenuation in a certain frequency band, the script displays the cheapest and lightest filter topology for given boundary conditions. In this section, the algorithm of the script will be introduced and some important details are discussed.

With the help of a graphical user interface (GUI) the boundary conditions for the configuration of the filter circuit are set. These are the minimum and maximum values of the used inductors, capacitors and resistors (L_{min}, L_{max}, C_{min}, C_{max}, R_{min}, R_{max}). Another input parameter is the attenuation within a given frequency band. Furthermore, it should be determined, whether there is a high ohmic or a low ohmic input and output impedance, because they influence the behavior of the filter circuit. Moreover, it is necessary to determine the number of the filter stages. For the evaluation, the program needs an input signal described in a text file. As output parameter of the script the required computing time or the calculated attenuation can be selected. In Fig. 4, the program flowchart of the main program is shown.

First of all, the entered boundary conditions are stored and initialized. In the next step, the various components of the different filter circuits are calculated. This is done using an optimization algorithm which is called Particle Swarm Optimization (PSO) (Brodersen, 2006). This process will be discussed later in this section. After calculating the components of each filter, the frequency characteristic of the different filter stages can be calculated throughout these components. For this purpose the $\underline{A}$-matrices of the filter circuits are used. After comparing the input signal with the selected minimum and maximum frequencies, the program controls whether the filters are generally useable. On the one hand it is important, that the filter fulfills the required attenuation in the given frequency band and on the other hand it is checked whether the filter matches with the given filter stage. If these two conditions are met, the respective control variable "Count" is set to the value 1 and a specific value for the weight and price for this filter is defined. If the first condition is not fulfilled, the control variable "Count" is set to 0, the specific cost and the specific weight of the filter are set to the value 100. In the next step, the cost and weight of the filter are sorted by size, starting with the lowest. After that, the five filters with the lowest weight and the five filters with the smallest price will be displayed. Filters with the control variable "Count" = 0 are not considered.

3.2 Studied filter circuits

For the program and the later practical applications, it is necessary to consider different filter circuits. In the simulation program the following passive filter circuits are implemented to the fifth order: LC-, RC-, LRC-, T- and π-filter. In active filters, operational amplifiers are used. The advantage of using operational amplifiers is the low weight and the low construction volume. The ideal operational amplifier, which is used in these simulations has the following characteristics: The internal impedance of the operational amplifier is infinitely large. The input current $\underline{I}_e$ is zero. The output impedance of the operational amplifier is zero. The voltage gain $A_{u,d}$ for differential signals is infinitely large. Therefore the voltage between the input terminals is zero, when the output signal is limited. The common-mode rejection is infinite large.

Previous studies (Rehfeld, 2012) have shown, that an active filter with an inductive coupling and capacitive decoupling has the best filter characteristics. In these previous studies different active filters with operational amplifiers were studied and the the active LC-filter in Fig. 5 was chosen. This filter is used for further considerations. It consists of a transformer, which decouples the interference signal via the inductance L_1 and transmits it to the inductance L_2, in the secondary circuit. There, the decoupled signal is inverted by an inverter circuit and is transfered by means of the capacitor C_1 back to the load circuit.

3.3 Filter Optimization algorithm

The optimization of a filter is a complex problem, because filters are described by several parameters, which in addition need to be in a certain solution space.

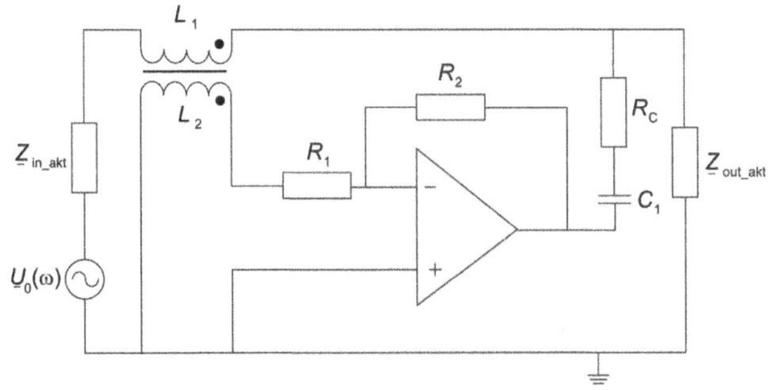

Fig. 5. Active Filter with inductive coupling and capacitance decoupling.

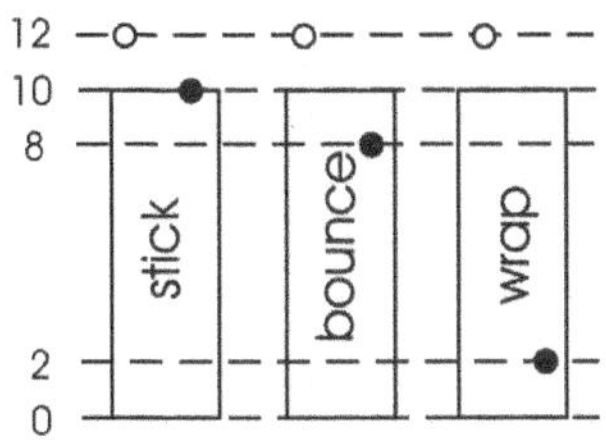

Fig. 6. Restriction condition for the PSO-algorithm (Brodersen, 2007).

3.3.1 Brute-Force-Method

The Brute-Force-Method tries each or a majority of all possible solutions meeting the boundary conditions, until a solution is found. The advantage of this method is the determination of an unique solution of the problem. However, the disadvantage is the requirement of high computing time due to an increasing number of variables in the solution space.

3.3.2 Ant-Colony-Algorithm

Another way to solve a complex problem is the Ant-Colony-Algorithm. Here, the natural behavior of ants when foraging is imitated. Each ant has a memory and also the entire ant colony has a collective memory. The ants have several possible routes to reach the feeding place. The result of the algorithm is the optimal way to the food source (Boysen, 2006). This method is used very often for the optimization of business processes. It would also be possible to design filters with this process.

3.3.3 Particle-Swarm-Optimization

Another solution method is the Particle-Swarm-Optimization (PSO). This process provided good results in the field of engineering, e.g. in the construction of bandpass filters or the design of gears and can be applied to the current problem. The algorithm will be described more in detail below and is adapted to the problem of filter design.

Mathematical description:

Basically, the i-th particle of a swarm is described by:

- $\boldsymbol{x}_i$: current position of the particle
- $\boldsymbol{v}_i$: current velocity of the particle
- $\boldsymbol{y}_i$: best position of the particle

In addition, a certain fitness function $f(\boldsymbol{x}_i)$ is assumed. From this function we obtain the so called fitness value, that indicates how far the current position is away from the target. From the fitness value the next best position results:

$$\boldsymbol{y}_i(t+1) = \begin{cases} \boldsymbol{y}_i(t) & \text{if } f(\boldsymbol{x}_i(t+1)) < f(\boldsymbol{y}_i(t)) \\ \boldsymbol{x}_i(t+1) & \text{if } f(\boldsymbol{x}_i(t+1)) \geq f(\boldsymbol{y}_i(t)) \end{cases} \tag{5}$$

The next important parameter of the PSO is the swarm. It consists of p particles. The swarm has a collective memory, which remembers the best position of the particle with the highest fitness value. This value is called the global best position. It is given by:

$$y_{\mathrm{gb}}(t+1) =$$

$$= \begin{cases} y_{\mathrm{gb}}(t) & \text{if } f\left(y_{0\ldots p}(t+1)\right) < f\left(y_{\mathrm{gb}}(t)\right) \\ \left(y_{0\ldots p}(t+1)\right)_{\max} & \text{if } f\left(y_{0\ldots p}(t+1)\right) \geq f\left(y_{\mathrm{gb}}(t)\right) \end{cases}$$

From the best position of each particle and the global best position, the new velocity of a particle is results:

$$\begin{gathered} \boldsymbol{v}_i(t+1) = w \cdot \boldsymbol{v}_i(t) + c_1 r_1(t) \cdot \\ \left[\boldsymbol{y}_i(t) - \boldsymbol{x}_i(t)\right] + c_2 r_2(t)\left[y_{gb}(t) - \boldsymbol{x}_i(t)\right] \end{gathered} \tag{6}$$

With the variables:

$$0 \leq c_1, c_2, w \leq 2 \tag{7}$$

The two functions $r_1(t)$ and $r_2(t)$ are random numbers (Brodersen, 2006). After determining the new velocity of each particle, the new current position of the particle results:

$$\boldsymbol{x}_i(t+1) = \boldsymbol{x}_i(t) + \boldsymbol{v}_i(t+1) \cdot \Delta t \tag{8}$$

This algorithm is repeated until (Wu, 2006):

a. a particle has found the optimal position

b. it is certain, that the optimal position for the solution space does not exist

It is possible that a particle leaves the predetermined solution space. There are three possible restriction conditions for this case. In Fig. 6 the conditions stick, bounce and wrap are

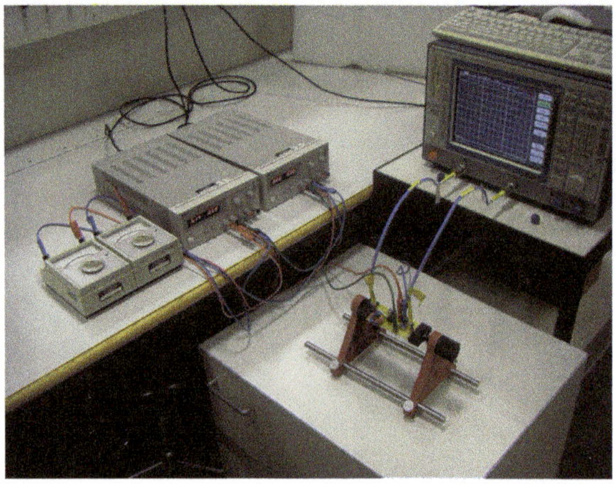

Fig. 7. Measurement setup for measuring the filter response.

shown. For the solution space the numbers from 0 to 10 are set. The white particle leaves the solution space. As a result, the black particle is created in the solution space, according to the following rules.

For the stick specification the particle remains at the maximum value of the solution space. The particle jumps back the amount in the solution space for which he has left it at the bounce condiction. For the wrap specification the particle starts at a lower limit, when leaving the solution space at the upper limit. It is increased by the amount, by which it has left the solution space at the upper limit.

PSO for filter design:

The PSO algorithm, is now applied to filter design. Since a filter consists of inductors, capacitors and resistors, these components can be regarded as the current position of a particle. Accordingly, the velocity of the particle also results as a vector consisting of a change of the inductance, capacitance and resistance. Thus, the best position of the particle is also a vector which consists of an inductor, a capacitor and a resistor.

$$\boldsymbol{x}_i = \begin{pmatrix} L_i(t) \\ C_i(t) \\ R_i(t) \end{pmatrix}, \quad \boldsymbol{v}_i = \begin{pmatrix} \Delta L_i(t) \\ \Delta C_i(t) \\ \Delta R_i(t) \end{pmatrix}, \quad \boldsymbol{y}_i = \begin{pmatrix} L_i(t) \\ C_i(t) \\ R_i(t) \end{pmatrix} \tag{9}$$

After the start of the algorithm for each particle a random position and velocity in the solution space are set. The specific attenuation is determined by means of the $\underline{\mathbf{A}}$-matrix process described in Sect. 2. With this attenuation the corresponding fitness value of the filter is determined. The fitness value is depending on the attenuation of the filter D in dB. The fitness value is 0, if the attenuation is smaller than the attenuation set value -10 dB. From this point the fitness value increases linearly. When the filter reaches the required attenuation D_{soll}, the fitness value will be 1.

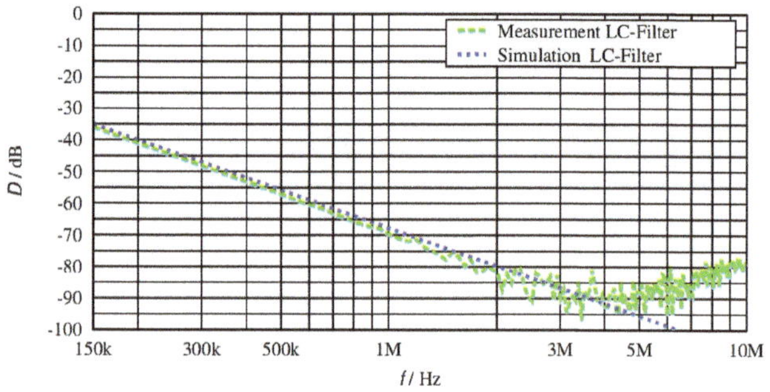

Fig. 8. Comparison of the simulation and measurement results of a passive LC filter.

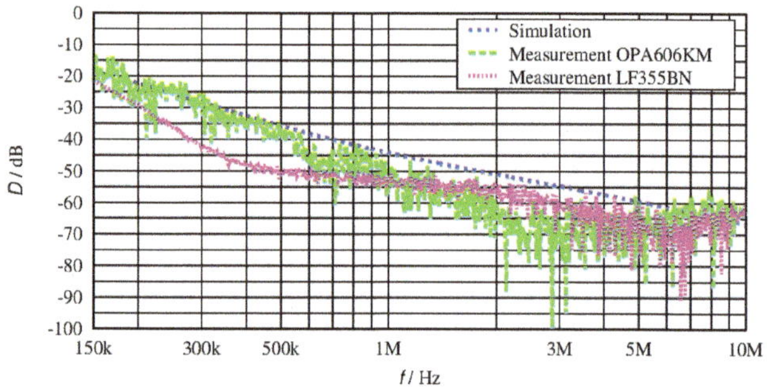

Fig. 9. Comparison of the simulation and measurement results of an active common-mode filter.

Through the fitness value the best position of a particle and the global best position $\boldsymbol{y}_{\mathrm{gb}}$ can be calculated. From these values the new velocity of the particle is determined. This process is repeated until, the optimal solution is found - in this case, the optimal attenuation of the filter.

4 Validation of the simulation by measurements

After different filter circuits have been simulated, these should be validated by measurements. For this purpose, different filters have been constructed and measured. The general measurement setup is shown in Fig. 7. The connections of the semi-rigid cables and the filter setup were done using high frequency SMA connectors. The components of the filter circuit were chosen small and with less tolerances so the measurement will be possible in the measured frequency range. The influence of the other cabling could be neglected, because the cabling did not change the measurement results noticeably.

In Fig. 8 the measurement and simulation results of a passive filter are shown. There the attenuation D in dB, is plotted over the frequency f in Hz.

Up to a frequency of 4 MHz, the filter responses are almost identical. From an attenuation of -80 dB the measurement is not longer accurate, since the signal-to-noise ratio becomes

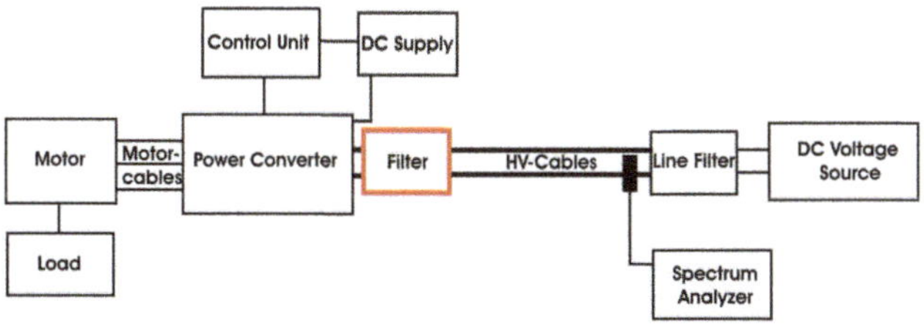

Fig. 10. Measurement setup of the electric powertrain including filter circuit.

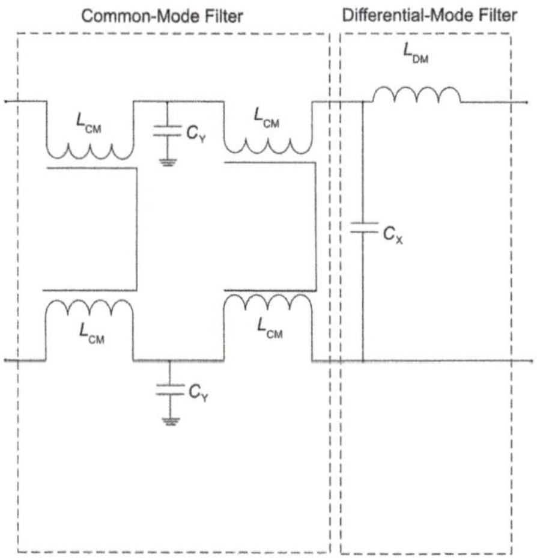

Fig. 11. Used filter circuit at the measurement setup of the electric powertrain.

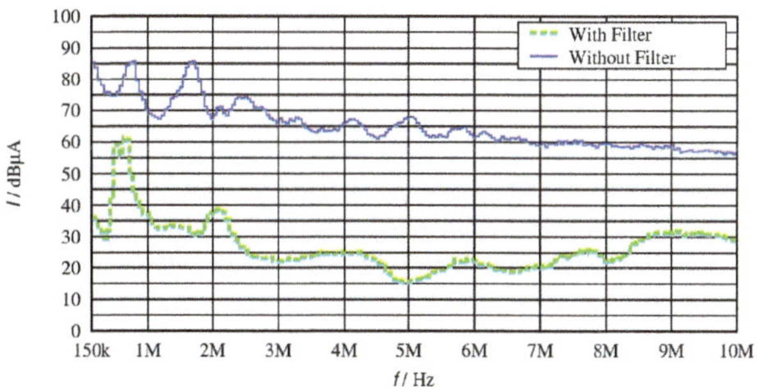

Fig. 12. Comparison of the measurement results with and without a filter.

too small. For this topology, parasitic effects are noticeable once the frequency exceeds about 4 MHz. This is explained by the parasitic inductance in series with the capacitor.

Another representative comparison shows the following diagram (Fig.9) of measurement and simulation results of the attenuation of an active common-mode filter. The characteristic of the simulated filter circuit is shown with big dotted blue line. For the measurement setup two different operational amplifiers were used. The reasons for the deviations from the simulations are on the one hand the limited bandwidth of the operational amplifiers and on the other hand the voltage drop between the inverting and noninverting input, which deviates from the simulation.

5 Practical setup of a filter for the electrical powertrain

After validating the simulated filter circuits by measurements, the filter is studied in a practical test setup of e.g. an electric powertrain. One possible measurement setup is shown in Fig. 10.

In Fig. 11 the used filter can be seen. The choice of the used components must take the real operating conditions of the filter into account. Magnetic components must be designed to avoid saturation. Moreover, insulation and capacitors must be electrically strong to withstand the occurring voltages. Please note that this is only one possible filter out of many different usable filter circuits. In this circuit Y-capacitors C_Y, common-mode chokes L_{CM} and a X-capacitor C_X as well as a differential-mode inductor L_{DM} are used.

The spectrum is measured with a current measuring clamp connected to a spectrum analyzer. In Fig. 12 a comparison of the interferences on the HV-cables with and without filters is shown.

The filter attenuates the interferences in the entire required frequency range. The lower attenuation at lower frequencies can be increased by using components with lower tolerance.

6 Conclusions

In the early development process of EMC filters, simulations are helpful. Due to the large number of possible filter variants, the manual determination is time-consuming. Therefore, a software tool has been developed which determines filters for DC applications quickly, efficiently and optimizes their weight, volume and cost. A further advantage is, that new filter concepts, such as hybrid filter structures can be implemented quickly and compared with conventional concepts. Here, an optimal filter design is determined by using a simulation tool including an effective optimization algorithm. The comparison of simulation and measurement results show good agreement. Investigations with an electrical powertrain setup verified the filter effect.

References

Boysen, N.: Ameisenalgorithmen, 2006. Universität Hamburg, 2006.

Brodersen, O.: Particle Swarm Optimization in der stochastischen Lagerhaltung, 2006. Gemeinsamer Workshop der Arbeitsgruppen "Entscheidungstheorie und -praxis" und "Fuzzy-Systeme, Neuronale Netze und Künstliche Intelligenz" der Gesellschaft für Operations Research e.V. (GOR) Kaiserslautern, 2006.

Brodersen, O. and Schumann, M.: Einsatz der Particle Swarm Optimization zur Optimierung universitärer Stundenpläne, 2007, Arbeitsbericht Nr. 05/2007. Universität Göttingen, 2007.

DIN-EN 50178 / VDE 0160: Ausrüstung von Starkstromanlagen mit elektronischen Betriebsmitteln, 1998, DIN Deutsches Institut für Normung e.V. und VDE Verband deutscher Elektrotechniker e.V., 1998.

Hagmann, G.: Grundlagen der Elektrotechnik, 2008, 13 Edn., AULA-Verlag, 2008.

Marti, O and Plettl, A.: Physikalische Elektronik und Messtechnik, 2007, Universität Ulm, 2007.

Rehfeld, M.: EMC-Studies for optimal Filter Design in Aircraft and Automotive Applications, 2012, Helmut Schmidt University, University of the Federal Armed Forces Hamburg, 2012.

Schinkel, M.: Entwurf und Simulation aktiver EMV-Filter für dreiphasige drehzahlveränderbare Antriebe, 2009, Technische Universität Berlin, 2009.

Schwab, A. J. and Kürner, W.: Elektromagnetische Verträglichkeit, 2011, 6 Edn., Springer-Verlag, 2011.

Wu, G. and Hansen, V. and Kreysa, E. and Gemünd, H.P.: Optimierung von FSS-Bandpassfiltern mit Hilfe der Schwarmintelligenz (Particle Swarm Optimization), 2006, URSI Adv. Radio Sci., 4, 65–71, 2006, http://www.adv-radio-sci.net/4/65/2006/.

31

A novel solution algorithm for nonlinearly loaded transmission lines inside resonating enclosures

R. Rambousky[1], **S. Tkachenko**[2], **and J. Nitsch**[2]

[1]Bundeswehr Research Institute for Protective Technologies and NBC Protection (WIS), Munster, Germany
[2]Otto-von-Guericke University Magdeburg, Magdeburg, Germany

Correspondence to: R. Rambousky (ronald.rambousky@ieee.org)

Abstract. Nonlinearly loaded lossless transmission lines inside a rectangular cavity are studied using the left- and right-hand Green's functions of the problem in time domain. These Green's functions are developed for a transmission line with quasi-matched loads. This ensures Green's functions of a short duration. Therefore, the amount of frequency data necessary to obtain time-domain Green's functions is quite limited. The time-domain Green's functions are finally convolved with the left- and right-hand line voltages. With this technique it is possible to treat arbitrarily loaded transmission lines in resonators. An example is presented to demonstrate the applicability of this technique to a transmission line with a simple diode as nonlinear load.

1 Introduction

Interconnecting cables play a major role in Electromagnetic Compatibility (EMC). These cables do not only transport information signals and energy in electronic systems, but they also act as effective collectors for unintended electromagnetic energy. Electronic components usually possess basic EMC hardness but often they are integrated in metallic housings and interconnected with other components. Therefore, the resulting configuration may differ significantly from the EMC test configuration of the single electronic component. Resonance effects of the housing, cable coupling in cable harnesses or nonlinear effects (e.g. demodulation) may effect electromagnetic compatibility issues and therefore the reliable operation of the electronic system.

Shielded rooms are designed to provide the electronic devices as a whole a sufficient electromagnetic protection. Since the shielding factor decreases for non-ideal screens with increasing frequencies, an intentional electromagnetic disturbance of a globally protected system will be expected more at the higher frequencies. If nonlinearities are excited by higher frequencies, then they also generate lower-frequency components in the noise spectrum by demodulation effects which may cause a number of problems for the electronic system, like, e.g., distortion of useful electronic signals, shift of operating points of transistor cascades of electronic circuits, or generation of dynamic instabilities in complex electronic systems.

Every electromagnetic field excitation of a shielded room, like a rectangular resonator, may generate resonance frequencies which are closely adjacent, in particular in the high frequency spectrum. Therefore, two-tone signals can be produced (Krauthäuser et al., 2002) that lead by field to cable coupling to demodulation effects in the current spectra in non-linear terminations. Other mechanisms which may also cause a two-tone high-frequency field inside the resonator is an exterior field excitation which enters the resonator through an aperture or is generated by an interior small antenna inside the resonator (Rambousky et al., 2013).

In this work the influence of nonlinear loads on cables (transmission lines) concerning transmitted current spectra is investigated. Transmission lines with nonlinear loads over PEC ground in free space show the expected demodulation and intermodulation effects. In resonators additionally there is a strong coupling of the transmission line with the resonance structure of the resonator and therefore the resulting spectra are much more comprehensive. In particular the resulting electromagnetic fields in the resonator increase significantly when the excitation frequency coincide with the resonance frequencies of the resonator.

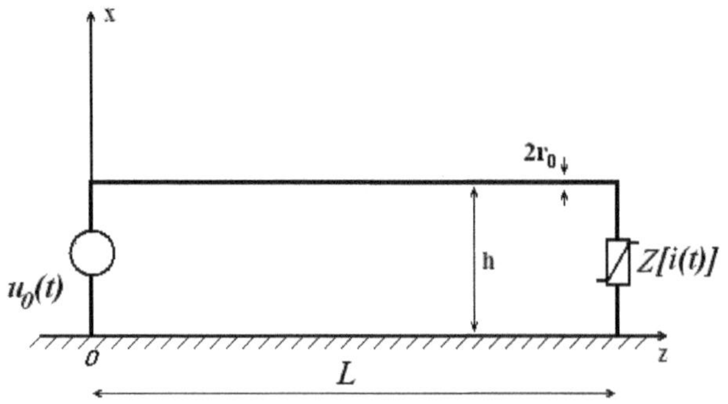

Figure 1. Geometry of the conductor above conducting ground.

Consider a classical transmission line which is fed/terminated at both ends by two active/passive sources $u_0(t)$ and $u_r(t)$, respectively. If one further assumes that the two corresponding left- and right-hand Green's functions $g_l(t)$ and $g_r(t)$ in time-domain are known then the current $i_L(t)$ in time-domain at the end of the transmission line at $z = L$ can be expressed as a sum of convolution integrals of these Green's functions and the two (voltage) sources:

$$i_L(t) = \int_{-\infty}^{t} g_l(L, t-\tau) u_0(\tau) d\tau + \int_{-\infty}^{t} g_r(L, t-\tau) u_r(\tau) d\tau \qquad (1)$$

Thus, with the knowledge of these two Green's functions one can calculate the current $i_L(t)$, for linear as well as for non-linear terminations of the transmission line. In this paper, the emphasis is laid on estimating these Green's functions.

2 Nonlinearly loaded transmission lines in free space above conducting ground

Consider first a linearly terminated transmission line above a perfectly conducting ground in the Agrawal form in free space (for the geometry see Fig. 1). The solution for this problem is known (Liu and Tesche, 1976) for the current at the end of the line ($z = L$) and can be represented as

$$I(L, \omega) = U_0(\omega) Y_l(L, \omega) + U_r(\omega) Y_r(L, \omega). \qquad (2)$$

Here $U_0(\omega)$ and $U_r(\omega)$ are the left- and right-hand sources of the conductor, respectively. The functions $Y_l(L, \omega)$ and $Y_r(L, \omega)$ are the corresponding complex admittance functions at the end ($z = L$) of the line. They are given in frequency domain by the expressions

$$Y_l(L, \omega) = \frac{2}{Z_C} \sum_{m=0}^{\infty} \exp(-\jmath(2m+1)kL) \qquad (3)$$

and

$$Y_r(L, \omega) = \frac{1}{Z_C} \left[1 + 2 \sum_{m=1}^{\infty} \exp(-\jmath 2mkL) \right], \qquad (4)$$

respectively. The quantity Z_C denotes the characteristic impedance of the line, $k = \omega / c_0$ is the wave number and $\jmath \equiv \sqrt{-1}$ the imaginary unit.

An inverse Fourier transformation of Eqs. (3) and (4) into time-domain yields

$$y_l(L, t) = \frac{2}{Z_C} \sum_{m=0}^{\infty} \delta\left(t - (2m+1)\frac{L}{c_0}\right) \qquad (5)$$

and

$$y_r(L, t) = \frac{1}{Z_C} \left[\delta(t) + 2 \sum_{m=1}^{\infty} \delta\left(t - 2m\frac{L}{c_0}\right) \right]. \qquad (6)$$

The dirac delta function is denoted by $\delta(t)$ and c_0 is the speed of light in vacuum. The convolution integral of Eq. (1) can simply be calculated and results in:

$$i(L, t) = \frac{2}{Z_C} \sum_{m=0}^{\infty} u_0\left(t - (2m+1)\frac{L}{c_0}\right) + \frac{1}{Z_C} \left[u_r(t) + 2 \sum_{m=1}^{\infty} u_r\left(t - 2m\frac{L}{c_0}\right) \right] \qquad (7)$$

This is the general solution in time-domain. Here the restriction to linear loads can be abandoned, and Eq. (7) now is valid for both, linear as well as nonlinear terminations. Assuming for the source at the right side of the conductor the relation

$$-\frac{u_r(t)}{Z_C} = \frac{i(L,t) Z(i(L,t))}{Z_C} =: \tilde{i}(t) \qquad (8)$$

one obtains from Eq. (7)

$$i(L, t) + \tilde{i}(t) = \frac{2}{Z_C} \sum_{m=0}^{\infty} u_0\left(t - (2m+1)\frac{L}{c_0}\right) - 2 \sum_{m=1}^{\infty} \tilde{i}\left(t - 2m\frac{L}{c_0}\right). \qquad (9)$$

If the voltage-current characteristic of the nonlinear load and the driving source $u_0(t)$ for $t \geq 0$ are known, Eq. (9) can be solved, in most cases numerically. Noting that the right-hand side of Eq. (9) depends only on past values of t, this equation may be solved for the unknown current $i(L, t)$ for later times using a standard root-finding algorithm.

3 Transmission line with nonlinear load inside a rectangular cavity

In this case the transmission line is positioned symmetrically inside the cavity (see Fig. 2). Before the actual problem

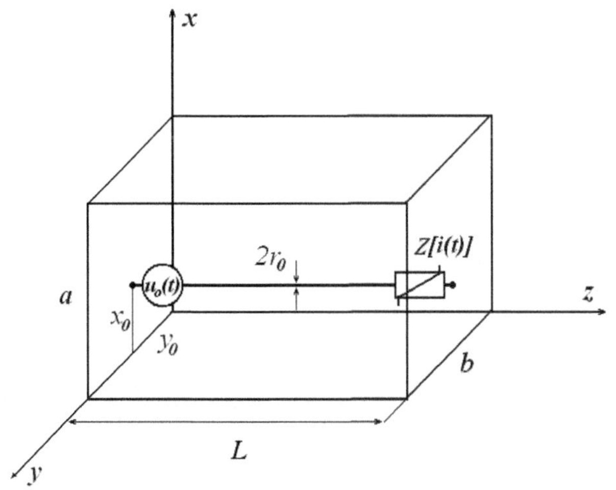

Figure 2. Schematic representation of the TL inside the cavity.

is treated, some important preliminary remarks are made. First, concerning the notations and results of a TL inside a rectangular resonator, reference is made to the publications (Tkachenko et al., 2011, 2013). The second observation relates to a numerical problem. Since a lossless transmission line is considered inside a cavity with a very high Q-value it is expected that the response of a once excited transmission line will exceed many transit times in duration. If, in addition, very high frequencies are required, one may end up not only in a computer-storage problem, but also needs very long computation time. Therefore, it would be desirable to reduce the duration of the Green's functions (here represented by the admittance functions) to only a few transit times. This, however, is possible to achieve, if the TL is matched. Following this idea, it is aimed to have a well matched line when computing the Green's functions (admittance functions). This goal is achieved in the following novel analysis.

For the moment the calculations take place in the frequency domain. During this time the terminations of the line are assumed to be linear. The explicit dependency of the frequencies, however, is suppressed in the quantities. In the rectangular resonator the admittance functions are given by:

$$Y_l(z) := \frac{jk}{\eta_0 L}\sum_{n_3=0}^{\infty}\frac{\epsilon_{n_3,0}\cos\left(\frac{n_3\pi z}{L}\right)}{\left(k_{\nu z}^2-k^2\right)S} \tag{10}$$

and

$$Y_r(z) := \frac{jk}{\eta_0 L}\sum_{n_3=0}^{\infty}\frac{\epsilon_{n_3,0}(-1)^{n_3}\cos\left(\frac{n_3\pi z}{L}\right)}{\left(k_{\nu z}^2-k^2\right)S} \tag{11}$$

Then the current along the line is expressed through

$$I(z) = U_0 Y_l(z) + U_r Y_r(z) = U_0 Y_l(z) - Z I(L) Y_r(z) \tag{12}$$

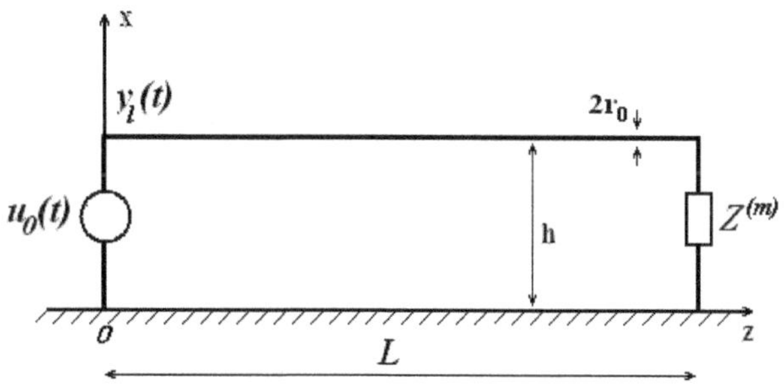

Figure 3. Matched TL inside the resonator with left-hand source.

and thus at the end of the line ($z = L$) one has

$$I(L) = \frac{U_0 Y_l(L)}{1 + Z Y_r(L)} \tag{13}$$

The impedance Z is now decomposed into two partial impedances, of which one $\left(Z^{(m)}\right)$ corresponds to the characteristic impedance (Djordjevic et al., 1986):

$$Z = Z^{(m)} + Z^{(nl)} \qquad \text{with} \qquad Z^{(m)} \equiv Z_C \tag{14}$$

Thereafter, an auxiliary circuit with $Z^{(m)}$ as terminating impedance as shown in Fig. 3 is examined. For this circuit one has:

$$I(z) = U_0 Y_l(z) - Z^{(m)} I(L) Y_r(z) \tag{15}$$

or solved for the current at the end of the line

$$I(L) = \frac{U_0 Y_l(L)}{1 + Z^{(m)} Y_r(L)} \tag{16}$$

Inserting Eq. (16) into Eq. (15) leads to

$$I(z) = U_0 \tilde{Y}_l^{\left(Z^{(m)}\right)}(z) \tag{17}$$

with the new admittance function

$$\tilde{Y}_l^{\left(Z^{(m)}\right)}(z) := Y_l(z) - \frac{Z^{(m)} Y_l(L) Y_r(z)}{1 + Z^{(m)} Y_r(L)} \tag{18}$$

In the next step, consider the same line from which the source $U_0(\omega)$ was removed. A source (active or passive) was instead inserted on the right side. The impedance at the end of the line is again $Z^{(m)}$ (see Fig. 4). Then, the following applies:

$$I(z) = U_r Y_r(z) - Z^{(m)} I(L) Y_r(z) \tag{19}$$

or at the point $z = L$

$$I(L) = \frac{U_r Y_r(L)}{1 + Z^{(m)} Y_r(L)} \tag{20}$$

Substituting Eq. (20) into Eq. (19) yields the current for the right-hand source

$$I(z) = U_r \tilde{Y}_r^{\left(Z^{(m)}\right)}(z) \tag{21}$$

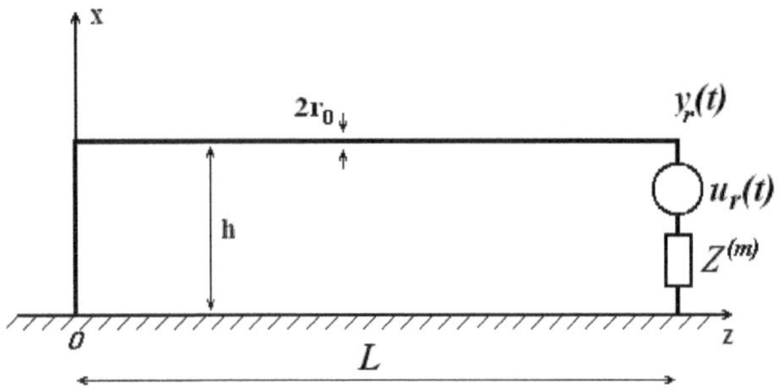

Figure 4. Matched TL inside the resonator with right-hand source.

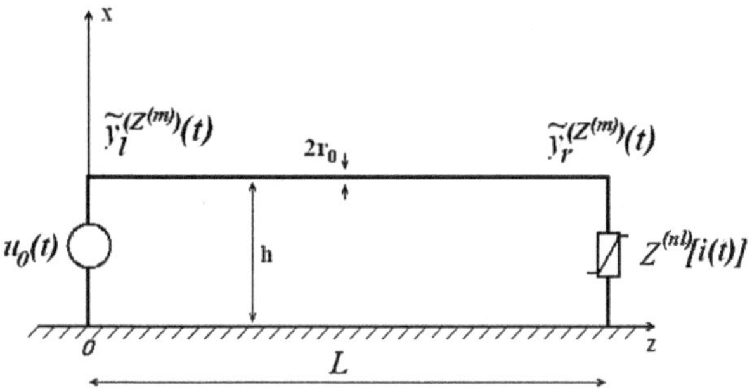

Figure 5. TL is fed by U_0 and terminated by $Z^{(nl)}$; current is estimated with the aid of renormalized admittance functions.

with

$$\tilde{Y}_r^{(Z^{(m)})}(z) := Y_r(z) - \frac{Z^{(m)}Y_r(L)Y_r(z)}{1+Z^{(m)}Y_r(L)} \tag{22}$$

With these new admittance functions (18) and (22) the above calculation is repeated for the following configuration depicted in Fig. 5. Now one obtains:

$$I(z) = U_0\tilde{Y}_l^{(Z^{(m)})}(z) - Z^{(nl)}I(L)\tilde{Y}_r^{(Z^{(m)})}(z) \tag{23}$$

In Eq. (23) U_r is assumed to be a passive source. Thus, the current at the end of the line results in:

$$I(L) = \frac{U_0\tilde{Y}_l^{(Z^{(m)})}(L)}{1+Z^{(nl)}\tilde{Y}_r^{(Z^{(m)})}(L)} \tag{24}$$

Using the renormalized admittance functions (18) and (22) in Eq. (24) it follows the expected result

$$I(L) = \frac{U_0Y_l(L)}{1+\left(Z^{(m)}+Z^{(nl)}\right)Y_r(L)} \tag{25}$$

But this is exactly the Eq. (13), taking into account the decomposition (14). This demonstrates that a given impedance can be decomposed into two parts where $Z^{(m)}$ can be suitably selected. In general, this will be the characteristic impedance of the line.

Now, by an inverse Fourier transform (23) is transformed into time-domain ($z = L$).

$$i_L(t) = u_0(t) * \tilde{y}_l^{(Z^{(m)})}(L,t) + u_r(t) * \tilde{y}_r^{(Z^{(m)})}(L,t) \tag{26}$$

The convolution integrals in Eq. (26) are indicated by '$*$'. In explicit form this reads:

$$i_L(t) = \int_{-\infty}^{t} \tilde{y}_l^{(Z^{(m)})}(L,t-\tau)u_0(\tau)d\tau + \int_{-\infty}^{t} \tilde{y}_r^{(Z^{(m)})}(L,t-\tau)u_r(\tau)d\tau \tag{27}$$

Here one clearly recognizes that the $\tilde{y}$-functions are the time-domain Green's functions, i.e., the responses of the terminated line in the resonator due to impulse excitations. Remember, these $\tilde{y}$-functions have been determined in frequency-domain for the linear part of the TL only (see Figs. 3 and 4). Equation (27) now is valid in time-domain. Therefore, also nonlinear loads can be chosen and there is no need to recompute frequency-domain data again if the excitation wave form or the (nonlinear) loads are changed.

Since causal functions are assumed ($u_0(t) \equiv 0$ for $t \leq 0$), and in the time domain the constraint to linear termination impedances now is abandoned, one arrives at the *temporary* result:

$$i_L(t) = \int_0^t \tilde{y}_l^{(Z^{(m)})}(L,t-\tau)u_0(\tau)d\tau - \int_0^t \tilde{y}_r^{(Z^{(m)})}(L,t-\tau)Z^{(nl)}(i_L(\tau))i_L(\tau)d\tau \tag{28}$$

In general, this equation has to be solved numerically. For this purpose it is advisable to rewrite Eq. (28) a little bit to

$$i_L(t) = i_{0L}(t) - \int_0^{t-\Delta} \tilde{y}_r^{(Z^{(m)})}(L,t-\tau)Z^{(nl)}(i_L(\tau))i_L(\tau)d\tau - \int_{t-\Delta}^{t} \tilde{y}_r^{(Z^{(m)})}(L,t-\tau)Z^{(nl)}(i_L(\tau))i_L(\tau)d\tau \tag{29}$$

with the abbreviation

$$i_{0L}(t) := \int_0^t \tilde{y}_l^{(Z^{(m)})}(L,t-\tau)u_0(\tau)d\tau. \tag{30}$$

After a change of variables in the second integral of Eq. (29) one arrives at a useful starting formula for numerical integration.

$$i_L(t) + \int_0^{\Delta} \tilde{y}_r^{(Z^{(m)})}(L,\tau) Z^{(nl)}(i_L(t-\tau)) i_L(t-\tau) d\tau$$

$$= i_{0L}(t) - \int_0^{t-\Delta} \tilde{y}_r^{(Z^{(m)})}(L,t-\tau) Z^{(nl)}(i_L(\tau)) i_L(\tau) d\tau \quad (31)$$

The integral on the left-hand side of Eq. (31) contains $\tilde{y}_r^{(Z^{(m)})}(L,\tau)$ – the right-hand admittance function – which is considered only over a short time-span $0 \leq \tau \leq \Delta$. During this short time Δ, the line still does not interact with the resonator. It behaves like a line in free space (see Eq. 6). Therefore, in this small period of time around $t = 0$, $\tilde{y}_r^{(Z^{(m)})}(L,\tau)$ is represented as a δ-function, or more precisely:

$$\tilde{y}_r^{(Z^{(m)})}(L,\tau) = \frac{1}{2Z_C}\delta(\tau) \quad \text{for} \quad -\Delta \leq \tau \leq \Delta \quad (32)$$

The δ-function here can be represented as a Gaussian-function:

$$\delta(\tau) \approx \frac{1}{\sqrt{\pi}\,\tau_0} \exp\left(-\frac{\tau^2}{\tau_0^2}\right) \quad (33)$$

using $\tau_0 = 0.1\,\text{ns}, \Delta = 4\tau_0, \Delta t = \frac{\tau_0}{40}$

Now one can perform the integration on the left-hand side of Eq. (31) and obtains

$$i_L(t)\left\{1 + \frac{1}{2Z_C} Z^{(nl)}(i_L(t)) \frac{\int_{-\Delta}^{\Delta} \exp\left(-\frac{\tau^2}{\tau_0^2}\right) d\tau}{\sqrt{\pi}\,\tau_0}\right\}$$

$$= i_{0L}(t) - \int_0^{t-\Delta} \tilde{y}_r^{(Z^{(m)})}(L,t-\tau) Z^{(nl)}(i_L(\tau)) i_L(\tau) d\tau \quad (34)$$

For the numerical integration the integral in Eq. (34) can be rewritten as a sum:

$$i_L(t_m)\left\{1 + \frac{1}{2Z_C} Z^{(nl)}(i_L(t_m)) \frac{\int_{-\Delta}^{\Delta} \exp\left(-\frac{\tau^2}{\tau_0^2}\right) d\tau}{\sqrt{\pi}\,\tau_0}\right\}$$

$$= i_{0L}(t_m)$$

$$\sum_{k=0}^{m-1} \tilde{y}_r^{(Z^{(m)})}(L, t_m - k\Delta t) Z^{(nl)}(i_L(k\Delta t)) i_L(k\Delta t)\Delta t \quad (35)$$

Here the sum is calculated at temporal sample points $t_m = m\Delta t$, where $m = 0, 1, ..., m_{\max}$. Similar to the situation of Eq. (9), also the right-hand side of Eq. (35) contains expressions of known functions or those that have already been calculated from previous times. Thus, Eq. (35) can be solved for each discrete time t_m using a root-finding algorithm.

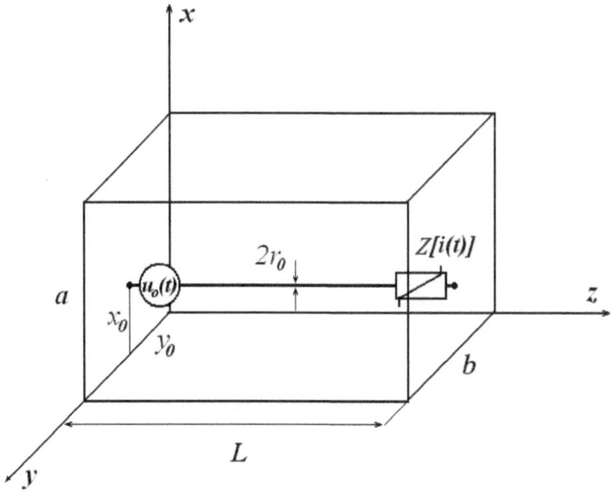

Figure 6. Parameters of the rectangular resonator: $a = 30$ cm, $b = 53$ cm, $L = 79$ cm; parameters of the TL: $x_0 = a/2$, $y_0 = b/2$, $r_0 = 1$ mm.

4 Generation and discussion of numerical results

In this section, a Schottky diode is chosen as a nonlinear load and selected results are presented and discussed. The dimensions of the used rectangular resonator and the position of the transmission line in the resonator parallel to the z axis is shown in Fig. 6. In the resonator the TL has a characteristic impedance of $Z_C = 333\,\Omega$. For the following calculations $Z^{(m)} = Z_C$ was used. The step size for numerical integration was set to $\Delta t = \tau_0/40$. A diode with the following characteristics

$$I(U) = I_S(\exp(U/U_t)) \quad \text{or} \quad U(I) = U_t \ln(1 + I/I_S) \quad (36)$$

was regarded and the typical values $U_t = 25$ mV and $I_S = 1\,\mu$A were used. A bi-linear approximation of the diode characteristic was choosen and is shown in Fig. 7. The bi-linear approximation of the diode characteristic was generated with the following formula and parameters:

$$U(I) = Z(I) \cdot I \quad (37)$$

$$Z(I) = \begin{cases} Z_1 \;, & I > 0 \\ Z_2 \;, & I < 0 \end{cases} \quad (38)$$

with $Z_1 = 2\text{k}\Omega, Z_2 = 0.1\,\text{M}\Omega$

In Figs. 8 and 9 the admittance functions $\tilde{y}_l^{(Z_C)}(L,t)$ and $\tilde{y}_r^{(Z_C)}(L,t)$ are shown in different time scales, respectively.

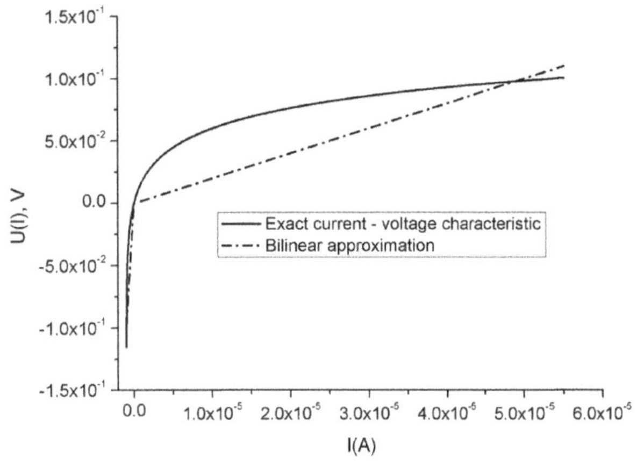

Figure 7. Characteristic of a generic diode and bi-linear approximation.

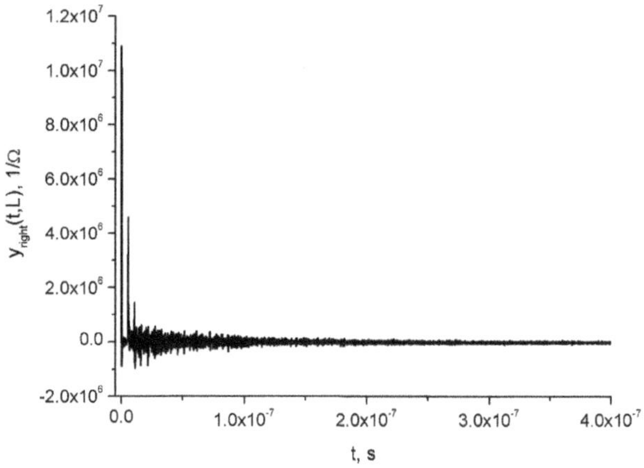

Figure 9. Right admittance function $\tilde{y}_r^{(Zc)}(L,t)$.

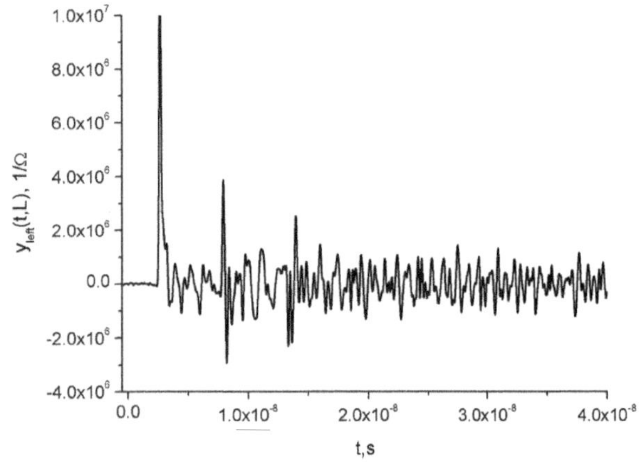

Figure 8. Left admittance function $\tilde{y}_l^{(Zc)}(L,t)$.

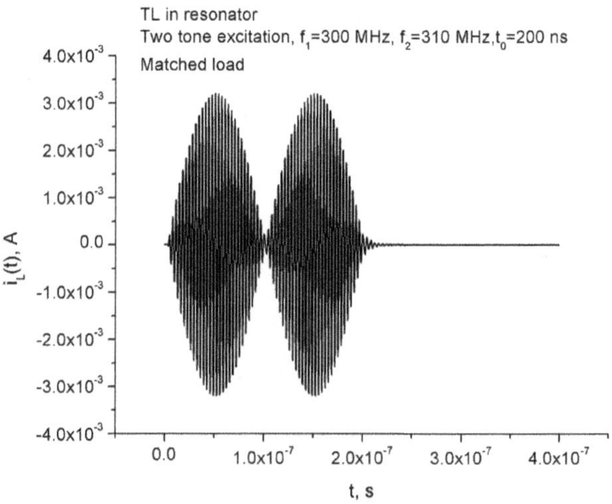

Figure 10. Current at the end of the TL in time domain (TL with linear and matched load in the rectangular resonator).

One can recognize the first three peaks resulting from the exciting δ-pulse along the TL. For $\tilde{y}_l^{(Zc)}(L,t)$ the peaks are the first multiples of L/c_0. As expected, for $\tilde{y}_r^{(Zc)}(L,t)$ the first peak is at $t=0$ and afterwards at time steps of $2L/c_0$. The excitation of the TL gets phased out for increasing time steps as can be seen in Fig. 9. Because the left- and right-hand Green's functions (admittance functions) are known, the current at the end of the TL can be calculated provided that the excitation $u_0(t)$ is known. The following *two tone* excitation is used:

$$u_0(t) = U_0\left[\sin(2\pi f_1 t) - \sin(2\pi f_2 t)\right]h(t)h(t_0 - t) \tag{39}$$

In Eq. (39) the Heaviside function is depicted by $h(t)$ and the following parameters are used: $U_0 = 1\,\mathrm{V}$, $f_1 = 300\,\mathrm{MHz}$, $f_2 = 310\,\mathrm{MHz} - 400\,\mathrm{MHz}$, $t_0 = 200\,\mathrm{ns}$.

Figure 10 shows the current at the end of the TL for the linear case with matched load. The frequency spectrum of the current is depicted in Fig. 11 where the two peaks at f_1 and f_2 can be clearly seen.

In Fig. 12 the current at the end of the TL in time domain is shown for the nonlinear load defined in Eq. (38). The second frequency f_2 of the two tone excitation was increased from 310 MHz to 400 MHz in steps of 10 MHz. The expected result can be seen in the frequency domain representation of the current at the end of the TL in the resonator. Figure 13 shows the result for the lowest frequency $f_2 = 310\,\mathrm{MHz}$ and therefore for the lowest frequency difference $\Delta f = f_2 - f_1 = 10\,\mathrm{MHz}$. Figures 14 and 15 show the currents in frequency domain for $f_2 = 350\,\mathrm{MHz}$ and $f_2 = 400\,\mathrm{MHz}$, respectively.

The expected results are clearly observable: At the difference of the two tone frequencies, peaks occur which are not present in the linear case. Additionally, in contrast to the linear load, now there are peaks at multiples of the two basic frequencies f_1 and f_2. Around these multiples there are smaller peaks representing multiples of $\Delta f = f_2 - f_1$. A moderate increase in peak heights with increasing Δf is observable.

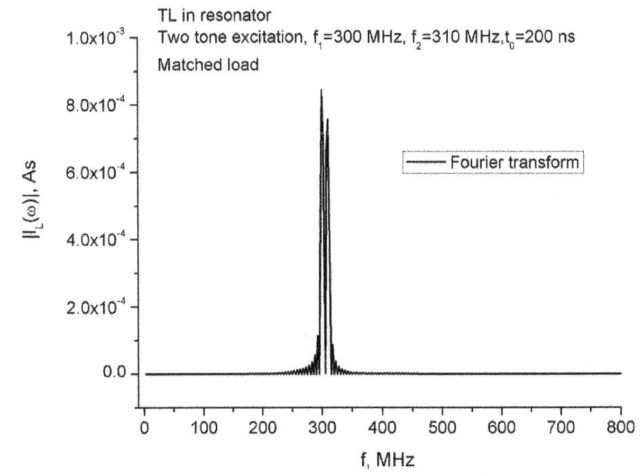

Figure 11. Current at the end of the TL in frequency domain (TL with linear and matched load in the rectangular resonator).

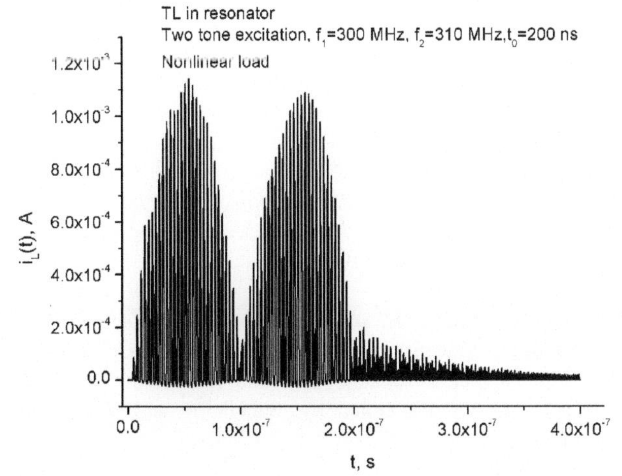

Figure 12. Current at the end of the TL in time domain (TL with nonlinear load in the rectangular resonator).

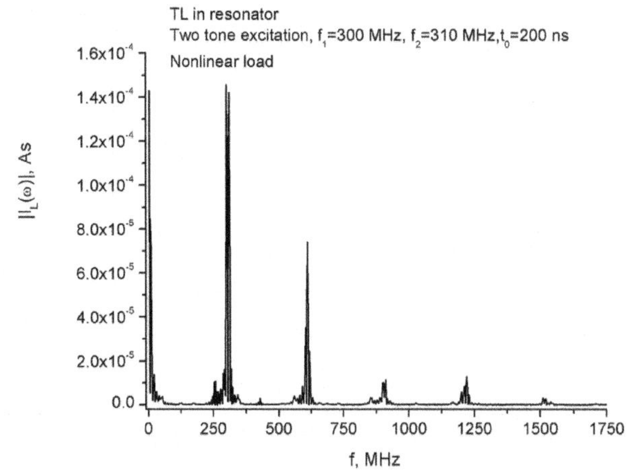

Figure 13. Current at the end of the TL in frequency domain (TL with nonlinear load in the rectangular resonator, $f_2 = 310\,\text{MHz}$, $\Delta f = 10\,\text{MHz}$).

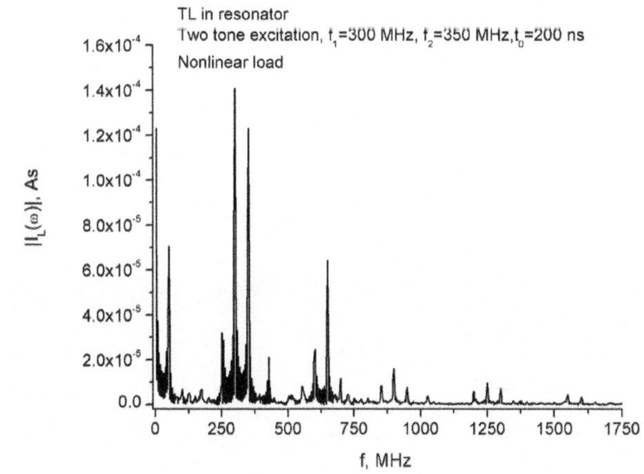

Figure 14. Current at the end of the TL in frequency domain (TL with nonlinear load in the rectangular resonator, $f_2 = 350\,\text{MHz}$, $\Delta f = 50\,\text{MHz}$).

However, it must be stated that the known phenomena (by measurement) of significant increase in current amplitude for excitation frequencies corresponding to one or more eigenfrequencies of the resonator are not seen in the presented simulations. Also the used Δf values do not correspond to the eigenfrequencies of the resonator. The reasons are: (a) The used resonator is quite flat and still behaves a bit like a *waveguide*. Additionally the TL has a high symmetry in the resonator. (b) Surely, the main reason is that the exciting frequencies were chosen too low. This had to be done because of performance limitations of the used standard computer. Going to higher excitation frequencies, the time steps have to be made smaller in time domain and the δ-function has to be defined narrower. In almost the same manner, the frequency steps have to be made smoother in frequency domain calculations. A loophole would be the use of high-performance computers or a modified solution using methods of perturbation theory.

So far there is no comparison of the presented analytical results with those obtained by other methods. However, for the simple example of linearly loaded transmission lines in free space a lot of comparisons with other methods (e.g. Method of Tesche, Reflexion Method) have been done and an excellent agreement was observed (Tesche et al., 1997).

5 Conclusion

In this work nonlinearly loaded lossless transmission lines inside a rectangular resonator were studied. Using the left- and right-hand Green's functions of the problem in time domain an analytic solution could be presented. The Green's functions were developed for linear matched loads ensuring short duration of a pulse-shaped excitation in time domain and therefore acceptable calculation time. An example was presented using a simplified diode as nonlinear load and

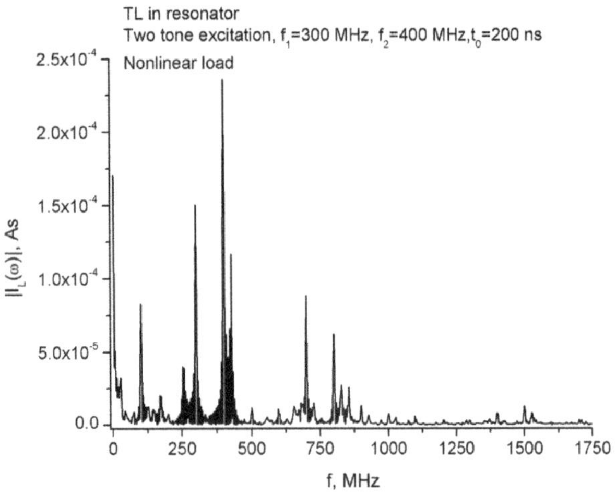

Figure 15. Current at the end of the TL in frequency domain (TL with nonlinear load in the rectangular resonator, $f_2 = 400\,\text{MHz}$, $\Delta f = 100\,\text{MHz}$).

two tone excitations. The demodulation and intermodulation effects could be clearly demonstrated. At multiples of the excitation frequencies and their sums and differences distinct resonances appeared. It is worth mentioning that the method presented above can be extended to non-uniform multiconductor transmission lines with weak losses. This extension may be performed on the basis of the two publications (Rambousky et al., 2014) and (Tkachenko et al., 2014) and references given therein.

Edited by: F. Sabath

References

Djordjevic, A., Sarkar, T., and Harrington, R.: Analysis of lossy transmission lines with arbitrary nonlinear terminal networks, IEEE T. Microw. Theory, 34, 660–666, 1986.

Krauthäuser, H., Tkachenko, S., and Nitsch, J.: The action of nonlinear effects in a resonator, in: XXVIIth General Assembly of the International Union of Radio Science, p. 2167, Maastricht, 2002.

Liu, T. and Tesche, F.: Analysis of antennas and scatterers with nonlinear loads, IEEE T. Antenn. Propag., 24, 131–139, 1976.

Rambousky, R., Tkachenko, S., and Nitsch, J.: Calculation of currents induced in a long transmission line placed symmetrically inside a rectangular cavity, in: IEEE International Symposium on Electromagnetic Compatibility (EMC), 796–801, Denver, 2013.

Rambousky, R., Nitsch, J., and Tkachenko, S.: Transmission-Line Super Theory as Antenna Theory for Linear Structures, in: Ultra-Wideband, Short-Pulse Electromagnetics 10, edited by: Sabath, F. and Mokole, E., Springer, Berlin, 2014.

Tesche, F., Ianoz, M., and Karlsson, T.: EMC Analysis Methods and Computational Models, John Wiley & Sons, Inc., New York, 1997.

Tkachenko, S., Nitsch, J., and Rambousky, R.: Electromagnetic Field Coupling to Transmission Lines Inside Rectangular Resonators, Interaction Notes, Note 623, http://www.ece.unm.edu/summa/notes/In/IN623.pdf, 2011.

Tkachenko, S., Rambousky, R., and Nitsch, J.: Electromagnetic Field Coupling to a Thin Wire Located Symmetrically Inside a Rectangular Enclosure, IEEE Trans. on EMC, 55, 334–341, 2013.

Tkachenko, S., Nitsch, J., and Rambousky, R.: Electromagnetic Coupling to Transmission Lines with Symmetric Geometry Inside Rectangular Resonators, in: Ultra-Wideband, Short-Pulse Electromagnetics 10, edited by: Sabath, F. and Mokole, E., Springer, Berlin, 2014.

32

Optimally Accurate Second-Order Time-Domain Finite-Difference Scheme for Acoustic, Electromagnetic, and Elastic Wave Modeling

C. Bommaraju, R. Marklein, and P. K. Chinta

Fachgebiet Theoretische Elektrotechnik, Fachbereich Elektrotechnik/Informatik, Universität Kassel (UNIK), 34109 Kassel, Germany

Abstract. Numerical methods are extremely useful in solving real-life problems with complex materials and geometries. However, numerical methods in the time domain suffer from artificial numerical dispersion. Standard numerical techniques which are second-order in space and time, like the conventional Finite Difference 3-point (FD3) method, Finite-Difference Time-Domain (FDTD) method, and Finite Integration Technique (FIT) provide estimates of the error of discretized numerical operators rather than the error of the numerical solutions computed using these operators. Here optimally accurate time-domain FD operators which are second-order in time as well as in space are derived. Optimal accuracy means the greatest attainable accuracy for a particular type of scheme, e.g., second-order FD, for some particular grid spacing. The modified operators lead to an implicit scheme. Using the first order Born approximation, this implicit scheme is transformed into a two step explicit scheme, namely predictor-corrector scheme. The stability condition (maximum time step for a given spatial grid interval) for the various modified schemes is roughly equal to that for the corresponding conventional scheme. The modified FD scheme (FDM) attains reduction of numerical dispersion almost by a factor of 40 in 1-D case, compared to the FD3, FDTD, and FIT. The CPU time for the FDM scheme is twice of that required by the FD3 method. The simulated synthetic data for a 2-D P-SV (elastodynamics) problem computed using the modified scheme are 30 times more accurate than synthetics computed using a conventional scheme, at a cost of only 3.5 times as much CPU time. The FDM is of particular interest in the modeling of large scale (spatial dimension is more or equal to one thousand wave lengths or observation time interval is very high compared to reference time step) wave propagation and scattering problems, for instance, in ultrasonic antenna and synthetic scattering data modeling for Non-Destructive Testing (NDT) applications, where other standard numerical methods fail due to numerical dispersion effects. The possibility of extending this method to staggered grid approach is also discussed. The numerical FD3, FDTD, FIT, and FDM results are compared against analytical solutions.

1 Introduction

Numerical methods in the time-domain suffer from artificial numerical dispersion. Standard numerical techniques, like Finite–Difference Time–Domain (FDTD) method (Taflove and Hagness, 2000) and Finite Integration Technique (FIT) (Marklein, 1997) provide estimates of the error of discretized numerical operators rather than the error of the numerical solutions computed using these operators (Geller and Takeuchi, 1995, 1998; Bommaraju and Marklein, 2004). Optimal accuracy means the greatest attainable accuracy for a particular type of scheme, e.g., second-order FD, for some particular grid spacing.

2 Modified operators for the 1-D case

Consider the homogeneous, source-free 1-D electromagnetic (EM) wave equation in the time domain

$$\left(\varepsilon_0 \frac{\partial^2}{\partial t^2} - \frac{1}{\mu_0}\frac{\partial^2}{\partial z^2}\right) E_x(z,t) = 0\,, \tag{1}$$

where E_x represents the electric field strength, ε_0 and μ_0 are the free-space permittivity and permeability, respectively. Let A_{n,n_t} be the discretized second-order Finite Difference (FD) operator for $\varepsilon_0 \frac{\partial^2}{\partial t^2}$, and K_{n,n_t} be the FD operator for $\frac{1}{\mu_0}\frac{\partial^2}{\partial z^2}$ and c_{n,n_t} represents the unknown discrete $E_x(N\Delta z, N_t\Delta t)$ component. The above equation can be represented as

$$\left(A_{n,n_t} - K_{n,n_t}\right) c_{n,n_t} = 0\,. \tag{2}$$

Following the FD3 method, A_{n,n_t} and K_{n,n_t} have the numerical dispersion of normal three point second derivative operators. The superscript "0" indicates the conventional FD (FD3) operators. Thus, omitting higher order terms, the operator error for the FD3 method read

$$\left(\delta A^0_{n,n_t} - \delta K^0_{n,n_t}\right) c_{n,n_t} = \varepsilon_0 \frac{(\Delta t)^2}{12}\left[\frac{\partial^4 E_x(z,t)}{\partial t^4}\right]\Bigg|_{n,n_t} - \frac{1}{\mu_0}\frac{(\Delta z)^2}{12}\left[\frac{\partial^4 E_x(z,t)}{\partial z^4}\right]\Bigg|_{n,n_t}. \tag{3}$$

The relation between the exact and conventional operators can be expressed as

$$A^{\text{exact}}_{n,n_t} = A^0_{n,n_t} - \delta A^0_{n,n_t} \tag{4}$$

$$K^{\text{exact}}_{n,n_t} = K^0_{n,n_t} - \delta K^0_{n,n_t}\,. \tag{5}$$

The transformation into the frequency domain yields

$$\left(\delta A_{n,n_\omega} - \delta K_{n,n_\omega}\right) c_{n,n_\omega} = \frac{\omega^2 (\Delta t)^2}{12}\left[\varepsilon_0 \omega^2 E_x(z,\omega)\right]\Bigg|_{n,n_\omega} + \frac{(\Delta z)^2}{12}\frac{\partial^2}{\partial z^2}\left[\frac{1}{\mu_0}\frac{\partial^2}{\partial z^2}E_x(z,\omega)\right]\Bigg|_{n,n_\omega}, \tag{6}$$

where $\delta A^0_{n,n_\omega}$, $\delta K^0_{n,n_\omega}$, c_{n,n_ω} are the frequency domain representations of $\delta A^0_{n,n_t}$, $\delta K^0_{n,n_t}$, and c_{n,n_t}, respectively. The quantity on the right-hand side of the above equation is the *basic error*. When E_x is an eigenfunction and ω approaches the corresponding eigenfrequency the basic error will not in general equal zero. Instead we have

$$\left(\delta A_{n,n_\omega} - \delta K_{n,n_\omega}\right) c_{n,n_\omega} = \frac{\omega^2(\Delta t)^2}{12}\left[\varepsilon_0\omega^2 E_x(z,\omega) + \frac{1}{\mu_0}\frac{\partial^2}{\partial z^2}E_x(z,\omega)\right]\Bigg|_{n,n_\omega} + \frac{(\Delta z)^2}{12}\frac{\partial^2}{\partial z^2}\left[\varepsilon_0\omega^2 E_x(z,\omega) + \frac{1}{\mu_0}\frac{\partial^2}{\partial z^2}E_x(z,\omega)\right]\Bigg|_{n,n_\omega}. \tag{7}$$

Observe that the equation in the square brackets is the reduced wave equation or Helmholtz equation. When E_x is an eigenfunction and ω is the corresponding eigenfrequency, the error will be zero as the homogeneous wave equation (inside square brackets) is zero. The criterion for obtaining *optimal accuracy* is that the net error of the discretized equation must be approximately zero whenever the operand is an eigenfunction and the frequency is equal to the corresponding eigenfrequency. Thus, when $E_x(z,t)=E_{xm}(z)\,\mathrm{e}^{-\mathrm{j}\omega_m t}$, $E_{xm}(z)$ is the mth eigenfunction of a mode with circular eigenfrequency ω_m, the bracketed term, the homogeneous wave equation, and hence its derivatives will be zero. Transforming above equation back into the time domain results in

$$\left(\delta A_{n,n_t} - \delta K_{n,n_t}\right) c_{n,n_t} = \frac{(\Delta t)^2}{12}\frac{\partial^2}{\partial t^2}\left[\varepsilon_0\frac{\partial^2}{\partial t^2}E_x(z,t) - \frac{1}{\mu_0}\frac{\partial^2}{\partial z^2}E_x(z,t)\right]\Bigg|_{n,n_t} + \frac{(\Delta z)^2}{12}\frac{\partial^2}{\partial z^2}\left[\varepsilon_0\frac{\partial^2}{\partial t^2}E_x(z,t) - \frac{1}{\mu_0}\frac{\partial^2}{\partial z^2}E_x(z,t)\right]\Bigg|_{n,n_t}. \tag{8}$$

The above equation shows that the basic error of the modified operators is given by the derivatives of the homogeneous wave equation. Omitting the details of the derivation, the modified operators that yield errors of the above form are as follows

$$A_{n,n_t} = \varepsilon_0 \frac{1}{(\Delta t)^2}\;\begin{array}{|c|ccc|}\hline t+\Delta t & 1/12 & 10/12 & 1/12 \\ t & -2/12 & -20/12 & -2/12 \\ t-\Delta t & 1/12 & 10/12 & 1/12 \\ \hline & z-\Delta z & z & z+\Delta z \\ \hline\end{array} \tag{9}$$

$$K_{n,n_t} = \frac{1}{\mu_0}\frac{1}{(\Delta z)^2}\;\begin{array}{|c|ccc|}\hline t+\Delta t & 1/12 & -2/12 & 1/12 \\ t & 10/12 & -20/12 & 10/12 \\ t-\Delta t & 1/12 & -2/12 & 1/12 \\ \hline & z-\Delta z & z & z+\Delta z \\ \hline\end{array}\,. \tag{10}$$

The homogeneous acoustic "A", elastic shear "E(S)", and elastic pressure "E(P)" 1-D wave equations read in the source-free case

$$\text{A} \qquad \left(\rho_{a0}\frac{\partial^2}{\partial t^2} - \frac{1}{\kappa}\frac{\partial^2}{\partial z^2}\right)p(z,t) = 0 \tag{11}$$

$$\text{E(S)} \qquad \left(\rho_{e0}\frac{\partial^2}{\partial t^2} - \mu\frac{\partial^2}{\partial z^2}\right)v_x(z,t) = 0 \tag{12}$$

$$\text{E(P)} \qquad \left(\rho_{e0}\frac{\partial^2}{\partial t^2} - (\lambda+2\mu)\frac{\partial^2}{\partial z^2}\right)v_z(z,t) = 0\,, \tag{13}$$

where in the "A" case p is the pressure, ρ_{a0} is the mass density at rest and κ is the adiabatic compressibility; in the "E(S)" and "E(P)" case, representing the shear and pressure waves, v_x, v_z are the particle velocities, ρ_{e0} is the mass density at rest and λ, μ are Lamé's constants. Due to the similarity between the wave equations, all results mentioned above for EM case are also valid for A, E(S) and E(P) cases.

3 Modified operators for the 2-D TM case

Consider the 2-D TM case in the xz plane, then the source-free electromagnetic wave equation read

$$\left(\varepsilon_0\frac{\partial^2}{\partial t^2} - \frac{1}{\mu_0}\left(\frac{\partial^2}{\partial x^2} + \frac{\partial^2}{\partial z^2}\right)\right)E_y(x,z,t) = 0\,. \tag{14}$$

Discretizing the unknown field strength

$$c_{p'r'N} = E_y(p'\Delta x, r'\Delta z, N\Delta t)\,, \tag{15}$$

where Δx, Δz represent the spatial grid spacing with respect to x and z, and Δt represents the temporal grid interval. The above equation can be written as

$$\left(A^1_{p'r'N} - K^1_{p'r'N} - K^3_{p'r'N}\right)c_{p'r'N} = 0\,. \tag{16}$$

$A^1_{p'r'N}$, $K^1_{p'r'N}$, $K^3_{p'r'N}$ are the temporal and spatial operators with respect to x and z, respectively. The index N represents the temporal grid points. The indices p', r' with $0 \le p' \le N_x$, $0 \le r' \le N_z$ represent the x and z directions, and N_x and N_z are the number of grid cells in x and z directions, respectively.

Using second-order central difference operators, the temporal operator error can be written as

$$\left(\delta A^1_{p'r'N}\right) = \frac{(\Delta t)^2}{12}\frac{\partial^2}{\partial t^2}\left(\varepsilon_0\frac{\partial^2}{\partial t^2}E_y(x,z,t)\right). \tag{17}$$

In order to have optimal accuracy the operator error should be

$$\delta A^{1'}_{p'r'N} = \left(\frac{(\Delta t)^2}{12} \frac{\partial^2}{\partial t^2} + \frac{(\Delta x)^2}{12} \frac{\partial^2}{\partial x^2} + \frac{(\Delta z)^2}{12} \frac{\partial^2}{\partial z^2} \right) \times \varepsilon_0 \frac{\partial^2}{\partial t^2} E_y(x, z, t) . \tag{18}$$

Making use of the relation stated in Eq. (4) yields

$$A^{1'}_{p'r'N} = \left(1 + \frac{(\Delta t)^2}{12} \frac{\partial^2}{\partial t^2} + \frac{(\Delta x)^2}{12} \frac{\partial^2}{\partial x^2} + \frac{(\Delta z)^2}{12} \frac{\partial^2}{\partial z^2} \right) \times \varepsilon_0 \frac{\partial^2}{\partial t^2} E_y(x, z, t) . \tag{19}$$

Neglecting higher order terms according to the second-order approximation the above equation can be approximated as

$$A^{1'}_{p'r'N} = \left(d_{p'r'N}\right) \left(f_{p'r'N}\right) \left(g_{p'r'N}\right) , \tag{20}$$

where d is the operator for $\left(1+\frac{(\Delta t)^2}{12}\frac{\partial^2}{\partial t^2}\right)\varepsilon_0\frac{\partial^2}{\partial t^2}E_y$, f is the operator for $\left(1+\frac{(\Delta x)^2}{12}\frac{\partial^2}{\partial x^2}\right)$ and g is the operator for $\left(1+\frac{(\Delta z)^2}{12}\frac{\partial^2}{\partial z^2}\right)$. By neglecting higher order terms, any function $u(z)$ can be approximated as

$$\frac{u(z+\Delta z) - 2u(z) + u(z-\Delta z)}{(\Delta z)^2} = \left(1 + \frac{(\Delta z)^2}{12} \frac{\partial^2}{\partial z^2}\right) \frac{\partial^2}{\partial z^2} u$$

$$\frac{u(z+\Delta z) + 10u(z) + u(z-\Delta z)}{12} = \left(1 + \frac{(\Delta z)^2}{12} \frac{\partial^2}{\partial z^2}\right) u . \tag{21}$$

Using the above equations, a modified temporal operator can be obtained. Following a similar approach, the spatial operators can also be obtained. Due to limited space, the modified operators are summarized at the end of this paper explicitly for an inhomogeneous medium.

In 2-D case and xz plane, the homogeneous acoustic "A" and elastic shear "E(S)" wave equation read in the source-free case

$$\left(\kappa \frac{\partial^2}{\partial t^2} - \frac{1}{\rho_{a0}} \left(\frac{\partial^2}{\partial x^2} + \frac{\partial^2}{\partial z^2} \right) \right) p_y(x, z, t) = 0 \tag{22}$$

$$\left(\rho_{e0} \frac{\partial^2}{\partial t^2} - \mu \left(\frac{\partial^2}{\partial x^2} + \frac{\partial^2}{\partial z^2} \right) \right) v_y(x, z, t) = 0 . \tag{23}$$

Due to the similarity between the wave equations, all results mentioned above for the electromagnetic waves are also valid for the acoustic and elastic waves.

4 Applicability to staggered grid approaches (FDTD, FIT)

In a homogeneous medium the first-order staggered grid approach consists of solving the following coupled first-order equations in differential form

$$\frac{\partial}{\partial t} H_y(z, t) = -\frac{1}{\mu_0} \frac{\partial}{\partial z} E_x(z, t) \tag{24}$$

$$\frac{\partial}{\partial t} E_x(z, t) = -\frac{1}{\varepsilon_0} \frac{\partial}{\partial z} H_y(z, t) - \frac{1}{\varepsilon_0} J_{ex}(z, t) . \tag{25}$$

For the sake of simplicity, subscripts are dropped from here onwards. Using a Taylor's series expansion in Eq. (24) yields

$$\frac{\partial}{\partial t} H(z, t) + \frac{(\Delta t)^2}{24} \frac{\partial^2}{\partial t^2} \frac{\partial}{\partial t} H(z, t) = -\frac{1}{\mu_0} \left(\frac{\partial}{\partial z} E(z, t) + \frac{(\Delta z)^2}{24} \frac{\partial^2}{\partial z^2} \frac{\partial}{\partial z} E(z, t) \right) . \tag{26}$$

To satisfy the general criterion for optimally accurate operators the Taylor's series should be

$$\frac{\partial}{\partial t} H(z, t) + \frac{(\Delta t)^2}{24} \frac{\partial^2}{\partial t^2} \left[\frac{\partial}{\partial t} H(z, t) + \frac{1}{\mu_0} \frac{\partial}{\partial z} E(z, t) \right] = -\frac{\frac{\partial}{\partial z} E(z, t) + \frac{(\Delta z)^2}{24} \frac{\partial^2}{\partial z^2} \left[\frac{\partial}{\partial z} E(z, t) - \mu_0 \frac{\partial}{\partial t} H(z, t) \right]}{\mu_0} . \tag{27}$$

The derivation of modified operators from the above equation is possible, but they apparently lead to an intractable implicit scheme. Thus, the modified operators formally exist but their use would lead to an apparently intractable implicit scheme.

5 Implementation

Let the following equation represent a discretized wave equation using modified operators

$$\left(A_{n,n_t} - K_{n,n_t}\right) c_{n,n_t} = f , \tag{28}$$

where here for the sake of completeness, even the source term is also considered. From Eqs. (9) and (10), the modified operator $(A_{n,n_t} - K_{n,n_t})$ has multiple non-zero elements for time $t+\Delta t$. Usage of these operators in the marching-on-in-time scheme to solve the FD equation results in an implicit scheme, rather than an explicit scheme. To obviate the need to solve a system of simultaneous linear equations at each time step, the modified operators are approximated by a *predictor-corrector scheme* based on the first-order Born approximation. The FD equation using the conventional operators can be written as

$$\left(A^0_{n,n_t} - K^0_{n,n_t}\right) c^0_{n,n_t} = f , \tag{29}$$

where f is the source term. First, predict the field at the next time step using the conventional operators A^0 and K^0. Compute δc, the correction to the displacement at time $t + \Delta t$, using the first-order Born approximation. To obtain the correction solve

$$\left(A^0_{n,n_t} - K^0_{n,n_t}\right) \delta c_{n,n_t} = -\left(\delta A_{n,n_t} - \delta K_{n,n_t}\right) c^0_{n,n_t} . \tag{30}$$

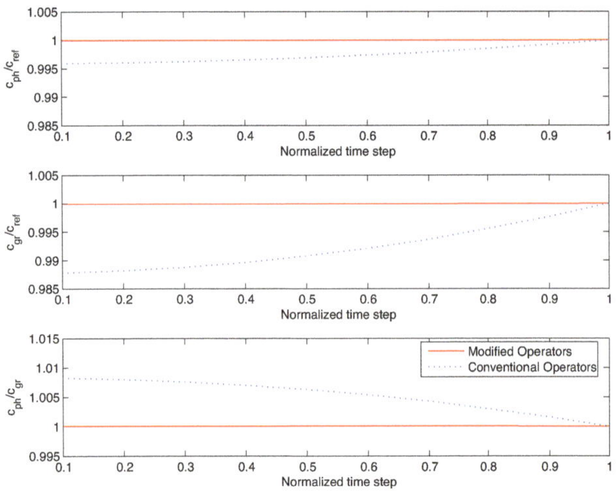

Fig. 1. Dispersion relation graphs.

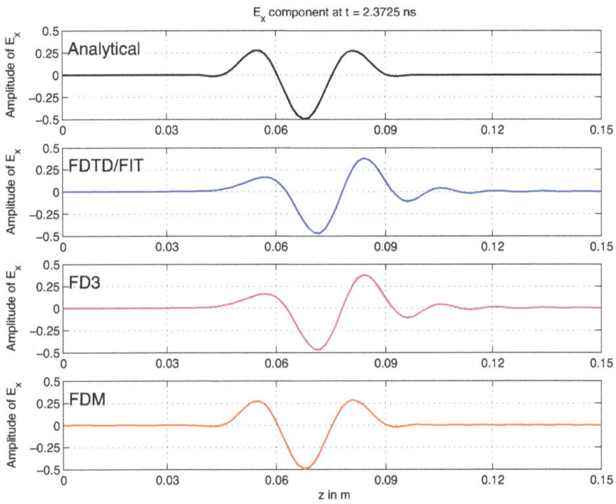

Fig. 2. Comparison between the analytical and numerical solutions.

The operators $\delta A_{n,n_t}$, $\delta K_{n,n_t}$ denote the difference between the modified operators A_{n,n_t}, K_{n,n_t} and the conventional operators A^0_{n,n_t}, K^0_{n,n_t}). The left-hand side of the above equation uses conventional operators, this means $\delta c_{n,n_t}$ can be obtained explicitly at time $t+\Delta t$. Compute the corrected field after each time step using c^0_{n,n_t} and $\delta c_{n,n_t}$ computed above, i.e., $c_{n,n_t+1}=c^0_{n,n_t+1}+\delta c_{n,n_t+1}$. Before advancing to the next time step redefine $c^0_{n,n_t+1} \leftarrow c_{n,n_t+1}$.

6 Simulation results

6.1 Dispersion relation graphs

Figure 1 shows the variation of numerical phase velocity and group velocity with respect to the normalized time step, $\widehat{\Delta t}$, for a given spatial resolution, number of cells per wavelength, in this case 20. Modified operators have less dependency on the normalized time step, and also have a sharper characteristics than standard operators. This is clearly visible as numerical phase and group velocities of modified operators tend to be always the free-space wave velocity, i.e., $c_{\text{ph}}=c_{\text{gr}}=c_0$.

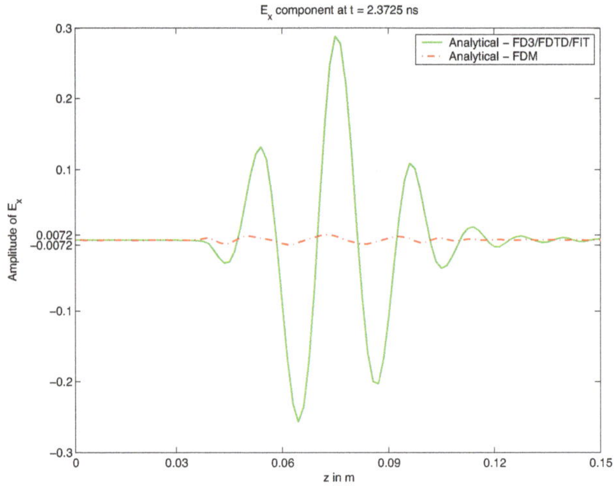

Fig. 3. Absolute difference between the analytical and numerical solutions.

6.2 1-D EM case: Free space modeling: PEC boundaries

All the simulations, see Figs. 2 and 3, are carried out in free-space of the length 1.5 m, 1000 cells, for 2.3725 ns, 950 time steps, with a raised cosine pulse of 2 cycles, RC_2 pulse, with a carrier frequency of 10 GHz as the source located at the center of the medium. The normalized time step $\widehat{\Delta t}$ is equal to 0.5, the reference cell size Δz and reference time step Δt are 1.5 mm and 5 ps, respectively. The spatial resolution is taken as 20. The results are plotted from z=0, . . . , 0.15 m. Figurre 2 shows a comparative plot of the analytical and numerical solutions. The numerical dispersion error is clearly visible in Fig. 3, displays the difference between the analytical and numerical solutions. The numerical dispersion error is reduced approximately by a factor of 40.

6.3 2-D TM case: inhomogeneous medium/plane wave boundaries

The 2-D modeling results in Figs. 4 and 5 are obtained by simulations carried out in an area of a length of 3 m, 2000 cells, and a width of 1.5 m, 1000 cells, for 10 ns, 4000 time steps, with a raised cosine pulse of 2 cycles, RC_2 pulse, with a carrier frequency of 10 GHz as source located along the top boundary of the medium. The normalized time step $\widehat{\Delta t}$ is equal to 0.5, the reference cell size Δz and reference time step Δt are 1.5 mm and 5 ps, respectively. The spatial resolution is taken as 20. A plane wave boundary condition is used to truncate the simulation. The first medium (0–2.25 m) is free space and the second medium (2.25–3 m) has an increase in permittivity by a factor of 4. A plane wave

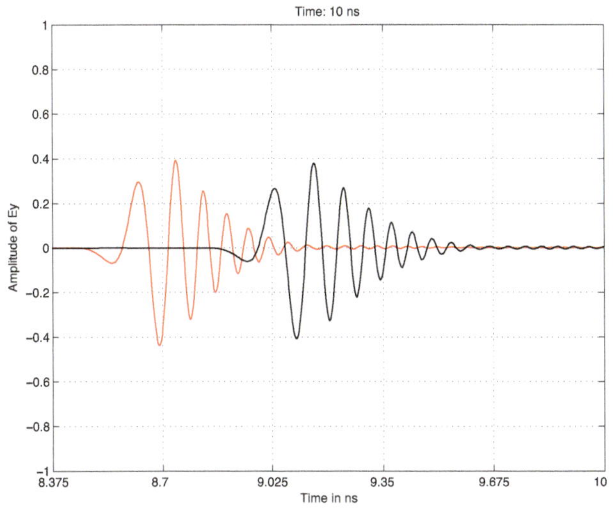

Fig. 4. Wave modeling: FIT.

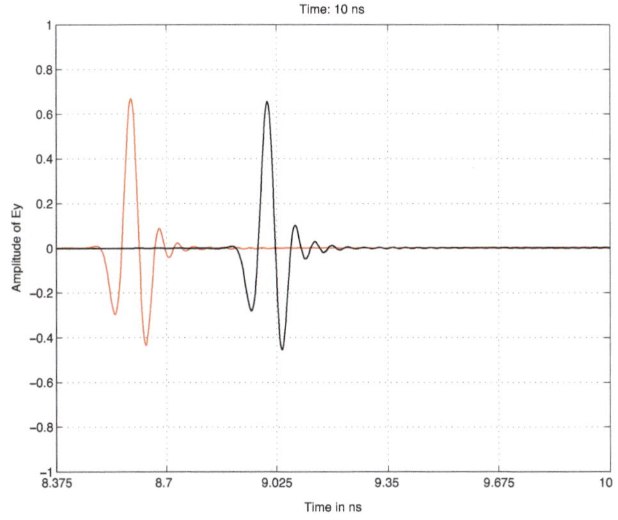

Fig. 5. Wave modeling: FDM.

is sent from the top of the medium and is observed as it passes through the second medium. Figures 4 and 5 shows the data collected by the receivers (inside the second medium). From these figures it clear that FDM outweighs FIT even in the case of an inhomogeneous problem.

6.4 2-D P-SV case: plexiglas/motion-free boundaries

Consider the results in Fig. 6 where the simulations are carried out for plexiglas with an area of length of 66.2266 mm, 1200 cells, and a width of 132.4535 mm, 2000 cells, for a time interval of 90.9707 μs, 6500 time steps, with a raised cosine pulse of 2 cycles, RC_2 pulse, with a carrier frequency of 5 MHz as source. The normalized time step $\widehat{\Delta t}$ is equal

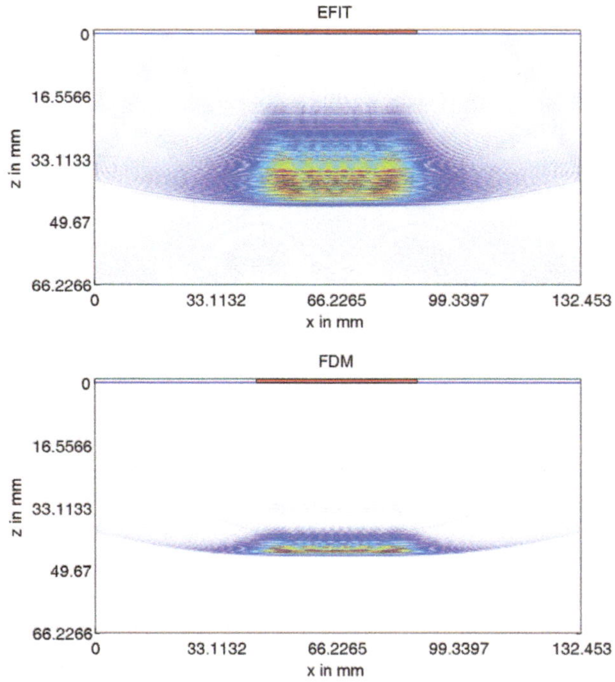

Fig. 6. Elastic P-SV wave modeling using EFIT and FDM.

to 0.5. The reference cell size Δz and reference time step Δt are 55.189 μm and 20.2157 ns, respectively. The spatial resolution, number of grid cells per wavelength, is taken as 9.8933. The motion-free boundary condition is used to truncate the simulation on all sides. A normal force is applied within a finite aperture at the top of the medium. The finite aperture is shown in red color. The EFIT and FDM snapshot of the wave field is taken at t=90.9707 μs. In this case, using FDM, accuracy is increased by a factor of 30 compared to EFIT.

7 Conclusions

The errors due to spatial and temporal discretization cannot be considered separately. The net error due to the combined effects of temporal and spatial discretization must be considered as a single quality. This error can be minimized by tuning the operators so that the errors due to spatial and temporal discretization come as close as possible to cancel each other. The error of the conventional operators is much worse than that of the modified operators, but approaches that of modified operators if and only if $\widehat{\Delta t}$ is equal to the Courant limit. The error of solutions obtained using modified operators is essentially uniform regardless of the choice of $\widehat{\Delta t}$ and spatial resolution. This is clear from the dispersion graphs. We make the following observations:

- a error reduction by a factor of about 40–100 for 2 times CPU requirement compared to FD3 method in 1-D case
- a error reduction by a factor of about 36 for 2.9 times CPU requirement compared to FD3 method in 2-D TM (SH) case, and
- a error reduction by a factor of about 30, for 3.5 times additional CPU requirement compared to FD3 method in 2-D P-SV case.

Please note that the reduction of error depends on the complexity of the problem, too. This method is straightly extendable to inhomogeneous problems and also to the 3-D problems. The ratio of required floating point operations will increase linearly with the dimension of the problem and a CPU time of about 5–8 times can be expected. An improvement of the results by a factor of 30 can be expected as the improvement in the accuracy is independent of the dimension of the problem. The P-SV case is not furnished here, but it can be treated in a similar way, which is part of our present research.

8 Appendix

$$K^{1'}_{p'r'N} = \left(\frac{1}{144(\Delta x)^2}\right) *$$

$t+\Delta t$	$z+\Delta z$	υ_{-0}	$-\upsilon_{-+}$	υ_{+0}
	z	$10\upsilon_{-0}$	$-10\upsilon_{-+}$	$10\upsilon_{+0}$
	$z-\Delta z$	υ_{-0}	$-\upsilon_{-+}$	υ_{+0}
		$x-\Delta x$	x	$x+\Delta x$
t	$z+\Delta z$	υ_{-0}	$-10\upsilon_{-+}$	υ_{+0}
	z	$10\upsilon_{-0}$	$-100\upsilon_{-+}$	$10\upsilon_{+0}$
	$z-\Delta z$	υ_{-0}	$-10\upsilon_{-+}$	υ_{+0}
		$x-\Delta x$	x	$x+\Delta x$
$t-\Delta t$	$z+\Delta z$	υ_{-0}	$-\upsilon_{-+}$	υ_{+0}
	z	$10\upsilon_{-0}$	$-10\upsilon_{-+}$	$10\upsilon_{+0}$
	$z-\Delta z$	υ_{-0}	$-\upsilon_{-+}$	υ_{+0}
		$x-\Delta x$	x	$x+\Delta x$

The indices p', r' represent the x and z directions respectively, $0 \le p' \le N_x$, $0 \le r' \le N_z$. N_x and N_z are the number of grid cells in x and z directions. In spatial operators, $\frac{1}{\mu}$ is represented as υ.

$$\upsilon_{-0} = \frac{\left(\upsilon_{(p'-1)r'} + \upsilon_{p'r'}\right)}{2}$$

$$\upsilon_{+0} = \frac{\left(\upsilon_{(p'+1)r'} + \upsilon_{p'r'}\right)}{2}$$

υ_{-+} represents the sum of υ_{-0} and υ_{+0}.

$$K^{3'}_{p'r'N} = \left(\frac{1}{144(\Delta z)^2}\right) *$$

$t-\Delta t$	$z+\Delta z$	υ_{0-}	υ_{0-}	υ_{0-}
	z	$-\upsilon_{-+}$	$-10\upsilon_{-+}$	$-\upsilon_{0-+}$
	$z-\Delta z$	υ_{0+}	υ_{0+}	υ_{0+}
		$x-\Delta x$	x	$x+\Delta x$
t	$z+\Delta z$	$10\upsilon_{0-}$	$10\upsilon_{0-}$	$10\upsilon_{0-}$
	z	$-10\upsilon_{-+}$	$-100\upsilon_{-+}$	$-10\upsilon_{-+}$
	$z-\Delta z$	$10\upsilon_{0+}$	$10\upsilon_{0+}$	$10\upsilon_{0+}$
		$x-\Delta x$	x	$x+\Delta x$
$t-\Delta t$	$z+\Delta z$	υ_{0-}	υ_{0-}	υ_{0-}
	z	$-\upsilon_{-+}$	$-10\upsilon_{-+}$	$-\upsilon_{0-+}$
	$z-\Delta z$	υ_{0+}	υ_{0+}	υ_{0+}
		$x-\Delta x$	x	$x+\Delta x$

$$\upsilon_{0-} = \frac{\left(\upsilon_{p'(r'-1)} + \upsilon_{p'r'}\right)}{2}$$

$$\upsilon_{0+} = \frac{\left(\upsilon_{p'(r'+1)} + \upsilon_{p'r'}\right)}{2}$$

Here υ_{-+} represents the sum of υ_{0-} and υ_{0+}.

$$A^{1'}_{p'r'N} = \left(\frac{\varepsilon_{00}}{144(\Delta t)^2}\right) *$$

$t+\Delta t$	$z+\Delta z$	1	10	1
	z	10	100	10
	$z-\Delta z$	1	10	1
		$x-\Delta x$	x	$x+\Delta x$
t	$z+\Delta z$	−2	−20	−2
	z	−20	−200	−20
	$z-\Delta z$	−2	−20	−2
		$x-\Delta x$	x	$x+\Delta x$
$t-\Delta t$	$z+\Delta z$	1	10	1
	z	10	100	10
	$z-\Delta z$	1	10	1
		$x-\Delta x$	x	$x+\Delta x$

where ϵ_{00} represents $\epsilon_{p'r'}$.

References

Bommaraju, C.: Optimally accurate second-order time-domain finite-difference scheme for the acoustic, electromagnetic and elastodynamic wave modeling, M.Sc. Thesis, University of Kassel, Kassel, Germany, 2004.

Bommaraju, C. and Marklein, R.: Optimally accurate second-order time-domain finite-difference scheme for the acoustic, electromagnetic and elastodynamic wave modeling: 1-D case, Proc. of IEEE Intern. Symposium on Antennas and Propagation and USNC/URSI National Radio Science Meeting, 2004.

Geller, R. J. and Takeuchi, N.: A new method for computing highly accurate DSM synthetic Seismograms, Geophysics Journal International, 123, 449–470, 1995.

Geller, R. J. and Takeuchi, N.: Optimally accurate second-order time-domain finite difference scheme for the elastic equation of motion: one-dimensional case, Geophysics Journal International, 135, 48–62, 1998.

Geller, R. J. and Takeuchi, N.: Optimally accurate second-order time-domain finite difference scheme for computing synthetic seismograms in 2-D and 3-D media, Physics of the Earth and Planetary Interiors, 119, 99–131, 2000.

Marklein, R.: Numerical Methods for the Modeling of Acoustic, Electromagnetic, Elastic and Piezoelectric Wave Propagation Problems in the Time Domain Based on the Finite Integration Technique (in German), Shaker Verlag, Aachen, Germany, 1997.

Taflove, A. and Hagness, S. C.: Computational Electrodynamics: The Finite-Difference Time-Domain Method, 2nd ed., Artech House, Boston, 2000.

33

Fast Multipole Method Applied to Volume Integral Equation Method

W. Hafla , A. Buchau, F. Groh, and W. M. Rucker

University of Stuttgart, Institute for Theory of Electrical Engineering, Pfaffenwaldring 47, 70569 Stuttgart, Germany

Abstract. The Volume Integral Equation Method (VIEM) has been used for the solution of three-dimensional non-linear magnetostatic field problems. The number of unknowns is minimal as only the magnetic material has to be discretized. For accurate solutions of problems where the magnetic field is small compared to the excitation field a difference field formulation has been developed. To reduce computational costs the fast multipole method is applied both on compression of the system matrix and during post processing. The efficiency of the formulation is demonstrated in several examples.

1 Introduction

The Boundary Element Method (BEM) has already proved to be very applicable when linear magnetostatic problems are solved since only surface integrals have to be carried out. For problems with non-linear $B(H)$ characteristics additional volume integrals occur. This generalization of the BEM is called Volume Integral Equation Method (VIEM). Its number of unknowns is minimal since only magnetized materials have to be discretized. Also the surrounding space is taken into account inherently, contrary to the Finite Element Method for example.

The main drawback of the method is its fully dense system matrix. It implies that memory requirements and the number of operations to solve the system of equations with an iterative solver grow by $O\left(N^2\right)$, where N is the number of unknowns. Also as integration is performed in three spatial directions, computational costs grow with third order with respect to the number of integration points.

A way to overcome these obstacles is to apply a suitable matrix compression technique. It can be used to compute the vector-by-matrix products during the iterative solving of linear system of equations. In Buchau et al. (2003a) it has been shown that with the fast multipole method (FMM) computational costs with BEM computations are reduced from $O\left(N^2\right)$ to approximately $O(N)$. As opposed to other compression techniques such as ACA (Buchau et al., 2003b) large reduction of computation time is also achieved with the FMM during post processing. This is why we decided to apply the FMM to the VIEM.

2 Formulation

The considered problem consists of free space, the magnetized material in the domain Ω_m with relative permeability μ_r and a solenoidal excitation field $\boldsymbol{H}_S$ due to free currents which can be computed with Biot-Savart's law. The total magnetic field can be split into the sum of the excitation field and the unknown induced field $\boldsymbol{H}_i$,

$$\boldsymbol{H}(\boldsymbol{r}) = \boldsymbol{H}_S(\boldsymbol{r}) + \boldsymbol{H}_i(\boldsymbol{r}). \tag{1}$$

If problems with highly permeably materials are solved $\boldsymbol{H}$ can become very small inside the magnetizable material compared to the excitation field. This is the case when $\boldsymbol{H}_S$ and $\boldsymbol{H}_i$ are almost anti-parallel. This is why huge cancellation errors usually occur. To overcome this problem a difference field formulation has been developed (Hafla et al., 2002) with which both linear and non-linear problems can be solved.

With Eq. (1) the basic equations for non-linear magnetostatic problems are

$$\operatorname{rot}\left(\boldsymbol{H}_S + \boldsymbol{H}_i\right) = 0 \tag{2}$$

$$\operatorname{div} \boldsymbol{B} = 0 \tag{3}$$

and the constitutive relation

$$\boldsymbol{B} = \mu_0 \boldsymbol{H} + \boldsymbol{M} = \mu_0 \mu_r (|\boldsymbol{H}|)\, \boldsymbol{H} \tag{4}$$

The induced field $\boldsymbol{H}_i$ can be represented by the integral formulation

$$\boldsymbol{H}_i(\boldsymbol{r}) = -\int_V \rho\left(\boldsymbol{r}'\right) \nabla G\left(\boldsymbol{r}, \boldsymbol{r}'\right) \mathrm{d}V' - \int_A \sigma\left(\boldsymbol{r}'\right) \nabla G\left(\boldsymbol{r}, \boldsymbol{r}'\right) \mathrm{d}A' \tag{5}$$

where σ and ρ are the equivalent surface and volume charges respectively and with Green's function of free space

$$G\left(\boldsymbol{r}, \boldsymbol{r}'\right) = \frac{1}{4\pi\mu_0} \frac{1}{\left|\boldsymbol{r} - \boldsymbol{r}'\right|}. \tag{6}$$

With this approach the irrotational condition Eq. (2) for $\boldsymbol{H}_i$ is always fulfilled. Considering induction's solenoidality Eq. (3) with Eq. (4) for points on the surface of Ω_m and within Ω_m gives two integral equations which have to be solved simultaneously (Babić et al., 2000; Kim et al., 2000).

- Integral equation from induction's solenoidality on $\partial\Omega_m$

Applying div $\boldsymbol{B}$=0 on $\partial\Omega_m$ yields the integral equation (McWhirter et al., 1982)

$$\frac{\sigma(\boldsymbol{r})}{2\lambda(\boldsymbol{r})}-\left[\int_A \sigma(\boldsymbol{r}')\,\partial_n G(\boldsymbol{r},\boldsymbol{r}')\,\mathrm{d}A'+\int_V \rho(\boldsymbol{r}')\,\partial_n G(\boldsymbol{r},\boldsymbol{r}')\,\mathrm{d}V'\right]$$
$$=\boldsymbol{n}(\boldsymbol{r})\cdot\boldsymbol{H}^S(\boldsymbol{r}) \tag{7}$$

with

$$\partial_n G(\boldsymbol{r},\boldsymbol{r}')=\boldsymbol{n}(\boldsymbol{r})\cdot\nabla G(\boldsymbol{r},\boldsymbol{r}'), \tag{8}$$

$$\lambda(\mathbf{r})=\frac{\mu_r(\boldsymbol{r})-1}{\mu_r(\boldsymbol{r})+1}. \tag{9}$$

The normal vector $\boldsymbol{n}$ points outwards from Ω_m. To avoid cancellation errors Eq. (7) can be solved in a first step for infinite permeability. In that case and if Ω_m is single-connected,

$$\boldsymbol{H}=\boldsymbol{H}^i\Big|_{\mu_r\to\infty}+\boldsymbol{H}^s=0, \tag{10}$$

so that Eq. (7) becomes an equation for the sources σ_∞ of the scalar magnetic potential of the excitation field,

$$\frac{1}{2}\sigma_\infty(\boldsymbol{r})+\int_A \sigma_\infty(\boldsymbol{r}')\,\partial_n G(\boldsymbol{r},\boldsymbol{r}')\,\mathrm{d}A'=\boldsymbol{n}(\boldsymbol{r})\cdot\boldsymbol{H}_S(\boldsymbol{r}). \tag{11}$$

Once Eq. (11) has been solved, it can be used to substitute the right-hand side in Eq. (7). This leads with the difference surface charge

$$\delta\sigma:=\sigma-\sigma_\infty \tag{12}$$

to the integral equation

$$\frac{\delta\sigma(\boldsymbol{r})}{2}+\lambda(\boldsymbol{r})\left[\int_A \delta\sigma(\boldsymbol{r}')\,\partial_n G(\boldsymbol{r},\boldsymbol{r}')\,\mathrm{d}A'+\int_V \rho(\boldsymbol{r}')\,\partial_n G(\boldsymbol{r},\boldsymbol{r}')\,\mathrm{d}V'\right]$$
$$=\frac{\sigma_\infty(\boldsymbol{r})}{2}(\lambda(\boldsymbol{r})-1) \tag{13}$$

From Eq. (10) follows

$$\boldsymbol{H}^S(\boldsymbol{r})=-\boldsymbol{H}^i(\boldsymbol{r})\Big|_{\mu_r\to\infty}=\int_A \sigma_\infty(\boldsymbol{r}')\,\nabla G(\boldsymbol{r},\boldsymbol{r}')\,\mathrm{d}A' \tag{14}$$

so the induced magnetic field can be computed with

$$\boldsymbol{H}(\boldsymbol{r})=-\int_V \rho(\boldsymbol{r}')\,\nabla G(\boldsymbol{r},\boldsymbol{r}')\,\mathrm{d}V'$$
$$-\int_A\left\{\begin{matrix}\delta\sigma(\boldsymbol{r}')\\ [\delta\sigma+\sigma_\infty](\boldsymbol{r}')\end{matrix}\right\}\nabla G(\boldsymbol{r},\boldsymbol{r}')\,\mathrm{d}A'+\left\{\begin{matrix}0\\ H_S(\boldsymbol{r})\end{matrix}\right\}$$
$$\text{if}\left\{\begin{matrix}\boldsymbol{r}\in\Omega_m\\ \boldsymbol{r}\notin\Omega_m\end{matrix}\right\}. \tag{15}$$

- Integral equation from induction's solenoidality in Ω_m

In cases where the relative permeability $\mu_r(H)$ depends on the magnetic field, solenoidality of induction within Ω_m leads to non-vanishing volume charges (Krstajić et al., 1992)

$$\rho(\boldsymbol{r})+\mu_0\boldsymbol{H}(\boldsymbol{r})\cdot\nabla\ln\mu_r(\boldsymbol{r})=0. \tag{16}$$

Substituting Eq. (15) into Eq. (16) for $\boldsymbol{r}\in\Omega_m$ gives

$$\rho(\boldsymbol{r})-\int_V \rho(\boldsymbol{r}')\,\nabla G(\boldsymbol{r},\boldsymbol{r}')\cdot\nabla\ln\mu_r(\boldsymbol{r})\,\mathrm{d}V'$$
$$-\int_A \delta\sigma(\boldsymbol{r}')\,\nabla G(\boldsymbol{r},\boldsymbol{r}')\cdot\nabla\ln\mu_r(\boldsymbol{r})\,\mathrm{d}A'=0. \tag{17}$$

- Solution Procedure

After Eq. (11) has been solved, the coupled system Eqs. (13), (17) can be solved. As it is in general non-linear an iterative procedure such as direct iteration has to be used where in every step Eq. (13), Eq. (17) is solved for a linear inhomogeneous distribution of μ_r. Between every non-linear iteration step the distribution of relative permeability must be updated. To enable convergence for problems where $\mu_r(H)$ characteristics have a maximum, we use

$$\mu_r^{(k)}=\begin{cases}B\left(H^{(k)}\right)/\left(\mu_0 H^{(k)}\right), & H^{(k)}<H_m\\ \tilde{B}^{(k)}/\left(\mu_0 H\left(\tilde{B}^{(k)}\right)\right), & H^{(k)}\geq H_m\end{cases} \tag{18}$$

where H_m is the field strength where this maximum occurs, k denotes the iteration step number, and $H(B)$ is the inverse magnetization characteristic $H(B)$ with $\tilde{B}^{(k)}=\mu_r^{(k-1)}H^{(k)}$.

3 Fast Multipole Method

The fully dense system matrix of the IEM arises from the fact that all elements interact with all the other. In most cases the matrix itself is not explicitly needed but it is only used to compute matrix-by-vector products, as for iteratively solving the linear system of equations for example. In the FMM algorithm, the matrix-by-vector product can be split into a near-field part due to elements that are close to each other and a far-field part for the remaining elements,

$$y=[A]\cdot\{x\}=[A_{near}]\cdot\{x\}+\{y_{far}\}.$$

The division of element interactions into a near-field and a far-field part is carried out by means of a hierarchical grouping scheme that is based on cubes, the so-called octree. The near-field matrix $[A_{near}]$ is a sparse matrix that equals the total matrix with the far-field interactions removed. Elements whose distance from each other is sufficiently large are considered group-wise, where the size of the group depends on the distance. Their interactions are given respect to in $\{y_{far}\}$. Therefore Green's function is approximated by a truncated series expansion into spherical harmonics Y_n^{-m}.

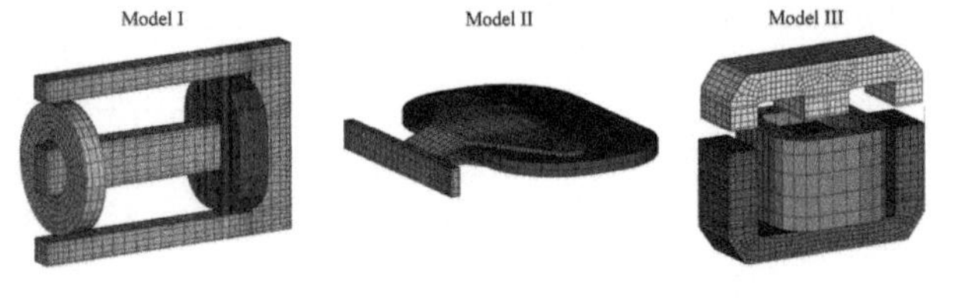

Fig. 1. Investigated models; finest meshes with problem oriented meshing strategy are shown.

The coefficients of the series expansion are called multipole coefficients M_n^m. For the presented formulations they can be calculated from the surface and volume charge densities,

$$M_n^m = \int_V \rho\left(\boldsymbol{r}'\right) \frac{1}{r'^{n+1}} Y_n^{-m}\left(\theta', \varphi'\right) \mathrm{d}V' + \int_A \sigma\left(\boldsymbol{r}'\right) \frac{1}{r'^{n+1}} Y_n^{-m}\left(\theta', \varphi'\right) \mathrm{d}A'. \tag{19}$$

We noticed that compared to BEM where only surface elements occur, it is much more important with the IEM to keep the number of near-field interactions as small as possible. This is because of two reasons. First, integrations with volume elements are in general much more time-consuming than with surface elements. Second, elements in the near-field are often located in all three spatial directions and not only in two as in the case of BEM. The latter not only leads to more computation time but also to a lower memory compression rate. A small near-field requires higher order L of series expansions. It was shown (Buchau et al., 2003a) that computational costs can be reduced with modified transformations of the multipole coefficients from $O\left(L^4\right)$ to $O\left(L^3\right)$.

Also, the different sizes of surface and volume elements have to be considered for the grouping scheme (Buchau et al., 2003c).

4 Numerical Results

The coupled system of Eqs. (13) and (17) has been discretized with Galerkin's method and second order isoparametric volume and surface nodal elements. GMRES with Jacobi preconditioner was used as linear solver. In the following the properties of the FMM with respect to the IEM are investigated. Then an example field problem is presented.

4.1 Fast Multipole Method

We investigated the application of the FMM to the IEM concerning three aspects that may influence compression rates. The first is the overall shape of the model. The second is the influence of the two different meshing strategies, problem oriented meshing and meshing with the aim to create as many as possible elements of the same size. And third, how the compression rate depends on the number of degrees of freedom (DOF).

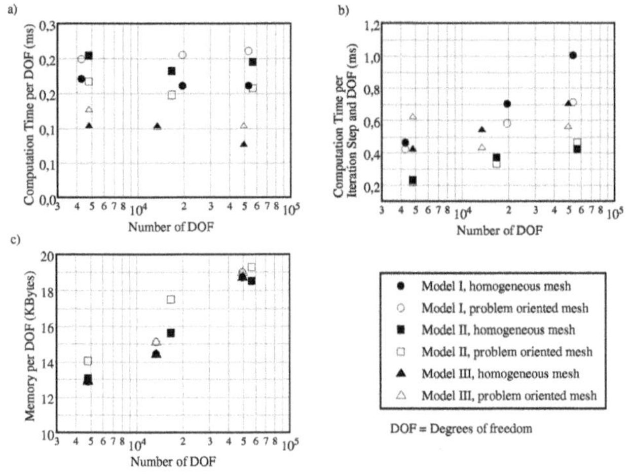

Fig. 2. Computational costs by application of the FMM.

The three different models shown in Fig. 1 were used for these investigations. Therefore all of them where meshed with both problem oriented and homogeneous meshes. Again for each of these two cases three different meshes with different number of elements where generated, so all in all 18 different meshes were used.

- Overall Model Shape

With BEM near-field interactions between surface elements are often found only in two dimensions. With volume elements near-interactions can occur up to all three dimensions depending on the overall model structure. Therefore we used the models in Fig. 1 with long and thin structures (Model I), flat structures (Model II) and bellied structures (Model III).

As it can be seen from Fig. 2a where the time per DOF for near-field assembling is shown, the time for model III is the lowest while the two other models are nearly the same. The time per DOF and iteration step during solving of the linear system of equations is displayed in Fig. 2b. In Fig. 2c memory requirements per DOF are shown. They depend roughly only on the number of DOF and not on the respective model.

- Meshing Strategy

When the FMM is applied to BEM calculations it becomes apparent that mesh homogeneity can have a significant influence on the compression rate. In conventional BEM one usually tries to use as few elements as possible. This mostly results in meshes with elements of largely varying sizes. The geometrical dimensions of the near-field of elements of large size are usually large. This means that if many small elements lie in the neighborhood of large elements, a high number of near-field integrations have to be computed which results in a lower compression rate. To investigate this we considered both homogeneous and problem oriented meshes.

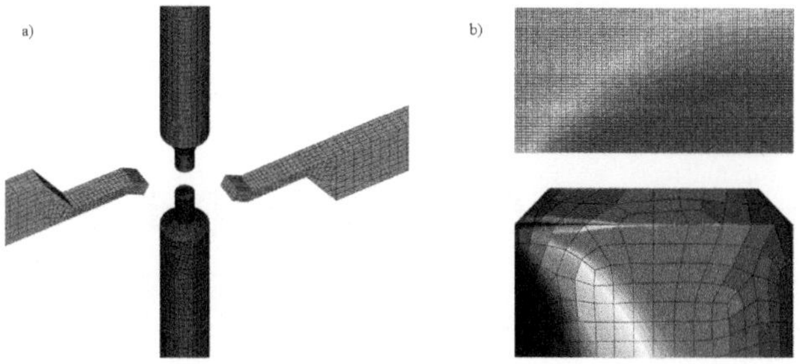

Fig. 3. Investigated arrangement of pole shoes for Magnetic Transmission X-ray Microscopy Project (6812 surface elements, 3130 volume elements, 36912 unknowns).

From Figs. 2a and 2c it can be seen that the increased size of the near-field of problem oriented meshes is reflected in increased computational costs for near-field assembly as well as an increase of total memory requirements.

- *Number of Degrees of Freedom*

For all three models different mesh sizes where used, both for homogeneous and problem oriented meshes. The assembly time per DOF of the near-field matrix in Fig. 2a is approximately independent of the number of DOF. Except for the coarse problem oriented mesh a logarithmic dependency between the number of DOF and computation time can be seen in Fig. 2b. The increase of memory requirements per DOF in Fig. 2c is believed to be caused by a growth of average near-field interactions due to the increasing number of volume elements.

4.2 Difference Field Formulation

For the Magnetic Transmission X-ray Microscopy Project at BESSY II (Eimüller, 2002) the field distribution in the air region between the pole shoes shown in Fig. 3a was investigated.

The flux lines run between the horizontal pole shoes. The influence of the passive vertical pole shoes was to be investigated especially at the sample location which is directly above the lower one. The vertical pole shoes don't saturate, so their relative permeability could be assumed constant, e.g. 1700 even for high accuracy calculation.

The stopping criterion of mean($\Delta\mu/\mu$) <2.5% of the non-linear solver was chosen to obtain accurate results in the region of interest. It was reached after 28 iterations and about 110 h. As the system matrix was compressed with the FMM only 729 MByte were required to solve the problem. Without the FMM about 10.1 GByte would have been required. This equals a compression rate of 93%.

Flux density was calculated at the nodes of the very fine evaluation mesh (19 101 nodes) shown in Fig. 3b. The calculation of the field at all 28 389 evaluation points took only 2 min.

5 Conclusions

A new indirect Volume Integral Equation Method formulation for non-linear magnetostatic problems was developed. Uniform treatment of surface and volume charges results in an elegant coupling of integral equations. Non-linear problems are tackled by consecutively solving linear problems. It has been shown that application of the fast multipole method leads to a large reduction of computational costs and therefore to an efficient and practicable alternative to e.g. Finite Element Method.

Acknowledgements. We would like to thank J. Konarev for the creation of discretized models and for the thorough compilation of the obtained results.

References

Babic, S., Akyel, C., and Gavrilovic, M.: Calculation improvement of 3d linear magnetostatic field based on fictious magnetic surface charge, IEEE Trans. Magn., vol. 36, no. 5, 3125–3127, Sept. 2000.

Buchau, A., Hafla, W., and Rucker, W. M.: Fast and efficient 3d boundary element method for closed domains, Symposium Reports of the Sixth International Symposium on Electric and Magnetic Fields, 211–214, Aachen, 2003a.

Buchau, A., Rucker, W. M., Rain, O., Rischmüller, V., Kurz, S., and Rjasanow, S.: Comparison between different approaches for fast and efficient 3d BEM computations, IEEE Trans. Magn., vol. 39, no. 3, 1107–1110, 2003b.

Buchau, A., Hafla, W., Groh, F., and Rucker, W. M.: Grouping schemes and meshing strategies for the fast multipole method, COMPEL, vol. 22, no. 3, 495–507, 2003c.

Eimüller, T.: Magnetic imaging of nanostructured systems with Transmission X-ray Microscopy, Dissertation, Universität Würzburg, 2002.

Hafla, W., Groh, F., Buchau, A., and Rucker, W. M.: Magnetostatic field computations by an integral equation method using a difference field concept and the fast multipole method, Proceedings of the 10th IGTE Symposium on Numerical Field Calculation in Electrical Engineering, 262–266, 2002.

Kim, D., Park, I., and Park, M., and Lee, H.: 3-d magnetostatic field calculation by a single layer boundary integral equation method using a difference field concept, IEEE Trans. Magn., vol. 36, no. 5, 3134–3136, Sept. 2000.

Krstajić, B., Andelić, Z., Milojković, S., and Babić, S.: Nonlinear 3d magnetostatic field calculation by the integral equation method with surface and volume magnetic charges, IEEE Trans. Magn., vol. 28, no. 2, 1088–1091, March 1992.

McWhirter, J. H., Oravec, J. J., and Haak, R. W.: Computation of magnetostatic fields in three-dimensions based on Fredholm integral equations, IEEE Trans. Magn., vol. MAG-18, no. 2, 456–460, March 1982.

34

Automotive Radar and Lidar Systems for Next Generation Driver Assistance Functions

R. H. Rasshofer and K. Gresser

BMW Group Research and Technology, Hanauer Str. 46, 80992 Munich, Germany

Abstract. Automotive radar and lidar sensors represent key components for next generation driver assistance functions (Jones, 2001). Today, their use is limited to comfort applications in premium segment vehicles although an evolution process towards more safety-oriented functions is taking place. Radar sensors available on the market today suffer from low angular resolution and poor target detection in medium ranges (30 to 60 m) over azimuth angles larger than ±30°. In contrast, Lidar sensors show large sensitivity towards environmental influences (e.g. snow, fog, dirt). Both sensor technologies today have a rather high cost level, forbidding their wide-spread usage on mass markets.

A common approach to overcome individual sensor drawbacks is the employment of data fusion techniques (Bar-Shalom, 2001). Raw data fusion requires a common, standardized data interface to easily integrate a variety of asynchronous sensor data into a fusion network. Moreover, next generation sensors should be able to dynamically adopt to new situations and should have the ability to work in cooperative sensor environments.

As vehicular function development today is being shifted more and more towards virtual prototyping, mathematical sensor models should be available. These models should take into account the sensor's functional principle as well as all typical measurement errors generated by the sensor.

1 Present status

1.1 Introduction

Radar and Lidar systems play an important role in modern premium segment vehicular technology (Schneider, 1999; Jones, 2001). Today, vehicle applications mainly represent comfort functions (e.g. ACC) although an evolution from comfort towards safety functions has started recently. Functional requirements for vehicular safety systems impose much more stringent requirements on next generation sensor systems than typically found in today's sensors. With state-of-the-art sensor technology available on the market today, a number of safety functions can only be realized with limited performance.

Today (2004), the over-all market penetration of long range vehicular perception sensors is very low and mainly restricted to vehicles of the premium segment. Most sensors are employed to realize simple ACC (adaptive cruise control) functions based on 77 GHz radar technology although some lidar-based ACC systems have shown up on the Japanese market recently. In ACC systems, commonly long range sensors with a detection range larger than 150 m on cars and an azimuth coverage of ±6 to ±8° are employed. These sensors are optimised for very smooth distance regulation and thus have rather poor dynamic performance. Typical latency values for new target detection are 250 to 500 ms, internal filtering commonly limits target dynamics to values below $\pm 4\,m/s^2$. Given these facts, ACC sensors available on the market cannot be used for dynamic cruise control applications like e.g. city traffic or stop-and-go without major adjustments in sensor firmware and hardware (Widmann, 2000).

1.2 New vehicle functions

Car manufacturers world wide have successfully implemented a number of new vehicle functions based on radar and lidar sensors recently. Most of these functions will be put to mass production in the next two years. Figure 1 shows an overview of these functions and a brief description of the detection ranges typically required along with the sensor technology likely to be used. As can be seen, today there is no clear trend towards using a specific sensor technology, radar and lidar systems are equally employed in a variety of applications. Sometimes a combination of multiple sensors is used to achieve the detection range required by the vehicle function. Usage of sensors available for mass production 2 to 3 years before the production start of the vehicle is typical for all applications listed here.

1.3 Limits of today's radar sensors

1.3.1 Azimuth measurement

Radar sensors available for production today typically have high measurement accuracy in range. The range resolution

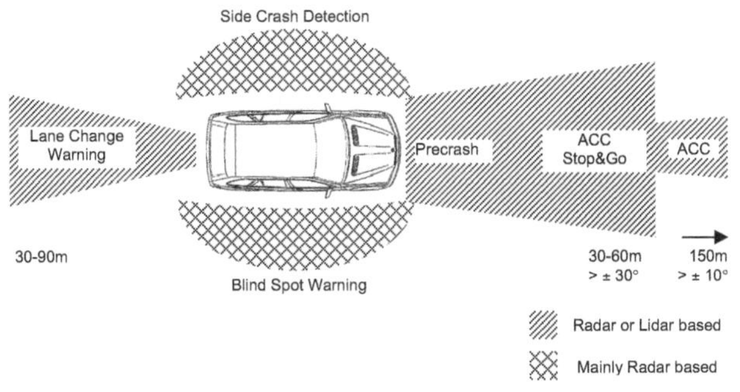

Fig. 1. New radar- and lidar-based vehicle functions likely to be available on the market in the next two years.

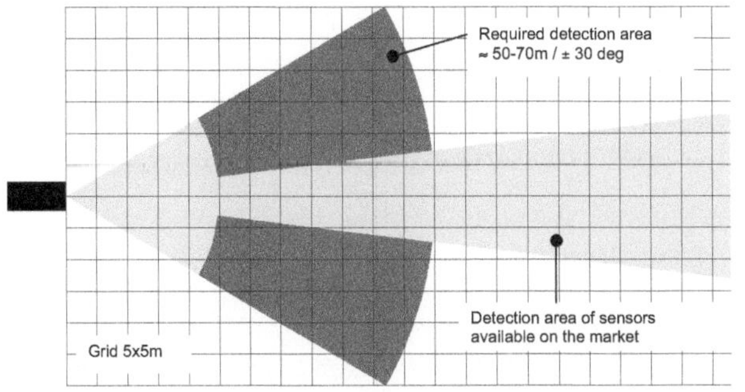

Fig. 2. Sensor coverage for front-looking applications like e.g. ACC stop & go: Requirements vs. radar sensors available.

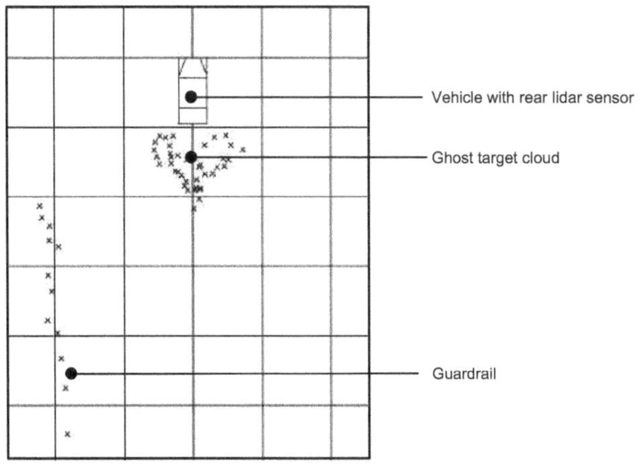

Fig. 3. Lidar ghost target cloud generated by water spray on a wet autobahn during heavy rain. Each point represents a target detected by the lidar. Grid: 5×5 m.

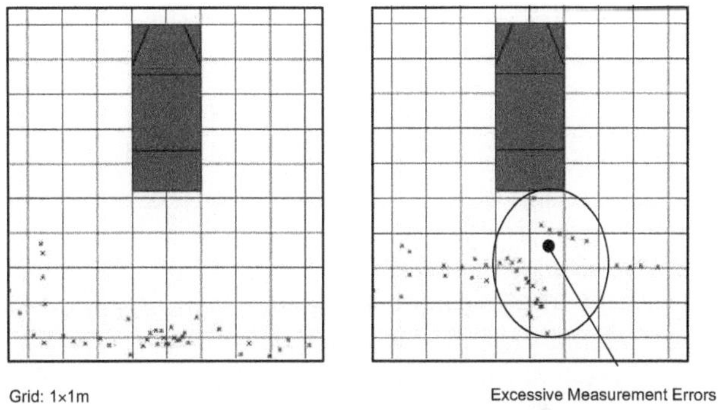

Fig. 4. Ghost target generation in a lidar system exceeding its dynamic range. Left: 4 m distance to highly reflective wall, no measurement errors. Right: 2 m distance to highly reflective wall, measurement errors increasing.

practically achieved is a factor of 1.3 to 2.0 larger than the physical limit implied by the signal bandwidth and usually is good enough for most applications. The main drawback of today's radar sensors is found in the measurement principles used for azimuth angle determination. As radar sensors employing mechanically scanning antennas – which have been used wide spread in the early days of automotive radar – have disappeared from the market almost completely, only electronic azimuth measurement principles like "sequential lobing" (Flacke, 1999) or "monopulse" (Li, 1999) are found in today's sensors. Both principles offer the possibility to measure a target's azimuth angle relatively to the sensor, but only work unambiguously in case of a single target. Generally speaking, both principles have a lack of angular resolution, i.e. cannot resolve two targets in the same range having different azimuth. Sensor manufacturers overcame this problem by means of tracking and filtering. As tracking and filtering methods rely on target model assumptions, critical detection errors tend to take place in certain situations, leading to target loss, wrong measurement of target's azimuth or even detection of ghost targets. These errors tend to appear rather seldom but with a finite probability so safety-critical decisions cannot only rely on information provided by today's radar sensors alone.

1.3.2 Detection area

Concerning front-looking applications like e.g. ACC stop & go, the detection of cutting-in vehicles represents a special challenge. To clearly classify the cutting-in case, target vehicles have to be detected in a distance of 50 to 70 m from the sensor on both adjacent lanes. This transfers to a desired detection azimuth of ±30° over 50 to 70 m distance which cannot be realized with any sensor on the market today. Figure 2 illustrates this situation. Next generation sensors should be enhanced in sensitivity to cover the important medium range area.

1.3.3 Frequency allocation

Well-known physical laws (1) relate radar signal bandwidth B to a radar's range resolution δR:

$$\delta R > c_0/2B \quad (1)$$

Typically, short range radars require high range resolution so ultra wide band sensors are commonly found in these applications (Rollmann, 2004, Gresham 2004). If cost-effective 24 GHz ISM band operation is desired, legal restrictions varying from country to country have to be taken into account. On a world-wide basis, only a few MHz of bandwidth are free to use. In the USA, FCC has allowed the use of ultra wideband radar sensors since 2002 (FCC, 2002) however in the EC, use of ultra wideband sensors is likely to be limited to a maximum of 7% fleet penetration and stop of new sensor deployment by end of 2012.

A long term solution to the limited bandwidth problem of 24 GHz near distance sensors will be the use of the recently allocated 79 GHz millimetre wave band spreading from 77 to 81 GHz. Sensors at these frequencies will be available not before 2015 in mass production since neither low-cost millimetre wave technology nor mass production facilities will be ready before that date.

1.4 Limits of today's lidar sensors

1.4.1 Lidar technology today

Today, lidar sensors can be divided into two groups: multibeam lidars and scanning lidars. While multibeam lidars employ an array of transmitting and receiving elements illuminating separate angular sections, scanning lidars usually use one transmitter and receiver with a mechanically scanned lens system (e.g. rotating mirror or prism). The azimuth of the rotating scan system directly represents the target azimuth in case of detection.

All lidars available on the market today operate in the near infrared. Their transmit output power level is limited by eye safety constrains. As the receiver sensitivity of a lidar sensor is limited by the detection diode (PIN diode or avalanche diode) operating close to the physical limits today, maximum sensor range is mainly influenced by lens aperture and hence by the size of the sensor.

1.4.2 Environmental conditions

Lidar sensors employ laser light for distance measurement. Atmospheric damping influences the maximum detection range, rain, snow and fog typically limit the lidar's detection range to the same performance a human eye has under these conditions. Given automotive environment, water spray of a preceding vehicle for front-looking applications or spray from the own vehicle for back-looking applications further limits the sensor performance by introducing ghost targets (see Fig. 3). Moreover, dirt and ice coverage on the sensor will stop its proper operation, however these effects can easily be detected by the sensors itself and thus can be signalled to the application.

1.4.3 Dynamic range

As object corners represent sharp boundaries for laser light, very large signal dynamic between adjacent lidar beams is

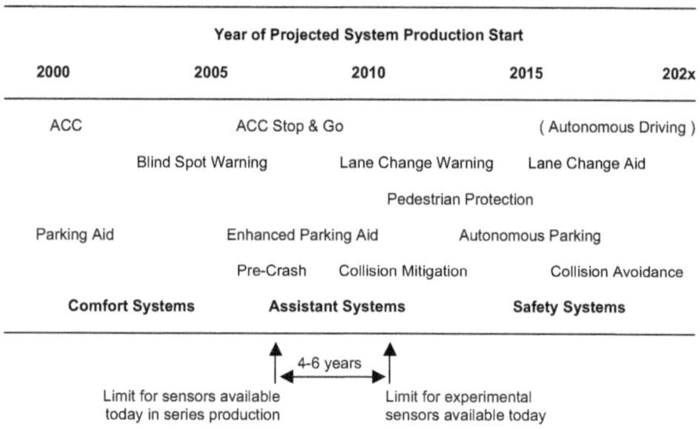

Fig. 5. General roadmap for future vehicle functions (not OEM specific).

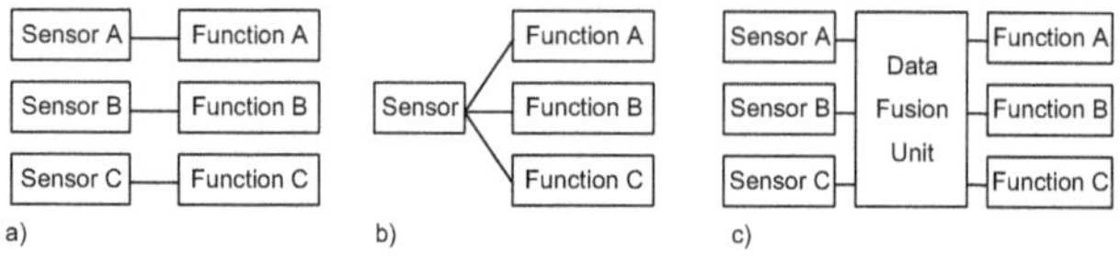

Fig. 6. Sensor system architectures. **(a)** Separate systems (present situation) **(b)** Multiple use of sensors, **(c)** Multi sensor data fusion.

noticed. To cope with this, dynamic intra-pulse gain control has to take place. If the maximum dynamic range of a lidar is exceeded, typically heavy generation of ghost targets along with real target losses starts due to non-linearity in the photo detector diode and compression of the video amplifier chain. Figure 4 shows an example of this phenomena.

2 Sensor requirements for next generation vehicular systems

2.1 General roadmap for future vehicle functions

Figure 5 shows a general roadmap for future vehicle functions valid for many car manufacturers. The roadmap displays the global trend of evaluation from comfort system via driver assistance functions towards pronounced safety features. The realization of these different functions today is mainly limited by the quality of sensors available. Due to this, the functions can be divided into three different groups:

1. Functions that can be realized with mass-production ready sensors available today
(e.g. parking aid or ACC)

2. Functions which can be realized as demonstrators using experimental high-end sensors (e.g. lane change warning or collision mitigation systems)

3. Functions that cannot be realized at all due to quality lacks of the corresponding sensor systems (e.g. collision avoidance).

Table 1. Typical strengths and weaknesses of automotive sensors available today.

	Short Range Radar	Long Range Radar	Lidar	Ultrasound	Video Camera	3D-Camera	Far IR Camera
Range Measurement < 2m	o	o	o	++	-	++	-
Range Measurement 2..30m	+	++	++	-	-	o	-
Range Measurement 30..150m	n.a.	++	+	- -	-	-	-
Angle Measurement < 10 deg	+	+	++	-	++	+	++
Angle Measurement > 30 deg	o	-	++	o	++	+	++
Angular Resolution	o	o	++	-	++	+	++
Direct Velocity Information	++	++	- -	o	- -	- -	- -
Operation in Rain	++	+	o	o	o	o	o
Operation in Fog or Snow	++	++	-	+	-	-	o
Operation if Dirt on Sensor	++	++	o	++	- -	- -	- -
Night vision	n.a.	n.a.	n.a.	n.a.	-	o	++

++ : Ideally suited / + : Good performance / o : Possible, but drawbacks to be expected;

- : Only possible with large additional effort / - - : Impossible / n.a. : Not applicable

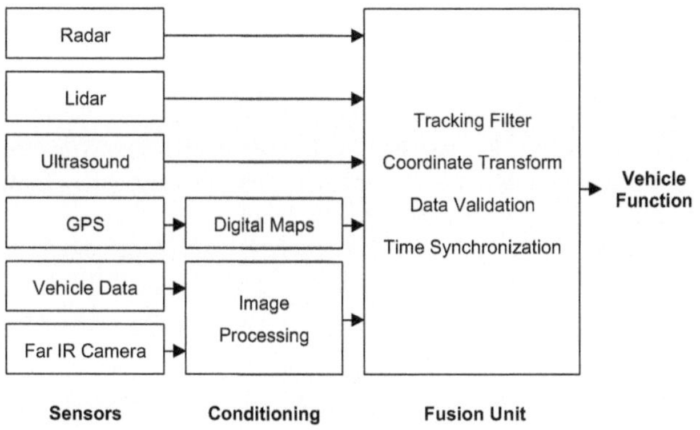

Fig. 7. Typical automotive sensor data fusion architecture.

Typically, proof of function has to take place 4 to 6 years before the start of vehicle production. By then, the step from an experimental sensor system towards a mass-production ready, automotive qualified part has to take place at the tier 1 supplier.

2.2 Trends in vehicle function architecture

Today, each vehicle application uses its own, functionally optimised sensor (Fig. 6a). In the future, this situation might change since the variety of new vehicle functions will not allow for separate sensor systems due to cost and space restrictions. One possible scenario will be the use of one sensor system by more than one vehicle function (e.g. re-use of ACC stop & go sensors for pre-crash application and pedestrian protection, see Fig. 6b). This scenario requires the same sensors to be operated in different applications sometimes having diverting requirements which usually imposes an enhanced quality level on the sensor.

As vehicle function development today mainly is limited by sensor quality, means to enhance sensor system performance have been discussed extensively. Related to their operation principle, different sensor technologies tend to show complementary strengths in measuring certain object parameters. Table 1 gives an overview of typical strengths and weaknesses for automotive sensors available today. As can be seen, strengths of different sensor types (e.g. radar and camera systems) can be combined to end up with an improved sensor system: This represents the concept of multi sensor data fusion shown in Fig. 6c. To practically perform data fusion, a multi step process has to take place. Figure 7 shows an overview of a typical multi sensor data fusion architecture. As far as the interface of the sensors to the fusion unit is concerned, certain restrictions apply. The sensors should provide a standardized object data interface which comprises only minimal and transparent data pre-processing and tracking. Moreover, the sensors should provide a precise measurement timestamp which enables the fusion unit to process asynchronous measurement data. Most sensors have the ability to internally judge a measured value with respect to the quality of measurement. This is obvious for radar and lidar where e.g. a good signal-to-noise ratio (SNR) usually indicates trustable measurements while a bad SNR makes measured values questionable. If these quality measures are available, they must be transmitted to the fusion unit to make probabilistic weighting of different measurements possible (Houles, 1989).

2.3 Sensor modelling

As vehicle functions emerge into more safety critical applications, function development and verification becomes increasingly more difficult. Testing the system to its limits during qualification and functional optimisation often means to provoke sensor or system failure. This might result in dangerous situations for the test engineers. If parameters should be tuned for the average driver, many non-skilled tests persons have to check the function during test sessions using experimental prototype systems. These tests often are impossible to be done in reality since they would endanger test persons or would require set up of artificial situations difficult to construct on test tracks. The solution to this problem is virtual function prototyping which is typically done in modern driving simulators. Although these simulators are highly developed with respect to vehicle dynamics and graphic simulation, functional development requires an accurate modelling of perception sensors which is not implemented in most simulators available up to now. Sensor modelling must include the sensors' measurement variance in range and angle, modelling of detection properties and description of sporadic sensor errors like generation of ghost targets and sporadic target losses. In the future, OEMs will require sensor manufacturers to provide virtual function models for their sensors well before first hardware samples become available. This will enable the OEM to speed up the prototyping process and to simplify the validation process.

Moreover, sensor error models provide very important input for sensor data fusion algorithms as they provide a-priori knowledge of the sensor's measurement noise covariance matrix **R** (Bar-Shalom 1989).

2.4 Vision: Cooperative sensors for automotive applications

On the way towards vehicle safety systems based on perception sensors, absolute reliability of sensor information becomes more and more crucial. For today's self-contained sensor systems, however, there is a limit of reliability given by the ambiguity of road scenarios combined with principle-inherent finite sensor measurement quality. A classic way to improve sensor information quality in radar and lidar is the use of cooperative targets which interact with an interrogating sensor. This concept has been successfully used in aviation (e.g. distance measurement equipment, DME) for decades and proved its feasibility to secure safety-critical decisions. Cooperative sensors could use e.g. communication properties of radar (Lindenmeier, 2003) to exchange object information between vehicles. This information could be used to check data plausibility and thus improve quality of self-contained measurements each vehicle did on its own. Moreover, communication to infrastructure (e.g. traffic lights) or beaconing of dangerous or vulnerable objects (e.g. pedestrians) could become possible.

3 Conclusions

Today, lidar and radar systems are mainly used in ACC comfort applications although a variety of new radar- and lidar-based vehicle functions will be introduced by OEMs in the next two years. Main drawback of today's radar sensors is their lack of angular resolution while today's lidar sensors show high sensitivity towards environmental influences. For next generation vehicle functions, quality of perception sensors is the main limiting factor. Ideally speaking, upcoming radar and lidar sensors should have improved angular resolution, should be free to use in any country and should be easily integrated into data fusion systems. Moreover, these sensors should have virtual function models available and should be able to interact with each other and with infrastructure to work in cooperative sensor environments.

References

Bar-Shalom, Y., Chang, K. C., and Blom, H. A. P.: Automatic Track Formation in Clutter with a Recursive Algorithm, Proceedings of the 28th Conference on Decision Control, Tampa, Florida, 1402–1408, December 1989.

Bar-Shalom, Y., Li, X., and Kirubarajan, T.: Estimation with Applications to Tracking an Navigation, John Wiley & Sons, Inc., New York, 2001.

FCC: First report and order, revision of part 15 of the comission's rules regarding ultra wideband transmission systems, FCC, Washington, D.C., ET Docket 98–153, 2002.

Flacke, J., Heckl, K., and Kaiser, B. : Verfahren zur Winkelbemessung mittels eines Mehrfachantennenradarsystems, European Patent EP 0 989 416 A2, June 1999.

Gresham, I., Jenkins, A., Egri, R., et.al.: : Ultra-Wideband Radar Sensors for Short-Range Vehicular Applications, IEEE Transactions on Microwave Theory and Techniques, Vol. 52, No. 9, 2105–2122, 2004.

Houles A. and Bar-Shalom, Y.: Multisensor Tracking of a Maneuvering Target in Clutter, IEEE Transactions on Aerospace and Electronic Systems, 176–189, 1989.

Jones, W. D.: Keeping Cars from Crashing, IEEE Spectrum, September 2001.

Li, D. D., Luo, S. C., Pero, C., Wu, X., and Knox, R. M.: Millimeter-Wave FMCW/Monopulse Radar Front-End for Automotive Applications, MTT-S Int. Microwave Symp.Dig., 277–280, 1999.

Lindenmeier, S., Boehm, K., and Luy, F.: A Wireless Data Link for Mobile Applications, IEEE Microwave and Wireless Components Letters, Vol. 13, No. 8, 326–328, 2003.

Rollmann, G., Knoll, P., Mekhaiel, M., Schmid, V., and Blöcher, H.-L.: Short Range Radar (SRR) – System for Automotive Application, IEEE MTT/AP German Newsletter, Vol. 8, No. 1, June 2004.

Schneider, R. and Wenger, J.: System aspects of future automotive radar, MTT-S Int. Microwave Symp. Dig., 293–296, 1999.

Widmann, G. R., Daniels, M. K., Hamilton, L., et.al.: Comparison of Lidar-Based and Radar-Based Adaptive Cruise Control Systems, SAE Technical Paper Series, 2000-01-0345, 2000.

Numerical simulation of transient electro-quasistatic fields using advanced subspace projection techniques

T. Steinmetz, G. Wimmer, and M. Clemens

Chair for Theory of Electrical Engineering and Computational Electromagnetics, Helmut-Schmidt-University, University of the Federal Armed Forces Hamburg, P.O. Box 700822 D-22008 Hamburg, Germany

Abstract. The transient simulation of electro-quasistatic fields requires a formulation that takes both dielectric effects as well as nonlinear conductive effects into account. The successive solution of large linear systems of equations with similar or even identical system matrices has to be performed repeatedly if this formulation is discretized by the Finite-Element method and an Implicit-Runge-Kutta method, respectively. The solution processes can be accelerated by using subspace recycling techniques and subspace projection extrapolation techniques. Numerical results for three-dimensional high-voltage applications are presented and the efficiency of these techniques is shown.

1 Introduction

For the analysis of technical devices operated by using slowly varying electromagnetic fields, electromagnetic wave propagation can be neglected if the characteristic dimension of the device under consideration is much smaller than the distance an electromagnetic wave covers during the characteristic time of the device. The characteristic time is the reciprocal of the highest frequency at which the device is operated (Dirks, 1996). This leads to the well-known quasistationary approximation to Maxwell's equations where changes in the current or charge distribution have effect instantly in space. This can be enforced in Maxwell's equations by discarding the solenoidal part of the displacement current in Ampere's law while keeping its irrotational part. Whereas in the static case, i.e. all time derivatives occuring in Maxwell's equations vanish ($\frac{\partial}{\partial t}\boldsymbol{D} \equiv \frac{\partial}{\partial t}\boldsymbol{B} \equiv 0$), the quasistationary approximation can further be classified in electro-quasistatic fields (EQS) and magneto-quasistatic fields (MQS) considering the governing electric or magnetic energy density w_e and w_m, respectively. If the electric energy density w_e exceeds the magnetic energy w_m for a given problem, the electro-quasistatic assumption is applicable, whereas the magneto-quasistatic assumption is applicable for $w_\mathrm{m} \gg w_\mathrm{e}$. The EQS assumption is expressed in Maxwell's equations by discarding the induction voltage in Faraday's law, $\frac{\partial}{\partial t}\boldsymbol{B} \equiv 0$. Typically, high electric energy densities and low magnetic energy densities occur if the dielectric relaxation time $\tau_\mathrm{relax} = \varepsilon/\kappa$ is much greater and the magnetic diffusion time $\tau_\mathrm{diff} = \mu\kappa d^2$ is much smaller than the device's characteristic time. Here, the permeability μ, the permittivity ε, the conductivity κ and the characteristic spatial dimension d are used.

2 Transient electro-quasistatic fields

Faraday's law under the EQS assumption yields

$$\operatorname{rot} \boldsymbol{E} = 0.$$

Hence, a description of electro-quasistatic fields is possible by a scalar potential function $\varphi(\boldsymbol{r})$ with $\boldsymbol{E}(\boldsymbol{r}) = -\operatorname{grad}\varphi(\boldsymbol{r})$. Looking at Ampere's law,

$$\operatorname{rot} \boldsymbol{H} = \boldsymbol{J} + \frac{\partial}{\partial t}\boldsymbol{D},$$

together with the material constitutive laws,

$$\boldsymbol{D}(\boldsymbol{r},t) = \varepsilon \boldsymbol{E}(\boldsymbol{r},t),$$
$$\boldsymbol{J}(\boldsymbol{r},t) = \kappa(\boldsymbol{E}(\boldsymbol{r},t))\,\boldsymbol{E}(\boldsymbol{r},t),$$

a conditional equation for the scalar potential φ under the EQS assumption follows,

$$\operatorname{div}(\kappa(\varphi)\operatorname{grad}\varphi) + \operatorname{div}\left(\varepsilon \operatorname{grad}\frac{\partial}{\partial t}\varphi(t)\right) = 0.$$

This equation is geometrically discretized using the Finite-Element method, which results in a nonlinear system of stiff ordinary differential equations in the time domain, which can be represented by the matrix equation

$$\mathbf{K}_{\kappa}(\Phi)\,\Phi + \mathbf{B}_{\varepsilon}\frac{\mathrm{d}}{\mathrm{d}t}\Phi = 0, \tag{1}$$

where the matrices **K** and **B** are finite-element stiffness matrices of the electrical conductivity and the electric permittivity, respectively. Moreover, Φ denotes the vector of the discrete scalar potential and $\Phi(t_0) = \Phi_0$ its initial value at time t_0.

3 Adaptive time integration

3.1 Runge-Kutta methods

In order to solve the resulting initial boundary value problem, time discretization based on Singly-Diagonal-Implicit-Runge-Kutta Methods (SDIRK) is performed (Clemens et al., 2004a; Lang, 2001). SDIRK methods are stiffly-accurate one-step methods with s internal stages which discretize equation (1) to systems of equations of the type

$$\left(\mathbf{K}_{\kappa}\left(\Phi_i^{(n)}\right) + \frac{1}{a_{ii}\Delta t_n}\mathbf{B}_{\varepsilon}\right)\Phi_i^{(n)} = \frac{1}{a_{ii}\Delta t_n}\mathbf{B}_{\varepsilon}\left(\Phi^{(n)} + \sum_{j=1}^{i-1} a_{ij}\Phi_j^{\prime(n)}\right),$$

with stage derivative values

$$\Phi_i^{\prime(n)} = \frac{\left(\Phi_i^{(n)} - \Phi^{(n)} - \sum_{j=1}^{i-1} a_{ij}\Phi_j^{\prime(n)}\right)}{a_{ii}}$$

and $\Phi_s^{(n)} = \Phi^{(n+1)}$. The time step length is denoted by Δt while a_{ij} denotes coefficients of the specific SDIRK method. In addition to the solution vector $\Phi^{(n+1)}$ at the time instant t_{n+1}, embedded SDIRK methods provide a second solution vector $\bar{\Phi}^{(n+1)}$. Both solution vectors differ since they belong to different orders of convergence p for $\Phi^{(n+1)}$ and $\bar{p}$ for $\bar{\Phi}^{(n+1)}$, respectively. Here, an SDIRK method with four internal stages and the orders of convergence $p=3$ and $\bar{p}=2$ is used.

3.2 Adaptive time step selection

The local truncation error of a time step can be estimated by

$$\left|\left|\Phi^{(n+1)}\right|\right|_{\mathrm{err}} = \sqrt{\frac{\left|\left|\Phi^{(n+1)} - \bar{\Phi}^{(n+1)}\right|\right|_{\infty}^{2}}{\left|\left|\Phi^{(n+1)}\right|\right|_{\infty}^{2} + \delta}},$$

with δ as an absolute error tolerance computed by $\delta = \theta \cdot \max_{t \le t^{(n)}}\left(||\Phi(t)||_{\infty}^{2}\right)$, $\theta \in \left[10^{-3}, 10^{-2}\right]$. This error estimate can be used twice: First, one can decide whether the time step leads to an acceptable local truncation error. Then the computed time step solution is accepted. If the local truncation error is greater than a given tolerance, the time step is

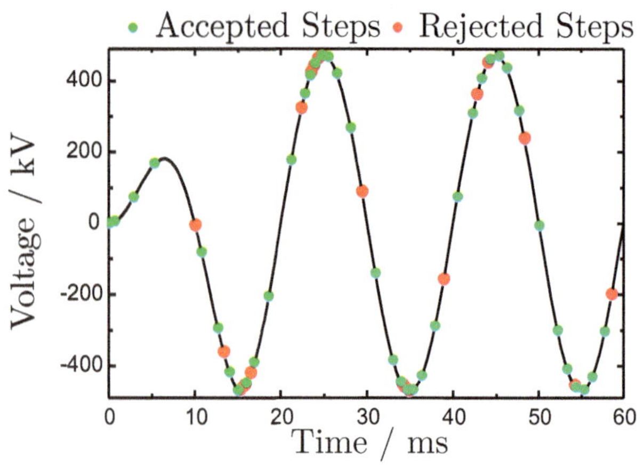

Fig. 1. Transient adaptive EQS simulation, displayed are accepted and rejected time steps.

rejected, i.e. a new time step is performed starting again from the previous time t_n. In both cases, the error estimate can be used to predict the next time step length

$$\Delta t_{n+1} = \nu\left(\frac{rtol}{\left|\left|\Phi^{(n+1)}\right|\right|_{\mathrm{err}}}\right)^{\frac{1}{p}}\Delta t_n,$$

using a closed loop model of an I-controller scheme (Gustafsson, 1994; Clemens et al., 2002) with $rtol$ as the required relative tolerance. Fig. 1 shows a transient simulation with adaptive choice of the times step length for a ramped sinusoidal excitation signal.

3.3 Solution of the algebraic systems of equations

In each stage of the time integration scheme a (non)linear algebraic system of equations of the form

$$\left(\mathbf{K}\left(\Phi_i^{(n)}\right) + \frac{1}{a_{ii}\Delta t^{(n)}}\mathbf{B}\right)\Phi_i^{(n)} = \mathbf{f}\left(\Phi^{(n)}, \ldots\right)$$

has to be solved. After a linearization by a Newton-Raphson method, the solution of a sequence of linear algebraic systems of equations $\mathbf{A}^l\mathbf{x}^l = \mathbf{b}^l$ with $\mathbf{A}^l = \mathbf{A}\left(\mathbf{x}^{l-1}\right)$, $l = 1, 2, \ldots$, and a start vector $\mathbf{x}_0$ is performed. In the following, the superscript l is omitted. As the system matrices $\mathbf{A}^l$ are symmetric, positive definite and sparse, a preconditioned conjugate-gradient method (PCG) is a suitable iterative solver for this type. Thereby, the PCG method generates iteration vectors

$$\mathbf{x}_i \in \mathbf{x}_0 + K_i\left\{\mathbf{M}^{-1}\mathbf{A}, \mathbf{M}^{-1}\mathbf{r}_0\right\}.$$

with $K_i\left\{\mathbf{M}^{-1}\mathbf{A}, \mathbf{M}^{-1}\mathbf{r}_0\right\}$ denoting the i-th Krylov space which is constructed iteratively within the PCG method. The preconditioning operator is denoted by **M** which is a spectral approximation of the system matrix **A**. Here an algebraic multigrid method (AMG) is used as preconditioner.

The speed of convergence of the PCG method depends on the condition number $\sigma(\mathbf{A})$ of the system matrix $\mathbf{A}$ at the one hand and on the choice of the start vector $\mathbf{x}_0$ at the other hand,

$$\|\mathbf{x}-\mathbf{x}_i\|_{\mathbf{A}} \leq 2\|\mathbf{x}-\mathbf{x}_0\|_{\mathbf{A}} \left(\frac{\sqrt{\sigma}-1}{\sqrt{\sigma}+1}\right)^{\sigma},$$

with $\|w\|_{\mathbf{A}} = \sqrt{w^{\mathrm{T}}\mathbf{A}w}$ and $\sigma(\mathbf{A})$ as the ratio of the highest and the lowest eigenvalue of the matrix $\mathbf{A}$, $\sigma(\mathbf{A}) := \lambda_n/\lambda_1$ (Golub and van Loan, 1996).

4 Subspace projection techniques

4.1 Subspace projection extrapolation

As shown above, a sequence of linear systems of equations has to be solved during a simulation. The generic way to do this is to treat these linear systems without any correlation. However, especially in the case of linear material behavior, the systems are very well correlated. Hence, the information gained in the solution process of one linear system can be reused in the following solution processes. In order to recycle this information, the proposed subspace projection extrapolation (SPE) approach in Clemens et al. (2004b) is to calculate the optimal linear combination of former solution vectors which is used to generate a start vector for the current solution process.

In this paper we recycle the information of the Krylov space which is spanned by the preconditioned residual vectors $\mathbf{z}_1, \ldots, \mathbf{z}_m$ constructed by the PCG method while solving the first linear system of equations during a transient simulation. This Krylov space contains approximations of eigenvectors of the matrix $\mathbf{A}$ associated with the smallest eigenvalues of $\mathbf{A}$. The corresponding matrix operator $\mathbf{Q}$ is built by arranging the Krylov basis vectors as columns vectors of $\mathbf{Q}$,

$$\mathbf{Q} = \{\mathbf{z}_1|\ldots|\mathbf{z}_m\} \in \mathbb{R}^{n\times m}.$$

Here, the application of a good preconditioner is crucial, because it is very important that the dimension of the matrix operator $\mathbf{Q}$ is low, otherwise the numerical costs of the following projection techniques may outweigh their profits. The matrix $\mathbf{Q}$ can then be used to solve the Galerkin-projected linear system

$$\mathbf{Q}^T\mathbf{A}\mathbf{Q}\mathbf{z} = \mathbf{Q}^T\mathbf{b}$$

instead of the originally system $\mathbf{A}\mathbf{x} = \mathbf{b}$ in order to compute a start vector

$$\mathbf{x}_0^{\mathrm{SPE}} := \mathbf{Q}\mathbf{z} = \mathbf{Q}\left(\mathbf{Q}^T\mathbf{A}\mathbf{Q}\right)^{-1}\mathbf{Q}^T\mathbf{b}$$

for the respective PCG iteration. Through this proceeding, the start vector $\mathbf{x}_0^{\mathrm{SPE}}$ already contains information associated with eigenvectors corresponding with the k smallest eigenvalues $\lambda_1, \ldots, \lambda_k$ of the system matrix $\mathbf{A}$. The PCG method will therefore converge with the effective condition number σ_{eff} instead of the condition number $\sigma(\mathbf{A})$ of the system matrix $\mathbf{A}$,

$$\sigma_{\mathrm{eff}}(\mathbf{A}) := \frac{\lambda_n}{\lambda_k} < \sigma(\mathbf{A}) := \frac{\lambda_n}{\lambda_1},$$

what can be understood as an implicit deflation of the PCG method.

4.2 Augmented PCG method

The solution processes can be accelerated additionally by not only generating a start vector reusing subspace information, but furthermore by modifying the PCG algorithm itself (Gosselet and Rey, 2002; Saad et al., 2000) in order to obtain residual vectors $\mathbf{r}_i$ which are orthogonal to the Krylov space spanned by the column vectors of $\mathbf{Q}$,

$$\mathbf{Q}^T\mathbf{r}_i = 0. \tag{2}$$

This can be achieved by augmenting the i-th Krylov space by the range of $\mathbf{Q}$ resulting in iteration vectors

$$\mathbf{x}_i \in \mathbf{x}_0 + K_i\left\{\mathbf{M}^{-1}\mathbf{A}, \mathbf{M}^{-1}\mathbf{r}_0\right\} \oplus \mathrm{Range}\{\mathbf{Q}\}.$$

Thus, using the augmented PCG method,

$$\mathbf{r}_i \perp K_i\left\{\mathbf{M}^{-1}\mathbf{A}, \mathbf{M}^{-1}\mathbf{r}_0\right\} \oplus \mathrm{Range}\{\mathbf{Q}\}$$

is valid. Defining the projector

$$\mathbf{P}_Q = \mathbf{I} - \mathbf{Q}(\mathbf{Q}^T\mathbf{A}\mathbf{Q})^{-1}\mathbf{Q}^T\mathbf{A}$$

which projects vectors onto the A-orthogonal complement of the space spanned by the column vectors of $\mathbf{Q}$, $\mathbf{Q}^T\mathbf{r}_i = 0$ is enforced if $\mathbf{P}_Q$ is applied to the preconditioned residual vectors in the Augmented PCG (AugPCG) iteration.

4.3 Combined algorithm

In Algorithm 1, the subspace projection extrapolation as well as the augmented PCG method are combined with $atol_{\mathrm{rhs}} = \max_{j=1,\ldots,l-1}\|\mathbf{b}^j\|$ as the maximum absolute value of the right-hand side vectors for all previous systems with $\Theta \in [10^{-3}, 10^{-2}]$ and an arbitrary start vector $\mathbf{x}_{00}$.

5 Numerical results

The first technical example shown in Fig. 2 is a high-voltage bushing which mainly consists of metal, ceramic insulators and insulation oil with a nonlinear electrical conducitvity characertistic $\kappa(\boldsymbol{E})$. The simulation with solely linear material behavior yields the result that the SPE-PCG with 13 column vectors in the matrix $\mathbf{Q}$ is more effective than the AugPCG method. This is due to the fact that the system matrices in the stages of the time integrator are nearly invariant.

As can be seen in Fig. 2, the plot of the scalar potential of the model with solely linear material behavior differs clearly

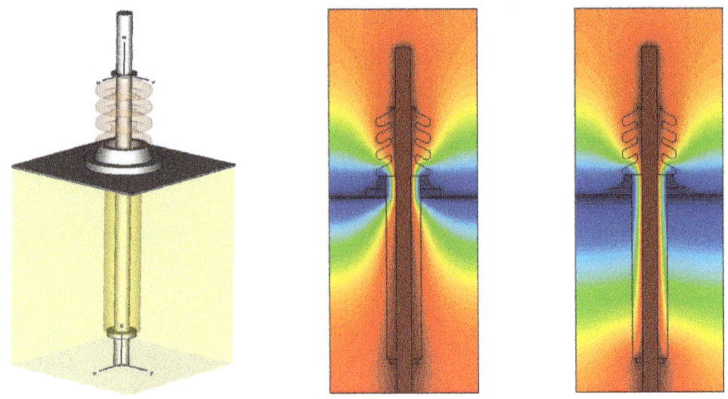

Fig. 2. HV Bushing: Geometry, plot of the scalar potential with linear and nonlinear electric conductivity (from left).

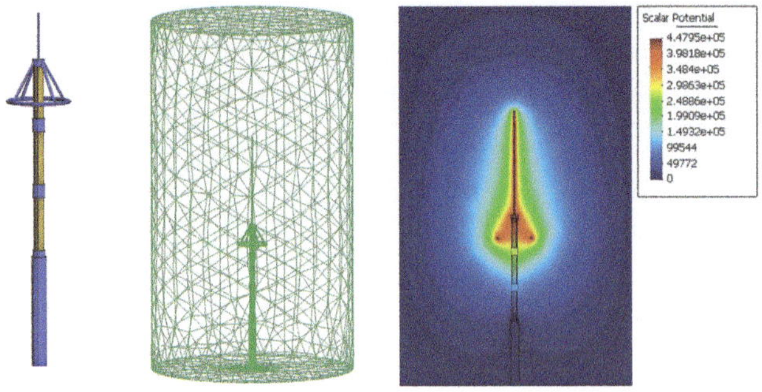

Fig. 3. HV surge arrester: Geometry, mesh and plot of the scalar potential (from left).

Algorithm 1 Combined Algorithm using subspace projection.

if SPE-PCG **then**
  $\mathbf{x}_0 = \mathbf{Q}(\mathbf{Q}^T\mathbf{A}\mathbf{Q})^{-1}\mathbf{Q}^T\mathbf{b}$
else if AugPCG **then**
  $\mathbf{x}_0 = \mathbf{Q}(\mathbf{Q}^T\mathbf{A}\mathbf{Q})^{-1}\mathbf{Q}^T\mathbf{b} + \mathbf{P}_Q\mathbf{x}_{00}$
end if
$\mathbf{r}_0 = \mathbf{b} - \mathbf{A}\mathbf{x}_0$
$r_{rel,0} = ||\mathbf{r}_0|| \,/\, (||\mathbf{b}|| + \Theta\ atol_{\text{rhs}})$
while $r_{rel,i} > \text{tol}_{\text{PCG}}$ **do**
  if SPE-PCG **then**
    $\mathbf{z}_i = \mathbf{M}^{-1}\mathbf{r}_i$
  else if AugPCG **then**
    $\mathbf{z}_i = \mathbf{P}_Q\mathbf{M}^{-1}\mathbf{r}_i$
  end if
  if $i = 0$ **then**
    $\mathbf{p}_i = \mathbf{z}_i$
  else
    $\beta_i = \mathbf{r}_i^T\mathbf{z}_i \,/\, \mathbf{r}_{i-1}^T\mathbf{z}_{i-1}$
    $\mathbf{p}_i = \beta_i\mathbf{p}_{i-1} + \mathbf{z}_i$
  end if
  $\alpha_i = \mathbf{r}_i^T\mathbf{z}_i \,/\, \mathbf{p}_i^T\mathbf{A}\mathbf{p}_i$
  $\mathbf{x}_{i+1} = \mathbf{x}_i + \alpha_i\mathbf{p}_i$
  $\mathbf{r}_{i+1} = \mathbf{r}_i - \alpha_i\mathbf{A}\mathbf{p}_i$
  $r_{rel,i+1} = ||\mathbf{r}_i|| / (||\mathbf{b}|| + \Theta\ atol_{\text{rhs}})$
  $i = i + 1$
end while

Table 1. Number of iterations and computation time for the HV bushing

Model	Alg. Var.	PCG-Iter.	Sol. time
Linear	PCG	2725	380.9 s
53502 DoF	SPE-PCG	49	19.6 s
Non-Linear	PCG	2748	391.4s
53502 DoF	AugPCG	1199	336.5 s
Non-Linear	PCG	3344	1361.5 s
162113 DoF	AugPCG	1531	1214.7 s

Table 2. Number of iterations and computation time for the HV surge arrester.

Model	Alg. Var.	PCG-Iter.	Sol. time
Non-Linear	PCG	4372	345.6 s
41100 DoF	AugPCG	1037	243.4 s
Non-Linear	PCG	7590	1846.7 s
150590 DoF	AugPCG	227	742.3 s

from the plot of the model with nonlinear electrical conductivity characteristic for the insulation oil. In that case, the AugPCG method, in which the projection matrix $\mathbf{P}_Q$ is updated by the system matrix $\mathbf{A}$ for each linear system, is more effective, as the system matrices differ from each other due to the nonlinearity. The number of column vectors in $\mathbf{Q}$ is 13 in a model with 53502 DoF and 16 in a model with 162113 DoF. Unfortunately, although the number of PCG iterations is reduced obviously, the numerical costs of the projection leads to smaller benefits in the computation time needed to solve all linear systems in a transient simulation. The number of the PCG iterations and the related computation times for the high-voltage bushing are shown in Table 1.

The second technical application is a high-voltage surge arrester shown in Fig. 3, a structure mainly consisting of metal, ceramic insulators and varistor material with a nonlinear electrical conductivity characteristic.

As in the first example, due to the nonlinearity the greatest acceleration in terms of CPU time is achieved by the AugPCG method in which the projection matrix with 21 column vectors for a model with 41100 DoF and 44 column vectors for the model with 150590 DoF, respectively, is updated for each linear system. The number of the PCG iterations and the related computation times for the high-voltage surge arrester are shown in Table 2.

The presented simulations are carried out on an double processor Intel Xeon (2.8 GHz, 2 GB RAM) with a constant time step length to gain a better comparability of the results.

6 Conclusion

A three-dimensional finite-element implementation based on a discrete formulation of electro-quasistatic fields was presented. Adaptive time integration based on an embedded Singly-Diagonal-Runge-Kutta method was performed. The repeated solution processes of the linear systems within the time integrator were accelerated using implicit and explicit subspace projection techniques. Their efficiency was shown in the transient simulation of technical applications.

References

Clemens, M., Wilke, M., and Weiland, T.: 3-D Transient Eddy Current Simulations Using FI^2TD with Variable Time Step Size Selection Schemes, IEEE Trans. Magn., 38, 605–608, 2002.

Clemens, M., Wilke, M., Benderskaya, G., De Gersem, H., Koch, W., and Weiland, T.: Transient Electro-Quasi-Static Adaptive Simulation Schemes, IEEE Trans. Magn., 240, 1294–1297, 2004a.

Clemens, M., Wilke, M., Schuhmann, R., and Weiland, T.: Subspace Projection Extrapolation Scheme for Transient Field Simulations, IEEE Trans. Magn., 40, 934–937, 2004b.

Dirks, K. H.: Quasi-Stationary Fields for Microelectronic Applications, Electrical Engineering, 79, 145–155, 1996.

Golub, G. and van Loan, C.: Matrix Computations, Johns Hopkins University Press, London, 1996.

Gosselet, P. and Rey, C.: On a Selective Reuse of Krylov Subspaces in Newton-Krylov Approaches for Nonlinear Elasticity, in: Proc. of the 14th Intern. Conf. on Domain Decomposition Methods, edited by Herrera, I., Keyes, D. E., Widlund, O. B., and Yates, R., 419–426, 2002.

Gustafsson, K.: Control-Theoretic Techniques for Stepsize Selection in Implicit Runge-Kutta Methods, ACM Trans. Math. Software, 20, 496–517, 1994.

Lang, J.: Adaptive Multilevel Solution of Nonlinear Parabolic PDE Systems: Theory, Algorithm and Application, Springer, Berlin, Heidelberg, New York, 2001.

Saad, Y., Yeung, M., Erhel, J., and Guyomarch, F.: A Deflated Version of the Conjugate Gradient Algorithm, SIAM J. Sci. Comput., 21, 1909–1926, 2000.

36

Exponential Modelling for Mutual-Cohering of Subband Radar Data

U. Siart, S. Tejero, and J. Detlefsen

Institute for High-Frequency Engineering, Technische Universität, Arcisstr. 1, 80333 München, Germany

Abstract. Increasing resolution and accuracy is an important issue in almost any type of radar sensor application. However, both resolution and accuracy are strongly related to the available signal bandwidth and energy that can be used. Nowadays, often several sensors operating in different frequency bands become available on a sensor platform. It is an attractive goal to use the potential of advanced signal modelling and optimization procedures by making proper use of information stemming from different frequency bands at the RF signal level. An important prerequisite for optimal use of signal energy is coherence between all contributing sensors. Coherent multi-sensor platforms are greatly expensive and are thus not available in general. This paper presents an approach for accurately estimating object radar responses using subband measurements at different RF frequencies. An exponential model approach allows to compensate for the lack of mutual coherence between independently operating sensors. Mutual coherence is recovered from the *a-priori* information that both sensors have common scattering centers in view. Minimizing the total squared deviation between measured data and a full-range exponential signal model leads to more accurate pole angles and pole magnitudes compared to single-band optimization. The model parameters (range and magnitude of point scatterers) after this full-range optimization process are also more accurate than the parameters obtained from a commonly used super-resolution procedure (root-MUSIC) applied to the non-coherent subband data.

1 Introduction

It is well known that the fundamental limit for radar range resolution is the signal bandwidth, provided there is no additional limitation through noise. Classical radar signal theory (see Cook and Bernfeld, 1993, for a comprehensive overview) relates range resolution to the effective bandwidth of the radar transmit signal. Super-resolution methods like autoregressive modelling (AR) or eigenanalysis methods (e.g. MUSIC) can achieve better resolution compared to the classical limits. They are basically model-based parameter estimators that use prior knowledge about the signal. Their performance and properties have been exhaustively investigated in the past (Pesavento et al., 2000; Zhang, 1998; Rao and Hari, 1989; Marple, 1987) just to mention a few).

In recent times and even more in the future there will be sensor platforms with not only one but several radar sensors designed for various purposes. Automotive radar equipment for parking aid, adaptive cruise control and a variety of driver assistance and comfort functions is an example from the low-cost sector. Also nowadays there are remote sensing platforms that gather information with more than one radar system. This rises the question for making optimum use of information that comes from different radar sensors and in different frequency areas. A prerequisite for optimum processing of multi-band radar information would be mutual coherency between the sensors in order to utilize also mutual phase information. For expenditure reasons, individual sensors on a platform are most often mutually incoherent so this condition is normally not fulfilled.

The purpose of this work is to investigate signal processing methods to restore coherency between the data from mutually incoherent radar sensors. The approach is different to sensor fusion techniques. It attempts to merge radar sensor information directly on the raw radar data level while sensor fusion techniques apply to fully processed sensor data for the sake of a more convenient object classification or multi-static radar measurements (e.g. Meinecke et al., 2003; Schiementz et al., 2003; Thomopoulos and Okello, 1993).

An important assumption behind the published approach is that the contributing sensors bring information about the same scenery. In a first approach, it is further assumed that the common scenery is observed from the same direction by two sensors as depicted in Fig. 1. The sensors generally operate at different carrier frequencies and with different signal bandwidths.

It is further assumed that the radar scenery is well approximated by an ensemble of real or effective discrete scattering centers and that the same scatterers are active in both sensors' frequency regimes. As a consequence, the radar response is considered to be mainly a multi-path response including P paths with identical round trip times τ_p in both sensors. Let us also assume that the radar data is given in frequency

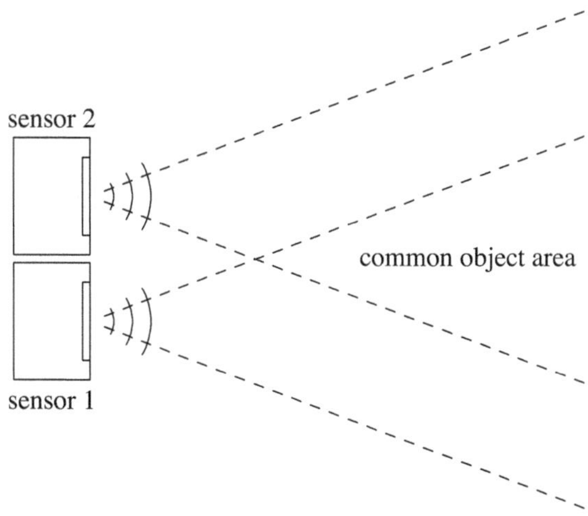

Fig. 1. Dual-band radar sensing of a common object area.

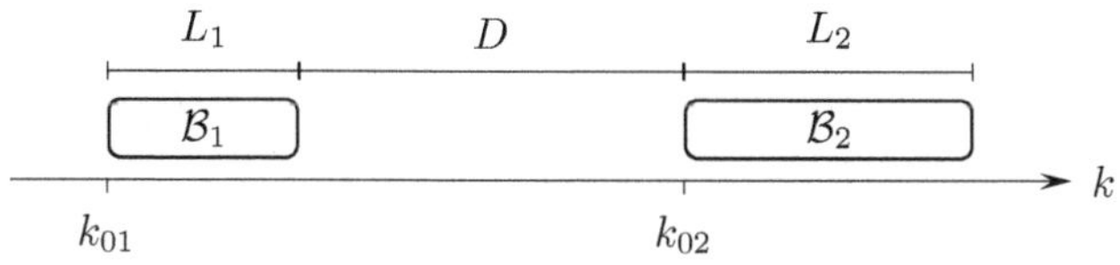

Fig. 2. Illustration of the parameters used to specify bandwidth and band gap.

domain. Both sensors deliver a discrete-frequency measure of the complex reflectivity, using the same angular frequency increment $\Delta\omega$. If the bandwidth of frequency band $\mathcal{B}_{1,2}$ is $B_{1,2}$, then the length of the data sequence is $L_{1,2}{=}B_{1,2}/\Delta\omega$, respectively. The distance in between bands $\mathcal{B}_1$ and $\mathcal{B}_2$ is also given in terms of angular frequency increments D. The start index of each band is $k_{01,2}{=}+-\omega_{01,2}/\Delta\omega$ where $\Delta\omega_{01,2}$ are the lower angular frequencies of each band. See Fig. 2 for illustration.

With these assumptions, the discrete-frequency complex reflectivity is

$$r[k] = \sum_{p=1}^{P} A_p \cdot e^{-j\Omega_p k} \tag{1}$$

where $\Omega_p{=}\Delta\omega{\cdot}\tau_p$ and $k{=}\omega/\Delta\omega$. If Eq. 1 is measured by a radar system no. n with angular frequency increment $\Delta\omega$, starting from the lower frequency ω_{0n}, and with bandwidth $B_n{=}L_n{\cdot}\Delta\omega$, then the noiseless data from this sensor is

$$r_n[k] = \sum_{p=1}^{P} A_p {z_p}^{k+k_{0n}} = \sum_{p=1}^{P} a_{pn} {z_p}^{k} \tag{2}$$

with $k = 0, 1, \ldots, L_n{-}1$. In this notation $z_p{=}e^{-j\Omega_p}$ and $a_{pn}{=}A_p e^{-j\omega_{0n}\tau_p}{=}A_p {z_p}^{k_{0n}}$. If the measurements are mutually coherent and Eq. 1 is valid in frequency bands $\mathcal{B}_1$ and $\mathcal{B}_2$, then necessarily

$$\frac{a_{p2}}{a_{p1}} = e^{-j\Omega_p(k_{02}-k_{01})} . \tag{3}$$

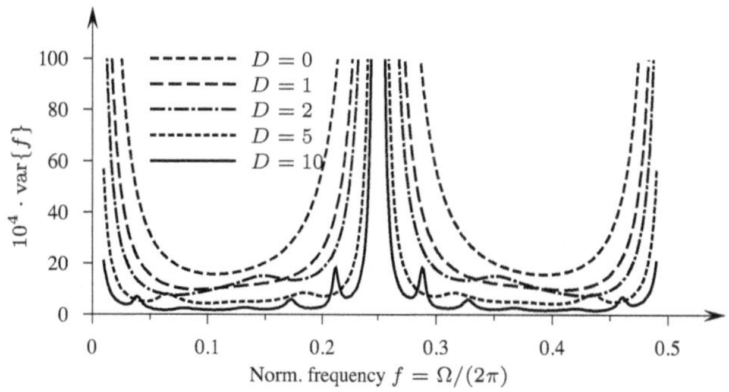

Fig. 3. Cramér-Rao lower bound for dual-band dual-frequency estimation.

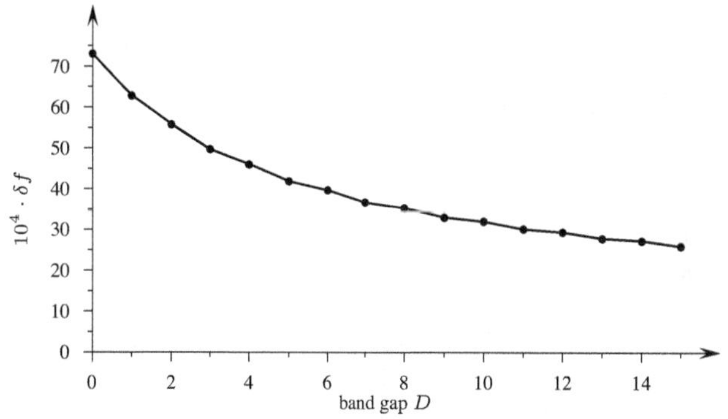

Fig. 4. Maximum achievable resolution versus band gap.

Equation 3 can be considered as the condition for mutual coherence between the data in $\mathcal{B}_1$ and $\mathcal{B}_2$.

2 Cramér-Rao lower bound

Estimating the range parameter of point scatterers observed by a radar system can be considered equivalent to estimating the frequency of a real or complex sinusoid in the frequency response of the radar scenery. With frequency modulated continuous wave (FMCW) or stepped-frequency radar this is task naturally appears. But the information delivered by any arbitrary impulse response measurement technique can in principal be processed in the frequency domain, where the round trip time of discrete echos corresponds to the frequencies of sinusoids. To investigate the maximum achievable resolution through dual-band estimation, the Cramér-Rao lower bound (CRLB) for frequency estimation of a sinusoid from dual-band data under the presence of a second sinusoid with equal magnitude is derived. It is assumed that the observed discrete data from bands $\mathcal{B}_{1,2}$ have the form

$$x_{1,2}[k] = s_{1,2}[k, \boldsymbol{\theta}] + w_{1,2}[k] \tag{4}$$

where the noiseless signals are

$$s_1[k, \boldsymbol{\theta}]{=}A\cos(2\pi f_1 k{+}\phi_1) + A\cos(2\pi f_2 k{+}\phi_2) \tag{5a}$$
$$s_2[k, \boldsymbol{\theta}]{=}A\cos(2\pi f_1 k{+}\phi_{12}){+}A\cos(2\pi f_2 k{+}\phi_{22}) \tag{5b}$$

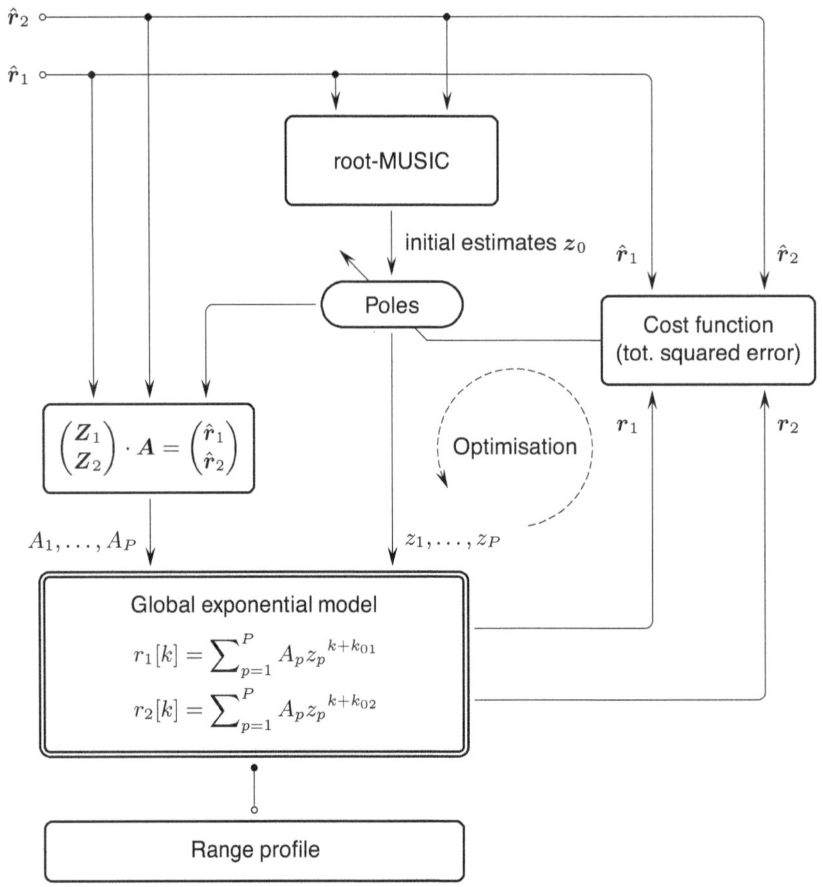

Fig. 5. Block diagram of the proposed global model fitting procedure.

and $w_{1,2}[k]$ is white Gaussian noise. Coherency between the bands implies that

$$\phi_{12} = \phi_1 - 2\pi f_1(L_1 + D) \tag{6a}$$
$$\phi_{22} = \phi_2 - 2\pi f_2(L_1 + D) \tag{6b}$$

and the parameter vector is $\boldsymbol{\theta}=[f_1\ f_2\ \phi_1\ \phi_2]^{\mathrm{T}}$. Assuming that the noise is white and Gaussian, the Fisher information matrix following Kay (1993) is

$$[\boldsymbol{I}(\boldsymbol{\theta})]_{ij} = \frac{1}{\sigma^2}\left(\sum_{\mathcal{B}_1} \frac{\partial s_1[k,\boldsymbol{\theta}]}{\partial \theta_i}\cdot\frac{\partial s_1[k,\boldsymbol{\theta}]}{\partial \theta_j} + \sum_{\mathcal{B}_2} \frac{\partial s_2[k,\boldsymbol{\theta}]}{\partial \theta_i}\cdot\frac{\partial s_2[k,\boldsymbol{\theta}]}{\partial \theta_j}\right). \tag{7}$$

From this, the CRLB for the i-th parameter is

$$\mathrm{var}(\theta_i) \geq [\boldsymbol{I}^{-1}(\boldsymbol{\theta})]_{ii}\,. \tag{8}$$

A plot of $[\boldsymbol{I}^{-1}(\boldsymbol{\theta})]_{ii}$ under the presumption that f_2=0.25 and ϕ_1=ϕ_2=0 is shown in Fig. 3. In this figure the distance D between the data windows varies while the window widths remain constant L_1=L_2=5. It is seen that estimation of frequency f_1 becomes inaccurate as f_1 approaches either DC or the Nyquist limit or if f_1 comes close to f_2. Furthermore it can be recognized that the increase of the CRLB nearby frequency f_2 is the slower, the wider the gap D.

Let us now define the minimum achievable resolution δf as the frequency difference $|f_2 - f_1|$ where the CRLB of f_1 exceeds the difference $|f_2 - f_1|$ that is to be resolved. Using this definition, the minimum resolvable frequency difference can be derived from Eq. 8 for any values of L_1, L_2, and D. As an example, the minimum achievable resolution (in this sense) is plotted in Fig. 4 versus the band gap D. In this figure, the length of the available data is L_1+L_2=20 and the signal-to-noise ratio is SNR=0 dB. The classical Rayleigh resolution limit Δ_{R} in this case is Δ_{R}=0.5/20=250·10^{-4}. It is seen that the resolution limit of an efficient estimator improves as the gap D between the frequency bands increases.

3 Model-based dual-band processing

Assuming that Eq. 1 is globally valid, then the signal model for the measured data $\hat{r}_{1,2}[k]$ in the subbands $\mathcal{B}_{1,2}$ is

$$r_1[k] = \sum_{p=1}^{P} A_p {z_p}^{k+k_{01}} \tag{9a}$$

$$r_2[k] = \sum_{p=1}^{P} A_p {z_p}^{k+k_{02}}\,. \tag{9b}$$

Table 1. Simulation parameters.

Quantity	*Value*
Length L_1 of subband 1	*32*
Length L_2 of subband 2	*32*
Number of complex sinusoids	*2*
Frequency difference	*variable*
Model order P	*one of 4/8*
Length D of interband gap	*one of 0/32/128*
Signal-to-noise ratio SNR	*variable*
no. of trials	*2 000*

This can be written in matrix notation

$$\boldsymbol{Z}\cdot\boldsymbol{A}=\begin{pmatrix}\boldsymbol{Z}_1\\ \boldsymbol{Z}_2\end{pmatrix}\cdot\boldsymbol{A}=\begin{pmatrix}\boldsymbol{r}_1\\ \boldsymbol{r}_2\end{pmatrix} \tag{10}$$

where $\boldsymbol{A}{=}[A_1,\ldots,A_P]^{\mathrm{T}}, \boldsymbol{r}_n{=}\big[r_n[0],\ldots,r_n[L_n-1]\big]^{\mathrm{T}}$ and

$$\boldsymbol{Z}_n=\begin{pmatrix} z_n{}^{k_{0n}+0} & z_2{}^{k_{0n}+0} & \cdots & z_P{}^{k_{0n}+0}\\ z_n{}^{k_{0n}+1} & z_2{}^{k_{0n}+1} & \cdots & z_P{}^{k_{0n}+1}\\ \vdots & \vdots & \ddots & \vdots\\ z_n{}^{k_{0n}+L_n-1} & z_2{}^{k_{0n}+L_n-1} & \cdots & z_P{}^{k_{0n}+L_n-1}\end{pmatrix}.$$

The purpose is to estimate the poles z_p under the boundary condition that the same values A_p apply to both subbands. Through $\tau_p{=}\Omega_p/\Delta\omega$ the pole angles Ω_p then yield the round trip times to be estimated. Generally, the amplitudes A_p may take any complex value and no prior knowledge is available about them. Given a set of poles z_p, optimum values A_p can be found by the least-squares solution of Eq. 10 when substituting $\boldsymbol{r}_{1,2}$ with the measured data $\hat{\boldsymbol{r}}_{1,2}$. With these parameters the signal model Eq. 9 is complete. As a measure for the modelling quality, the total square deviation

$$J=\sum_{\mathcal{B}_1}\left|\hat{r}_1[k]-r_1[k]\right|^2+\sum_{\mathcal{B}_2}\left|\hat{r}_2[k]-r_2[k]\right|^2 \tag{11}$$

between the signal model $\boldsymbol{r}_{1,2}$ and the measured data $\hat{\boldsymbol{r}}_{1,2}$ is used. This clarifies the exponential model fitting procedure under consideration. See Fig. 5 for a block diagram. A set of poles z_p forms the degrees of freedom while the corresponding amplitudes A_p follow from a least-squares comparison between P complex sinusoids with frequencies Ω_p and the measured data $\hat{\boldsymbol{r}}_{1,2}$. This global exponential model can be optimized iteratively by varying the pole angles.

Tuning the pole angles is a highly ambiguous problem, especially when the band gap D is large. Therefore it is required that the initial estimates for the pole angles are as accurate as possible. Super-resolution methods that are based on auto-covariance estimation are principally sufficient for processing mutually incoherent multi-band data. They do not employ mutual phase information even if the subband data were coherent, however. Phase information is introduced by the global signal model Eq. 1 that leads to the subband representation Eq. 2 and the phase condition Eq. 3.

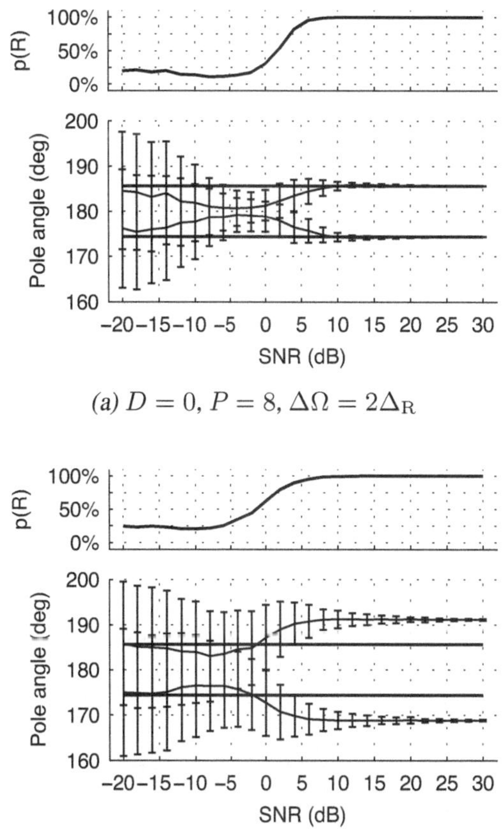

(a) $D=0,\ P=8,\ \Delta\Omega=2\Delta_{\mathrm{R}}$

(b) $D=32,\ P=8,\ \Delta\Omega=2\Delta_{\mathrm{R}}$

Fig. 6. Resolution properties of an autoregressive model (covariance method).

In Fig. 5 the input measurement data is represented by the row vectors $\boldsymbol{r}_{1,2}$. To achieve initial estimates for the poles z_p the root-MUSIC algorithm is chosen. The according complex amplitudes A_p are determined by the minimum-least-squares solution of Eq. 10 where the measured data is substituted in the row vector on the right side. In the following the total squared deviation between the signal model and the measured data is minimized through an iterative nonlinear least-squares minimization algorithm based on the Levenberg-Marquardt method. During this optimization $|z_p|{=}1$ so the pole angles are the actual optimization parameters. In general, the difference $k_{02}{-}k_{01}$ is another degree of freedom when the data $\mathcal{B}_1$ and $\mathcal{B}_2$ are not coherent. It is used to set up the matrix $\boldsymbol{Z}$ in Eq. 10. In the following simulations it is assumed to be known, however. It should be noted that without any knowledge about the phase angles $\arg\{A_p\}$ only the difference $k_{02}{-}k_{01}$ can be estimated. The values $k_{01,2}$ are ambiguous because multiplication of $\boldsymbol{A}$ with $[z_1{}^m,\ldots,z_P{}^m]^{\mathrm{T}}$ in Eq. 10 with an arbitrary value m can always be compensated by adding $-m$ to all exponents in $\boldsymbol{Z}$.

Once the model parameters are optimized, the signal

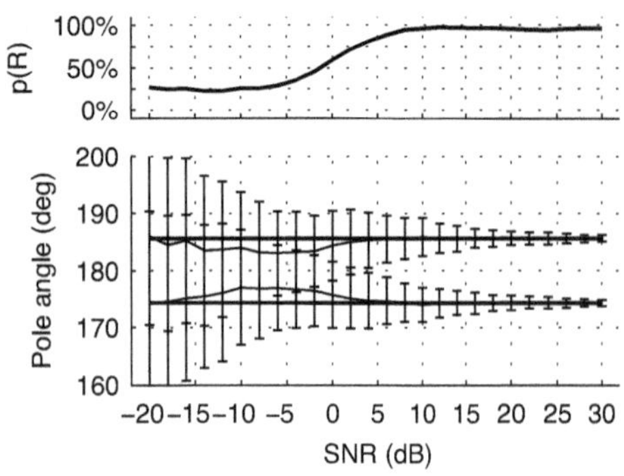

(a) $D = 0$, $P = 8$, $\Delta\Omega = 2\Delta_{\mathrm{R}}$

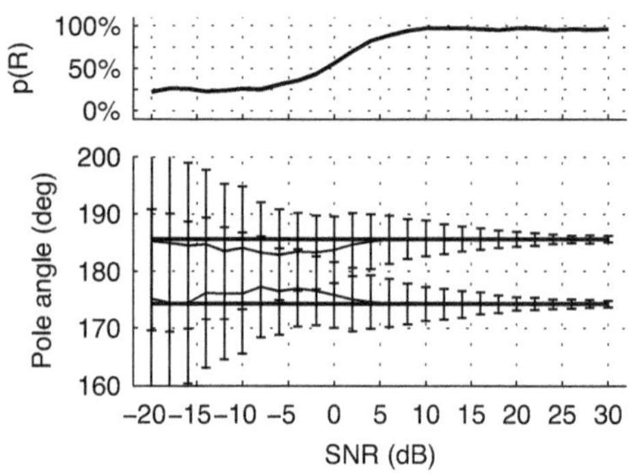

(b) $D = 32$, $P = 8$, $\Delta\Omega = 2\Delta_{\mathrm{R}}$

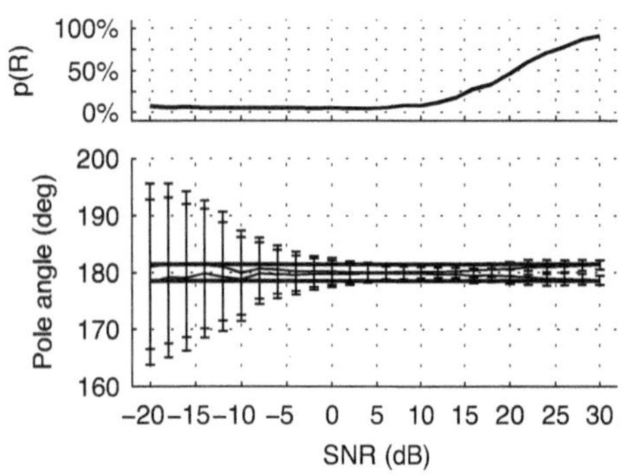

(c) $D = 0$, $P = 8$, $\Delta\Omega = 0.5\Delta_{\mathrm{R}}$

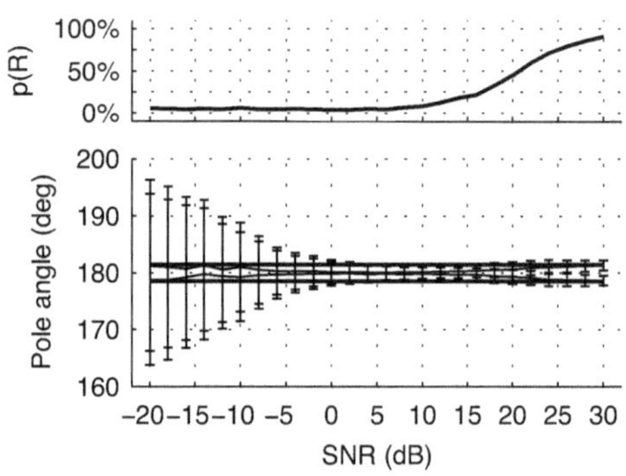

(d) $D = 32$, $P = 8$, $\Delta\Omega = 0.5\Delta_{\mathrm{R}}$

Fig. 7. Resolution properties of the root-MUSIC algorithm.

model can be used to compute the range profile. This can be done by either evaluating the pole angles directly or by using the model to compute an arbitrary number of interpolated and extrapolated samples along the frequency axis to achieve a closed data vector without any missing samples. Such a data series is then sufficient to provide a discrete Fourier transform (DFT) range profile. Although the signal model allows to extrapolate the measured data to any desired wide bandwidth, leading to arbitrary narrow pulses in the time/range domain, this bandwidth extrapolation yields neither additional information nor increased resolution. Radar pulses become narrow through bandwidth extrapolation but no additional targets or reflections will appear.

4 Simulation results

In the following simulations the mean and variance of pole angle estimates in the single-band and in the dual-band case are investigated. The length of the subband data is $L_1 = L_2 = 32$ and the signal $s[k]$ consists of two complex sinusoids with equal magnitude. Their normalized frequency difference is given in comparison to the Rayleigh resolution bound $\Delta_{\mathrm{R}} = 1/(L_1 + L_2)$. The mean and variance estimates are based on 2000 trials and they are plotted versus the signal-to-noise ratio of the input data. A survey on the simulation parameter setup is given in Table 1.

From the resulting pole angle estimates the probability of resolution $p(\mathrm{R})$ is computed. Since the total number of poles is given by the chosen model order, a criterion for proper detection and resolution of the targets based on the actual pole angles has to be introduced. The following procedure was used to plot $p(\mathrm{R})$ in Figs. 6 to 8: Provided that Ω_i and Ω_j are the true pole angles to represent the two complex sinusoids then first find z_i of which $\arg\{z_i\}$ is closest to Ω_i. Next find z_j of which $\arg\{z_j\}$ is closest to Ω_j. The sinusoids are considered to be resolved if, and only if $|i-j| = 1$.

Figure 6 shows the resolution properties and the quality of the pole angle estimates using an autoregressive (AR) model. This figure compares the single-band case (D=0, Fig. 6a to the dual-band case (D=32, Fig. 6b. The model order is P=8 and the frequency distance between the two sinusoids is twice the Rayleigh resolution bound. Resolution is achieved also in the dual-band situation but the estimates are biased, however. Therefore, AR estimation is not sufficient to provide accurate initial values for the exponential model optimizer although AR modelling can be applied to dual-band data.

As a comparison, the same approach using the root-MUSIC algorithm are shown in Fig. 7, additionally for the case where the frequency distance is only half the Rayleigh bound Δ_{R} (Figs. 7c, d. It is seen that the root-MUSIC result is not influenced by missing samples and that root-MUSIC

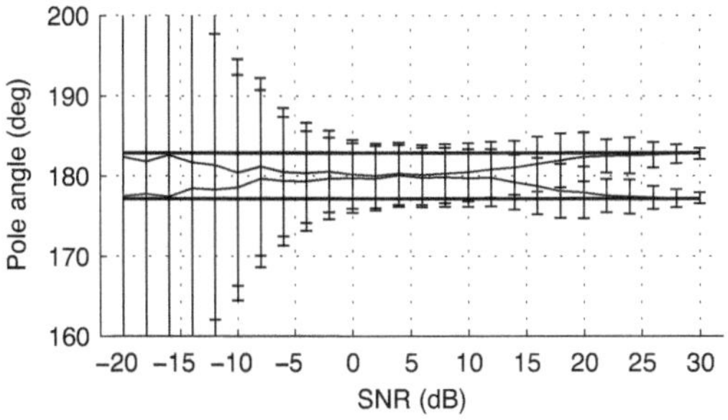

(a) $D = 0$, $P = 4$, $\Delta\Omega = \Delta_R$

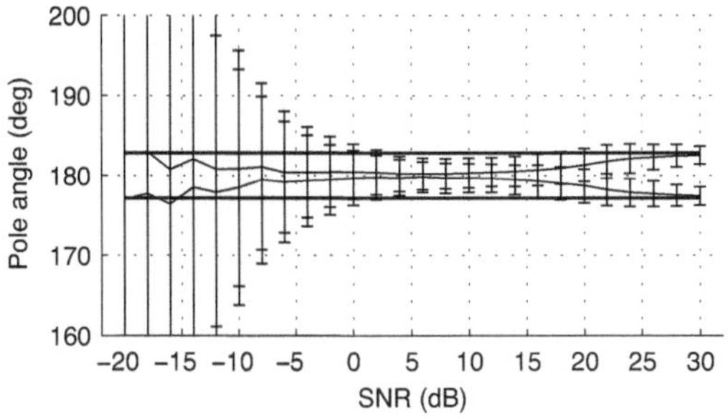

(b) $D = 128$, $P = 4$, $\Delta\Omega = \Delta_R$

Fig. 8. Variance of the pole angle estimates after the global multi-band model fit.

still achieves resolution below the Rayleigh bound, provided the signal-to-noise ratio is sufficiently high. For these properties, root-MUSIC was chosen as the initial pole angle estimator.

In Fig. 8 the statistical properties of the two poles closest to the true pole angles after the optimization procedure explained in section 3 are shown. The model order in this simulation is P=4 and the frequency distance is equal to the Rayleigh resolution bound. In Fig. 8a the optimization result using a single band (D=0) is shown in comparison to a dual-band optimized model in Fig. 8b. It is seen that in both cases the pole angle variance decreases considerably when SNR>22 dB. The variance also increases rapidly for SNR<0 dB. In the region 0 dB<SNR<22 dB the variance of the two poles closest to the true positions is considerably smaller when using dual-band data for the global model fitting procedure.

5 Conclusion

A model-based signal processing procedure for dual-band radar range estimation has been proposed. The algorithm basically recovers coherency between the distinct radar data by making use of the prior knowledge that identical scatterers form the radar range profile in both subbands. The Cramér-Rao lower bound indicates a smaller range estimation variance for the dual-band case compared to equal bandwidth single-band estimation, yielding a better resolution. A root-MUSIC algorithm is used to achieve initial range estimates from incoherent subband data. These initial estimates are further improved by making use of mutual phase information. This is achieved by fitting a global exponential model to the observed subband data and presuming that the exponential components are identical in both subbands. Simulation results show the improved accuracy of the coherent dual-band estimates compared to the result of conventional super-resolution methods.

References

Bresler, Y. and Macovski, A.: Exact Maximum Likelihood Parameter Estimation of Superimposed Exponential Signals in Noise, IEEE Trans. Acoust., Speech, Signal Processing, ASSP-34, 1081–1089, 1986.

Bruckstein, A. M., Shan, T.-J., and Kailath, T.: The Resolution of Overlapping Echos, IEEE Trans. Acoust., Speech, Signal Processing, ASSP-33, 1357–1367, 1985.

Chan, Y. T., Lavoie, J. M. M., and Plant, J. B.: A Parameter Estimation Approach to Estimation of Frequencies of Sinusoids, IEEE Trans. Acoust., Spcech, Signal Processing, ASSP-29, 214–219, 1981.

Cook, C. E. and Bernfeld, M.: Radar Signals, An Introduction to Theory and Applications, Artech House, Boston, 1993.

Gu, H.: Estimating the Number of Signals and Signal Resolution, IEEE Trans. Signal Processing, SP-46, 2267–2270, 1998.

Kay, S. M.: Estimation Theory, vol. 1 of Fundamentals of Statistical Signal Processing, Prentice Hall, Upper Saddle River, 1993.

Kumaresan, R. and Tufts, D. W.: Estimating the Parameters of Exponentially Damped Sinusoids and Pole-Zero Modeling in Noise, IEEE Trans. Acoust., Speech, Signal Processing, ASSP-30, 833–840, 1982.

Liu, S. and Xiang, J.: Novel Method For Super-Resolution in Radar Range Domain, IEE Proc. Radar, Sonar and Navigation, Vol. 146, 40–44, 1999.

Marple, S. L.: Digital Spectral Analysis, Prentice Hall, Englewood Cliffs, 1987.

Meinecke, M.-M., Obojski, M. A., M., T., Dörfler, R., Marchal, P., Letellier, L., Dariu, G., and Morris, R.: Approach for Protection of Vulnerable Road Users Using Sensor Fusion Techniques, in: Proc. Int. Radar Symposium (IRS), DGON, Dresden, 125–130, 2003.

Pesavento, M., Gershman, A. B., and Haardt, M.: Unitary Root-MUSIC with a Real-Valued Eigendecomposition: A Theoretical and Experimental Performance Study, IEEE Trans. Signal Processing, SP-48, 1306–1314, 2000.

Rao, B. D. and Hari, K. V. S.: Performance Analysis of Root-MUSIC, IEEE Trans. Acoust., Speech, Signal Processing, ASSP-37, 1939–1949, 1989.

Schiementz, M., Fölster, F., and Rohling, H.: Angle Estimation Techniques for different 24 GHz Radar Networks, in: Proc. Int. Radar Symposium (IRS), DGON, Dresden, 405–410, 2003.

Thomopoulos, S. C. A. and Okello, N. N.: Design and experimental validation of a multiband radar data fusion system, in: Proc. SPIE, Boston, MA, 26–36, 1993.

Zhang, Q. T.: A Statistical Resolution Theory of the AR Method of Spectral Analysis, IEEE Trans. Signal Processing, SP-46, 2757–2766, 1998.

37

Accelerating the numerical computation of indirect lightning effects by means of vector fitting

J. Anatzki and F. Gronwald

Institute of Electromagnetic Theory, Hamburg University of Technology, Harburger Schloss Str. 20, 21079 Hamburg, Germany

Abstract. In the context of numerical computation of indirect lightning effects it is customary to use volume-discretizing methods in time domain, such as the Finite Difference Time Domain (FDTD) method, the Finite Integration Technique (FIT), or the Transmission Line Matrix (TLM) method. If standard lightning electromagnetic pulses (LEMPs) of tenths of microseconds duration are used as excitations, these methods require long computation times, as implied by the Courant criterion. It is proposed to use shorter pulse forms and to compare the transfer functions obtained by different pulse durations by means of macromodels that are obtained from the vector fitting method. Numerical computation of lightning related transfer functions of a canonical structure indicate that the duration of the exciting pulse can typically be shortened by at least one order of magnitude if compared to a standard pulse.

1 Introduction

Modern computer-based computation techniques make it possible to simulate lightning related effects on aircraft structures (Apra et al., 2008). These techniques are defined either in time or frequency domain. In case of time domain computations long computation times usually are required. This is a consequence of the Courant criterion which relates the minimum discretisation length of the geometric model to the maximum allowed time step (Hoffman, 2001). Typically, if geometric details in the range of millimeters need to be resolved, the maximum allowed time step will be of the order of picoseconds. Then the simulation of the lightning strike over its whole duration of several tenths of microseconds usually is a time consuming task.

In this contribution an adaptive macromodelling technique is introduced which is based on the idea to determine the LEMP transfer function of a system by the use of excitation pulses that are shorter than standard lightning pulses. This method is based on an adaptive stopping criterion which has been used in the context of time domain characterization of microwave components (Deschrijver et al., 2009). In the present case of lightning analysis, several simulations with comparatively short excitation pulses have to be performed. The resulting transfer functions are characterized by a small number of parameters which are obtained from the vector fitting procedure and represent macromodels of the system (Gustavsen and Semlyen, 1999; Gustavsen, 2006). These macromodels can be used to quantitatively compare the transfer functions that are obtained by means of the different excitation pulses.

In principle, transfer functions calculated from different excitations should be the same if they relate to the same system. However, due to numerical inaccuracies of the time domain data and rounding errors the use of short pulses is limited. On the other hand, the use of comparable long excitation pulses will result in long computation times. Therefore a compromise between these two constraints needs to be found. To this end, comparatively short pulses are used at the beginning of an adaptive procedure. The duration of the first pulse can be chosen to be about a hundredth of the duration of a regular LEMP pulse, yielding a first transfer function with an associated macromodel. Then the duration of the excitation pulses is successively increased, yielding further transfer functions with associated macromodels. The differences between the parameters of different macromodels are observed until they fall below a predefined limit and an approximate convergence has been achieved, thus finally yielding an acceptable transfer function.

In the following Sect. 2, the methodology of the proposed method is described in more detail. In Sect. 3, a numerical example illustrates the method by means of a canonical tank model and Sect. 4 provides the conclusion.

2 Methodology

In lightning analysis, LEMP transfer functions are the main quantities of interest. They relate a lightning current $i_{\text{in}}(t)$ as relevant input variable to an observable which often is given by an induced voltage $v_{\text{out}}(t)$. In case of numerical lightning analyis in time domain an observable, such as $v_{\text{out}}(t)$, is calculated from the excitation $i_{\text{in}}(t)$ for a number of discrete points in time. Then a discrete transfer function $H_D(k)$ is obtained by Fourier or Laplace transformation according to

$$H_D(k) = \frac{V_{\text{out}}(k)}{I_{\text{in}}(k)} = \frac{\mathfrak{F}\{v_{\text{out}}(m)\}}{\mathfrak{F}\{i_{\text{in}}(m)\}}, \quad (1)$$

where k and m denote discrete indices in frequency and time domain, respectively. The vector fitting method allows to approximately fit a rational function to the discrete transfer function (Gustavsen and Semlyen, 1999; Gustavsen, 2006). This rational function can be written as a residue-pole expansion of the form

$$H_R(s) = \sum_{j=1}^{N} \frac{r_j}{s - p_j} + d + s \cdot e. \quad (2)$$

The advantage of $H_R(s)$ is that, as a result, in the context of lightning analysis LEMP transfer functions usually are characterized by only a small number of parameters r_j, p_j, d, and e. As will be seen below, realistic LEMP transfer functions can already be represented by a first order approximation with $N = 1$.

The proposed method can now be illustrated by the block diagram which is shown in Fig. 1. Starting from a comparatively short excitation pulse, the desired observable is obtained from a numerical time-domain calculation. This allows to calculate the desired transfer function $H_D(k)$ and a corresponding macromodel which is expressed by $H_R(s)$. A problem that is associated to the use of short pulses is that the discrete Fourier transform of such a signal will tend to have a sparse set of samples in the low frequency range. Due to dominant quasistationary contributions of typical lightning currents, the main frequency range of interest is actually located within this sparsely populated range of low frequencies. However, the sparsity makes the determination of a related transfer function susceptible to numerical errors. Therefore it is necessary to check the stability of results by using longer excitation pulses. It is then possible to compare the transfer functions $H_R(s)$ that are obtained by excitation pulses of different lengths. If an increase of pulse duration does not lead to a significant change of $H_R(s)$ it is assumed that the desired transfer function is obtained with sufficient accuracy.

To make the notion of a short excitation pulse more precise, in Fig. 2 a standard double exponential lightning pulse of the form

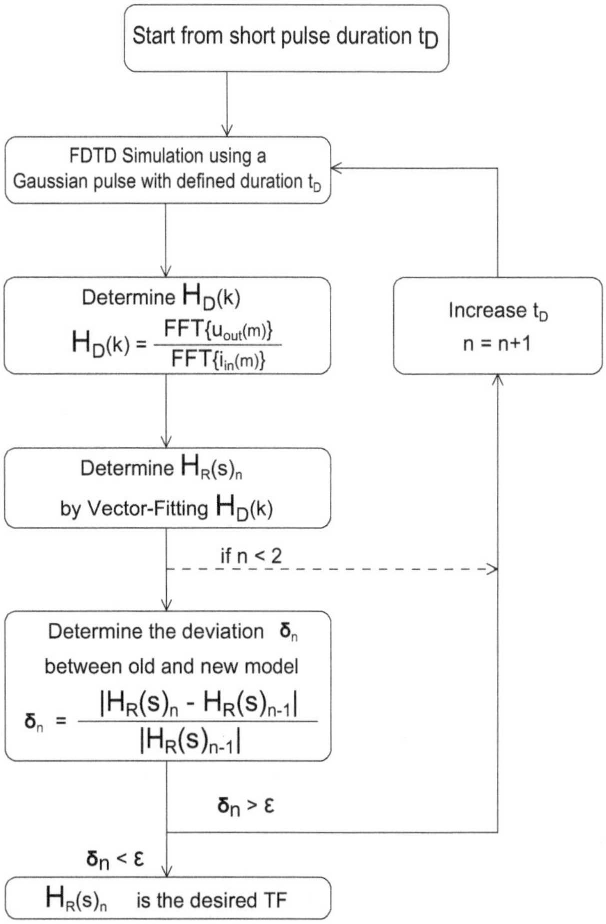

Fig. 1. Block diagram of the proposed adaptive macromodelling routine.

$$i(t) = I_0(e^{-\alpha t} - e^{-\beta t}) \quad (3)$$

is contrasted to a unipolar Gaussian pulse of the form

$$i(t) = I_1 \cdot e^{-\frac{(t-t_0)^2}{2 \cdot T^2}}. \quad (4)$$

The parameters of the pulses are adjusted such that they have a similar rise time but the Gaussian pulse decays much faster if compared to the double exponential pulse. The amplitude spectrum of both pulses is shown in Fig. 3. For the double exponential pulse, the amplitude spectrum is well-known and characterized by dominant low-frequency contributions (Lee, 1995). The amplitude spectrum of the unipolar Gaussian pulse is characterized by dominant low frequency contributions as well and thus the Gaussian pulse represents a suitable excitation to obtain LEMP transfer functions from time-domain analysis.

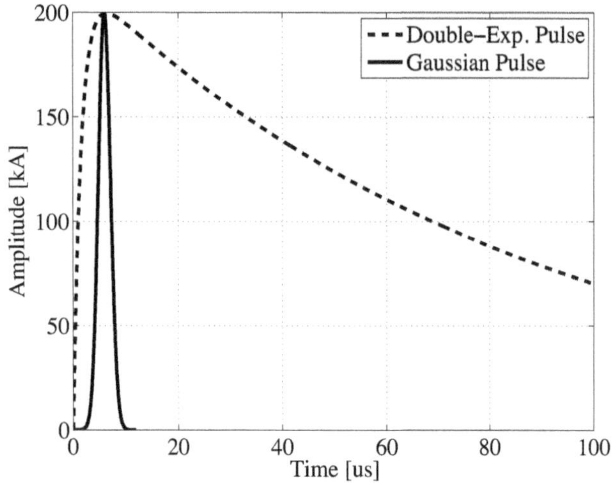

Fig. 2. Graphical comparison between a double exponential pulse and a Gaussian pulse. Both pulses have the same maximum value and are characterized by a similar rise rime. The double exponential pulse is a standard pulse, defined within the standard (SAE 5412A, 2005), where the time from beginning to peak value 200 kA is 6.4 us, the time from beginning to decay to 50 percent peak value, that is 100 kA, is 69 us.

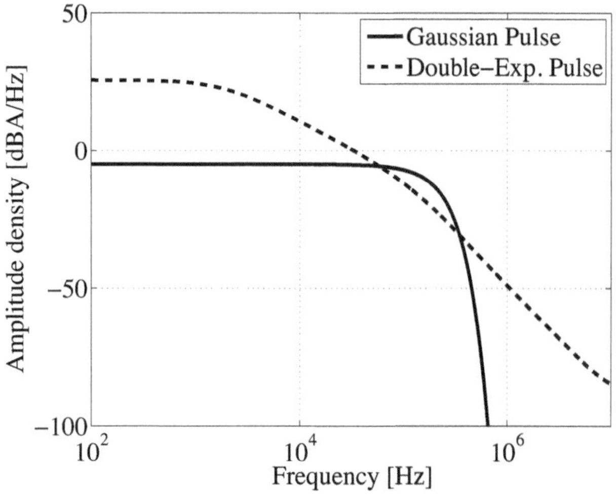

Fig. 3. Amplitude spectrum of both double exponential pulse and Gaussian pulse, as displayed in Fig. 2.

The duration of a Gaussian excitation is determined by the parameter T that is introduced in Eq. (4). It is inversely proportional to the frequency F_{max} which is defined as the frequency where the amplitude spectrum of the Gaussian pulse has decayed by -20 dB if compared to its maximum value,

$$T \sim \frac{1}{F_{\mathrm{max}}}. \tag{5}$$

Therefore a longer pulse duration implies a smaller value for F_{max} and vice versa. This circumstance is illustrated in Fig. 4 where Gaussian pulses corresponding to $F_{\mathrm{max}} =$

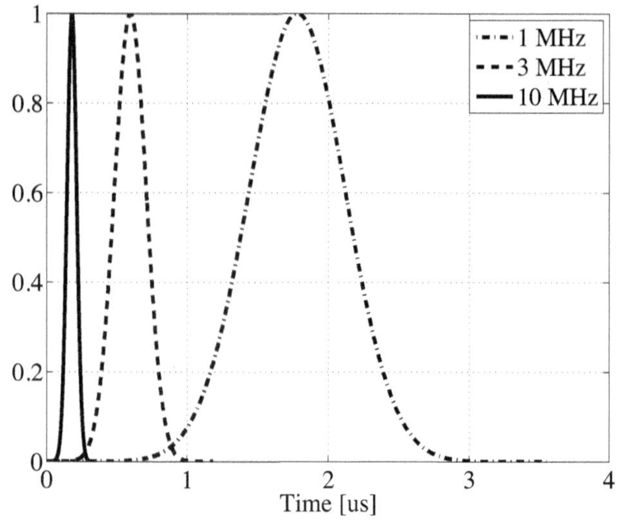

Fig. 4. Gaussian pulses corresponding to $F_{\mathrm{max}} = 1$ MHz, 3 MHz, and 10 MHz. A larger value of F_{max} implies a shorter pulse duration.

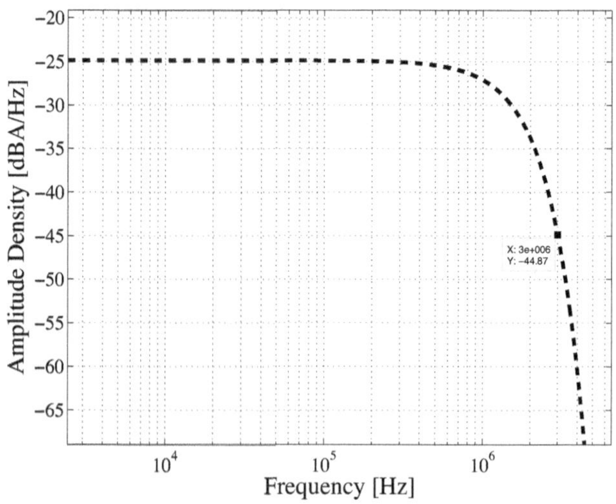

Fig. 5. Amplitude spectrum of a Gaussian pulse with $F_{\mathrm{max}} =$ 3 MHz. At 3 MHz its value has decayed by 20 dB if compared to the maximum value.

1 MHz, 3 MHz, and 10 MHz are displayed. Additionally, the amplitude spectrum of the Gaussian pulse with $F_{\mathrm{max}} =$ 3 MHz is shown in Fig. 5.

It is suggested to use Gaussian excitations of different parameters F_{max}, corresponding to different time durations, within the macromodelling routine of Fig. 1. In the following section this routine will be illustrated by a numerical example.

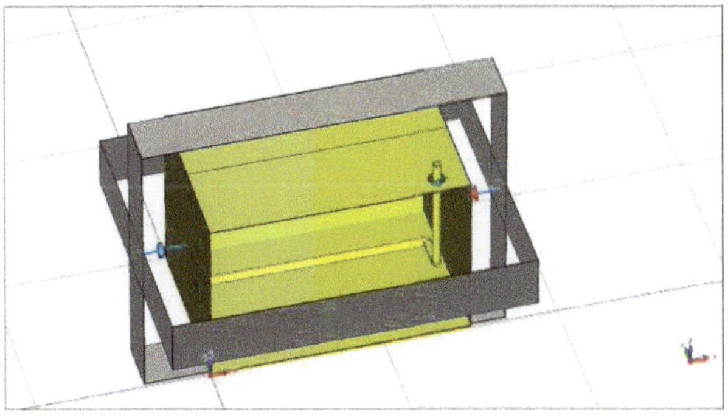

Fig. 6. Canonical tank geometry with interior tank pipe. For better visibility the front side of the tank is removed. The outer metallic frame serves as a return conductor.

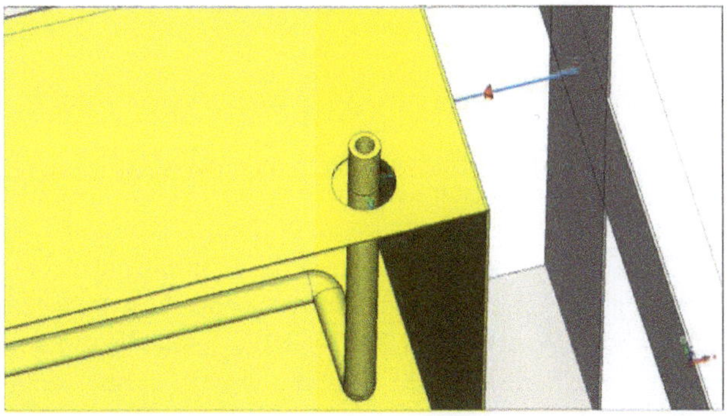

Fig. 7. Enlarged view of tank pipe and circular exit hole. The arrows between tank pipe and boundary of the circular hole indicate voltage monitors which record the voltage $v_{\text{out}}(t)$ during the numerical simulation.

3 Numerical example: LEMP transfer function of a canonical tank model

As a specific example a canonical tank model is considered. The tank model consists of a rectangular box which contains a fuel pipe. The fuel pipe is electrically bonded to one side of the interior of the tank and leaves the tank through a circular hole as shown in Fig. 6. The tank is subject to a lightning current source $i_{\text{in}}(t)$ and the induced voltage $v_{\text{out}}(t)$ between the fuel pipe and the boundary of the circular hole is the observable of interest, compare Fig. 7 . Then a LEMP transfer function is defined according to Eq. (1).

The tank model has been created within the CST Microwave Studio software which also is used to run the simulation model and to generate the time domain data $v_{\text{out}}(t)$ from the input variable $i_{\text{in}}(t)$ by means of the Finite Integration Technique (Clemens and Weiland, 2001). The result of a sample calculation is shown in Figs. 8 and 9. A mainly inductive response is observed, where voltage and current approximately are related by $u(t) = L\frac{di(t)}{dt}$. This observation has been discussed in more detail in (Gronwald, 2010). As

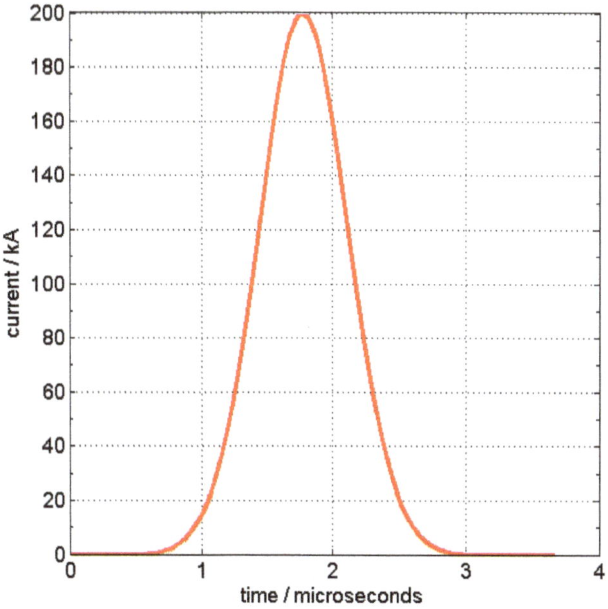

Fig. 8. Sample Gaussian current excitation $i_{\text{in}}(t)$.

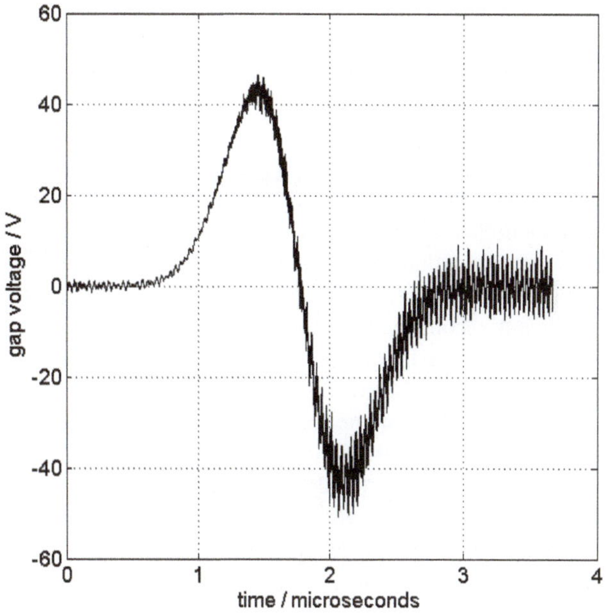

Fig. 9. Induced voltage $v_{\text{out}}(t)$ which is due to the current excitation of Fig. 8.

a general rule, LEMP transfer functions of aircraft structures are mainly characterized by inductive and resistive contributions. This is due to the quasistationary nature of the lightning current which makes radiation contributions negligible. Also capacitive effects can be neglected as long as the metallic parts of aircraft structures are properly bonded (Fisher et al., 1990). Therefore, in the present case, it is meaningful to consider the transfer admittance

$$Y_R(s) = \frac{1}{R+s\cdot L} = \frac{1/L}{s+R/L} \triangleq H_R(s)\Big|_{N=1} = \frac{r}{s-p} \qquad (6)$$

as an appropriate first order macromodel for the (inverse) transfer function of the canonical tank model. For details on macromodels of elementary networks it is referred to (Antonini, 2003).

To calculate the parameters L and R which, according to (6), determine the LEMP transfer admittance $Y_R(s)$, it is first necessary to perform a time domain calculation for a given excitation $i_{\text{in}}(t)$ to compute $v_{\text{out}}(t)$. Then the corresponding discrete transfer function $H_D(k)$ can be obtained from (1). Next, the inverse of this transfer function is fitted by means of the vector fitting method to the first order macromodel $Y_R(s)$ which is given by (6). This yields the pole p and residue r which both determine the parameters L and R.

Following the procedure described in the previous paragraph, macromodels for the LEMP transfer function of the canonical tank model have been obtained for nine different Gaussian excitations. The results are given in Tab. 1. The chosen pulse durations vary by a factor of 80, that is, the computation times to compute v_{out} vary by a factor of about 80 as well. The computations were performed on a single PC with 3.0 GHz CPU, 4 GB RAM, using the CST Microwave Studio 2010 software on a 32-bit Windows platform.

In order to judge the resulting parameters L and R one should note that the material of the tank model is modelled as highly conducting aluminium ($\sigma = 3.72 \cdot 10^7$ S/m). Therefore the voltage response is dominantly inductive, such that the resistive response is somewhat hidden and does not significantly contribute to the voltage response. To judge the quality of the fitted macromodels the relative root mean square error is included in Tab. 1 which quantifies the deviation between the numerically calculated transfer function $H_D^{-1}(k)$ and the fitted first order LEMP transfer admittance $Y_R(s)$. It is seen that this deviation is small, thus justifying the assumption of a simple first order macromodel. Towards shorter pulse durations the fit error increases. This indicates that the possibility to shorten the pulse duration is limited due to numerical inaccuracies that accompany the computation of the Fourier transformation in order to obtain $H_D^{-1}(k)$. However, the fit error increases towards longer pulse durations as well. To explain this observation it is noted that for long computation times numerical instabilities tend to build up, at least for the present numerical model which only contains minor losses. The smallest fit error occurs for the Gaussian pulse of time duration 1.83 us.

For further validation, it is of interest to compare the results that were obtained from the Gaussian excitations to an analogous calculation using a standard double exponential pulse as displayed in Fig. 2. However, as it turns out, the long pulse duration leads to strong oscillations that are interpreted as numerical instabilities and make it not feasible to

Table 1. Resulting macromodels for various Gaussian excitation pulses of different durations.

$F_{\max}$ MHz	Duration us	L nH	R uΩ	RMSE 10^{-6}	Solver Time sec
8.0	0.46	0.5216	2.512	24.79	535
6.0	0.61	0.5214	2.982	15.09	716
4.0	0.91	0.5216	2.226	6.072	1073
2.0	1.83	0.5220	1.633	1.211	2114
1.0	3.66	0.5225	0.907	3.605	4279
0.8	4.58	0.5227	0.765	5.584	5328
0.6	6.11	0.5230	0.619	8.732	7015
0.3	12.2	0.5242	0.334	31.64	14077
0.1	36.6	0.5262	0.251	116.6	42537

extract a meaningful LEMP transfer function. As an alternative, an independent calculation was performed using the Method of Moment code CONCEPT-II which is defined in frequency domain (Brüns et al., 2011). To this end, a surface discretized model of the canonical tank was created and simulations at 161 discrete frequency samples between 1 kHz and 8 MHz were performed. From the resulting data a corresponding first order macromodel was obtained with parameters $L = 0.5192$ nH, $R = 2.8887$ uΩ, and a relative root mean square error RMSE$= 0.0023$, thus supporting the results of Tab. 1.

4 Conclusions

Standard lightning pulses are characterized by durations of tenths of microseconds, thus leading to long computation times for time domain calculations of lightning current induced quantities. However, in order to obtain LEMP transfer functions it is also possible to use shorter excitation pulses, involving less computation time. Typically, LEMP transfer functions are characterized by only a small number of parameters. Therefore the vector fitting procedure is useful to obtain these parameters which, in turn, can be used to quantitatively compare transfer functions that are obtained from different excitation pulses. This makes it possible to quantify whether a short excitation pulse is sufficient to determine a LEMP transfer function with sufficient accuracy.

References

Apra, M., D'Amore, M., Gigliotti, K., Sarto, M. S. and Volpi, V.: Lightning Indirect Effects Certification of a Transport Aircraft by Numerical Simulation, IEEE Trans. Electromagn. Compat., 50(3), 513–523, 2008.

Antonini, G.: Spice Equivalent Circuits of Frequency-Domain Responses, IEEE Trans. Electromagn. Compat., 45(3), 502–512, 2003.

Brüns, H.-D., Freiberg, A., and Singer, H.: CONCEPT-II Manual of the Program System, (Hamburg University of Technology, February 2011), see also http://www.tet.tu-harburg.de/concept/, 2011.

Clemens, M. and Weiland, T.: Discrete electromagnetism with the finite integration technique, Progress In Electromagnetics Research, PIER, 32), 65–87, 2001.

Deschrijver, D., Pissoort, D., and Dhaene, T.: Adaptive Stopping Criterion for Fast Time Domain Characterization of Microwave Components, IEEE Microw. Wireless Compon. Lett., 19(12), 765–767, 2009.

Fisher, F. A., Perala, R. A., and Plumer, J. A.: Lightning Protection of Aircraft, (Lightning Technologies, Inc., Pittsfield, USA), 1990.

Gronwald, F.: Classification and quantification of lightning-current induced voltages within aircraft structures, Proc. of the 9th Int. Symp. on EMC, EMC Europe 2010, Wroclaw, Poland, 559–564, 13–17 September, 2010.

Gustavsen, B. and Semlyen, A.: Rational Approximation of Frequency Domain Responses by Vector Fitting, IEEE Trans. Power Delivery, 14(3), 1052–1061, 1999.

Gustavsen, B.: Improving the Pole Relocating Properties of Vector Fitting, IEEE Trans. Power Delivery, 21(3), 1587–1592, 2006.

Hoffman, J. D.: Numerical Methods for Engineers and Scientists, 2nd ed., (Marcel Dekker Inc., New York), 2001.

Lee, K. S. H. (ed.): EMP Interaction: Principles, Techniques, and Reference Data, revised printing, Taylor & Francis, Washington D.C. 1995.

SAE Aerospace Recommended Practice (ARP) 5412A, Aircraft Lightning Environment and Related Test Waveforms, February 2005.

Coupling path influence on the conducted emission measurement results by the EMC semiconductor test board

M. Frick[1], R. Eidher[1], and R. Weigel[2]

[1]Engineering Electromagnetic Compatibility, Robert Bosch GmbH, Reutlingen, Germany
[2]Institute for Electronics Engineering, Friedrich-Alexander-University of Erlangen-Nuremberg, Germany

Abstract. The fact of reduced development time in the automotive sector requires a change in the electromagnetic compatibility (EMC) validation of electronic components. To increase the development time of automotive components, e.g. sensors or electronic control units (ECUs), it is mandatory to verify the EMC behavior of the integrated circuit (IC) itself in an early development stage simultaneous to the development process of the entire ECU into which it is embedded. The conducted IC EMC tests emission as well as immunity is realized with a specific test printed circuit board (PCB) which is built according to the guidelines of the Bosch, Infineon, Siemens VDO Specification (BISS, 2007). The measurement results of these EMC tests are affected by the test PCB itself. To deduce the application requirements of the component from the application circuit of the test PCB it is necessary to consider this influence by the printed circuit board. The text at hand gives an overview on how far the emission results are adulterated by the used test PCB. The presented method also grants a validation of the disturbance voltage directly at the IC pin considering the application circuit.

1 Introduction

To accelerate the development process of automotive components it is impossible to verify the complete new product, e.g. a sensor or ECU, at the end of the development process respecting the EMC. The tight schedule demands a systematic EMC validation of each complex sub module which is built into the component, e.g. an IC or microcontroller (μC) at an early development stage. A consistent EMC validation of each of these complex sub modules will usually decrease the number of EMC based redesigns and promises a good EMC performance of the entire component.

As usually at EMC tests, the conducted IC EMC tests are divided into emission and immunity tests, too. Both test versions are accomplished by a specific test PCB which has been particularly developed for each IC and includes the respective application circuits. In both emission and immunity case the test PCBs are usually populated with the same application circuits. Only the radio frequency (RF) coupling networks, which are used to carry the electro magnetic energy to the test receiver (emission) or the IC pin (immunity), differ. In case of the emission test the coupling network, as shown in Fig. 4, is used to connect the IC pin with the test receiver and matches the IC pin impedance to the 50 Ω measurement impedance as well as possible. At the immunity test, in contrast, the coupling network is only used to connect the RF immunity test setup with the IC pin via a DC block capacitor. The application circuit and the coupling/decoupling network influence the measurement results most. An efficient method to determine and consider the influence of the EMC semiconductor test PCB during the conducted emission test is given in this paper.

2 Conducted emission of integrated circuits

To verify an integrated circuit regarding its EMC behavior it is necessary to solder the device under test (DUT), including the application circuits, to a specific test PCB as shown in Figs. 2 and 3. By means of this test PCB it is possible to set the IC into different operating modes, like it is used in the ECU, and measure the conducted emission of any IC pin. Figure 1 illustrates the result of such a conducted emission measurement. The diagram shows the disturbance voltage of a supply pin which is measured at the RF connector of the test PCB by using an electromagnetic interference (EMI) test receiver. The frequency range of these measurements is

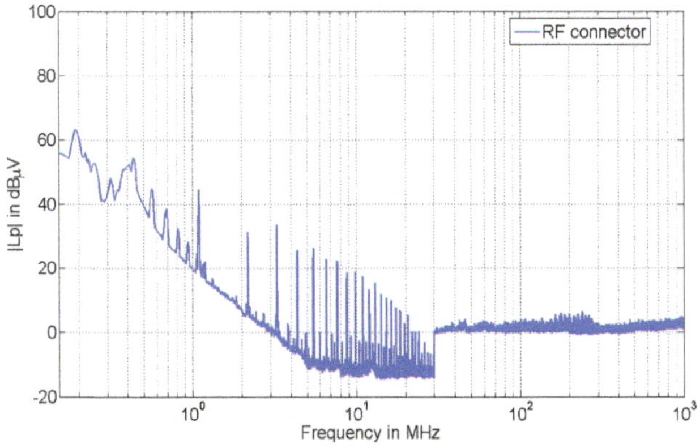

Fig. 1. Measured disturbance voltage at the RF-connector.

Fig. 2. EMC test PCB, top layer.

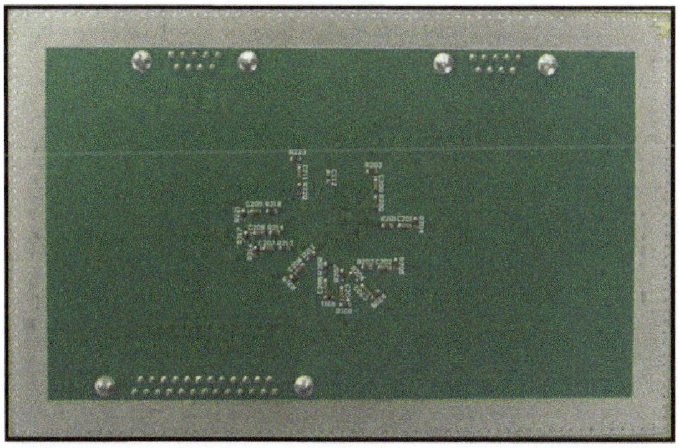

Fig. 3. EMC test PCB, bottom layer.

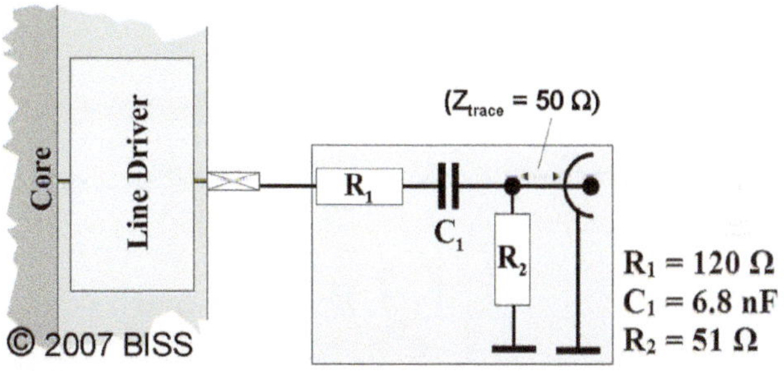

Fig. 4. Coupling network of the conducted emission measurement (BISS, 2007).

required in the BISS (2007) paper. It starts at 150 kHz an ends at 1 GHz. One significant item in this diagram is the amplitude offset at 30 MHz. This offset does not arise from the DUT itself but from the altered measurement bandwidth of the test receiver. The peaks at 1.1 MHz and integer multiples which rise up out of the noise floor derive from a charge pump which is embedded in the IC.

3 EMC test PCB

Figures 2 and 3 show the PCB which is used to determine the conducted emission of the IC. The circuit board consists of four layers with a dimension of 160 × 100 mm and is manufactured out of FR4 PCB material. The connectors are mounted onto the top layer, as shown in Fig. 2, to supply the IC and monitor the signals during the measurements. The IC and the application circuits are mounted on the top layer, too.

Regarding the bottom layer, which is shown in Fig. 3, it is only the RF coupling networks that are mounted onto the surface to match the IC pin impedance appropriate to the 50 Ω impedance of the measurement equipment. The coupling network consists of a 120 Ω series resistor on the IC side of the coupling network, followed by a 6.8 nF coupling capacitor and a 51 Ω parallel resistor to ground as shown in Fig. 4.

3.1 Characterization of the EMC test PCB

To characterize the scattering parameters of the test PCB a prober measuring station is used. The main constituent of this measuring station is a four port vector network analyzer (VNA) and a manual probe system. The VNA decreeds about a test frequency range from 9 kHz up to 8.5 GHz and a dynamic range of more than 120 dB. The manual probe system offers a front and back side microscope that enables a connection of the top and bottom layer concurrently.

Figure 5 shows the calibration substrate which is used to calibrate the entire measurement setup including the probes. By means of this substrate, a full two port calibration of the measurement setup is realized in the open, short, load and thru standard. After the measurement setup is calibrated the PCB characterization can be started. In Fig. 6 the top and bottom layer including the test points, which are used to characterize the coupling path of the supply pin, are shown.

To detect the complex scattering parameters of the test board any RF coupling path is considered as an independent two-port.

Fig. 5. Calibration measurement setup.

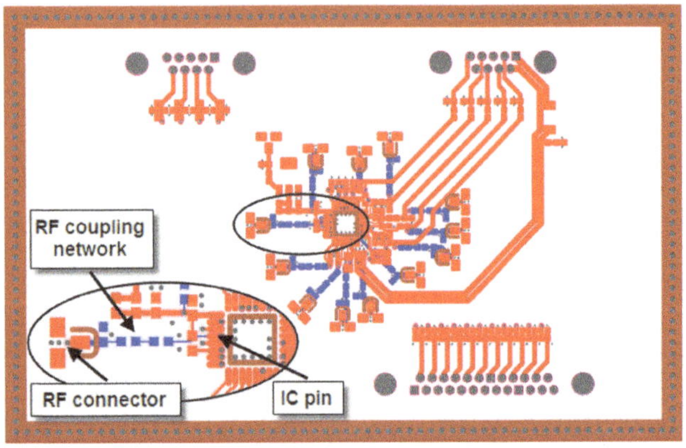

Fig. 6. Test points of the coupling path characterization.

Therefore the first port of the VNA is connected to the IC side of the coupling path and the second port is connected to the RF connector of the test PCB as it is shown in Fig. 7. During this measurements neither the RF connector nor the IC is mounted on the test PCB.

The results of this scattering parameter characterization is given in Fig. 8. The blue line illustrates the reflection coefficient S_{11} at the IC side of the coupling path. The worse progression of the curve is a result of the impedance mismatch between the 50 Ω port of the VNA and the coupling network. The cyan curve in this diagram shows the reflection coefficient S_{22} at the RF connector side of the coupling path. The progression of the curve gives a feedback for the well matching impedance of the VNA and the measured device under test (DUT). The forward transmission parameter S_{21} is congruent with the backward transmission parameter S_{12}, this confirms the reciprocity of the two-port.

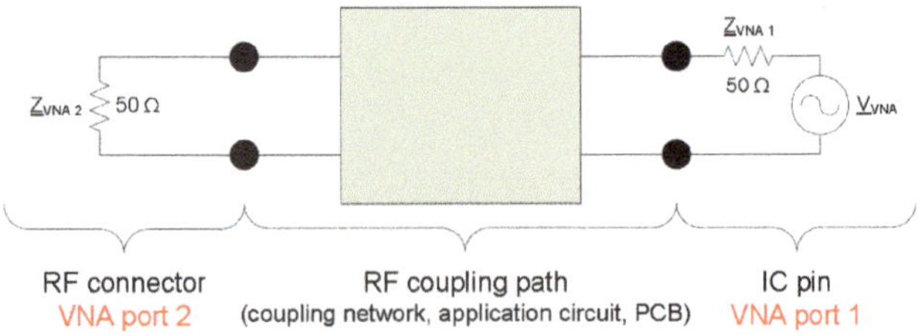

Fig. 7. Equivalent circuit of the scattering parameter measurement.

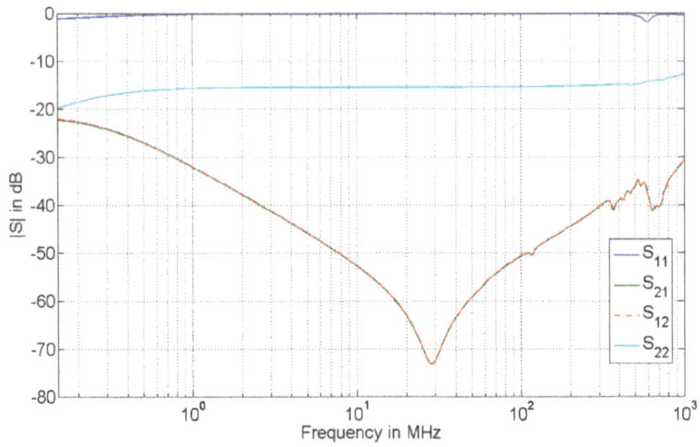

Fig. 8. Scattering parameter of the two-port.

4 Applied method

To determine the disturbance voltage directly at the IC pin, the influence of the used test PCB on the measurement results has to be considered. Therefore this particular PCB problem has to be transformed into a basic framework which could be used to describe an EMC problem in general (Paul, 2006). This framework divides all included elements of that problem into three main aspects like:

- Source (emitter)
- Transfer (coupling path)
- Receptor (receiver)

In case of the conducted emission the IC pin is considered as a source of the electromagnetic energy. After the energy is produced it will be transferred by using the RF coupling path of the EMC test PCB. Finally the transmitted RF energy will be absorbed by the EMI test receiver, as shown in Fig. 9. The IC pin, which generates the disturbance voltage, could be understood as a Thévenin equivalent circuit which is including a RF voltage source and an internal resistor.

The coupling path is represented by the influence of the coupling network, the application circuit and the printed circuit board. The third and last part of the given EMC basic decomposition is the used EMI test receiver which is interpreted as the receptor at the two-port output. In order to determine the disturbance voltage at the IC pin, the applied method is separated into the following items:

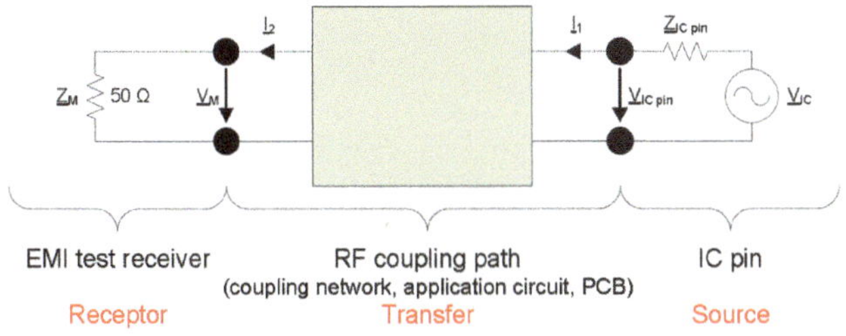

Fig. 9. Equivalent circuit of the IC pin emission measurement.

1. Measuring the disturbance voltage V_{M} at the RF connector of the test PCB
2. Observing any RF coupling path as an independent two-port
3. Determining the voltage gain A_{v} of any coupling path
4. Calculating the disturbance voltage $V_{\mathrm{IC\,pin}}$ straight at the IC pin by using the voltage gain of the coupling path

After the complex scattering parameters of the EMC test PCB are measured, the voltage gain of each coupling path can be calculated by using the following formula (Michel, 1981):

$$A_{\mathrm{v}([\mathbf{S}],\boldsymbol{\Gamma}_{\mathrm{M}})} = \frac{\hat{V}_{\mathrm{M}}}{\hat{V}_{\mathrm{IC\,pin}}} = \frac{-S_{21}(\boldsymbol{\Gamma}_{\mathrm{M}}+1)}{(S_{11}S_{22}-S_{12}S_{21}+S_{22})\boldsymbol{\Gamma}_{\mathrm{M}}-S_{11}-1} \quad (1)$$

As it is shown in Eq. (1) the voltage gain of the coupling path is depending on the scattering parameters [**S**] of the test PCB on the one hand, and the reflection coefficient $\boldsymbol{\Gamma}_{\mathrm{M}}$ which is appearing at the RF connector of the test PCB on the other hand. Section 3.1 describes how to measure the scattering parameters. However, the reflection coefficient cannot be measured during the emission test and therefore remains unknown. For this reason the consequence of change in this parameter is investigated in the next few chapters.

4.1 Output impedance of the considered two-port

The reflection coefficient at the RF connector of the printed circuit board results from the impedance mismatch between the RF connector itself and the output impedance of the two-port, as shown in Fig. 10. For this reason the output impedance of the two-port has to be calculated first.

In addition the two-port, as shown in Fig. 11, which includes the behavior of the coupling network, the application circuit and the printed circuit board, is considered as a Thévenin equivalent circuit including the elements of the input circuit (Berroth, 2009). Now the internal resistance of the Thévenin equivalent circuit, which is identical to the output impedance of the two-port, has to be calculated.

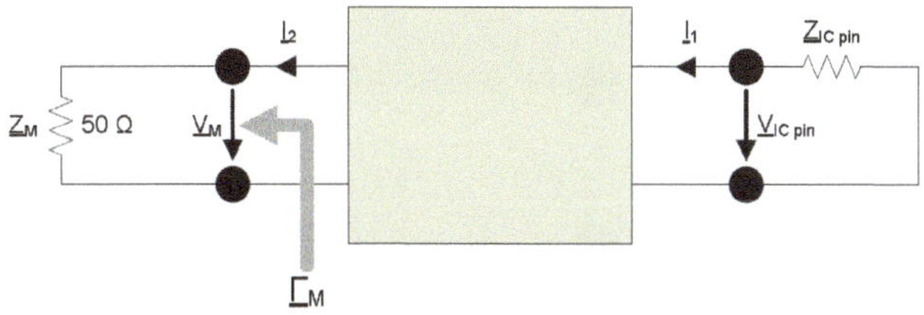

Fig. 10. Reflection coefficient at the RF connector of the test PCB.

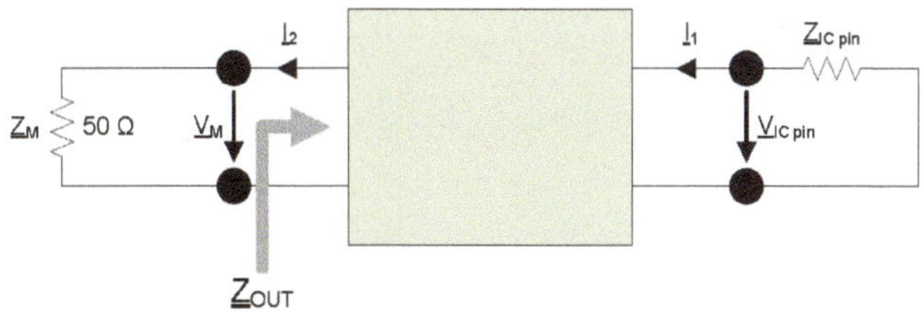

Fig. 11. Output impedance of the two-port.

Figure 12 shows this new equivalent circuit which is used to determine the two-port output impedance.

The output impedance calculation of the two-port is realized by the following equation (Berroth, 2009), in which [**A**] is the chain parameter (ABCD-parameter) matrix of the two-port and $\mathbf{Z}_{\mathrm{IC\,pin}}$ the impedance of the IC pin.

$$\mathbf{Z}_{\mathrm{eq}} = \mathbf{Z}_{\mathrm{OUT}} = \left.\frac{A_{22}\mathbf{Z}_{\mathrm{IC\,pin}}+A_{12}}{A_{21}\mathbf{Z}_{\mathrm{IC\,pin}}+A_{11}}\right|_{V_{\mathrm{IC}}=\mathrm{shorted}} \quad (2)$$

With Eq. (3) to Eq. (6) (Meinke and Gundlach, 1992) the chain parameters of the two-port can be calculated by the measured scattering parameters of each coupling path.

$$A_{11} = \frac{(1+S_{11})(1-S_{22})+S_{12}S_{21}}{2S_{21}} \quad (3)$$

$$A_{12} = \frac{Z_0((1+S_{11})(1+S_{22})-S_{12}S_{21})}{2S_{21}} \quad (4)$$

$$A_{21} = \frac{(1-S_{11})(1-S_{22})-S_{12}S_{21}}{Z_0 2S_{21}} \quad (5)$$

$$A_{22} = \frac{(1-S_{11})(1+S_{22})+S_{12}S_{21}}{2S_{21}} \quad (6)$$

Equation (2) in combination with Eqs. (3) to (6) demonstrates the influence of different parameters on the two-port output impedance like scattering parameters, reference impedance ($Z_0 = 50\ \Omega$) and the IC pin impedance. The scattering parameters and the reference impedance are now well known as shown above. The only missing parameter to calculate the output impedance is the unknown IC pin impedance. Since it is not possible neither to measure the pin impedance during the emission test nor to determine it with the available measurement results, both worst cases, open circuit and short circuit of the IC pin impedance, are investigated. Figure 13

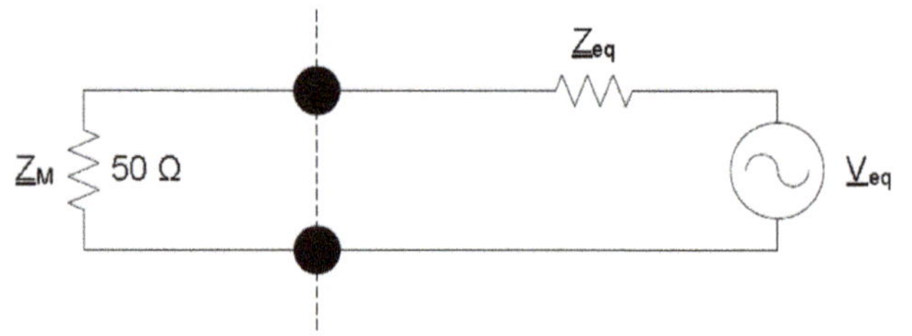

Fig. 12. Thévenin equivalent circuit of the two-port output (Berroth, 2009).

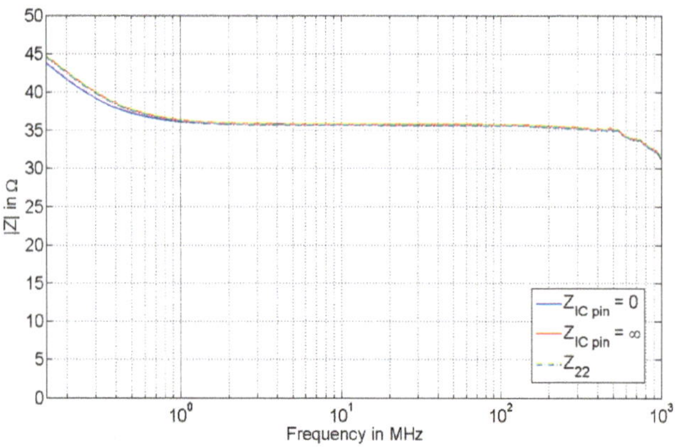

Fig. 13. Output impedance of the two-port by varying the IC pin impedance.

illustrates the marginal influence of a varying pin impedance between 0 and ∞ on the output impedance of the two-port. The green dash-dot line shows the progression of the secondary open loop impedance of the two-port. This graph matches completely with the output impedance in case the pin impedance is ∞.

4.2 Reflection coefficient at the RF connector

Both worst case values which have previously been determined for the open circuit and short circuit case of an IC pin will be used to calculate the reflection coefficient at the RF connector of the EMC test PCB. Equations (7) and (8) display the interrelationship between the reflection coefficient and the varying output impedance of the two-port.

$$\Gamma_{M_{min}} = \frac{Z_M - Z_{OUT_{max}}}{Z_M + Z_{OUT_{max}}} \tag{7}$$

$$\Gamma_{M_{max}} = \frac{Z_M - Z_{OUT_{min}}}{Z_M + Z_{OUT_{min}}} \tag{8}$$

As shown in Fig. 14, the variation of the IC pin impedance between the both worst cases 0 and ∞ is hardly visible. Both curves are nearly congruent over the entire frequency range.

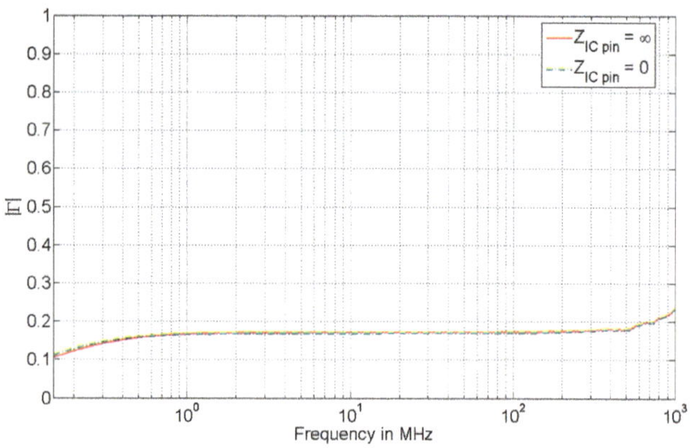

Fig. 14. Reflection coefficient at the RF connector of the test PCB by varying the IC pin impedance.

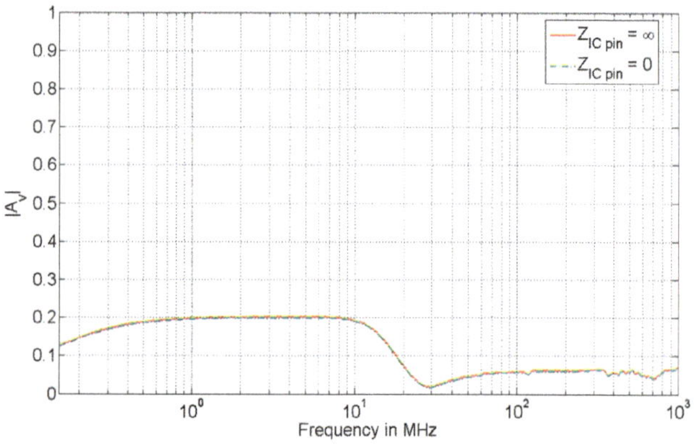

Fig. 15. Voltage gain of the coupling path by varying the IC pin impedance.

4.3 Voltage gain of the coupling path

The reflection coefficient, whose performance range is calculated in the last chapter, is the last missing parameter of Eq. (1). Now the voltage gain of the coupling path can be determined for both worst case reflection coefficient values by using the measured scattering parameters of the EMC test PCB. Figure 15 shows the results of the voltage gain by varying the IC pin impedance.

The red line illustrates the performance for an open circuit inside the IC pin ($Z_{IC\,pin} = \infty$) which is accompanied by a minimum reflection coefficient at the RF connector. The green dash-dot line represents a short circuit inside the IC pin ($Z_{IC\,pin} = 0$) which is accompanied by a maximum reflection coefficient at the connector.

4.4 Disturbance voltage straight at the IC pin

After the voltage gain of the coupling path is determined the disturbance voltage straight at the IC pin can be calculated by the following equation (Michel, 1981):

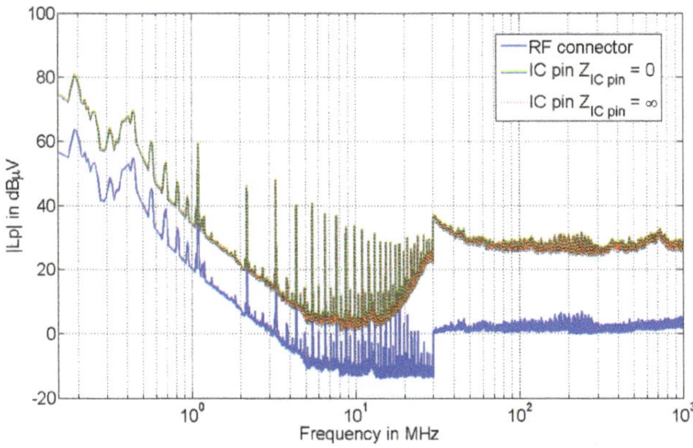

Fig. 16. Influence of the RF coupling path to the conducted emission measurement results.

$$\hat{V}_{\mathrm{IC\,pin}} = \frac{\hat{V}_{\mathrm{M}}}{\boldsymbol{A}_{\mathrm{v}([\mathbf{S}],\boldsymbol{\Gamma}_{\mathrm{M}})}} \tag{9}$$

The influence of the EMC semiconductor test PCB on the conducted emission EMC test is shown in Fig. 16. The blue line illustrates the disturbance voltage of a supply pin, measured at the RF connector of the test PCB which is accompanied by the output voltage at the two-port. The green respectively red dotted curve represents the determined disturbance voltage directly at the IC pin of both worst cases.

As in case of the voltage gain the deviation between a short circuit inside the IC pin ($\boldsymbol{Z}_{\mathrm{IC\,pin}} = 0$) and an open circuit inside the IC pin ($\boldsymbol{Z}_{\mathrm{IC\,pin}} = \infty$) is also very low and calculated to less than 0.1 dB. This robustness of the RF coupling path regarding its transmission behavior against the varying IC pin impedance is caused by the dominant external circuit devices. The difference between the blue and the green respectively red dotted line, which could be up to 36 dB in case of this investigated supply pin, is caused by the EMC test PCB.

5 Conclusions

The RF coupling path influence on the conducted emission measurements due to the EMC semiconductor test PCB is presented in this paper. By using this method a pin selective validation of the IC regarding its conducted emission, excluding the external circuit devices is possible. The robustness of the RF coupling path against the varying IC pin impedance, regarding both, its transmission behavior and the voltage gain of the investigated IC pin, has been proved. This ruggedness of the two-port is caused by the dominant external application circuit of the supply pin. A more dominant external circuit increases the robustness of the coupling path against the varying IC pin impedance. After a pin selective validation of the IC regarding its conducted emission is realized, the application requirements of the component (ECU/sensor) can be deduced from the application circuit of the EMC test PCB.

References

Berroth, M.: Vorlesungsskript, Theorie der Schaltungen III, Kapitel 4: Grundzuege der Vierpoltheorie, Institut fuer Elektrische und Optische Nachrichtentechnik, Universitaet Stuttgart, http://jfz.50g.com/files/tdslll-kapitel_4.pdf, October 2009.

BISS; Bosch, Infineon, Siemens VDO Specification: Generic IC EMC Test Specification 2004–2007.

Meinke, H. and Gundlach, F. W.: Taschenbuch der Hochfrequenztechnik, Hrsg. von Lange, K., Loecherer, K.-H., ISBN-10:3-540-54714 -2, Springer-Verlag, Fuenfte Auflage, 1992.

Michel, H.-J.: Zweitor-Analyse mit Leistungswellen, Studienbuecher Elektrotechnik, ISBN-10:3-519-06112-0, B. G. Teubner Stuttgart, 1981.

Paul, C. R.: Introduction to electromagnetic compatibility, ISBN-13: 978-0-471-75500-5, John Wiley & Sons Inc., second edition, 2006.

Coherent and non-coherent processing of multiband radar sensor data

S. Tejero, U. Siart, and J. Detlefsen

Institute for High Frequency Engineering, Technische Universität München, D-80290 München, Munich, Germany

Abstract. Increasing resolution is an attractive goal for all types of radar sensor applications. Obtaining high radar resolution is strongly related to the signal bandwidth which can be used. The currently available frequency bands however, restrict the available bandwidth and consequently the achievable range resolution. As nowadays more sensors become available e.g. on automotive platforms, methods of combining sensor information stemming from sensors operating in different and not necessarily overlapping frequency bands are of concern. It will be shown that it is possible to derive benefit from perceiving the same radar scenery with two or more sensors in distinct frequency bands. Beyond ordinary sensor fusion methods, radar information can be combined more effectively if one compensates for the lack of mutual coherence, thus taking advantage of phase information.

At high frequencies, complex scatterers can be approximately modeled as a group of single scattering centers with constant delay and slowly varying amplitude, i.e. a set of complex exponentials buried in noise. The eigenanalysis algorithms are well known for their capability to better resolve complex exponentials as compared to the classical spectral analysis methods. These methods exploit the statistical properties of those signals to estimate their frequencies. Here, two main approaches to extend the statistical analysis for the case of data collected at two different subbands are presented. One method relies on the use of the band gap information (and therefore, coherent data collection is needed) and achieves an increased resolution capability compared with the single-band case. On the other hand, the second approach does not use the band gap information and represents a robust way to process radar data collected with incoherent sensors. Combining the information obtained with these two approaches a robust estimator of the target locations with increased resolution can be built.

1 Introduction

Increasing resolution is an attractive goal for all types of radar sensor applications. Obtaining high radar resolution is strongly related to the signal bandwidth which can be used. The currently available frequency bands however, restrict the available bandwidth and consequently the achievable range resolution. As nowadays more sensors become available e.g. on automotive platforms, methods of combining sensor information stemming from sensors operating in different and not necessarily overlapping frequency bands are of concern. Beyond ordinary sensor fusion methods, radar information can be combined more effectively if one compensates for the lack of mutual coherence, thus taking advantage of phase information.

In this article, the possibility to obtain a benefit in the radar performance by exploiting the availability of sensors operating at different frequency bands will be analyzed. We will concentrate on the dualband case, i.e. two sensors which observe the same radar scenario from the same aspect angle.

2 Radar signal model

In the common frequency bands for radar systems the targets can be considered to be in the high frequency range, as they usually are greater than a wavelength. Some very popular high frequency approximations for solving scattering problems are the Geometrical Optics (GO) theory and one of its extensions, the Geometrical Theory of Diffraction (GTD) (Hansen, 1981; James, 1986; McNamara et al., 1990). Both approximations are based in the decomposition of a scatterer in a set of individual scattering centers with a specific frequency dependence. These scattering centers corresponds to specular reflections, diffraction at wedges or tips or creeping wave terms. Assuming that the high frequency approximation is valid, the radar response of a target or a group of them can be represented as a sum of individual scattering centers with constant delay:

$$X(t) = \sum_{p=1}^{P} A_p \delta(t - \tau_p) \quad (1)$$

where each Dirac's delta represents a scattering center at distance $c_0 \tau_i / 2$ with amplitude α_i. In the high-frequency range this model is valid for a wide frequency band, so that the expression for the frequency domain results in

$$x(f) = \sum_{p=1}^{P} A_p e^{j2\pi f \tau_p}. \quad (2)$$

This radar response is sampled by a system working at a frequency band $\mathcal{B}_l$ and can be expressed as:

$$x_l[n] = \sum_{p=1}^{P} A_p z_p^{k_{0l}+n} = \sum_{p=1}^{P} a_{pl} z_p^n \quad (3)$$

where

$$z_p = e^{j\Omega_p}$$
$$\Omega_p = \Delta\omega\tau_p$$
$$a_{pl} = A_p e^{j\omega_{0l}\tau_p}$$

and τ_p is the round trip time, $\Delta\omega$ the angular frequency increment and ω_{0l} is the angular start frequency for band $\mathcal{B}_l$.

3 Radar signal processing

Analyzing Eqs. (2) and (3) it can be noted that the radar problem of finding the position of the scattering centers in the frequency radar response is equivalent to the estimation of the frequencies of complex exponentials in a time domain signal or the angles of the poles z_p. Therefore, spectral or frequency estimation algorithms can be applied. In this paper, the dualband analysis is based on the Multiple Signal Classification (MUSIC) algorithm, but the results can be extended to other parametrized spectral estimation techniques, like Autoregressive (AR) algorithms, as well.

3.1 Single-band MUSIC

A short review of the MUSIC algorithm for a single-band dataset is given here, further information can be found in the literature (Marple, 1987; Schmidt, 1986). Assume the radar frequency response of Eq. (3) is received in the presence of complex white Gaussian Noise (CWGN) $w[n]$ in the frequency band $\mathcal{B}$

$$y[n] = \sum_{p=1}^{P} a_p z_p^n + w[n], \quad n = 0, \ldots, N-1$$
$$y[n] = x[n] + w[n]. \quad (4)$$

With the samples of the data sequence $x[n]$, the so called data matrix $\mathbf{X}$ of order L –equal to the data matrix in the AR covariance method– can be built and decomposed into the matrices $\mathbf{B}$ and $\mathbf{C}$:

$$\mathbf{X} = \begin{bmatrix} x[L-1] & x[L-2] & \cdots & x[0] \\ x[L] & x[L-1] & \cdots & x[1] \\ \vdots & \vdots & \ddots & \vdots \\ x[N-1] & x[N-2] & \cdots & x[N-L] \end{bmatrix} = \mathbf{B} \cdot \mathbf{C} \quad (5)$$

$$\mathbf{B} = \begin{bmatrix} a_1 z_1^{L-1} & a_2 z_2^{L-1} & \cdots & a_P z_P^{L-1} \\ a_1 z_1^{L} & a_2 z_2^{L} & \cdots & a_P z_P^{L} \\ \vdots & \vdots & \ddots & \vdots \\ a_1 z_1^{N-1} & a_2 z_2^{N-1} & \cdots & a_P z_P^{N-1} \end{bmatrix} \quad (6)$$

$$\mathbf{C} = \begin{bmatrix} 1 & z_1^{-1} & \cdots & z_1^{-L+1} \\ 1 & z_2^{-1} & \cdots & z_2^{-L+1} \\ \vdots & \vdots & \ddots & \vdots \\ 1 & z_P^{-1} & \cdots & z_P^{-L+1} \end{bmatrix} \quad (7)$$

An eigenvalue decomposition is applied to the autocorrelation-like matrix $\mathbf{X}^H\mathbf{X}$, so that

$$\mathbf{X}^H\mathbf{X} = \mathbf{V}\Lambda\mathbf{V}^H \quad \text{with} \quad (8)$$

$$\mathbf{V} = \begin{bmatrix} v_1[1] & v_2[1] & \cdots & v_L[1] \\ v_1[2] & v_2[2] & \cdots & v_L[2] \\ \vdots & \vdots & \ddots & \vdots \\ v_1[L] & v_2[L] & \cdots & v_L[L] \end{bmatrix} \quad (9)$$

$$\Lambda = \begin{bmatrix} \lambda_1 & 0 & \cdots & 0 \\ 0 & \lambda_2 & \cdots & 0 \\ \vdots & \vdots & \ddots & \vdots \\ 0 & 0 & \cdots & \lambda_L \end{bmatrix} \quad (10)$$

and $\lambda_k = 0$ for $k = P+1, \ldots, L$ as the matrix $\mathbf{X}^H\mathbf{X}$ is of rank P. It can be shown, that any principal eigenvector $\mathbf{v}_k$ ($k = 1, \ldots, P$) of the matrix $\mathbf{X}^H\mathbf{X}$ is a linear combination of the columns of $\mathbf{C}^H$ (e.g. Marple, 1987), composed by the signal poles:

$$\mathbf{C}^H = \begin{bmatrix} 1 & 1 & \cdots & 1 \\ (z_1^*)^{-1} & (z_2^*)^{-1} & \cdots & (z_P^*)^{-1} \\ \vdots & \vdots & \ddots & \vdots \\ (z_1^*)^{-L+1} & (z_2^*)^{-L+1} & \cdots & (z_P^*)^{-L+1} \end{bmatrix}$$

Therefore, the columns of $\mathbf{C}^H$ are orthogonal to the non-principal eigenvectors $\mathbf{v}_k$ ($k = P+1, \ldots, L$).

In the presence of noise, the autocorrelation-like matrix of the signal plus noise can be approximated by

$$\mathbf{Y}^H\mathbf{Y} \approx \mathbf{X}^H\mathbf{X} + \sigma^2\mathbf{I} = \mathbf{X}^H\mathbf{X} + \sum_{k=1}^{L} \sigma^2 \mathbf{v}_k \mathbf{v}_k^H$$
$$= \sum_{k=1}^{P} (\lambda_k + \sigma^2) \mathbf{v}_k \mathbf{v}_k^H + \sum_{k=P+1}^{L} \sigma^2 \mathbf{v}_k \mathbf{v}_k^H. \quad (11)$$

with σ^2 the noise variance. The principal eigenvectors $\mathbf{v}_k$ ($k=1,\ldots,P$) expand the signal plus noise subspace and the eigenvectors $\mathbf{v}_k$ ($k=P+1,\ldots,L$) the noise subspace.

The MUSIC algorithm exploits the orthogonality between the noise subspace eigenvectors $\mathbf{v}_k$ and the columns of $\mathbf{C}^H$, the signal vectors

$$\mathbf{s}(z_p)=\left[1\ (z_p^*)^{-1}\ \cdots\ (z_p^*)^{-L+1}\right]^T \tag{12}$$

in different ways. The Spectral MUSIC looks for the peaks in the pseudospectrum $X(e^{j\omega})$

$$\begin{aligned} X(e^{j\omega}) &= \frac{1}{D(e^{j\omega})} \\ &= \frac{1}{\mathbf{s}^H(e^{j\omega})\left(\sum_{k=P+1}^{L}\mathbf{v}_k\mathbf{v}_k^H\right)\mathbf{s}(e^{j\omega})} \end{aligned} \tag{13}$$

which are at the signal frequencies and the Root-MUSIC algorithm obtains directly the roots of the null spectra polynomial $D(z)$:

$$D(z)=\sum_{k=P+1}^{L} V_k(z)V_k^*(1/z^*) \quad \text{with} \tag{14}$$

$$V_k(z)=v_k[1]+v_k[2]z^{-1}+\cdots+v_k[L]z^{-(L-1)} \tag{15}$$

Next, two possible extensions of this algorithm to the dual-band case are presented.

3.2 Multiband radar signal processing

The radar response is sampled at two different subbands $\mathcal{B}_1$ and $\mathcal{B}_2$:

$$y[n]=\sum_{p=1}^{P} a_p z_p^n + w[n] \qquad \text{with} \tag{16}$$

$$n=\begin{cases} N_{10},\ldots,N_{10}+N_1-1 & \text{for } \mathcal{B}_1 \\ N_{20},\ldots,N_{20}+N_2-1 & \text{for } \mathcal{B}_2 \end{cases}$$

Two main approaches to treat the dualband case can be identified, the non-coherent approach, where no information on the band gap between the subbands is used and the coherent approach, where the information about the band gap is used in the algorithm.

3.2.1 Non-coherent multiband MUSIC

Using the data of the two subbands, a new data matrix is built by the superposition of the two subband matrices:

$$\mathbf{X}_{\text{nc}}=\begin{bmatrix}\mathbf{X}_1\\ \mathbf{X}_2\end{bmatrix} \qquad \text{where} \tag{17}$$

$$\mathbf{X}_1=\begin{bmatrix} x[N_{10}-L_1+1] & \cdots & X[N_{10}] \\ \vdots & \ddots & \vdots \\ x[N_{10}+N_1-1] & \cdots & x[N_{10}+N_1-L_1] \end{bmatrix} \tag{18}$$

$$\mathbf{X}_2=\begin{bmatrix} x[N_{20}-L_2+1] & \cdots & X[N_{20}] \\ \vdots & \ddots & \vdots \\ x[N_{20}+N_2-1] & \cdots & x[N_{20}+N_2-L_2] \end{bmatrix} \tag{19}$$

and $L_1=L_2=L$. As in the single-band case, the matrix $\mathbf{X}_{textnc}$ can be decomposed into the product of the matrices $\mathbf{B}_{textnc}$ and $\mathbf{C}$, with

$$\mathbf{B}_{textnc}=\begin{bmatrix}\mathbf{B}_1\\ \mathbf{B}_2\end{bmatrix} \tag{20}$$

and $\mathbf{C}$ defined as in Eq. (7). As the matrix $\mathbf{C}$ has the same structure as in the single-band case, the same procedure can be applied directly. No information of the band gap $N_{20}-N_{10}$ is used, therefore, no coherence between the two subbands is required.

3.2.2 Coherent multiband MUSIC

To exploit the relative position between the two subbands, a data matrix is built including the band-gap information in its structure.

$$\mathbf{X}_{textc}=\left[\mathbf{X}_2\ \mathbf{X}_1\right] \tag{21}$$

where $\mathbf{X}_1$ and $\mathbf{X}_2$ are defined as in Eqs. (18) and (19), $L_1+L_2=L$ and $N_2-N_1=L_2-L_1$. The new data matrix $\mathbf{X}_{textc}$ is decomposed as the product of the matrices $\mathbf{B}$ and $\mathbf{C}_{textc}$, with $\mathbf{B}$ defined as in Eq. (6) and

$$\mathbf{C}_{textc}=\begin{bmatrix} 1 & \cdots & z_1^{-L_2+1} & z_1^{-\Delta N} & \cdots & z_1^{-\Delta N-L_1+1} \\ \vdots & \ddots & \vdots & \vdots & \ddots & \vdots \\ 1 & \cdots & z_P^{-L_2+1} & z_P^{-\Delta N} & \cdots & z_P^{-\Delta N-L_1+1} \end{bmatrix} \tag{22}$$

with $\Delta N=N_{20}-N_{10}+N_2-N_1$.

Comparing the structure of the data matrices for the single-band- $\mathbf{X}$, for the non-coherent dualband- $\mathbf{X}_{textnc}$ and for the coherent dualband-algorithm $\mathbf{X}_{textc}$, can be observed the difference of the approaches for analyzing the data. In the first two approaches, a window of length L is moved through the data set to characterize the relation between neighbor data samples with a maximum distance L between them. In the third approach, to build the data matrix $\mathbf{X}_{textc}$ the observation window is split into two sub-windows of length L_1 and L_2, each of them containing data from a different subband. This

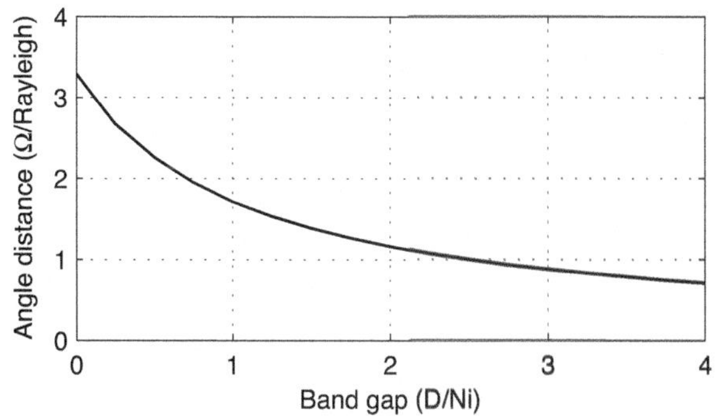

Fig. 1. Maximum resolvable angle distance using the non-coherent and coherent pole combination. Rayleigh angle distance is defined as $\frac{2\pi}{N_1+N_2}$.

window analyzes the relation between the neighbor samples in the same subband and the samples of the other subband.

Now, the noise subspace eigenvectors $\mathbf{v}_k$ are orthogonal to the columns of $\mathbf{C}^H_{textc}$, which are composed by the signal poles evaluated at the two subbands simultaneously

$$\mathbf{s}(z_p) = \left[1 \cdots (z_p^*)^{-L_2+1} (z_p^*)^{-\Delta N} \cdots (z_p^*)^{-\Delta N - L_1 + 1}\right]^T \quad (23)$$

The pole vectors which are orthogonal to the noise subspace, are identified again with two main procedures. The spectral MUSIC algorithm looks for the peaks in the pseudospectrum $X(e^{j\omega})$

$$X(e^{j\omega}) = \frac{1}{D(e^{j\omega})} = \frac{1}{\mathbf{s}^H(e^{j\omega})\left(\sum_{k=P+1}^{L} \mathbf{v}_k \mathbf{v}_k^H\right)\mathbf{s}(e^{j\omega})}$$

and the root-MUSIC obtains the roots of the null spectra polynomial $D(z)$

$$D(z) = \sum_{k=P+1}^{L} V_k(z) V_k^*(1/z^*)$$

with the polynomial $V_k(z)$ evaluated at the two subbands:

$$\begin{aligned} V_k(z) &= v_k[1] + \cdots + v_k[L_2] z^{-L_2+1} \\ &\quad + v_k[L_2+1] z^{-\Delta N} + \cdots + v_k[L] z^{-\Delta N - L_1 + 1} \end{aligned} \quad (24)$$

It is clear from the structure of the signal vectors $s(z_p)$ or the noise polynomials $V_k(z)$, that the information of the band gap is used and therefore, coherence is required. As the gap information ΔN is included directly in the null-spectra polynomial, the number of zeros of polynomial, i.e. the number of signal poles orthogonal to the noise subspace vectors, increases linearly with the band gap. This makes the distinction of the real signal poles from the spurious polynomial zeros difficult.

To overcome this problem, the information of the non-coherent and coherent algorithms is combined. Only the zeros of the coherent polynomial surrounding the zeros of the incoherent polynomial are analyzed. It is assumed, that the detection of a group of scatterers is achieved with the non-coherent approach, while with the coherent approach only an increase of resolution is expected. It may occur that two signal poles are present where the non-coherent approach detects only one. Assuming that the coherent polynomial zeros are approximately distributed uniformly along the unit circle, the maximum search distance to avoid spurious zeros is fixed by the total number of zeros, which is approximately equal to ΔN. This approach implies a limit in the maximum distance between two poles which can be still resolved with this algorithm as can be seen in Fig. 1.

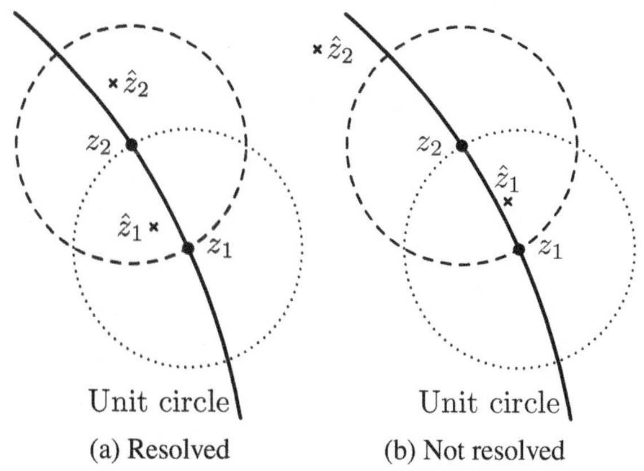

Fig. 2. Illustration of the resolution criteria: two targets are said to be resolved if the distances from the estimated poles ($\hat{z}_p$) to the true ones (z_p) is lower than the distance between the true positions.

4 Simulations

The resolution capability of the dualband processing has been assessed by means of simulations. Monte Carlo analysis of the probability of resolution, i.e. the frequency of experiments for which two targets are resolved over the number of experiments where two targets are present, have been carried out. A signal composed by two poles on the unit-circle with angular distance $\Delta\Omega$ between them and equal amplitude A buried in CWGN has been used as radar signal. The signal is sampled at two subbands with equal number of samples $N_1 = N_2 = 32$ and with a variable band gap of D samples between them. Both dualband root-MUSIC algorithms with $L = 8$ and $P = 2$ have been applied. As the number of expected targets P is assumed to be known in advance for the algorithm the number of detected targets can not be used as resolution criterion. The definition of resolution proposed here is illustrated in Fig. 2, two targets are said to be resolved if the distance between the estimated poles and the true ones is smaller than the distance between the true positions.

In Fig. 3 results of the Monte Carlo analysis for different angular distances between the poles $\Delta\Omega$ and signal to noise

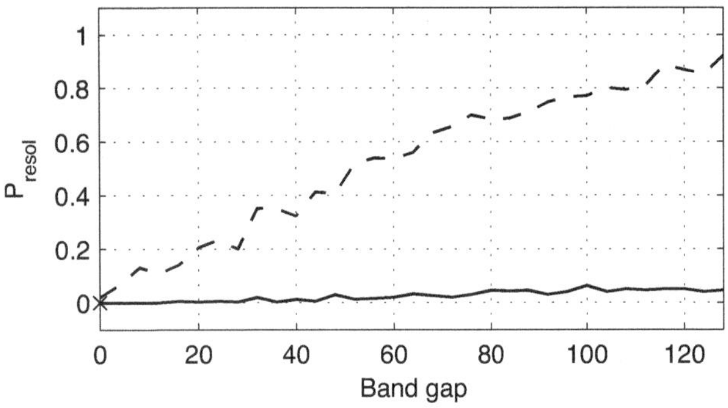

(a) $\Delta\Omega = 0.15\ \frac{2\pi}{N_1+N_2}$, SNR=30 dB

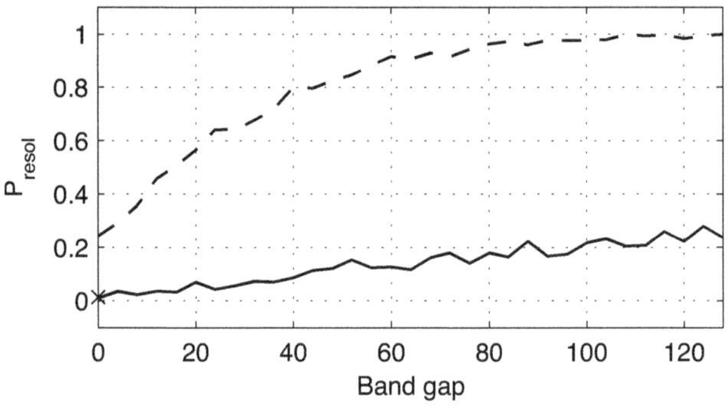

(b) $\Delta\Omega = 0.15\ \frac{2\pi}{N_1+N_2}$, SNR=35 dB

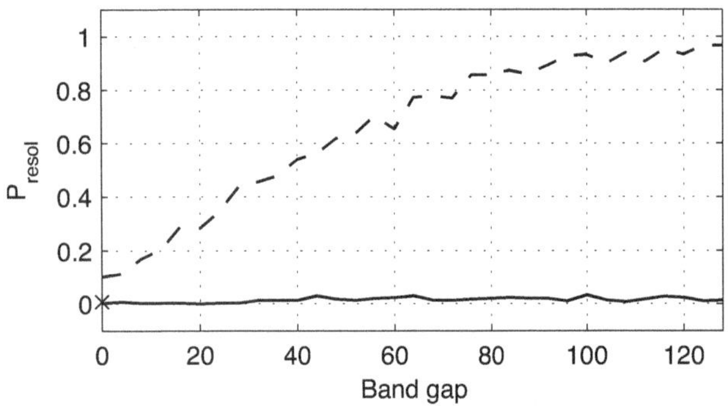

(c) $\Delta\Omega = 0.25\ \frac{2\pi}{N_1+N_2}$, SNR=20 dB

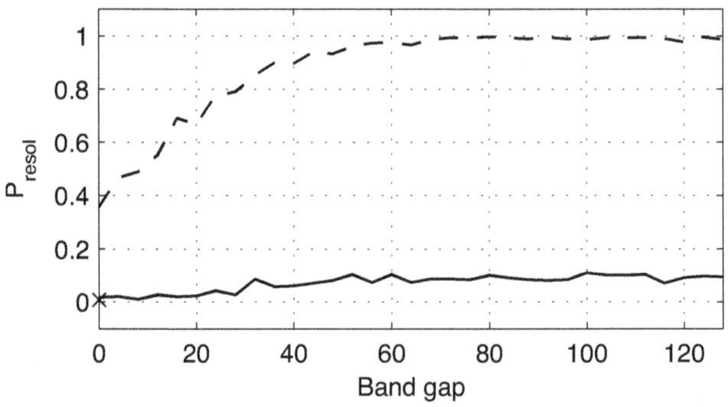

(d) $\Delta\Omega = 0.25\ \frac{2\pi}{N_1+N_2}$, SNR=25 dB

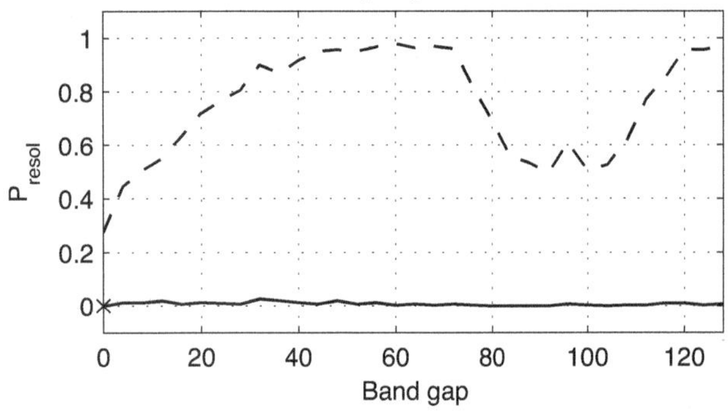

(e) $\Delta\Omega = 0.5\ \frac{2\pi}{N_1+N_2}$, SNR=10 dB

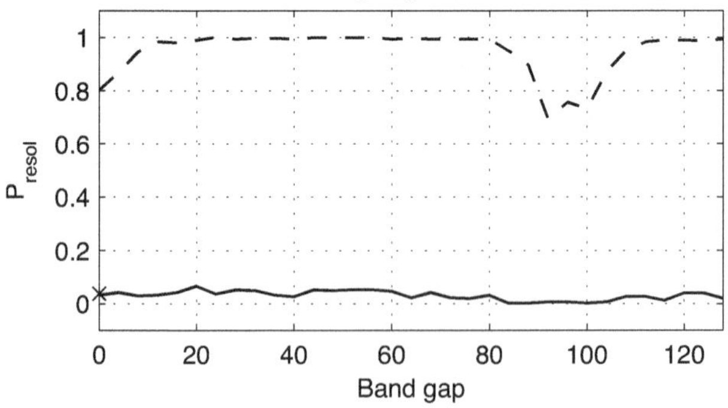

(f) $\Delta\Omega = 0.5\ \frac{2\pi}{N_1+N_2}$, SNR=15 dB

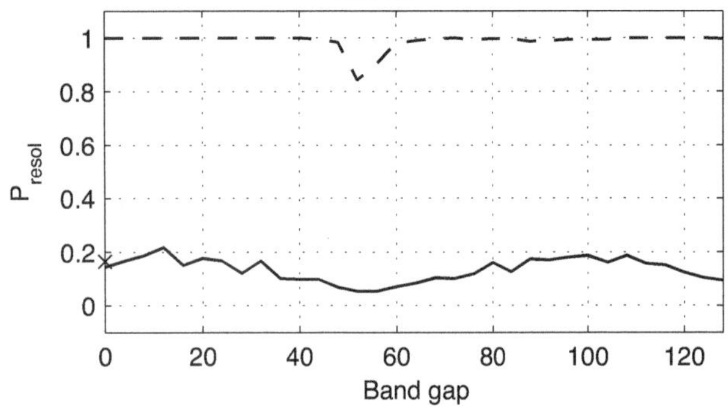

(g) $\Delta\Omega = 0.75\ \frac{2\pi}{N_1+N_2}$, SNR=15 dB

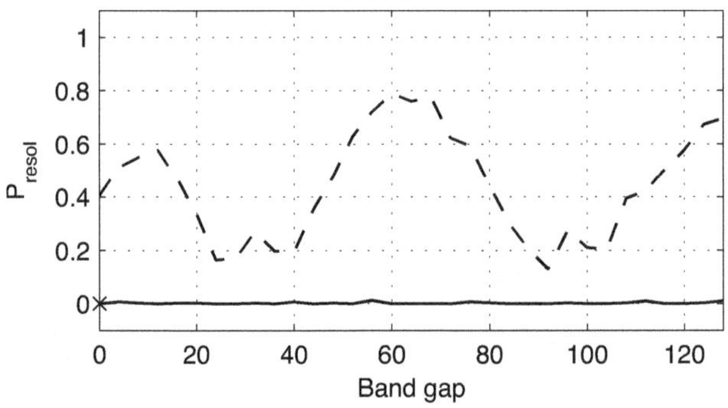

(h) $\Delta\Omega = 1.00\ \frac{2\pi}{N_1+N_2}$, SNR=0 dB

Fig. 3. Probability of resolution Monte Carlo analysis. Radar signal composed by two poles on the unit-circle with angular distance between them of $\Delta\Omega$, sampled at two subbands with $N_1 = N_2 = 32$ and with a band gap between them of D samples. Applied algorithms: Dualband root-MUSIC non-coherent (solid line), coherent (dashed line) with $L = 8$ and $P = 2$ and single-band ($\times$) with $N = N_1 + N_2 = 64$. The resolution event is defined as in Fig. 2.

ratio, defined as $\mathrm{SNR} = A^2/\sigma^2$, are shown. Also several single-band simulations with $N = N_1 + N_2 = 64$ have been carried out. The cases with a probability of resolution around 50% have been selected for demonstration.

It can be seen, e.g. Figs. 3f or 3g, that the dualband non-coherent approach has resolution performance similar to the single-band case. The non-coherent approach is therefore a robust way to exploit the whole bandwidth of a signal, also if the signal information is split in different, non-adjacent and mutually incoherent subbands $\mathcal{B}_{\mathrm{eq}} = \sum \mathcal{B}_i$.

The coherent approach achieves a higher probability of resolution. For low angle distances, $\Delta\Omega < 0.5 \frac{2\pi}{N_1 + N_2}$ from Figs. 3a to 3d, the improvement in the dualband coherent algorithm increases continuously with the band gap. Also an increase is observed in the dualband non-coherent approach (Fig. 3a). For greater pole distances however, the probability of resolution shows a periodic behavior with period $\approx \frac{2\pi}{\Delta\Omega}$. This is again observed in both dualband approaches. The origin of this periodicity effect will be subject of further investigations.

5 Summary and outlook

A signal model for the multiband radar response based on GO and GTD has been presented. Based on this model, it can be seen that the range estimation using frequency-domain radar information is analog to the spectral or frequency estimation techniques for time-domain signals. The MUSIC algorithm for spectral estimation has been extended to obtain two ways to process the multiband case: non-coherent and coherent. The non-coherent algorithm does not require coherency between the subbands. It does not use the band gap information and the resolution performance is similar to the single-band case using the sum of the bandwidths. It is therefore a robust approach to exploit the total bandwidth of a signal, also if the signal information is split in different, non-adjacent and mutually incoherent subbands. The coherent algorithm exploits the band gap information and therefore, coherent data sets are required. An increase in the resolution performance compared to the non-coherent algorithm is achieved. Also a periodic effect with period $\approx \frac{2\pi}{\Delta\Omega}$ for $\Delta\Omega > 0.5 \frac{2\pi}{N_1+N_2}$ has been observed, which will be subject of further investigations.

References

Hansen, R. C.: Geometric Theory of Diffraction, IEEE Press, New York, 1981.

James, G. L.: Geometrical Theory of Diffraction for Electromagnetic Waves, Peter Peregrinus, Stevenage, Herts, England, revised third edn., 1986.

Marple, S.: Digital Spectral Analysis, Prentice Hall, Englewood Cliffs, New Jersey, 1987.

McNamara, D. A., Pistorius, C., and Malherbe, J.: Introduction to the Uniform Geometrical Theory of Diffraction, Artech House, Boston London, 1990.

Schmidt, R. O.: Multiple Emitter Location and Signal Parameter Estimation, IEEE Transactions on Antennas and Propagation, AP-34, 1986.

Efficient computation of source magnetic scalar potential

W. Hafla, A. Buchau, and W. M. Rucker

Institute for Theory of Electrical Engineering, Stuttgart, Germany

Abstract. Magnetic field problems are often excited by a known source field which itself is caused by free currents. Some formulations to solve the problem require knowledge of the source magnetic scalar potential whose gradient is the source field. In general, it has to be computed as a line integral. An approach for efficient computation of this potential has been developed that is based on an algorithm that gives short integration paths as well as on application of the fast multipole method. Numerical results indicate the efficiency of this approach especially when the number of current-carrying volume elements or the number of observation points are high.

1 Introduction

Magnetic field problems often consist of a known exciting source field due to free currents $\boldsymbol{J}$ and an unknown induced field. Many formulations to solve the problem have the magnetic scalar potential, its normal derivative or its sources as unknowns. Some of them require only the source field $\boldsymbol{H}_J$ as known quantity which can be obtained by evaluation of Biot-Savart law,

$$\boldsymbol{H}_J(\boldsymbol{r}) = \frac{1}{4\pi} \int\limits_V \boldsymbol{J}(\boldsymbol{r}') \times \nabla' G(\boldsymbol{r}, \boldsymbol{r}') dV'$$

where $\nabla' G$ is the gradient of Green's function of free space,

$$\nabla' G(\boldsymbol{r}, \boldsymbol{r}') = \frac{\boldsymbol{r} - \boldsymbol{r}'}{\left|\boldsymbol{r} - \boldsymbol{r}'\right|^3}. \tag{1}$$

Other formulations require knowledge of the magnetic scalar potential ψ_J whose gradient is the source field $\boldsymbol{H}_J$. In general, it can be computed as a line integral, If ψ_J has to be computed at the nodes $\boldsymbol{r}_i$ of a mesh line integrals

$$\psi_J(\boldsymbol{r}_i) = \int\limits_{C_i} \boldsymbol{H}_J\left(\boldsymbol{r}'_i\right) \cdot d\boldsymbol{r}' \tag{2}$$

have to be carried out where all integration paths C_i begin at the same arbitrary start node $\boldsymbol{r}_S$ of a given reference potential ψ_S and reach to every mesh node $\boldsymbol{r}_i$. Direct evaluation of Eq. (2) suffers from two drawbacks. First, errors accumulate during the numerical integration. Integration paths should therefore be as short as possible. We therefore constructed an algorithm that spans a tree over the mesh nodes and tries to keep integration path lengths short. Second, computation time can become large when the line integral is evaluated. We therefore reduced computation time by applying the fast multipole method (FMM) to evaluate Biot-Savart law.

2 Algorithm for integration path finding and for integration

There are some requirements for good integration paths. Of course, already computed potentials should be reused. Also, the sum over all integration path lengths as well as the maximum length should be as short as possible, i.e. the tree should be balanced. The basic idea we pursued was to regard the considered mesh on whose nodes the potential has to be computed as a graph as described in Gross and Yellen (2004). It is an undirected edge-weighted graph whose nodes equal a set V of vertices. The problem of finding paths C_i is equivalent to finding a spanning tree T whose vertices are V. The weights of the edges of T correspond to the length of the edges.

In order to build T, adjacency relationships between the vertices have to be determined first. Thus, three major steps have to be performed to compute the potential at the nodes:

1. Determine adjacency relationships between the nodes.

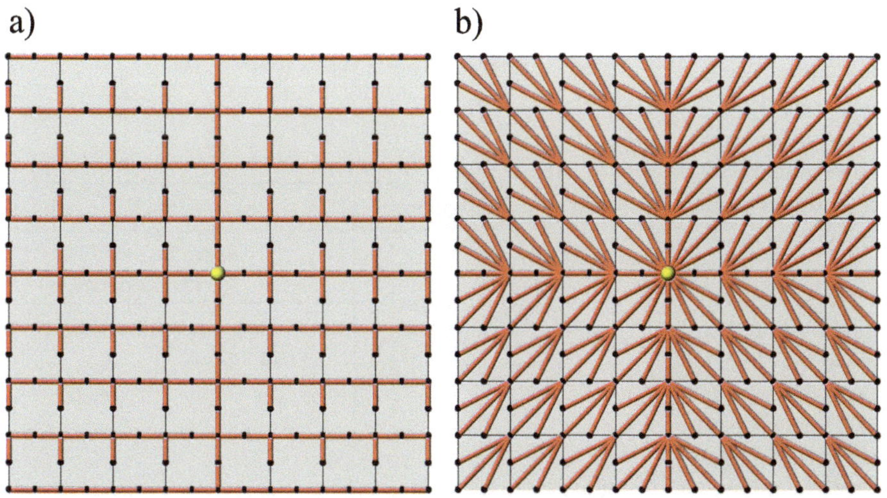

Fig. 1. Quadrate meshed with 64 eight-noded quadrilaterals. The integration tree (red) with its root node at the centre (yellow) connects all nodes of the mesh (black).

buildTree()

- Define a mesh node as root vertex and set its potential ψ_0
- *current level* = 0
- Do loop over all vertices of *current level*
  - Do loop over all neighbour vertices of current vertex
    - Add all neighbour vertices that do not exist yet within tree
- Until (no vertices have been created in *current level*)

Fig. 2. Algorithm for tree building.

2. Generate a spanning tree whose root is the start node $\boldsymbol{r}_S$.

3. Find and integrate C_i along the paths by recursively going from all leafs to the root of the tree.

Step 1 can be processed by use of the elements of the mesh. Dependant which nodes within an element are regarded as adjacent, different realizations of T are obtained. For example, in Fig. 1a only direct neighbour nodes along element edges were defined as adjacent whereas in Fig. 1b all nodes within each element that shares the node were considered so. The adjacency relationships between the N nodes can be stored in a sparse matrix, the so-called adjacency list.

For the construction of an algorithm that generates the spanning tree of step 2 we found it to be crucial to grow the tree level-wise. With this approach realizations of T are mostly good balanced which again can be seen from Fig. 1. For that we solely use the information stored in the adjacency list, i.e. node coordinates are ignored. Starting from the tree that contains at its level 0 only the start node at $\boldsymbol{r}_S$, its neighbour vertices are determined with the adjacency list. They are added to T's next level 1. Then all neighbour vertices of

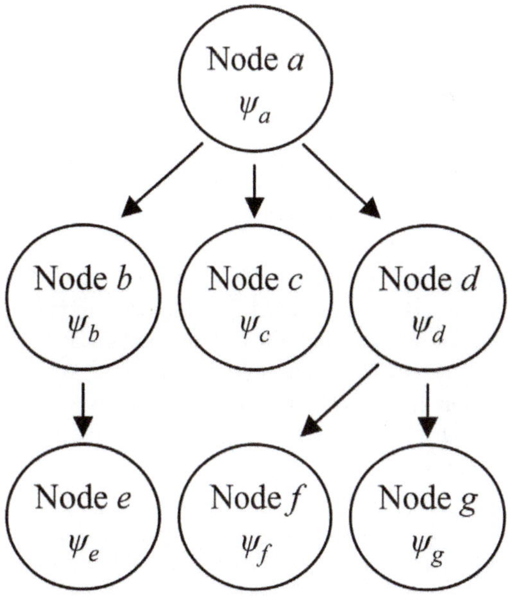

Fig. 3. Example of an integration tree.

level 1 are added at level 2 if they are not part of the tree yet and so on. The complete algorithm is shown in Fig. 2.

All line integrals along C_i in step 3 run from the root of T to its leafs. An efficient way to reuse already computed parts of the integral is to determine paths C_i by recursively starting at the leafs of the tree and moving upwards. For example, consider T displayed in Fig. 3. If the potentials ψ_a, ψ_d and ψ_f at the nodes a, d and f have been computed,

$$\psi_g = \psi_d + \int_{\boldsymbol{r}_d}^{\boldsymbol{r}_g} \boldsymbol{H}_J\left(\boldsymbol{r}'_i\right) \cdot d\boldsymbol{r}', \tag{3}$$

i.e. to compute ψ_g only integration between the ancestor of vertex g and vertex d is necessary. If integration is started at the leafs of T, ancestor vertices can be successively called until a vertex is reached whose potential has already been computed. From there on, integration between all vertices to the regarded leafs are computed and stored. In Fig. 4 the respective diagram that realizes the algorithm is displayed.

As described by Binns, Lawrenson and Trowbridge (1992) it has to be ensured that C_i does not intersect current-carrying regions. Since the exciting field is not irrotational it is also necessary to prevent the potential from being multi-valued. This can occur if the region where the potential is computed is multiply connected as with transformer problems. It therefore has been proposed by Simkin and Trowbridge (1979) to introduce suitable cuts to make the region simply connected. As the adjacency list is built with the aid of mesh elements, cuts can be defined easily by detaching elements that adjoin the cut. This means that nodes that lie on the cut have to be duplicated and assigned to elements of one side of the cut. That way the integration tree does not span over the cut which allows the potential to be discontinuous,

Vertex::computePotential()

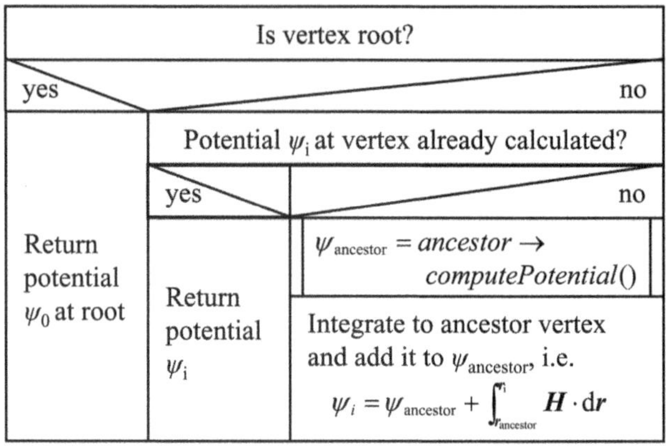

Fig. 4. Algorithm for line integration along edges of the tree.

3 Fast multipole method

Volume integrations are generally time-consuming, therefore computation of Eq. (1) is often slow. This may also hold for determination of $\boldsymbol{J}$ if it is not interpolated but computed at every integration point.

Greengard and Rokhlin (1987) already stated that the FMM can in principle be used for efficient computation of Eq. (1) but as far as it is known to the authors it has never been realized for three dimensional problems. We therefore developed an approach that allows application of the FMM to Eq. (1). It can be easily integrated in an existing FMM implementation.

Whilst in the following all derivations are valid for free currents in general, only magnetostatic problems are considered where free currents occur only in exciting coils. For simplicity volume elements that carry free currents are therefore denoted as coil elements.

In the FMM algorithm, field computation is split into a near-field part $\{\boldsymbol{H}_{\text{near}}\}$ due to observation points that are close to coil elements and a far-field part $\{\boldsymbol{H}_{\text{far}}\}$ for the remaining elements,

$$\{\boldsymbol{H}_J\} = \{\boldsymbol{H}_{\text{near}}\} + \{\boldsymbol{H}_{\text{far}}\} . \tag{4}$$

The division of element interactions into a near-field and a far-field part is carried out by means of an hierarchical grouping scheme that is based on cubes, the so-called octree. The near-field part has to be computed directly using Eq. (1). Coil elements with large distances from the considered observation point are considered group-wise in $\{\boldsymbol{H}_{\text{far}}\}$, where the size of the group depends on the distance. Therefore field sources are approximated by a truncated series expansion into spherical harmonics Y_n^{-m}. The coefficients of the series expansion are called multipole coefficients M_n^m which have been derived by Greengard and Rokhlin (1987) for single layer integral operators,

$$M_n^m = \int_A \sigma\left(\boldsymbol{r}'\right) r'^n Y_n^{-m}\left(\theta', \varphi'\right) dA' . \tag{5}$$

It was shown by Buchau et al. (2004) that multipole expansions are also possible for the double layer integral operator whose kernel is the gradient of Green's Function. Contrary to these operators the integrand of Eq. (1) has vector character. Also the differential operator cannot be put in front of the integral as in the case of field calculation with the FMM. However, the usual substitution of Green's function of its multipole expansion is still possible. Starting from this approach insertion of

$$G\left(\boldsymbol{r}, \boldsymbol{r}'\right) = \sum_{n=0}^{L} \sum_{m=-n}^{n} \frac{r'^n}{r^{n+1}} Y_n^m(\theta, \varphi)\, Y_n^{-m}\left(\theta', \varphi'\right) \tag{6}$$

into Eq. (1) gives

$$\boldsymbol{H}_J(\boldsymbol{r}) = \frac{1}{4\pi} \sum_{n=0}^{L} \sum_{m=-n}^{n} \boldsymbol{M}_n^m \frac{1}{r^{n+1}} Y_n^m(\theta, \varphi) \tag{7}$$

with the coefficients

$$\boldsymbol{M}_n^m = \int_V \boldsymbol{J}\left(\boldsymbol{r}'\right) \times \nabla_{\boldsymbol{r}'} \left[r'^n Y_n^{-m}\left(\theta', \varphi'\right)\right] dV' . \tag{8}$$

While Eq. (6) is exact if the order of the series expansion L is infinite, for numerical calculations it has to be chosen $L < \infty$, e.g. we chose $L = 9$. With the FMM algorithm the multipole coefficients of each cube of the octree are transformed to the so-called local coefficients $\boldsymbol{L}_n^m$ from which the magnetic field can be computed,

$$\boldsymbol{H}_{\text{far}}(\boldsymbol{r}) = \frac{1}{4\pi} \sum_{n=0}^{\infty} \sum_{m=-n}^{n} \boldsymbol{L}_n^m r^n Y_n^m(\theta, \varphi) . \tag{9}$$

This approach allows a standard FMM implementation to be used when each component of Eq. (9) is treated as a scalar

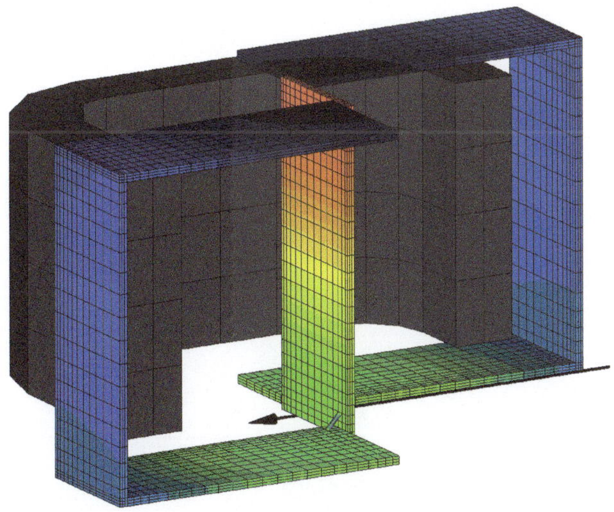

Fig. 5. Investigated mesh of TEAM Workshop Problem 13. Some elements of the coil are omitted for better graphical representation. On the right-hand side obtained integration paths are shown from the dashed section on the left.

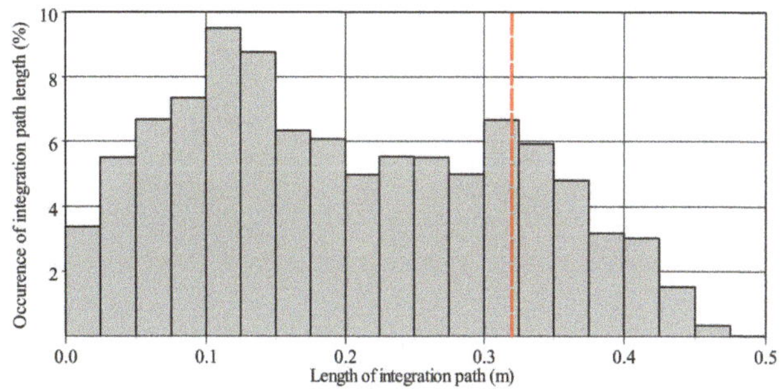

Fig. 6. Histogram plot of occurring integration path lengths.

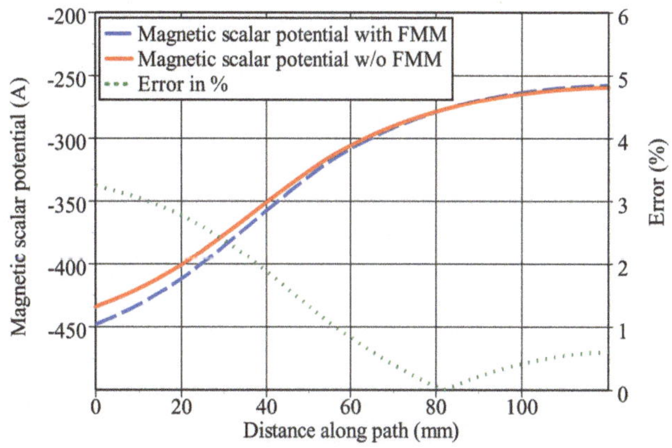

Fig. 7. Comparison of magnetic scalar potential obtained by direct evaluation of Biot-Savart law and by application of FMM. The evaluation path is shown as arrow in Fig. 5.

potential. Threefold execution of the algorithm, one for each component gives the three components of $\boldsymbol{H}_{\text{far}}$. The contribution of the near-field $\boldsymbol{H}_{\text{near}}$ needs to be computed only once and can be easily performed during one of the steps to compute a far-field component.

4 Numerical results

4.1 TEAM workshop problem 13

A mesh of the model from TEAM Workshop Problem 13 as published by Nakata et al. (1990) has been used to investigate the integration paths obtained by the presented algorithm. In Fig. 5 a contour plot of the computed magnetic scalar potential is displayed. As there is an air gap that separates the middle plate from the outer ones, auxiliary meshes that connect the parts were introduced. To keep the potential single-valued this was only done at the bottom of the middle plate but not on the top. As a results the potential is discontinuous between the upper air gaps. An enlargement of the auxiliary mesh within the dashed box is shown on the right in Fig. 5. For generation of the adjacency list only nodes lying beneath element edges were considered adjacent as in Fig. 1a. An approach as in Fig. 1b gives approximately the same accuracy but takes more computation time. Statistics of the lengths of integration paths are shown in the histogram plot in Fig. 6. The theoretical maximum integration length is the direct line within the mesh that connects the start node with the node of farest distance. It is drawn as red line within Fig. 6. This theoretical maximal length is in practice always exceeded as integration paths have to run between mesh nodes.

To compare the accuracy between direct evaluation of Biot-Savart law and with application of the FMM the potential has been computed along the arrow shown in Fig. 5. Results are shown in Fig. 7. The order of the series expansions was chosen to $L = 9$ and second order hexahedrons were used as coil elements. As results obtained by direct evaluation can be considered accurate, the error with FMM is about 3.5%. We believe this is because different to direct evaluation, we interpolated the magnetic field between mesh nodes when applying the FMM.

4.2 Toroidal coils

In Fig. 8 different meshes of toroidal coils are shown that were used to investigate the efficiency of the FMM with respect to computation time and memory requirements. The source magnetic scalar potential has been computed on the toroidal cores of the coils. In Fig. 8 mesh A, the computed potential and on the other meshes the magnetic field magnitude is shown. As the cores are multiply connected cuts had to be introduced to make the potential discontinuous which can bee seen in Fig. 8 mesh A.

In Table 1 details of the computational costs by application of the FMM as well as without its application are shown.

Fig. 8. Investigated meshes of toroidal coils. The contour plot on Mesh A represents scalar magnetic potential, the other meshes the magnitude of magnetic field.

Table 1. Comparison between application of FMM and classical approach regarding memory requirements and computation time for source magnetic scalar potential computation.

Mesh	Coil Elements	Observation points	Memory with FMM/MBytes	Time with FMM	Time w/o FMM
A	4 393	30 197	196	9 min 23 s	1 h 43 min
B	24 024	30 197	642	42 min 48 s	9 h 25 min
C	24 024	128 077	952	46 min 20 s	39 h 56 min
D	12 012	52 637	396	36 min 21 s	8 h 12 min
E	8 605	30 197	307	12 min 55 s	3 h 22 min

Observation points refer to the mesh nodes of the toroidal core. There is a large reduction of computation time especially when the number of coil elements or the number of observation points is large. The calculations were run on a 3.4 GHz Intel Pentium IV computer.

5 Conclusions

The calculation of source magnetic scalar potential is realized by performing line integrations over the magnetic field. Integration paths can be represented as edges of a spanning tree over the nodes of the mesh. Results indicate that the presented algorithm for tree generation gives balanced trees with short integration paths. Numerical errors that accumulate during line integration are therefore low. To accelerate magnetic field computation the FMM has been applied to Biot-Savart law. Computation time is thereby significantly reduced by the cost of extra memory requirements.

References

Binns, K. J., Lawrenson, P. J., and Trowbridge, C. W.: The Analytical and Numerical Solution of Electric and Magnetic Fields, John Wiley and Sons, 1992.

Buchau, A., Hafla, W., and Rucker, W. M.: Fast and efficient 3-D boundary element method for closed domains, COMPEL, 23, 859–865, 2004.

Greengard, L. and Rokhlin, V.: The rapid evaluation of potential fields in three dimensions, Lecture Notes in Mathematics, 1360, edited by Anderson, C. and Greengard, C., Berlin, Germany: Springer-Verlag, 121–141, 1987.

Gross, J. L. and Yellen, Y.: Handbook of Graph Theory, CRC Press LLC, 2004.

Nakata, T., Takahashi, N., Fujiwara, K., Olszewski, P., Muramatsu, K., Imai, T., and Shiraki, Y.: Numerical analysis and experiments of 3-d non-linear magnetostatic model, Proc. Int. Symp. and TEAM Workshop, Supplement A, Okayama, Japan, 308–310, 1990.

Simkin, J. and Trowbridge, C.: On The use of the Total Scalar Potential in the Numerical Solution of Field Problems in Electromagnetics, Intern. J. for Num. Methods in Engineeering, 14, 423–440, 1979.

Proposal for scalable models in EMC simulation

Á. Leibinger and Á. Hajdu

Automotive Electronics, Robert Bosch GmbH, Reutlingen, Germany

Abstract. In this paper a method "Component Series Modeling" (CoSeMod) is presented. This allows fast and easy implementation of scalable model generations for passive component series based on measurement data or specification provided by manufacturer. These can be used in circuit models for fast EMC analysis and optimization, especially in frequency ranges where conducted emission and susceptibility dominate. EMC tasks require high precision equivalent circuit models of components. Models generated with CoSeMod provide in many cases as high a quality as original (static) models do. One feature of scalability is that new netlisting is not needed after component changes. The process of model creation is based on similarities of the components of the same model series (packaging, manufacturing process, material etcetera). Required equations of the relationship between nominal and parasitic values are calculated by nonlinear regression. Model generation for unknown components of a known series is possible with interpolation. Implementation is possible with relatively simple actions made in circuit simulator Saber. An EMC application example of the implemented model is also shown in this paper.

1 Introduction

High quality but scalable models of components are needed in EMC optimization tasks to achieve realistic optimization results. Manufacturers of electronic components nowadays provide datasheets for the components or, in some cases, simulation models are available in standard format e.g. SPICE, MAST, VHDL-AMS. These models are 'static' in terms of representing only one component from a series according to the requirement of the customers. The parameters of the parasitic effects are, in the equivalent circuit, not dependent on other values. This means for an EMC behavioural optimization that there are two possibilities to use the existing models.

- If the parasitic effects are not considered in the optimization and only the main parameter of the component is optimized, the result of the optimization can be very implausible. Parasitic effects can significantly influence the spectral and transient character of a circuit. Usage of this method is restricted for that reason.
- If the models of every individual component with parasitic effects are considered, the models have to be swapped in the circuit schematic during optimization. A new netlist has to be generated after every iteration step and bias point values have to be calculated again. This is an appropriate method to obtain exact optimization results. Due to swaps and bias calculations the realization is very resource consuming and complicated.

Models delivered by manufacturers are not specified for HF or EMC analysis and they focuse on behavioural modelling of the component. It is recommended to use our own scalable models for whole component series in case of EMC optimization tasks.

Scalable models can be realized in circuit simulators from datasheet or measured data (e.g. Cojocaru et al., 2002). One possible realization is a subcircuit model with lookup table (LUT). LUT models have accurate results but several disadvantages downgrade usability: no good overview of relations, no adjustability, complicated realization, and no possibility to statistically correct modelling. Another possible solution is to build up the model by using simulators hardware description language. It is very time-consuming as well and it can lead – without enough coding experience – to errors.

The most appropriate modelling for EMC optimization can be, in many cases, the CoSeMod scalable modelling for component series. In this method an equivalent circuit is realized as a subcircuit. Values of the circuit elements are calculated from the nominal value of the given component with a fitted regression model (Seber and Wild, 1989). In this paper models for four different series are investigated with 0603 or 0805 size and with X7R or C0G dielectric material.

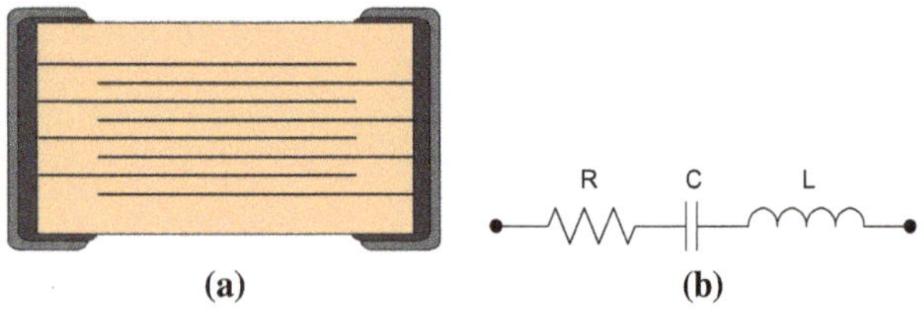

Fig. 1. (**a**) MLCC (multi-layer ceramic chip capacitor) structure (**b**) simplified equivalent circuit of MLCCs.

2 Theoretical background

The function of this method will be demonstrated on capacitors because of the relative simplicity of their construction but it is not restricted to them. Best modelling can be achieved if an accurate equivalent circuit is known and the parasitic element values are in strong correlation with the nominal value.

2.1 Modeling of capacitors

In monolithic ceramic capacitors the plates are stacked up in a comb-like structure in multiple layers as shown in Fig. 1 so the surface-to-volume ratio is large, and a ceramic material acts as the dielectric medium. Even if an accurate equivalent circuit of a MLCC involving parasitics is complex, it can be lumped as the simplest series or parallel circuit model, which represents the real and imaginary (resistive and reactive) parts of the total equivalent circuit impedance. Which circuit model should be used and which parasitic elements are determinative or negligible depends on the environment of operation, namely the applied frequency range and application. The goal was, in this case, to use the models in frequency ranges where conducted electromagnetic (EM) effects dominate. Usually, the frequency band for conducted emission measurements is 150 kHz–108 MHz in European countries (see in CISPR SC/D, 2008). However, modelling frequency range spans up to 1 GHz depending on the required accuracy. A realistic impedance characteristic is expressed with three-element RLC equivalent circuit parameters according to the required simplicity and fidelity. This structure is sufficiently accurate in most optimization cases; however the attenuation capability of the capacitor can be underestimated in noise reduction tasks, as described in Smith and Hockanson (2001).

2.2 Measuring impedance characteristic

Measurements were realized with a network analyzer equipped with 2000 μm HF probes. The DUT was placed on a rubber mat to avoid the influence of copper pad inductivity and to define a substrate for the measurements. In practical usage the substrate influence has to be considered also as in Lakshminarayanan et al. (2000). The test is demonstrated in Fig. 2. One port *S11* scattering parameter (Kurokawa, 1965) measurement was carried out and the data was converted to get the impedance curve of the component based on the following equation:

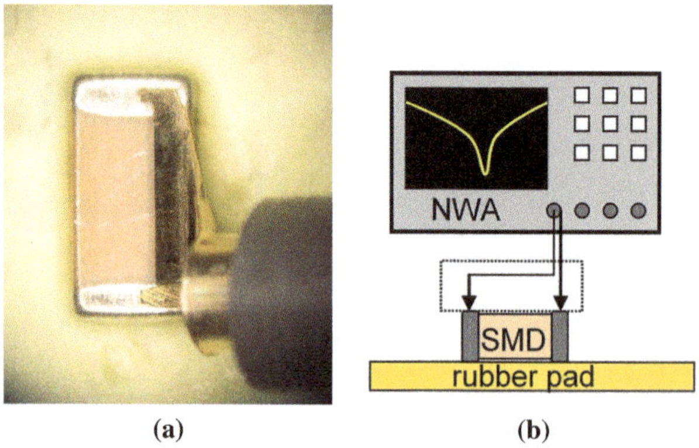

Fig. 2. (**a**) Measurement pins of HF probe on 0805 MLCC. (**b**) Test setup for impedance curve measurement with NWA.

$$Z_{11} = Z_0 \frac{1+S_{11}}{1-S_{11}} \tag{1}$$

Five pieces of the same component were measured. Components of one series were chosen by size, dielectric material, voltage limit, termination structure.

2.3 Model parameter extraction from measured impedance curves

The impedance curve can be split in three main regions. The capacitor behaves in the first region under series resonance frequency (SRF) as an ideal capacitor. So the impedance is determined by the following equation:

$$\log\left|Z_{(f<<\mathrm{SRF})}\right| = \log\frac{1}{2\pi f C} = \log\frac{1}{2\pi C} - \log f \tag{2}$$

At the resonance frequency there is no imaginary part of the impedance value so that the impedance is equal with the resistance of the circuit.

$$\log\left|Z_{(f=\mathrm{SRF})}\right| = R \tag{3}$$

Over the resonant point the inductive behavior dominates:

$$\log\left|Z_{(f>>\mathrm{SRF})}\right| = \log 2\pi f L = \log 2\pi L + \log f \tag{4}$$

From Eqs. (2) and (4) follows that a line with a slope of $\pm\log f$ has to be fitted on the measured data. L and C can be calculated from the y-value at the interception point of the y-axis.

As was mentioned previously, the investigated impedance curves of each capacitance value are an average of five measurements on five pieces of the same component. To perform a successful and reasonable regression on the dataset, the deviation of each curve can not be too large from the averaged curve. Therefore the equivalent circuit parameters were extracted from each impedance curve and their standard deviation was expressed from the average value, on which the curve fitting will be performed later. The standard deviation of the interception parameter of the fitted straight lines during the parameter extraction was not taken into account since its value was negligible (less than 0,01%). Also the variability of values around the average is considered to follow approximately a Gaussian distribution. The investigation was performed on the smallest and the largest value for each capacitance range. The standard deviation of the ESC value is small and fits into the capacitance tolerance range given by the manufacturers (±5% for COG and ±10% for X7R). Also the equivalent inductance of the components shows an acceptable result, the standard deviation values are below ±5% in all cases. Notable differences can be observed only in the case of the ESR values. The measurement of the equivalent series resistance is less accurate for higher capacitance values with this measurement method, as described in Hajdu (2010).

2.4 Curve fitting with non-linear regression

Before performing the nonlinear regression, the appropriate model has to be selected. By taking a look at the shape or tendency of the data points, the main curve category that fits the results can be determined. However, at this step a trade-off has to be made between accuracy and simplicity. After defining the best model a first initial value for the model parameters is required to generate the first curve. To identify the fidelity of the model, the sum-of-squares is calculated. To reach predefined quality, the variables have to be adjusted to make the curve come closer to the data points. The goal is to minimize the sum of squared residuals:

$$\phi(x) = \frac{1}{2}\sum_{i=1}^{m} f_i^2(x) = \frac{1}{2}\|F(x)\|^2 \tag{5}$$

Where $F : \mathbf{R^n} \rightarrow \mathbf{R^m}$ and $m \geq n$. Difference of the observed data point y_i and the calculated curve point at x_i is denominated as f_i where $\boldsymbol{\beta}$ is a vector of adjustable parameters.

$$f_i(x) = y_i - f(x_i; \boldsymbol{\beta}) \tag{6}$$

The most commonly used algorithm is the Levenberg-Marquardt method (Moré, 1978), which we also used. Some fitting quality indicators were used so that the quality of different fittings was comparable. The two indicators we used were adjusted R^2 and χ^2. The coefficient of determination, R^2 , can be calculated with the following expression with the mean of all observed data, $\bar{y}$:

$$R^2 = 1 - \frac{\sum_i (y_i - f(x_i; \boldsymbol{\beta}))^2}{\sum_i (y_i - \bar{y})^2} \tag{7}$$

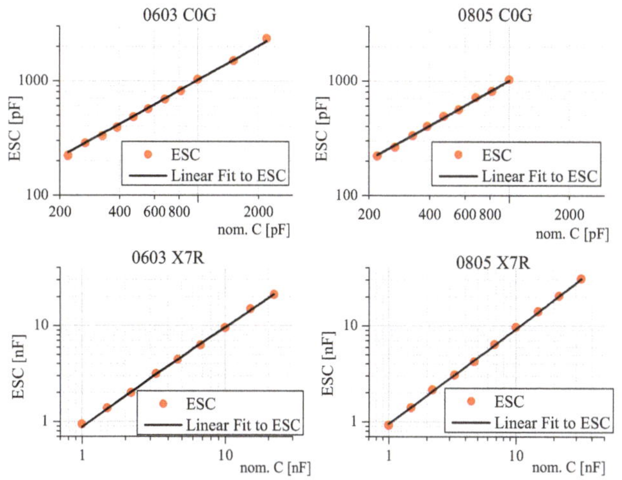

Fig. 3. The extracted ESC values and the results of curve fitting.

R^2 can be modified to adjusted R^2 to improve informative value with the sample size n and the total number of regressors k without the constant term in this form:

$$\text{adj.}R^2 = 1 - \left(1 - R^2\right)\frac{n-1}{n-k-1} \tag{8}$$

The errors can be assumed to have a normal distribution so that a reduced χ^2 distribution can be also used to test the quality of fit. The variance σ related to the measurement error for y_i is needed for the calculation of the reduced Chi-square.

$$\chi^2_{\text{red}} = \frac{\chi^2}{\nu} = \frac{1}{\nu}\sum_i \frac{(y_i - f(x_i; \boldsymbol{\beta}))^2}{\sigma^2} \tag{9}$$

The degrees of freedom (ν) can be calculated as $n - k$.

3 Fitting results

3.1 Equivalent series capacitance

The nominal capacitance of the measured DUT is dominant at lower frequencies. Therefore the data points resulting from the parameter extraction are expected to be very close to the nominal capacitance value. Figure 3 shows the corresponding curve fitting results for all the four types of capacitors with a linear regression. The outcome meets the previous expectations. The relation between the nominal capacitance of the DUT and the measured ESC values is linear.

The data points are plotted on a logarithmic scaled x- and y-axis, therefore the fitted curves also look linear.

3.2 Equivalent series resistance

To understand the series resistance of the capacitance we have to consider the structure of the component. The simplest interpretation is that the series resistance is the sum

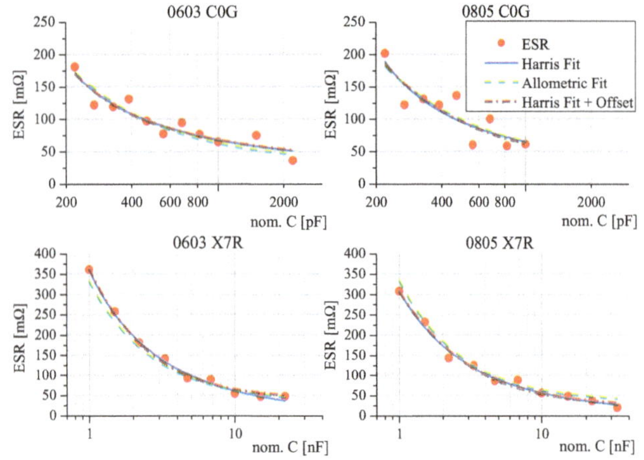

Fig. 4. The extracted ESR values and the results of curve fitting.

of the termination resistances and electrode resistances. A more sophisticated explanation can be found in Coda et al. (1976). Higher capacitance values can be reached with enlarging the overlapping area of the electrode plates, increasing the number of the plates, or with decreasing the distance between two electrodes. The last has negligible influence on the equivalent resistance; the other two reduce the resistance value. Thus one can expect that the resistance decreases with increasing nominal capacitance value. This expectation is confirmed by the measurement results in Fig. 4.

As there is no information about the internal structure of the capacitor, two standard fitting models were used for curve fitting: Allometric, Harris. They provide good quality and simplicity at the same time. The Harris fit gave a better result. An important aspect has to be taken into consideration. In the case of the two power form functions a reasonable meaning is behind their equations. The so-called Harris fit has a maximum value that it cannot exceed when the variable goes to zero. On the other hand, it does not have a minimum boundary so theoretically a DUT with large capacitance value can have zero ESR. This is obviously impossible. Therefore a modification on the Harris fit was necessary. By introducing a fourth parameter, the resulting function has boundaries for both extreme values. The outputs of the regression analysis for Harris fit with offset are collected in Table 1.

3.3 Equivalent series inductance

The most difficult parasitic component to measure and investigate is the equivalent series inductance. First of all it is performed at frequencies usually above 100 MHz so the imperfections in the measurement setup, the unwanted structural and external effects have extended influence on the accuracy. In addition, the value of the ESL in case of chip multilayer ceramic capacitors is quite small mainly due to the SMD packaging, and is usually around 0.5–1.5 nH.

Table 1. Curve-fitting results for the ESR values of the four different capacitors with Harris + offset.

	0603 C0G	0805 C0G	0603 X7R	0805 X7R
Eq.		$y=a+(b+cx_d)-1$		
a	15	30	40	20
b	−0.02	-5×10^{-4}	5×10^{-4}	5×10^{-4}
c	0.01	3×10^{-5}	0,002	3×10^{-3}
d	0.23	1	1.2	1
Red. χ^2	230.9	585.3	78.2	128.7
Adj. R^2	0.85	0.73	0.99	0.98

The investigation of the measured results is also more complicated. To find out the source of ESL, again the structure of the chip multilayer capacitor shown in Fig. 1 should be inspected. The electrodes can be considered as flat metallic plates, whose thickness is negligible compared to the other two dimensions. Each elemental capacitor is in series with itself not only its own self-inductance, but also the mutual inductance of all the other plates, the magnitude of which will diminish with the plate spacing. As these spacings are generally small, these mutuals will approach, though never exceed, the self-inductance of any one plate. The inductance in series with each elemental capacitor will therefore not exceed nL. With n capacitors with L inductance in parallel, and since these can now be considered fully decoupled, the effective inductance of the whole capacitor will be the inductance of any one of them divided by n. This means that in case of the two extreme conditions, the resultant inductance does not change as the number of layers increases if the spacing between them is small enough. Or it decreases by $1/n$ if the electrodes are far from each other. Here the influence of mutual inductances between the terminals and the end of the opposite electrodes or between two parallel but not overlapping plates are neglected, since they are a lot smaller in value compared to those between the overlap region of the plates. Also the inductance of the terminals themselves has to be taken into account. To simplify the calculation, they can be considered to be constant for the investigated series of capacitances since the geometry of the SMD packaging does not differ to a great extent. As a consequence of the previously mentioned relationships, the ESL value of a chip multilayer ceramic capacitor is expected to decrease as the nominal capacitance increases, but because of the many influential parameters (plate geometry and number of layers) the correlation might be more complicated.

The previous expectations seem to be valid in most of the cases if we take a look at the fitted curves in Fig. 5 for all the four different capacitor types.

The distribution of the data points in case of the X7R type dielectric seems to follow a periodic behaviour. The origin of this behaviour is in a structure change between the values

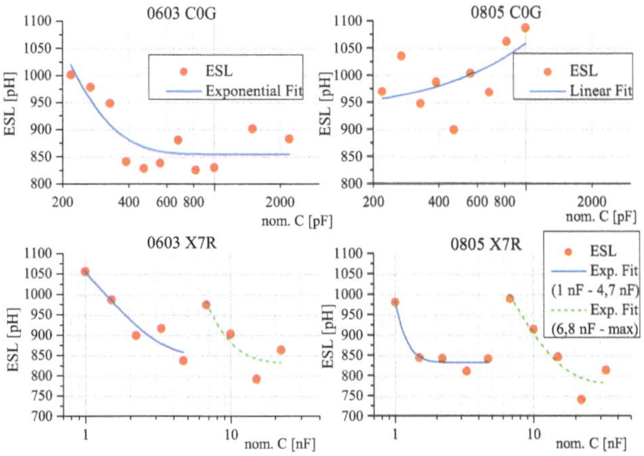

Fig. 5. The extracted ESL values and the results of curve fitting.

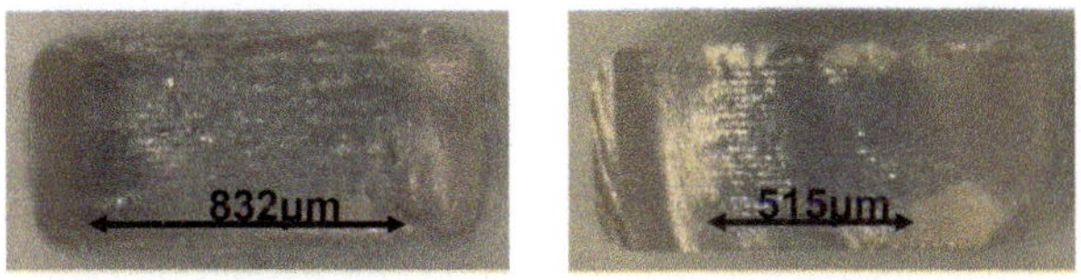

(a) **(b)**

Fig. 6. Cross section view of 0805 X7R capacitances with **(a)** 4.7 nF and **(b)** 6.8 nF.

4.7 nF and 6.8 nF. The increased capacitance value is reached with an increased number of plates but their dimensions are lesser. This is depicted in Fig. 6.

Unfortunately no such behavior could be observed regarding the components with COG type dielectric, although in the case of the 0603 sized DUT the data points are still not randomly scattered, so here the distribution was approximated also by an exponential function. For the 0805 sized COG type capacitor, a reasonable fitting function could not be found, so a simple linear one was used.

4 Implementation of the model

To test the previously expressed equations of the scalable models they have to be implemented in a circuit simulator. The implementation was realized as a subcircuit in Saber. A little trick was needed because the compiler of Saber does not understand every mathematical operation. Therefore the implemented equations look a bit more complex, because the power operation was substituted by the combination of exponential and logarithm functions. The step in the ESL curve of X7R capacitors between 4.7 nF and 6.8 nF, which has been explained in Sect. 3.3, was realized by the use of the *int* command, which rounds the operand to the nearest integers towards minus infinity. Realization for the 0805 X7R capacitor (right bottom corner in Figs. 3–5) is in Fig. 7, where *C_nom* is the nominalvalue of the capacitor in nF.

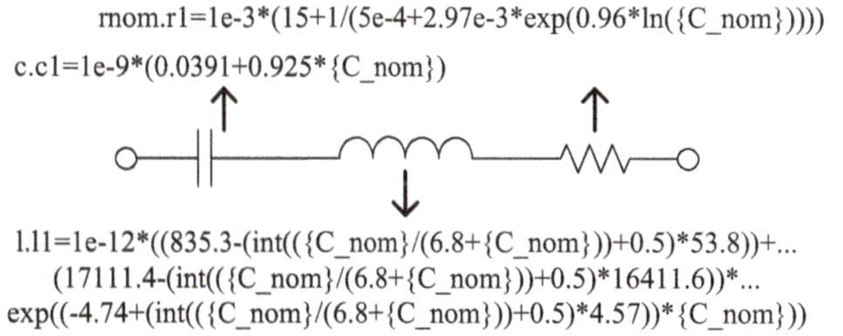

Fig. 7. Subciruit realization of the scalable model for one 0803 size X7R capacitor.

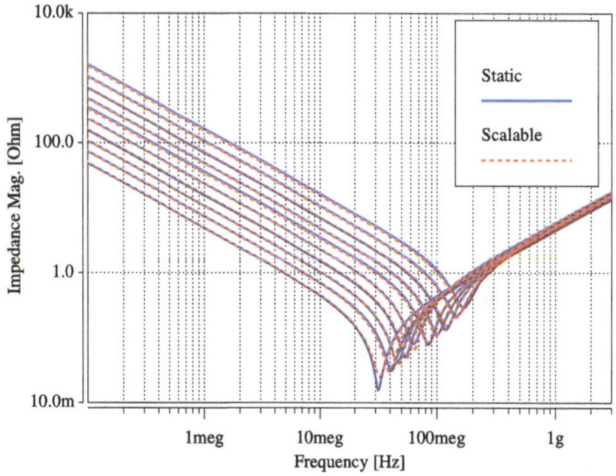

Fig. 8. Comparison of scalable and static model impedance magnitude for 0805 X7R capacitance series.

The scalable models were compared with the static models to validate them. The goal was to show that they are almost as good a quality as static models are. The comparison of the model shown in Fig. 7, is in Fig. 8.

The difference between the simulation and the measurement is quite small in the case of this series, but to have a better overview, the accuracy of each equivalent circuit parameter has to be tested. Therefore results of those capacitances were collected where the difference between the measured value and the fitted curve was the biggest. In this way the maximal margin can be determined and so the worst case situation can be investigated. The results are collected in Table 2. The accuracy is very high in all cases except in case the of the ESR. The measurement uncertainty of the resistance value also influences this result (see Sect. 3.2). This value is responsible for the amplitude in the spectrum which is measured usually in dB. For that reason this deviation is also acceptable for EMC purposes.

5 Possible usage in EMC

Scalable models are not avoidable if the goal is an automated EMC optimization, described in Leibinger (2010).

Table 2. Maximal modelling failures.

Parameter	ESC	ESR	ESL
0603 COG	3.20%	15%	5.60%
0805 COG	3.30%	22.10%	9.10%
0603 X7R	3.20%	14.40%	5.80%
0805 X7R	3.40%	13.00%	6.70%

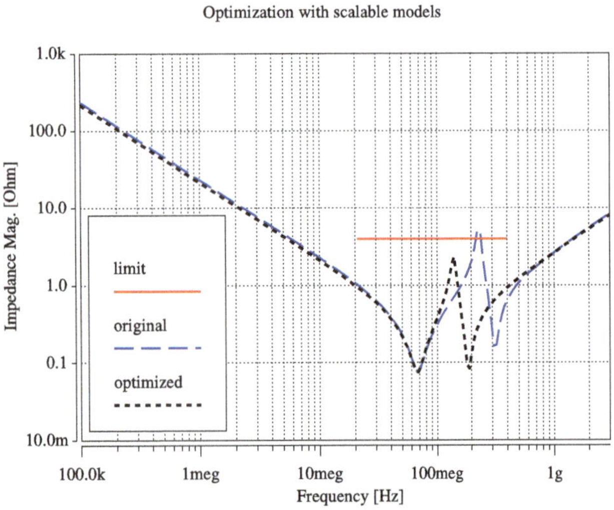

Fig. 9. Optimization of an input filter of two parallel SMD capacitors.

One typical task is the parameter optimization for input filters. In Fig. 9 is an example of two parallel 0603 size scalable capacitor models with X7R (1–22 nF) and C0G (220–2200 pF) dielectric for input filter optimization.

The advantageous behaviour of the scalable models is that they are easily extendable. With new functions within the subcircuit, more optimization goals can be defined simultaneously e.g. thermal, cost. The thermal or cost functions can be determined with the same method as the parasitics above.

6 Conclusions

Scalable models can be used in simulations, when a rapid optimization is desired. They reduce the number of parameters to be changed and therefore save computing capacity. Validation of the models must be proved with statistical measurement data if possible. A very good model approximationwith interpolation is possible for unknown components of one model series, which is presented with only few known components. This can be considered as a time-saving possibility for fast modelling. If a sufficient level of maturity is achieved in such models, these models can provide as good a quality as static models do for many component types. Good scalable modelling can be achieved with components in series that have little scattering regarding nominal value and parasitic values. SMD components are suitable for such a modelling.

References

CISPR SC/D: CISPR 25 Vehicles, boats and internal combustion engines – Radio disturbance characteristics – Limits and methods of measurement for the protection of on-board receivers, Standard, IEC, 2008.

Coda, N. and Selvaggi, J. A.: Design Considerations for High-Frequency Ceramic Chip Capacitors, IEEE Trans on Parts, Hybrids, Packag, 12(3), 206–212, 1976.

Cojocaru, V., Markell, D., Capwell, J., Weller, T., and Dunleavy, L.: Accurate Simulation of RF Designs Requires Consistent Modeling Techniques, High Frequency Electronics, 1(2), 46–51, 2002.

Hajdu, A.: High-Frequency Characterization and Modeling of Electronic Components, Diploma Thesis, 2010.

Kurokawa, K.: Power Waves and the Scattering Matrix, IEEE Trans. on MTT, MTT-13(2), 194–202, 1965.

Lakshminarayanan, B., Gordon Jr., H. C., and Weller, T. M.: A Substrate-Dependent CAD Model for Ceramic Multilayer Capacitors, IEEE Trans. on Microwave Theory and Techniques, 48(10), 1687–1693, 2000.

Leibinger, A.: Automatisierungsmglichkeiten der EMV Optimierung in der Netzwerksimulation, AmE 2010 Conf., Dortmund, 2010.

Moré, J. J.: The Levenberg-Marquardt algorithm: Implementation and theory In Numerical Analysis, edited by: Watson, G. A., Lecture Notes Math., 630, 105–116, Springer, 1978.

Seber, G. A. F. and Wild, C. J.: Nonlinear Regression, John Wiley & Sons, Canada, 1989.

Smith, L. D. and Hockanson, D.: Distributed SPICE Circuit Model for Ceramic Capacitors, IEEE Electronic Components and Technology Conference, 2001.

42

A low-noise high dynamic-range time-domain EMI measurement system for CISPR Band E

C. Hoffmann[1] **and P. Russer**[2]

[1]Technische Universität München, Lehrstuhl für Hochfrequenztechnik, Munich, Germany
[2]Technische Universität München, Lehrstuhl für Nanoelektronik, Munich, Germany

Abstract. In this paper, a broadband time-domain EMI measurement system for measurements from 9 kHz to 18 GHz is presented that allows for compliant EMI measurements in CISPR Band E. Combining ultra-fast analog-to-digital-conversion and real-time digital signal processing on a field-programmable-gate-array (FPGA) with ultra-broadband multi-stage down-conversion, scan times can be reduced by several orders of magnitude in comparison to a traditional heterodyne EMI-receiver. The ultra-low system noise floor of 6–8 dB and the real-time spectrogram allow for the characterisation of the time-behaviour of EMI near the noise floor. EMI measurements of electronic consumer devices and electric household appliances are presented.

1 Introduction

Electromagnetic interference (EMI) and electromagnetic compatibility (EMC) are traditionally measured on heterodyne EMI-receivers (Hagenhaus, 1942). The sequential measurement of typically several thousand frequencies yields long scan times on the order of hours or even days. Time-domain measurement systems can reduce scan times by several orders of magnitude. After broadband sampling of the EMI signal, the spectrum is calculated via the Fast-Fourier-Transform (FFT) and detectors are applied digitally. With the presented system, an EMI measurement over the complete band from 9 kHz to 18 GHz with an IF-bandwidth of 9 kHz takes less than 3 min. Using a traditional EMI-receiver with a dwell-time of 100 ms, the measurement of roughly 2×10^6 frequency points would take over 55 h.

The standardization and development of new communication technologies like WIMAX is attended by an increasing usage of higher frequency bands in the high microwave range. This development implies the need for fast and comprehensive EMI measurements in the frequency range above 1 GHz.

The presented time-domain EMI measurement system allows for the CISPR-compliant measurement of EMI in CISPR Band E up to 18 GHz. The real-time capability permits the characterisation of the time-behaviour of a device's emission. Spectrogram measurements of the data transmission between two modern smartphones and the non-stationary radiated emission of a microwave oven are shown.

2 Time-Domain EMI measurement system

The block diagram of the presented time-domain EMI measurement system is shown in Fig. 1. The EMI-signal is received by a broadband antenna for radiated emission or a line impedance stabilization network (LISN) for conducted emission. For measurements from 9 kHz to 1.1 GHz, the input signal is filtered by a lowpass filter to prevent spectral overlap due to a violation of Shannon's theorem. The filtered signal is sampled by the floating-point analog-to-digital converter (Braun and Russer, 2005) with a sampling rate of around 2.6 Gs/s. An FPGA computes the signal spectrum via the FFT and weights the spectrum. The amplitude spectrum is displayed.

2.1 Fast-Fourier-Transform

To compute the EMI signal spectrum, the digitized EMI-signal is transformed by FFT. The FFT is an efficient algorithm for the computation of the Digital-Fourier-Transform (DFT). The FFT exploits symmetry and repetition properties and is defined as (Oppenheim and Schafer, 2009)

$$X[k] = \sum_{n=0}^{N-1} x[n] e^{\frac{-j2\pi kn}{N}}, \qquad (1)$$

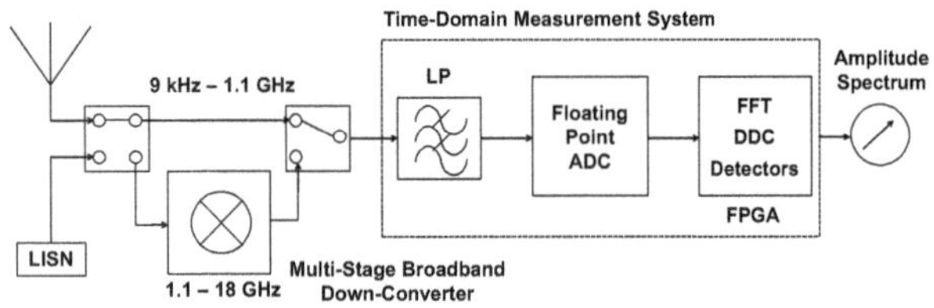

Fig. 1. Time-domain EMI measurement system.

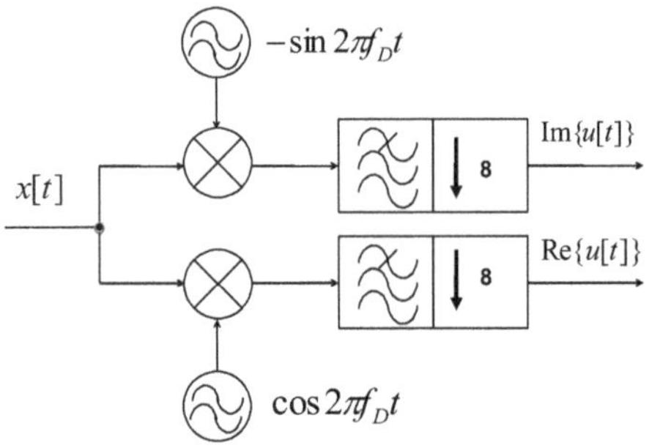

Fig. 2. Digital down-conversion, taken from Braun et al. (2009).

where $X[k]$ is the discrete amplitude spectrum of the discrete time signal $x[n]$.

2.2 Short-Time-Fast-Fourier-Transform

The Short-Time-Fast-Fourier-Transform (STFFT) is defined as an FFT over a limited time-interval. A Gaussian window function $w[n]$ is applied, corresponding to the IF-filter of a conventional measurement receiver. By application of the STFFT, a spectrogram is calculated. The spectrogram is a FFT taken over a time-interval of the sampled time-domain signal. It depends on the discrete time coordinate τ of the window and the discrete frequency k. The STFFT is calculated by (Oppenheim and Schafer, 2009)

$$X[\tau,k] = \sum_{n=0}^{N-1} x[n+\tau]w[n]e^{\frac{-j2\pi kn}{N}}. \qquad (2)$$

2.3 Digital Down-Conversion (DDC)

In order to avoid data overflow in the signal processor and to enable real-time processing of the signal, the frequency range from DC to 1.1 GHz is subdivided into eight subbands with a bandwidth of 162.5 MHz each. For the in-phase and quadrature channel, a polyphase decimation filter reduces the sampling frequency in order to fulfill the Nyquist criterion. Every subband is digitally down-converted to the baseband and the subbands are processed sequentially (Braun and Russer, 2005). The output sampling frequency is 325 MHz, while the bandwidth is 162.5 MHz. The block diagram is shown in Fig. 2.

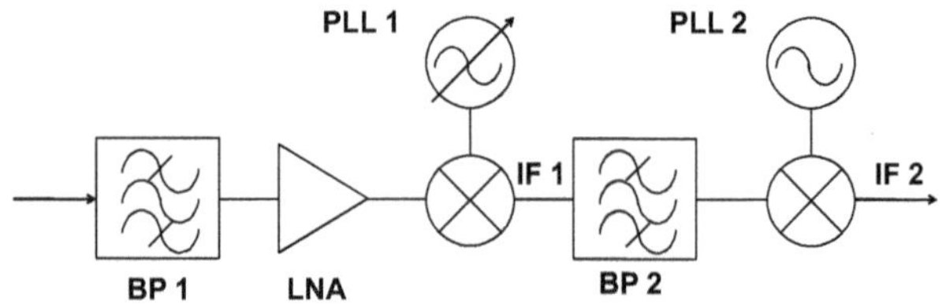

Fig. 3. 1.1–6 GHz down-converter.

3 Multi-stage broadband down-converter

To overcome the limitations imposed by available analog-to-digital converters, an ultra-broadband multi-stage down-converter is used to extend the frequency range up to 18 GHz. The upper frequency limit of the time-domain measurement system was increased to 3 GHz (Braun et al., 2009), 6 GHz (Braun et al., 2010) and 18 GHz (Hoffmann et al., 2010) by the addition of the multi-stage broadband down-converter. The EMI signal from 1.1–6 GHz is down-converted to the frequency range below 1.1 GHz by the 1.1–6 GHz down-converter. The down-converted signal is sampled and the spectrum is calculated. The frequency band from 6–18 GHz is in a first step down-converted to the input range of the 1.1–6 GHz down-converter. In a second step, it is down-converted to the range below 1.1 GHz and sampled.

3.1 1.1–6 GHz down-converter

The block diagram of the 1.1–6 GHz down-converter is shown in Fig. 3. The EMI signal from 1.1–6 GHz is down-converted by two mixer stages. The input band is divided into 16 subbands with a bandwidth of 325 MHz each. The inherent overlap of the subbands is used to eliminate unwanted mixer spurious from the intermediate frequency band by a slight frequency shift of the subbands. The subbands are sequentially up-converted to a first intermediate frequency band IF 1 above the input band using a broadband mixer. The intermediate frequency band is filtered by the narrowband bandpass filter BP 2 and down-converted to the frequency range below 1.1 GHz by the second mixer. The local oscillator frequencies are generated by low-noise PLL-synthesizers.

Because of the nonlinear characteristics of the mixer, a large number of mixing products are generated at its output. These frequencies f_{IF} are determined by (Vendelin et al., 2005)

$$f_{\mathrm{IF}}^{m,\pm n} = |m \cdot f_{\mathrm{LO}} \pm n \cdot f_{\mathrm{RF}}|\ ,\ m,n \in \mathbb{N}, \qquad (3)$$

where f_{RF} is the RF input frequency and f_{LO} is the local oscillator frequency.

If only the fundamental frequencies of f_{LO} and f_{RF} are taken into consideration, i.e. $m,n = 1$, we obtain two frequency components $f_{\mathrm{RF}\,1,2}$ according to Eq. (3) (Maas, 1993)

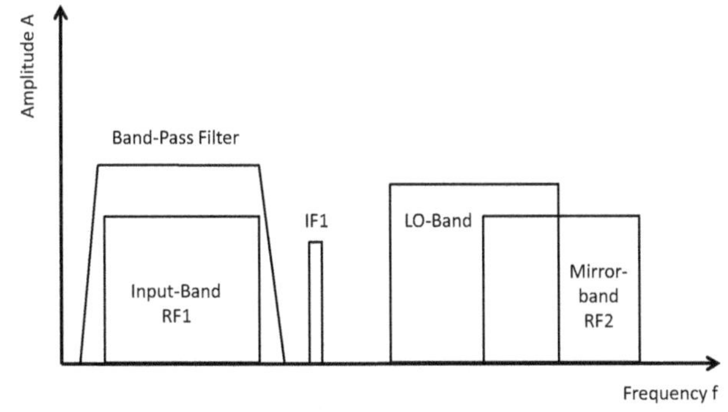

Fig. 4. Up-conversion frequency bands.

$$f_{\mathrm{RF}\,1,2} = |f_{\mathrm{LO}} \pm f_{\mathrm{IF}}|. \tag{4}$$

The frequency conversion yields two sidebands. The image-frequency signal is converted to the same intermediate frequency as the desired signal. Because the input band and the image band are not overlapping spectrally, the image band is sufficiently suppressed by the preselection bandpass filter BP 1 in this two-stage mixing scheme as illustrated in Fig. 4.

The preselection band-pass filter also increases the spurious-free dynamic range of the system by suppressing low-frequency out-of-band EMI signal components. A high-gain, low-noise InGaP/GaAs MMIC preamplifier increases system sensitivity.

3.2 6–18 GHz down-converter

The block diagram of the 6–18 GHz down-converter is shown in Fig. 5. The input band is divided into 3 ultra-wide subbands: band 1 from 6–9 GHz, band 2 from 9–13 GHz and band 3 from 13–18 GHz. Broadband, low-insertion loss, single-input, triple-output (SP3T) PIN-diode switches are used to switch between the bands. Preselection filters increase system dynamic range. The EMI-signal is amplified by a low-noise amplifier and down-converted to the frequency band from 1.1–6 GHz via a broadband mixer with low conversion loss. The local oscillator frequencies are generated by a low-noise PLL-Synthesizer and an active multiplier. Because the local oscillators fundamental is not sufficiently attenuated by the mixer's LO-IF isolation, a fifth order planar bandpass filter is used to suppress the fundamental by over 75 dB.

3.3 Noise behaviour

With the signal-to-noise-ratios $\mathrm{SNR_{in}}$ at the input and and $\mathrm{SNR_{out}}$ at the output of a system, the system's noise figure F is given by (Davenport and Root, 1987)

$$F = \frac{\mathrm{SNR_{out}}}{\mathrm{SNR_{in}}}. \tag{5}$$

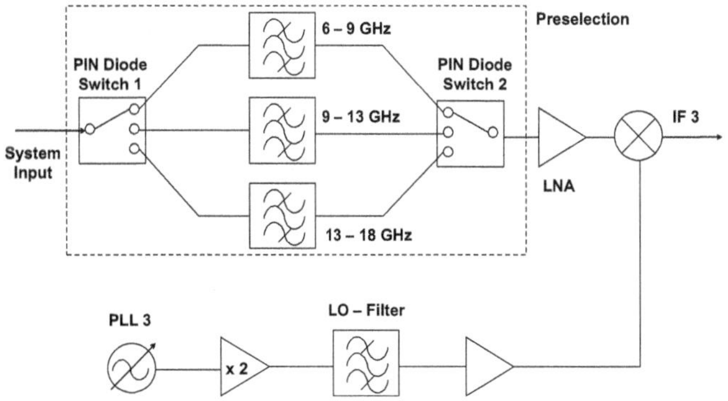

Fig. 5. 6–18 GHz GHz down-converter.

The available noise power P_N at the output of a system with a noise figure F and unity gain is defined by

$$P_N = kT_0 B_{\mathrm{ENB}}, \tag{6}$$

where k is the Boltzmann constant, T_0 is the ambient temperature and B_{ENB} is the equivalent noise bandwidth. The noise power at the output of a system, that exhibits the noise figure F can be calculated by

$$P_N = FkT_0 B_{\mathrm{ENB}}. \tag{7}$$

For the calculation of the system noise figure above 1 GHz, the noise figures of the multi-stage broadband down-converter components up to the first mixer have been considered. The noise figure F of a cascaded system with N stages is determined by

$$F = 1 + F_1 + \frac{F_2 - 1}{G_1} + \frac{F_3 - 1}{G_1 G_2} + \ldots + \frac{F_N - 1}{\prod_{k=1}^{N-1} G_k}, \tag{8}$$

where G_i is the available power gain of stage i and F_i is the noise figure of stage i. With Eq. (8), the system's noise figure was calculated to be less than 6 dB in the range from 1–6 GHz and less than 8 dB from 6–18 GHz.

The theoretical average system noise floor can be calculated using Eq. (7). The equivalent noise bandwidth B_{ENB} of the IF-filters with Gaussian characteristic is obtained with

$$B_{\mathrm{ENB}} = \int_{-\infty}^{\infty} |H(f)|^2 df, \tag{9}$$

using the filter's transfer functions $H(f)$. This yields an equivalent noise bandwidth $B_{\mathrm{ENB},9\,\mathrm{kHz}}$ of 6.8 kHz for the 9 kHz digital IF-filter. With the 9 kHz IF-filter we obtain a theoretical average noise floor of $-22.5\,\mathrm{dB}\,\mu\mathrm{V}$ from 1–6 GHz and $-20.5\,\mathrm{dB}\,\mu\mathrm{V}$ from 6–18 GHz. The measured system noise floor is higher by about 2–3 dB though, because of cable attenuation, the minimum attenuation of the variable attenuator and mismatch effects.

Fig. 6. SP3T PIN-diode switch.

3.4 Dynamic range

EMI measurement systems have to exhibit a sufficient dynamic range to be able to display transient emissions correctly. The dynamic range of a system is limited by the system noise floor on the lower end and the maximum input level without distortion of the input signal on the upper end. For spectrum analyzers, the maximum level is typically the 1 dB compression point ($P_{1\mathrm{dB}}$) of the system. Heterodyne EMI-receivers add a preselection to increase the dynamic range by suppressing out-of-band broadband or narrowband signals and Gaussian noise.

The preselection filter with the equivalent impulse bandwidth $B_{\mathrm{imp,pre}}$ increases the dynamic range for pulse signals by

$$\Delta D_{\mathrm{pulse}} = 20\log_{10}\frac{B_{\mathrm{imp,sys}}}{B_{\mathrm{imp,pre}}}, \tag{10}$$

with the equivalent impulse bandwidth of the measurement receiver $B_{\mathrm{imp,sys}}$.

For Gaussian noise, the dynamic range is increased by

$$\Delta D_{\mathrm{noise}} = 10\log_{10}\frac{B_{\mathrm{noise,sys}}}{B_{\mathrm{noise,pre}}}, \tag{11}$$

where $B_{\mathrm{noise,sys}}$ and $B_{\mathrm{noise,pre}}$ are the equivalent noise bandwidths of the measurements receiver and preselection filter.

Out-of-band narrowband signals with the frequency f_n are attenuated by the preselection filter's transfer function $H_{\mathrm{pre}}(f)$. This yields an increase of dynamic range according to

$$\Delta D_n = 20\log_{10} H_{\mathrm{pre}}(f_n). \tag{12}$$

Pulse-modulated signals are used to determine the IF dynamic range of EMI measurement systems above 1 GHz. CISPR 16-1-1 (CISPR 16-1-1 Ed. 3.1 Am. 1, 2010) demands an IF dynamic range of at least 40 dB for EMI measurements above 1 GHz.

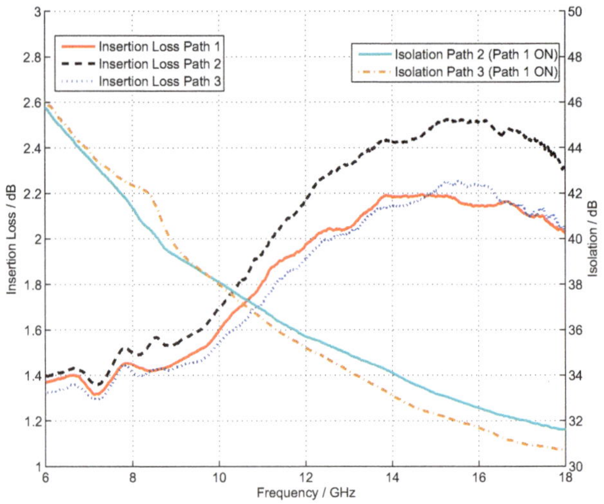

Fig. 7. Measurement of the SP3T PIN-diode switch.

4 Hardware implementation

As explained in Sect. 3.3, a low-noise figure of the first components in the RF Frontend is vital for maintaining a low system noise figure. PIN-diodes are commonly used in microwave and millimeter wave switches and attenuators.

By variation of the stored charge $Q = I_F \cdot \tau$ in the intrinsic region of a PIN-diode, its series resistance R_S can be changed by several orders of magnitude (Sze, 1988) according to

$$R_S = \frac{W^2}{I_F \cdot \tau(\mu_n + \mu_p)}, \tag{13}$$

where W is the width of the intrinsic region, I_F is the forward bias current, τ is the minority carrier lifetime and μ_n and μ_p are the electron and hole mobilities.

PIN-diode switches with single series diodes exhibit low insertion loss in the ON-state, but cannot provide the needed isolation in the OFF-state. To increase isolation, a series-shunt configuration was implemented, where dual shunt-diodes short-circuit the paths in the OFF-state in addition to the series diode in reverse polarity. The switches were fabricated on a glass reinforced hydrocarbon/ceramic RF-substrate and are shown in Fig. 6.

The performance characteristics of the switches were measured with a vector network analyzer (VNA). The measured scattering parameters are shown in Fig. 7. The insertion loss for Path 1 and Path 3 is below 2.2 dB from 6–18 GHz. At 2.5 dB, the insertion loss of the middle path is around 0.3 dB higher, because of coupling to the other two paths. The isolation exceeds 30 dB at 18 GHz.

5 Measurement results

For the characterisation of the system IF dynamic range, a pulse generator fed a pulse modulated sinusoidal signal with

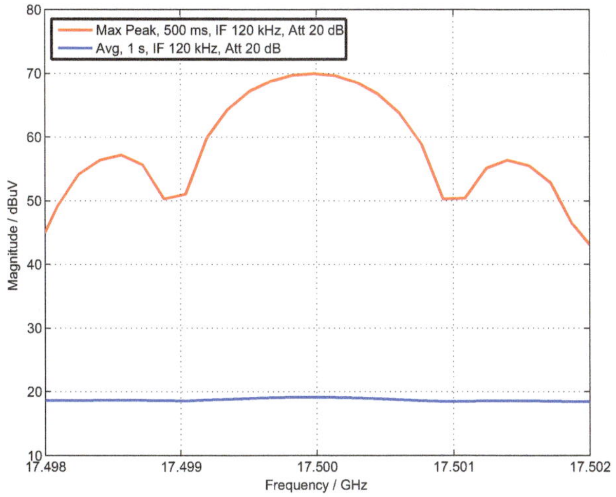

Fig. 8. Measurement of a pulse-modulated signal.

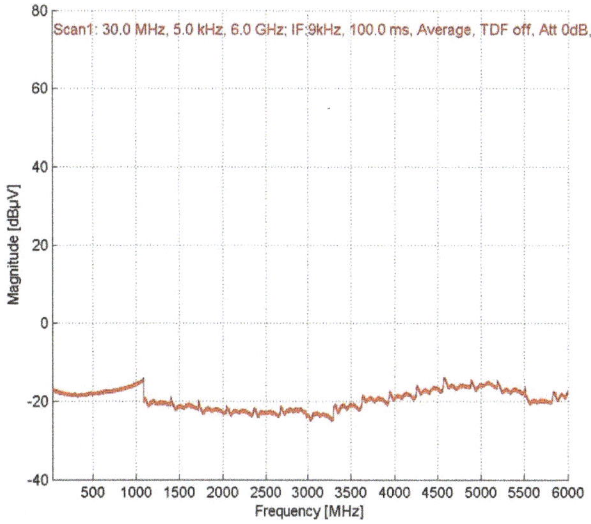

Fig. 9. System noise floor from 30 MHz to 6 GHz.

a frequency of 17.5 GHz to the system input. The signal pulse width was set to 1 μs and the pulse period to 40 ms. The spectrum was weighted by peak and average detectors and is shown in Fig. 8. With this pulse period, the average detector already shows the system noise floor. The difference in level between the peak and average detector measurements is defined as the IF dynamic range. The measurements indicate an IF dynamic range of over 60 dB for an IF-bandwidth of 1 MHz, exceeding CISPR 16-1-1 requirements by over 20 dB.

Figure 9 shows the ultra-low system noise floor measured from 30 MHz to 6 GHz. The system exhibits a noise floor of below $-15\,\mathrm{dB}\,\mu\mathrm{V}$ in this range. To quantify the noise floor above 6 GHz, the noise voltages V_N were measured at distinct frequencies using the average detector and a matched input and the respective noise powers P_N were calculated. As the bandwidth of the used IF-filter with a Gaussian characteristic was 120 kHz, the resulting powers were normalized to an IF-bandwidth of 1 Hz, using

$$P_{\mathrm{NF,AV}} = P_N / B_{\mathrm{ENB,IF}}, \tag{14}$$

where $P_{\mathrm{NF,AV}}$ is the average power of the system's noise floor, and $B_{\mathrm{ENB,IF}}$ is the equivalent noise bandwidth of the IF-filter. The results are given in Table 1 and show that the system noise floor power spectral density from 6-18 GHz stays below $-160\,\mathrm{dBm\,Hz}^{-1}$.

Measurements of the radiated emission of electronic consumer devices and electric household appliances were performed in a full anechoic chamber. A broadband quad-ridged horn antenna with a bandwidth from 1.7–20 GHz was used (Datasheet RF-Spin QRH20, 2010). To compensate for cable losses and to give the electric field strength of the EMI, the corresponding transducer factors and the antenna factor were applied.

Figure 10 shows a spectrogram of the WI-FI data between two modern smartphones. The two phones formed a wireless network and files were transmitted via the file-transfer-protocol (FTP). The antenna was placed in a distance of 3 m to the test subjects. The real-time spectrogram allows for the examination of the time-behaviour. While the WI-FI network was idle for the most part, the transmission of a file, lasting from 2 s to 6 s, is visible.

Table 1. System noise floor from 6 GHz to 18 GHz.

Frequency/MHz	Power spectral density/dBm Hz^{-1}
6000	−164.5
7000	−164.4
8000	−164.7
9000	−163.1
10 000	−163.2
11 000	−163.4
12 000	−162.8
13 000	−162.8
14 000	−162.9
15 000	−162.6
16 000	−161.4
17 000	−161.4
18 000	−160.7

For the measurement shown in Fig. 11, a microwave oven was placed in a distance of 3 m to the antenna. The spectrum shows the magnetron's fundamental at around 2.45 GHz and several higher harmonics up to 18 GHz. The scan time using an IF-filter bandwidth of 9 kHz and a frequency resolution of 50 kHz took around 120 s.

The magnetron's output frequency is non-stationary. The time-domain system's real-time capability allows for the examination of the time-behaviour of the magnetron's 6th harmonic. In Fig. 12, the spectrogram shows the time-behaviour

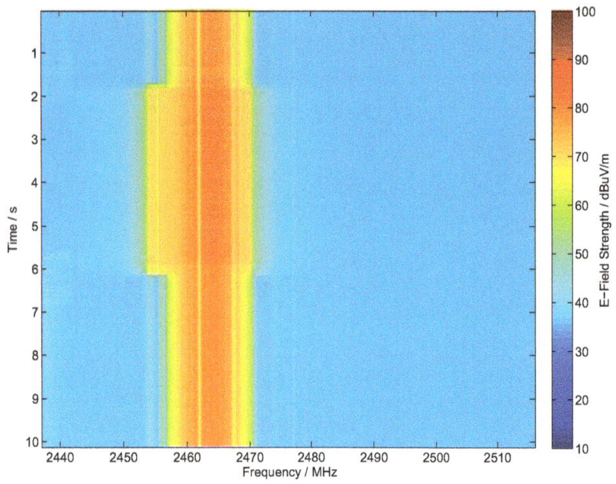

Fig. 10. Spectrogram of WI-FI communication between two mobile phones.

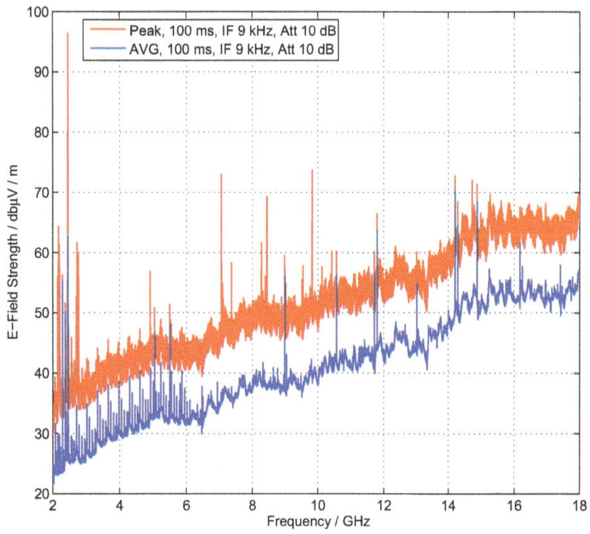

Fig. 11. Emission spectrum of a microwave oven.

of the magnetron's 6th harmonic over a 20 s time-period. In a medium power setting, the magnetron turns on at around 2 s in time and turns off at 8 s. The change in frequency of the free-running oscillator is visible. The high sensitivity of the system and the real-time operation with an IF-bandwidth of 9 kHz enables the measurement of the 6th harmonic, which is located around 20 dB above the system noise floor.

6 Conclusions

A highly sensitive time-domain EMI measurement system for the frequency range from 9 kHz to 18 GHz was presented. The system provides for EMI measurements in CISPR Band E in accordance to the requirements stated in

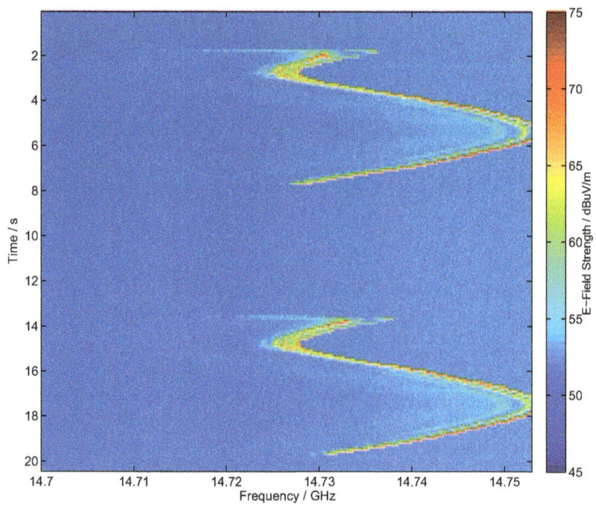

Fig. 12. Spectrogram of the 6th harmonic of a microwave oven.

CISPR 16-1-1. In comparison to traditional EMI-receivers, scan times can be decreased by several orders of magnitude. The high sensitivity, due to a low system noise floor of 6–8 dB, allows for the broadband characterisation of narrowband interference near the noise floor. Measurements of electric household appliances and consumer electronic devices were presented, including measurements of the subjects time-behaviour via the spectrogram.

Acknowledgements. The authors would like to thank the Bayerische Forschungsstiftung and GAUSS Instruments GmbH, Munich, Germany for funding this project and the EMV GmbH, Taufkirchen, Germany for supplying the antenna for the measurements. The authors are indebted to Stephan Braun and Arnd Frech from GAUSS Instruments GmbH for numerous helpful discussions.

References

Braun, S. and Russer, P.: A low-noise multiresolution high-dynamic ultra-broad-band time-domain EMI Measurement System, IEEE Transactions on Microwave Theory and Techniques, 53, 3354–3363, 2005.

Braun, S., Hoffmann, C., and Russer, P.: A Realtime Time-domain EMI Measurement System for Measurements above 1 GHz, IEEE EMC Society Symposium on Electromagnetic Compatibility, Austin, USA, 2009.

Braun, S., Hoffmann, C., and Russer, P.: Emissionsmessung im Zeitbereich oberhalb 1 GHz, EMV 2010, Düsseldorf, Germany, 2010.

CISPR 16-1-1, Ed. 3.1 Am. 1: Specification for radio disturbance and immunity measuring apparatus and methods Part 1-1: Radio disturbance and immunity measuring apparatus – Measuring apparatus, International Electrotechnical Commission, 2010.

Datasheet RF Spin: Broadband Quad-Ridged Horn Antenna QRH20, Data Sheet, 2010.

Davenport, W. B. and Root, D. L.: An Introduction to the Theory of Random Signals and Noise, John Wiley & Sons, 1987.

Hagenhaus, K.: Die Messung von Funkstörungen, Elektrotechnische Zeitschrift, 63, 182–187, 1942.

Hoffmann, C., Braun, S., and Russer, P.: A Broadband Time-Domain EMI Measurement System for Measurements up to 18 GHz, EMC Europe 2010, Wroclaw, Poland, 34–37, 2010.

Maas, S. A.: Microwave Mixers, Artech House, 1993.

Oppenheim, A. V. and Schafer, R. W.: Discrete-Time Signal Processing, 3rd Edition, Pearson Prentice Hall, 2009.

Sze, S. M.: Modern Semiconductor Device Physics, John Wiley & Sons, 1988.

Vendelin, G. D., Pavio, A. M., and Rohde, U. L.: Microwave Circuit Design using Linear and Nonlinear Techniques, second edition, John Wiley & Sons, 2005.

43

A multilevel fast spectral domain algorithm for electromagnetic analysis of infinite periodic arrays with large unit cells

T. F. Eibert

Institute of Radio Frequency Technology, University of Stuttgart, Germany

Abstract. A multilevel fast spectral domain algorithm (MLFSDA) is introduced for the efficient evaluation of the matrix vector products due to the boundary integral (BI) operator within a hybrid finite element - BI (FEBI) method for the analysis of infinite periodic arrays. The MLFSDA utilizes the diagonalization property of the spectral domain (SD) BI representation and handles the large numbers of Floquet modes required for large (with respect to wavelength) periodic unit cells by similar hierarchical techniques as applied in the multilevel fast multipole method/algorithm (MLFMM/MLFMA). With the capability of the MLFSDA to handle very large periodic unit cells, it becomes possible to model finite antennas and scatterers with the infinite periodic array model. For a cavity-backed antenna element and for a semi-finite array of 4 cavity-backed antenna elements in the finite direction, the dependence of the input impedances on the unit cell sizes is investigated and it is found that array resonances disappear for reasonably large unit cell dimensions. Finally, a semi-finite array of antipodal Vivaldi antenna elements is considered and simulation results for infinite periodic, finite, and semi-finite array configurations are compared to measured data.

1 Introduction

The infinite periodic array model is still an important approach for analysis and design of large finite arrays. With periodic excitation, the field solution in the array becomes also periodic and Floquet's theorem can be employed to reduce the computational domain down to a single array element (unit cell), thus significantly speeding up analysis and reducing memory requirements. In the context of integral equation (IE) formulations, periodic Green's functions are constructed, where especially spectral domain (SD) representations amenable to multilayered configurations are used (Mittra, 1988; Aroudaki et al., 1995). Arbitrarily inhomogeneos array elements can be modelled by the hybrid finite element - boundary integral (FEBI) technique (McGrath and Pyati, 1994; Lucas and Fontana, 1995; Eibert et al., 1999, 2000, 2003). Most approaches work with simple half-space Green's functions in SD (McGrath and Pyati, 1994; Lucas and Fontana, 1995), whereas a spatial domain Green's function computed via the Ewald transformation is used in Eibert et al. (1999). Also, the SD half-space Green's function has been extended to multilayer configurations above and below the FE mesh in Eibert et al. (2003). The approach in Eibert et al. (2000, 2003) evaluates the SD BI very efficiently by a so-called fast spectral domain algorithm (FSDA), but the approach works efficiently only for relatively small unit cell dimensions (with respect to wavelength), since the necessary Floquet mode numbers increase with the size of the pertinent unit cells.

In non-periodic IE and FEBI approaches, the use of fast integral solvers such as the fast multipole method (FMM) and its multilevel versions (MLFMM/MLFMA) (Chew et al., 2001) or the adaptive integral method (AIM) (Bleszynski et al., 1996) (similarly pre-corrected FFT methods Gedney et al., 2003) have become standard over the past years. Periodic IE and FEBI formulations can also be accelerated by fast integral methods. However, large problems with large unit cells require to work with large numbers of Floquet modes (or other types of series terms in spatial domain or mixed spatial/spectral representations (Ewald transformation)) and this leads to considerably increasing computational effort, even with the use of conventional fast integral methods.

In FSDA (Eibert et al., 2000), it was already realized that the standard SD formulation leads to diagonalization of the pertinent IE and BI operators. In this paper, it is shown that the standard SD formulation can be cast in an FMM-like representation. Thus, FSDA and multilevel FSDA (MLFSDA) approaches can be derived by similar techniques

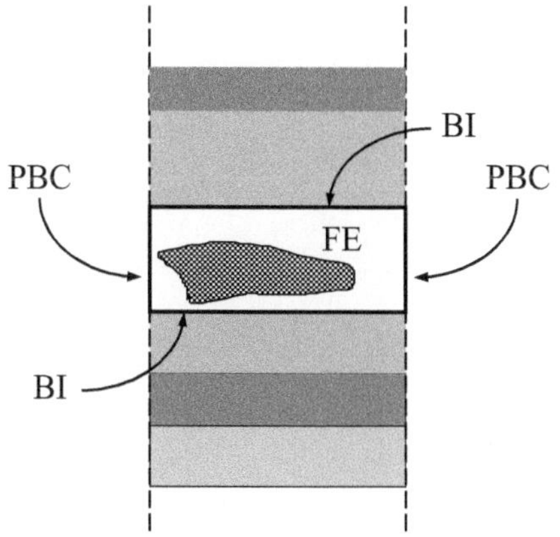

Fig. 1. Cross section of array unit cell with indication of BI and periodic boundary condition (PBC) surfaces.

as employed for FMM and MLFMM/MLFMA. Compared to MLFMM/MLFMA, MLFSDA does not need any near-coupling corrections and all interactions are computed on the coarsest level with only one group. For the considered periodic problems, the summation over the Floquet wavenumbers represents an exact integration rule and also the translation operator evolves directly from the SD representation. In this paper, MLFSDA is employed to accelerate a FEBI technique working with tetrahedral volume and triangular BI elements.

With MLFSDA, unit cells with dimensions of many wavelengths can be efficiently handled. Thus, it is possible to model multilayer FSS with dissimilar periodicities by the supercell approach, where a period common to all individual layers is defined and several array elements in the various layers are placed in the computational unit cell. This approach is probably not as efficient as the cascading technique in Ma et al. (2005), but has increased modelling flexibility.

With the capability to handle large unit cells, it becomes possible to place entire finite configurations in a periodic unit cell, which is large enough to decouple the periodic images. Thus, the advantages of the periodic modelling approach (Floquet mode series instead of SD integral) can also be used to compute finite configurations such as finite antenna arrays.

The paper is organized as follows. After presenting some basic equations, the MLFSDA technique and the modelling of finite and semi-finite configurations are discussed. In the results section, the transmission coefficient of an 8-layer FSS with dissimilar periodicities is computed and compared to measured data. For the examples of a cavity-backed patch antenna element and a corresponding quadruple patch array, it is investigated which unit cell sizes are necessary to decouple the periodic images in finite and semi-finite configurations, respectively. Finally, a semi-finite array of antipodal Vivaldi antenna elements is investigated.

2 Formulation

2.1 Hybrid FEBI formulation

The hybrid FEBI formulation for infinite periodic array configurations employing Floquet's theorem to reduce the problem to a single unit cell of the array (see Fig. 1) can be found in many references such as McGrath and Pyati (1994); Lucas and Fontana (1995); Eibert et al. (1999, 2003). The periodicity in the xy–plane is defined by the lattice vectors $\boldsymbol{\rho}_a$, $\boldsymbol{\rho}_b$ via the shifting relation

$$\boldsymbol{\rho}_{mn} = m\,\boldsymbol{\rho}_a + n\,\boldsymbol{\rho}_b\,, \quad \boldsymbol{r} = \boldsymbol{\rho} + z\hat{\boldsymbol{z}} = x\hat{\boldsymbol{x}} + y\hat{\boldsymbol{y}} + z\hat{\boldsymbol{z}}\,, \tag{1}$$

where m and n are integers. Assuming the periodicity conditions given in Eibert et al. (1999), electromagnetic analysis can be carried out by enforcing stationarity of a standard FE functional in terms of the electric field intensity $\boldsymbol{E}$ with BI termination for the FE portion of the unit cell, shown in Fig. 1.

2.2 Multilevel fast spectral domain algorithm

In a standard implementation of the hybrid FEBI method, the BI contribution results in fully populated matrix blocks for the unknowns in the top and bottom BI surfaces and therefore it is desirable to evaluate the BI expression in an alternative manner. It is well known that the standard SD formulation diagonalizes the BI operator and this property was already utilized for the formulation of an FSDA in Eibert et al. (2000). In this section, an MLFSDA is derived which allows very fast evaluation of the BI contributions during the computation of matrix-vector products in an iterative equation solver, even for large unit cells.

We start with the SD representation of the spatial Green's function in the BI

$$\bar{\mathbf{G}}_\mathrm{p}(\boldsymbol{\rho},\boldsymbol{\rho}') = \int_{k_x}\int_{k_y} \tilde{\bar{\mathbf{G}}}_\mathrm{p}(k_x,k_y) \times \mathrm{e}^{-jk_x(x-x')}\mathrm{e}^{-jk_y(y-y')}\,dk_x dk_y\,, \tag{2}$$

where

$$\tilde{\bar{\mathbf{G}}}_\mathrm{p}(k_x,k_y) = \frac{1}{A}\sum_{p=-\infty}^{\infty}\sum_{q=-\infty}^{\infty} \tilde{\bar{\mathbf{G}}}(k_x,k_y)\,\delta(\boldsymbol{k}_t - \boldsymbol{k}_{tpq}) \tag{3}$$

in the case of an infinite periodic array problem with

$$A = |\boldsymbol{\rho}_a \times \boldsymbol{\rho}_b|,\ \boldsymbol{k}_t = k_x\hat{x} + k_y\hat{y},\ \boldsymbol{k}_{tpq} = \boldsymbol{k}_{t00} + \frac{2\pi}{A}\left[p(\boldsymbol{\rho}_b \times \hat{z}) + q(\hat{z}\times\boldsymbol{\rho}_a)\right],$$

and $\tilde{\bar{\mathbf{G}}}$ the corresponding non-periodic SD Green's function (see Eibert et al., 1999, 2003) and introduce the decomposition

$$\boldsymbol{\rho} - \boldsymbol{\rho}' = (\boldsymbol{\rho} - \boldsymbol{\rho}_g) + (\boldsymbol{\rho}_g - \boldsymbol{\rho}'_g) - (\boldsymbol{\rho}' - \boldsymbol{\rho}'_g) \tag{4}$$

based on a regular grouping of the basis functions in the BI surfaces. $\boldsymbol{\rho}_g$ is the group center close to observation point $\boldsymbol{\rho}$ and $\boldsymbol{\rho}'_g$ is the group center close to the source point $\boldsymbol{\rho}'$. Substituting these expressions into the discretized BI contribution (e.g. see Eibert et al., 1999) gives

$$\iint_S \boldsymbol{b}_m(\boldsymbol{\rho}) \cdot \iint_S \bar{\mathbf{G}}_\mathrm{p}(\boldsymbol{\rho}, \boldsymbol{\rho}') \cdot \boldsymbol{b}_n(\boldsymbol{\rho}')\, ds' ds$$

$$= \int_{k_x} \int_{k_y} \tilde{\boldsymbol{b}}_m^*(k_x, k_y) \cdot \bar{\mathbf{T}}_{mg,ng}(k_x, k_y) \cdot \tilde{\boldsymbol{b}}_n(k_x, k_y)\, dk_x dk_y \quad (5)$$

with

$$\tilde{\boldsymbol{b}}_n(k_x, k_y) = \iint_S \boldsymbol{b}_n(\boldsymbol{\rho}) \mathrm{e}^{jk_x(x-x_{ng})} \mathrm{e}^{jk_y(y-y_{ng})}\, ds \quad (6)$$

and the corresponding translation operator

$$\bar{\mathbf{T}}_{mg,ng}(k_x, k_y) = \tilde{\bar{\mathbf{G}}}_\mathrm{p}(k_x, k_y) \times \mathrm{e}^{-jk_x(x_{mg}-x_{ng})} \mathrm{e}^{-jk_y(y_{mg}-y_{ng})}. \quad (7)$$

Also, $\boldsymbol{b}_n$ are the usual Rao-Wilton-Glisson basis functions used to descretize the BI. Equation (5) has the same structure as diagonalized FMM representations (Chew et al., 2001), except that the FMM integration over the Ewald sphere is replaced by SD lateral wavenumber integration, i.e. a truncated Floquet mode series in the case of the considered infinite periodic problem (see Eq, (3)). Thus, by applying the same principles as done in FMM or MLFMM/MLFMA, fast and multilevel fast spectral domain algorithms (FSDAs and MLFSDAs) can be realized to evaluate the BI. However, some important differences are observed. Most importantly, the translation operator in Eq. (5) is known analytically and the whole representation Eq. (5) is exact for all interactions, including near- and self-terms. Moreover, for infinite periodic problems the integration rule for the computation of the SD integral in Eq. (5) is also exactly given by the Floquet mode representation of the periodic Green's function (see Eq. (3)). Consequently, a particular error analysis is not necessary for FSDA and MLFSDA, since the algorithms give the same results as conventional SD implementations. In MLFSDA, we must only guarantee that the required interpolation and anterpolation steps are carried out without introducing additional errors.

FSDA has already been presented in Eibert et al. (2000). Due to the exact validity of Eq. (5), this algorithm does not work with any grouping and computes all interactions among all basis functions at once. This algorithm is well suited for periodic problems with small unit cells, however, becomes inefficient for large unit cells, where an increasing number of Floquet modes is needed. In MLFSDA, large unit cells are covered by a regular square grid of groups and a hierarchical grouping structure is introduced. Since FSDA and MLFSDA do not suffer from low-frequency break-down, the size of the groups on the finest level can be chosen on the same order as the sizes of the discretization subdomains. On higher levels, lower level groups are combined until there is only one group on the coarsest level L. All interactions according to Eq. (5) are computed for the single group on the coarsest level, where the SD sample density is chosen according to the size of this group. One level below, every other sample is left out and so on until the finest level is reached. The sampling density on the lower levels is determined by the sizes of the groups on these levels. The sample densities on the various levels are determined according to the Shannon sampling theorem, where sufficient oversampling must be provided for the interpolation and anterpolation procedures to work.

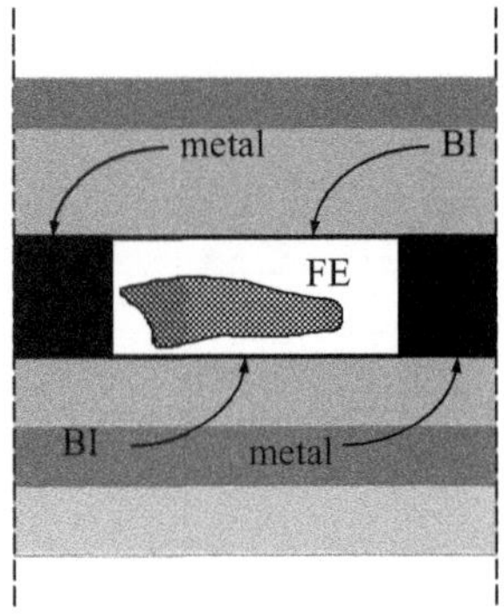

Fig. 2. Cross section of array unit cell for finite array modelling with indication of BI surfaces.

2.3 Computation of finite and semi-finite configurations by increasing the infinite array period

With the MLFSDA, it is possible to compute periodic array problems with large unit cells. For many array problems, this is not important, since the period is often on the order of a half wavelength. However, large unit cells are required if multilayer FSSs with dissimilar periodicities shall be computed by the supercell approach and we can even place an entire finite or semi-finite configuration in one unit cell. Since the influence of the periodic unit cell images on the finite configuration must be small in this case, the period in the computational periodic array model must be chosen very large and a model according to Fig. 2 evolves from the original model in Fig. 1. Due to the use of a Green's function with metallic backing in the BI, the finite array in the unit cell is embedded in a metallic plate with thickness according to the height of the FE domain. In practical computations, we found that the influence of the periodic images on the finite array vanishes for reasonably large periodic array periods, especially if little (negligible for finite configuration) losses are assumed in some of the material layers modelled in the BI.

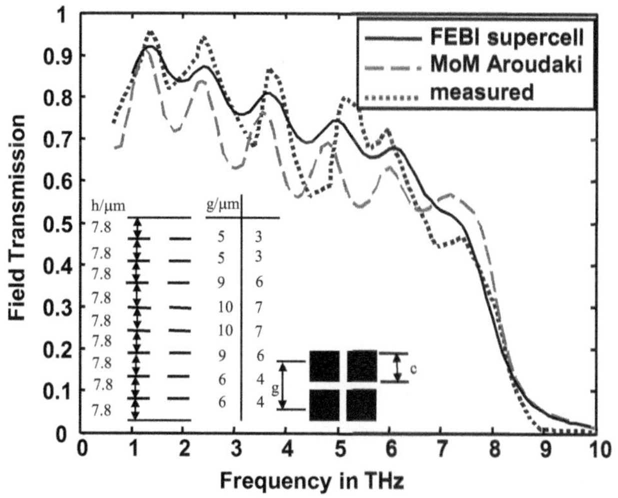

Fig. 3. Near-infrared 8-layer low-pass FSS with dissimilar periodicities, $\varepsilon_r = 2.31$, $\tan\delta = 0.035$ in FEBI computations, incidence angle 30° with respect to normal, power average of TE and TM.

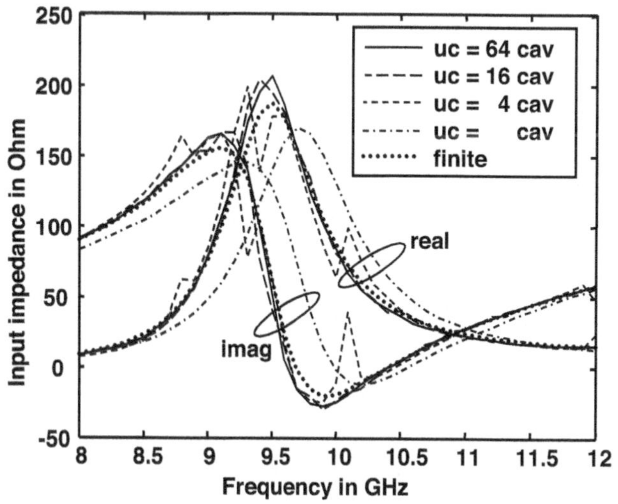

Fig. 5. Input impedance of cavity-backed patch antenna element according to Fig. 4 for infinite periodic array configurations with increasing unit cell dimensions.

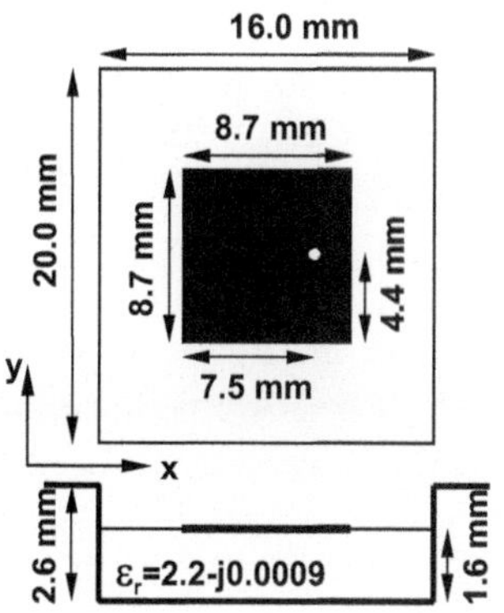

Fig. 4. Cavity-backed patch antenna element.

3 Results

The first example is the near-infrared 8-layer FSS with dissimilar periodicities depicted in Fig. 3 and already considered in Aroudaki et al. (1995). For a valid infinite periodic model, the supercell approach must be applied for this problem. The dimensions of the used (smallest) supercell are 90 μm by 90 μm resulting in up to 18 by 18 patch elements in the upper two layers.

Unfortunately, material losses and fabrication tolerances are not very well known for this FSS. Our FEBI-MLFSDA computations were performed for a dielectric substrate with $\varepsilon_r = 2.31 - j0.08$. A truncated Floquet mode series ranging from –7 to 7 was used for the representation of the basis function radiation patterns on the finest level and the Floquet mode series representing the Green's function on the coarsest level ranged from –53 to 53. The discretization model consisted of about 3.6 Mio volume unknowns including 20223 BI unknowns. The utilized RAM was about 2.8 GByte including about 1.8 GByte for storing the search vectors in the GMRES solver. Field transmission results are given in Fig. 3 and compared to measured results from Aroudaki et al. (1995) and approximate MoM results from Aroudaki et al. (1995), obtained using only one patch in every layer. Especially near the passband edge of the low-pass, our FEBI-MLFSDA supercell results capture the measured behavior more accurately.

The next problem is intended to demonstrate the feasibility of the approach discussed in Sect. 2.3 for the treatment of finite configurations. A simple cavity-backed patch antenna element excited by a probe current as depicted in Fig. 4 is considered. Reference computations were carried out by a 3-D hybrid FEBI code for finite problems, where the cavity-backed patch antenna element was placed in a thick finite metallic ground plane of size 160 mm by 200 mm with height 5 mm. Infinite periodic array computations with the FEBI-MLFSDA code were performed for equally increasing unit cell sizes in the two array dimensions in order to reduce the coupling between the unit cell patch and its periodic images. The input impedance results in Fig. 5 contain clearly visible array resonances for unit cell sidelengths (uc) of 4 and 16 times the cavity sidelengths (cav), but the curves for uc=64 cav do not show anymore obstruction by array resonances and agree pretty well with the finite ground plane reference results. When the unit cell has the same size as the cavity, the input impedance is also smooth but shows a noticeable frequency shift compared to the finite ground plane results. Due to anterpolation of the translation operator in the setup phase, the sample density required on the coarsest level could be kept very small (-6 to +6), no matter how large the unit cell was chosen.

Next, a finite by infinite array consisting of 4 patch ele-

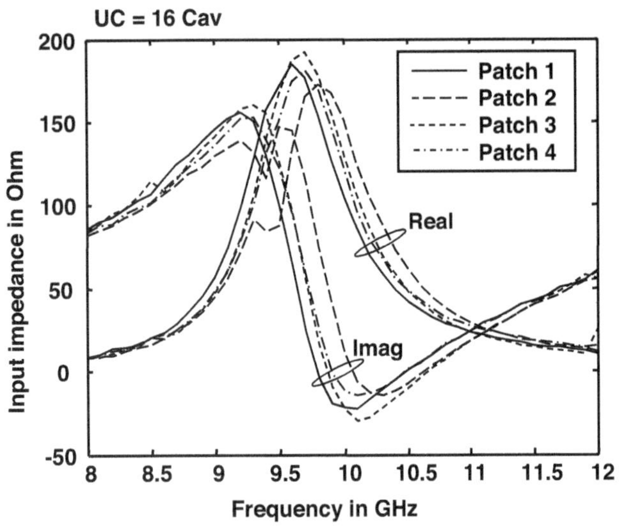

Fig. 6. Input impedances of finite by infinite patch array (4 elements in finite direction) consisting of patch elements according to Fig. 4 obtained with infinite periodic array model with a unit cell 16 times larger than the size of the array in the finite direction.

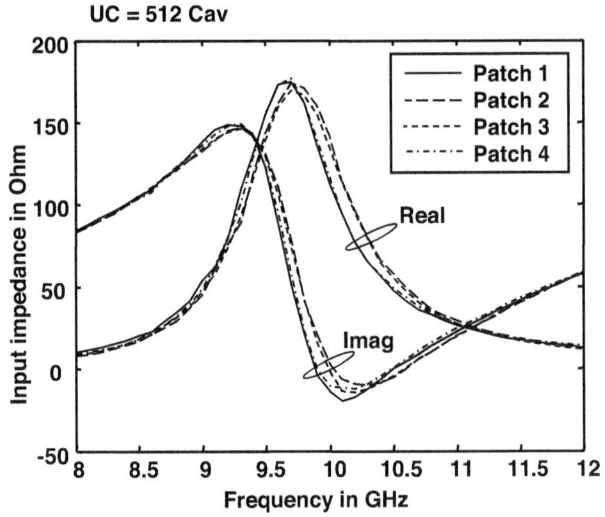

Fig. 7. Input impedances of finite by infinite patch array (4 elements in finite direction) consisting of patch elements according to Fig. 4 obtained with infinite periodic array model with a unit cell 512 times larger than the size of the array in the finite direction.

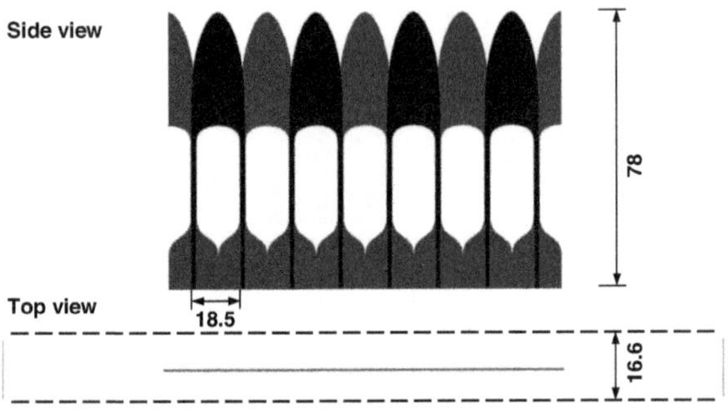

Fig. 8. Front (black) and rear (dark grey) side metallizations of 8-element column of antipodal Vivaldi elements. The dashed lines in the top view represent the periodic boundaries, the through lines at left and right sides indicate the electric boundaries in the FE domain for the semi-finite simulations (dimensions in mm, substrate: Rogers Duroid, $\varepsilon_r = 2.2$, 0.5 mm thick).

ments according to Fig. 4 in the finite direction is considered. In the infinite y direction, the patch elements are separated by metallic walls, whereas no metallic walls are present in the finite x direction. Figs. 6 and 7 show the input impedances of the 4 patch elements within the unit cell chosen for the infinite periodic FEBI-MLFSDA computations. The results in Fig. 6 computed with a periodic unit cell sidelength (uc) of 16 times the size of the array in the finite direction are clearly influenced by array resonances due to the periodic images in the computational model. However, after increasing the sidelength of the periodic unit cell to 512 times the array size in the finite direction, no more array resonances can be observed in the input impedance results in Fig. 7.

The final example is a semi-finite phased-array antenna of antipodal Vivaldi elements as illustrated in Fig. 8 and discussed in Wilden et al. (2004). The array consists of 8 equally-fed elements on one substrate in the finite direction and up to 256 active elements in the other direction (periodic boundaries indicated by dashed lines in the figure). The advantage of the antipodal Vivaldi elements is that they can be fed by microstrip lines. However, a 180° phase shift is required between neighboring feed ports. The fabricated antenna elements are fed by 1 to 8 power dividers, where the necessary phase shifts are realized by rat-race couplers. In our simulations, each Vivaldi element was fed by a short piece of coaxial line (excited by impressed line currents in the FE mesh) and forward and backward travelling waves on the coax line were extracted by the matrix pencil method in order to determine the pertinent S parameters. First simulations were performed for a double element of the antipodal Vivaldi radiators. Figure 9 shows simulation results for a free-standing double element computed by a hybrid FEBI code for finite configurations and compares it to the passive reflection coefficient directly measured at the input of a middle element of an 8-element column, where the remaining elements were terminated by 50 Ω plugs. The agreement of simulated and mesured results is very reasonable.

Figure 10 compares simulated infinite periodic results (obtained by FEBI-MLFSDA according to periodic array dimensions as indicated in Fig. 8) and measured active reflection results. The measured active reflection coefficient was found for a middle element of a finite array of 9 8-element columns by considering the coupling to all elements in the finite array. Given the relatively small size of the measured array, it is remarkable that the simulated array resonance peak at about 10 GHz is also found in the measured data. Simulated array resonances at other frequencies are not that obvious in the measured results.

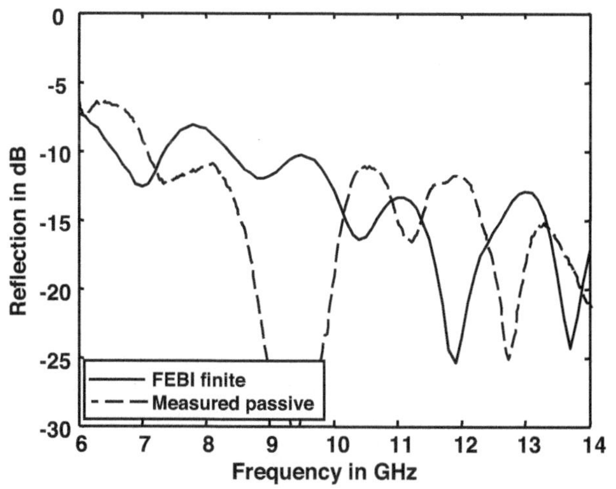

Fig. 9. Simulated and measured reflection coefficients for antipodal Vivaldi antenna element.

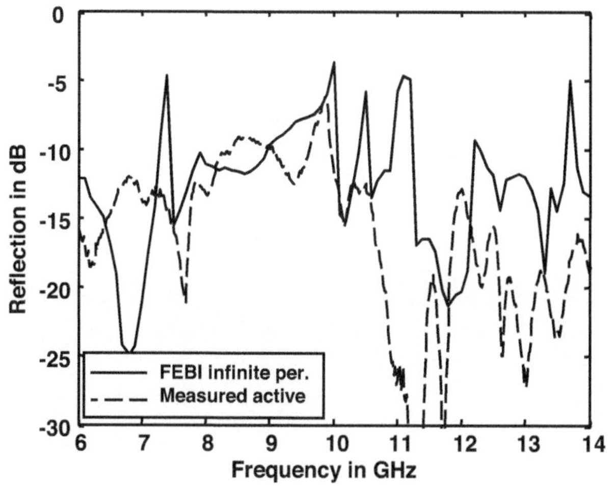

Fig. 10. Simulated and measured reflection coefficients for antipodal Vivaldi antenna array.

Further FEBI-MLFSDA simulations were performed for the semi-finite array configuration (see Fig. 8). The computed reflection coefficients of the 8 elements in the column were combined according to the 1 to 8 power divider used to feed the fabricated antenna. The resulting active reflection coefficient (FEBI semi-finite (sum)) is shown in Fig. 11 and compared to measured results obtained at the input port of the power divider (Measured active (sum)). Again, the coupling to 9 columns was considered to determine measured active results. For comparison, also the simulated infinite periodic element results are included. Interesting to note is, that the simulated semi-finite sum results show considerably weaker array resonances than the infinite periodic element results. In the measured active sum results determined at the input port of the power divider, almost no array resonances are found and the active reflection coefficient is below -10 dB

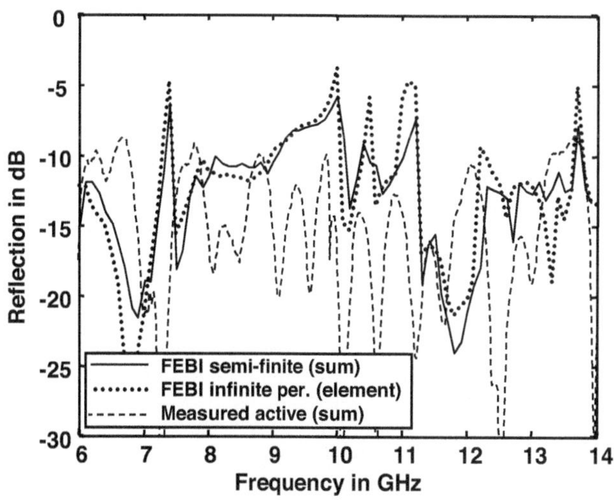

Fig. 11. Simulated and measured reflection coefficients for antipodal Vivaldi antenna array.

in most parts of the considered frequency range. The averaging among the 8 array elements in the finite array direction obviously suppresses the array resonances.

4 Conclusions

The hybrid finite element - boundary integral (FEBI) method for the analysis of infinite periodic arrays was extended by a multilevel fast spectral domain algorithm (MLFSDA) which accelerates the matrix vector product evaluation for the BI operator. The MLFSDA utilizes the diagonalization property of the spectral domain BI representation and employs similar multilevel procedures as applied in the multilevel fast multipole method/algorithm (MLFMM/MLFMA). With MLFSDA, large (with respect to wavelength) periodic unit cells can be treated by the FEBI method, which are for instance necessary for the modelling of multilayer frequency selective surfaces (FSSs) with dissimilar periodicities by the supercell method. Moreover, it is now possible to solve finite electromagnetic antenna and scattering problems by the infinite periodic approach und utilize its simplified evaluation of the BI operator (Floquet mode series instead of spectral integral). The finite antenna or scatterer is placed in the periodic unit cell and the dimensions of the unit cell are chosen large enough to reduce coupling between the periodic images to a negligible level. For a finite and a semi-finite cavity-backed patch antenna problem, the necessary sizes of the computational unit cells were studied. Finally, a semi-finite array of antipodal Vivaldi antenna elements was investigated. Since the presented FEBI method works with mutilayered Green's functions including metallic backing in the BI surfaces, the approach is very well suited for antennas and scatterers embedded in a thick metallic plate and not so well for free-standing configurations. However, the MLFSDA has the potential so speed-up the evaluation of various infinite periodic integral equation operators (e.g. electric field integral equation for planar circuits and antennas in multilayered media).

References

Mittra, R., Chan, C. H., and Cwik, T.: Techniques for Analyzing Frequency Selective Surfaces — A Review, Proc. IEEE 76, 1593–1615, 1988.

Aroudaki, H., Hansen, V., Gemünd, H.-P., and Kreysa, E.: Analysis of Low–Pass Filters Consisting of Multiple Stacked FSS's of Different Periodicities with Applications in the Submillimeter Radioastronomy, IEEE Trans. Antennas Propagat., 43, 1486–1491, 1995.

McGrath, D. T. and Pyati, V. P.: Phased Array Antenna Analysis with the Hybrid Finite Element Method, IEEE Trans. Antennas Propagat., 42, 1625–1630, 1994.

Lucas, E. W. and Fontana, T. W.: A 3-D Hybrid Finite Element/Boundary Element Method for the Unified Radiation and Scattering Analysis of General Infinite Periodic Arrays, IEEE Trans. Antennas Propagat., 43, 145–153, 1995.

Eibert, T. F., Volakis, J. L., Wilton, D. R., and Jackson, D. R.: Hybrid FE/BI Modeling of 3-D Doubly Periodic Structures Utilizing Triangular Prismatic Elements and a MPIE Formulation Accelerated by the Ewald Transformation, IEEE Trans. Antennas Propagat., 47, 843–850, 1999.

Eibert, T. F. and Volakis, J. L.: Fast Spectral Domain Algorithm for Hybrid Finite Element/Boundary Integral Modeling of Doubly Periodic Structures, IEE Proceedings Microwaves, Antennas Propagat., 147, 329–334, 2000.

Eibert, T. F, Erdemli, Y. E., and Volakis, J. L.: Hybrid Finite Elemente - Fast Spectral Domain Multilayer Boundary Integral Modeling of Doubly Periodic Structures, IEEE Trans. Antennas Propagat., 51, 2517–2520, 2003.

Chew, W. C., Jin, J.-M., Michielssen, E., and Song, J.: Fast and Efficient Algorithms in Computational Electromagnetics, Boston: Artech House, 2001.

Bleszynski, E., Bleszynski, M., and Jaroszewicz, T.: AIM: Adaptive Integral Method Compression Algorithm for Solving Large–Scale Electromagnetic Scattering and Radiation Problems, Radio Science, 31, 1225–1251, 1996.

Gedney, S. D., Zhu, A., Tang, W.-H., Liu, G., and Petre, P.: A fast, high-order quadrature sampled pre-corrected FFT for electromagnetic scattering, Microwave Opt. Technology Letters, 36, 343–349, 2003.

Ma, J.-F., Mittra, R., and Huang, N. T.: Analysis of Multiple FSS Screens of Unequal Periodicity Using an Efficient Cascading Technique, IEEE Trans. Antennas Propagat., 53, 1401–1414, 2005.

Wilden, H., Poppelreuter, B., Saalmann, O., Brenner, A., and Ender, J.: Design and Realisation of the PAMIR Antenna Frontend, EUSAR, 493–496, 2004.

44

Frequency versus time domain immunity testing of Smart Grid components

F. Gronwald

Institute of Electromagnetic Theory, Hamburg University of Technology, Harburger Schloss Str. 20, 21079 Hamburg, Germany

Correspondence to: F. Gronwald (gronwald@tuhh.de)

Abstract. Smart Grid components often are subject to considerable conducted current disturbances in the frequency range 2–150 kHz and, as a consequence, it is necessary to provide reliable immunity test methods. The relevant basic standard IEC 61000-4-19 that is currently under discussion focusses on frequency domain test methods. It is remarked in this contribution that in the context of frequency domain testing the chosen frequency spacing is related to the resonance response of the system under test which, in turn, is characterized in terms of resonance frequencies and quality factors. These notions apply well to physical system but it is pointed out by the example of an actual smart meter immunity test that smart grid components may exhibit susceptibilities that do not necessarily follow a resonance pattern and, additionally, can be narrowband. As a consequence it is suggested to supplement the present frequency domain test methods by time domain tests which utilize damped sinusoidal excitations with corresponding spectra that properly cover the frequency range 2–150 kHz, as exemplified by the military standard MIL-STD-461.

1 Introduction

In view of Electromagnetic Compatibility (EMC) of smart electrical energy networks, the so-called "smart grids", there are strong concerns about the frequency range 2–150 kHz. It has been observed that smart grid components, such as smart energy meters, can be subject to conducted disturbances that leads to malfunction, even though the components comply to the mandatory EMC tests (Kirchhof, 2010; CENELEC SC 205A, 2010, 2013). This situation is mainly due to insufficient EMC standards in the frequency range 2–150 kHz and has been identified as a major problem for already some time (Jäkel, 2004; Strzelecki and Benysek, 2008; Smolenski, 2012). As a remedy, a new basic standard IEC 61000-4-19 is currently under discussion which describes test procedures for immunity to conducted, differential mode disturbances in the frequency range 2–150 kHz (IEC 61000-4-19, 2014). The test procedures are defined in frequency domain and utilize predefined frequency steps in order to pass from one test frequency to another. This follows the usual pattern of EMC testing in frequency domain.

In this paper we contribute to the current discussion and point out that testing in time domain is a meaningful addition to the frequency domain test methods that already are suggested. It is noted that the parameters of frequency domain testing are chosen according to physical resonance models. Smart grid components, however, are mainly communication devices that, due to their digital nature, do not necessarily follow resonance patterns. They may exhibit narrowband susceptibilities that can be missed if, during frequency testing, the frequency steps are chosen as too large. In contrast, time domain pulses exhibit frequency spectra which continuously cover frequency intervals, such that narrowband susceptibilities are less likely to be missed.

In the following we will focus on smart meter devices and review in Sect. 2 corresponding EMC standards that at present are in effect. In Sect. 3 the relation of frequency domain testing to physical models is pointed out. By the example of an actual smart meter test it is shown that the susceptibility of a smart meter can be very narrowband. As alternative test pulses in time domain, the well-known damped sinusoidal pulses are introduced in Sect. 4. In contrast to civil standardization, military standards, such as MIL-STD 461, already suggest to apply these pulses in the low frequency regime 2–150 kHz. As a conclusion, it is recommended to follow MIL-STD 461 and to apply damped sinusoidal pulse testing also in the low-frequency regime to smart meters.

2 EMC standards for smart meter immunity testing

Smart meters, as an important class of smart grid component, have to comply to the product standard IEC 62052-11. This standard refers to the following EMC basic standards (IEC 62052-11, 2003):

- Standard IEC 61000-4-2, related to immunity testing with respect to electrostatic discharge (IEC 61000-4-2, 2008).
- Standard IEC 61000-4-3, related to immunity testing with respect to external radiated electromagnetic fields in the frequency range 80 MHz to 2 GHz (IEC 61000-4-3, 2010).
- Standard IEC 61000-4-4, related to immunity testing with respect to fast transients and bursts, as produced by fast switching processes (IEC 61000-4-4, 2012).
- Standard IEC 61000-4-5, related to immunity testing with respect to transients of a double exponential form, as produced by lightning or slow switching processes (IEC 61000-4-5, 2005).
- Standard IEC 61000-4-6, related to immunity testing with respect to conducted disturbances in the frequency range 150 kHz to 80 MHz (IEC 61000-4-6, 2008).
- Standard IEC 61000-4-12, related to immunity testing with respect to non-repetitive damped sinusoidal transients that also are called "ring waves". The test frequencies of these sinusoidal transients are given by 100 kHz and 1 MHz (IEC 61000-4-12, 2006).

Concerning immunity testing with respect to conducted disturbances, it is noted that the relevant basic standard IEC 61000-4-6 only covers the frequency range 150 kHz to 80 MHz. A closer inspection of this standard additionally reveals that only common mode signals are applied to the test objects, while in typical energy distribution networks it is the differential mode which also is relevant (Strzelecki and Benysek, 2008; Paul, 2006).

To also take into account conducted differential mode disturbances in the frequency range below 150 kHz, the draft standard IEC 61000-4-19 "Testing and measurement techniques – Test for immunity to conducted, differential mode disturbances in the frequency range from 2 kHz to 150 kHz, at a.c. ports" is currently under discussion. To the time of writing it is not known when this standard will be finalized and whether it will be made obligatory for smart meter devices, such that at present there is no standardized test procedure in effect which guarantees immunity of smart meter devices with respect to conducted disturbances in the frequency range 2–150 kHz.

3 Immunity tests in frequency domain and the quality factor of a complex system

From the viewpoint of EMC modelling, a smart meter device is not a canonical structure, such as a simple transmission line or antenna, but rather a complex system. If linearity and time-invariance can be assumed, the immunity of a complex system is characterized by a transfer function $H(s)$ which relates an output quantity, such as an observable $O(s)$ of an operational state, to an input quantity, such as an applied current $I(s)$. In the resonance region, transfer functions are properly discussed in the framework of the singularity expansion method (SEM) and written in the general form (Baum, 1976)

$$H(s) = \frac{O(s)}{I(s)} \approx \sum_{m=1}^{M} \left(\frac{R_m}{s - s_m} + \frac{R_m^*}{s - s_m} \right) \tag{1}$$

where the approximation sign indicates that in general the relation is not a mathematically exact equality but of sufficient accuracy if M dominant poles s_m of the system are taken into account. These poles represent the frequencies where the system considered is susceptible to the input quantity $I(s)$. The variable s denotes the Laplace variable in the complex plane and the variable R_m denotes the residue associated to the mth resonance pole

$$s_m = \sigma_m + j\omega_m \,. \tag{2}$$

The real part σ_m of the pole, for passive systems, is negative and represents losses of the system, the imaginary part ω_m is the resonance frequency of the pole. If the representation (1) is compared to the transfer function of a damped harmonic oscillator, such as an *RLC* resonant circuit, it is seen after some analysis that the variables σ_m and ω_m can also be expressed in terms of an Eigenfrequency $\omega_{0,m}$ and a quality factor Q_m. The corresponding relations are given by (Chauveau et al., 2007)

$$\omega_{0,m} = |s_m|, \qquad Q_m = -\frac{\omega_{0,m}}{2\sigma_m} \tag{3}$$

These notions and concepts are important in order to correctly interpret immunity tests in frequency domain (Gronwald et al., 2012).

As an example, in Fig. 1 the frequency dependency of the received power of a small dipole antenna is displayed in the vicinity of resonance, serving as an elementary model of an EMC victim. At resonance, the received power attains a maximum and the susceptibility of this system becomes, in this frequency range, largest as well. For reliable immunity testing it is decisive to hit the maxima of susceptibility and this, in turn, requires choosing the frequency steps accordingly. If

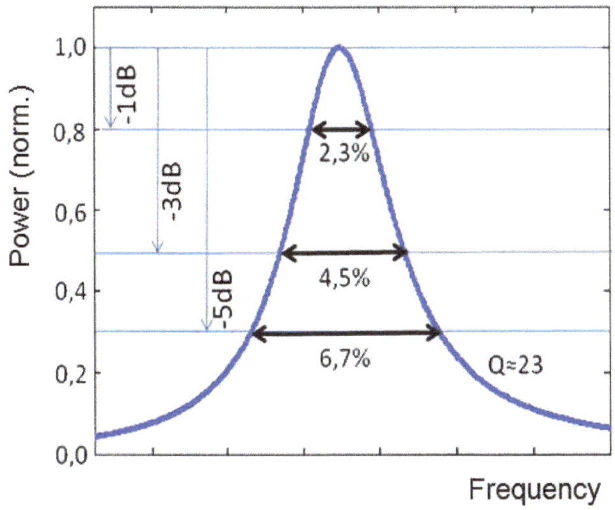

Figure 1. The diagram shows the received power of a matched, small dipole antenna of quality factor $Q = 23$. It can be noted that the quality factor is given by the reciprocal of the percentage of the 3 db-bandwidth relative to the resonance frequency, i.e., $Q = 23 \approx 1/(0.045)$

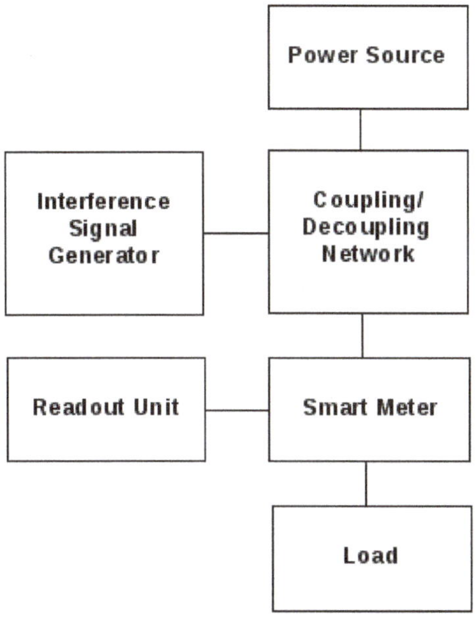

Figure 2. General setup of a smart meter immunity test. It follows the usual pattern of conducted immunity testing (Paul, 2006). The draft standard IEC 61000-4-19 gives more details on how to specify the necessary components that are shown in the figure.

testing with a margin of 1 dB with respect to maximum susceptibility is desired this requires, in the example of Fig. 1, to choose frequency steps according to a 2.3 % rule, that is

$$f_{n+1} = f_n \cdot 1{,}023 \tag{4}$$

as can be seen from Fig. 1 where the 2.3 % bandwidth is indicated. This measure is connected to the quality factor $Q = 23$ of the system, the reciprocal of which is equal to the 3 dB-bandwidth. This is also indicated in Fig. 1.

In the draft standard IEC 61000-4-19 frequency steps are prescribed according to a 2 % rule. This allows reliable testing of systems up to a quality factor of $Q = 26$ if a 1 dB margin is desired. For higher quality factors it is then possible to miss frequencies where susceptibility is at a maximum and, as a consequence, immunity at a minimum.

For smart meter devices it is not immediate to determine a quality factor on the basis of physical models. Clearly, smart metering involves digital signal processing and this is not directly accessible to electromagnetic theory. Within a smart metering device analog signals are readily converted to digital signals and then further processed. Disturbances in the frequency range 2–150 kHz may affect this process due to insufficient low-pass filtering and aliasing effects but there is no corresponding general rule which applies to any smart meter.

As an example, an actual smart meter which passed the current EMC standards but turned out to be susceptible during operation has been investigated by means of an immunity test setup as sketched in Fig. 2. A rather small frequency step width of about 0.4 %, which is well below the standard frequency step width 2 %, was chosen during the test, yielding the result curve of Fig. 3.

The test result exhibits narrowband susceptibilities that exceed acceptable accuracy limits by order of magnitudes. In this case, the frequency step width of 0.4 % turned out to be necessary to recognize the corresponding peaks. It follows from this example that the detection of susceptibilities of actual smart meter devices can require frequency steppings that are below 2 %, i.e., below the step width suggested in the draft standard IEC 61000-4-19.

4 Time domain test procedures and damped sinusoidal pulses

Time domain test procedures by means of damped sinusoidal pulses have been accomodated in a number of EMC standards, such as in the general MIL-STD-461 (MIL-STD-461F, 2007). In this standard the damped sinusoidal pulse mathematically is described in the test procedure CS116 by

$$i(t) = I_0 e^{-\pi f_0 t/D} \sin(2\pi f_0 t) \tag{5}$$

where D indicates a damping factor which has to assume a value of 15 ± 5. For a frequency $f_0 = 10\,\text{kHz}$, a damping factor of $D = 15$, and initial current amplitude $I_0 = 1\,\text{A}$ the resulting current function $i(t)$ is plotted in Fig. 4. The corresponding current spectrum is expressed by the Fourier

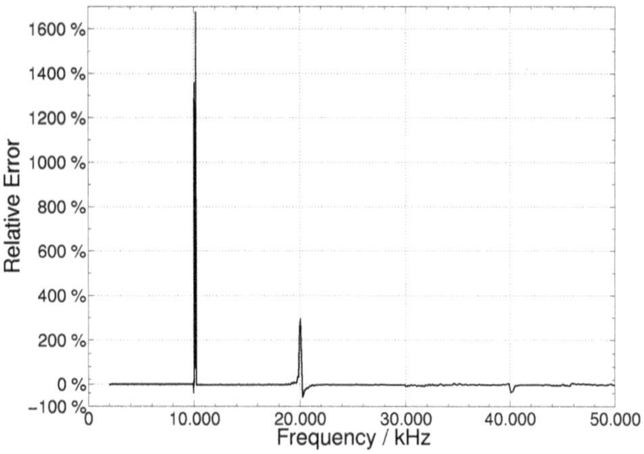

Figure 3. Result of immunity testing of a smart meter device that passed the current EMC tests but nevertheless failed during operation in an actual energy network. Two pronounced peaks at 10 kHz and 20 kHz show unacceptable susceptibilities in the frequency range 2–150 kHz, where sinusoidal differential mode test signals were applied.

transform

$$I(f) = \frac{1}{2\pi f_0} \frac{I_o}{1 + \left(\frac{1}{2D} + j\frac{f}{f_0}\right)^2}. \tag{6}$$

This spectrum can be put on a logarithmic scale according to

$$I(f)(\mathrm{dB_{As}}) = 20\log(|I(f)|/\mathrm{As}) \tag{7}$$

and is shown in Fig. 4 as well.

It is seen from Eq. (6) and Fig. 4 that the spectrum peaks at the frequency f_0 and smoothly decays towards lower and higher frequencies. Clearly, there are no frequency gaps in the spectrum such that is is less likely that narrowband susceptibilities are missed.

In the standard MIL-STD-461 compliance of equipment under test shall be demonstrated in the frequency range 10 kHz to 100 MHz at selected frequencies (MIL-STD-461F, 2007). The civil standard (IEC 61000-4-12, 2006) that applies to smart meters only requires damped sinusoidal pulses at test frequencies 100 kHz and 1 MHz. For these pulses the corresponding amplitude spectra decay towards lower frequencies, compare Eq. (6), such that the frequency interval 2–150 kHz is not reasonably covered towards its lower end. In view of both the presence of considerable disturbances in the complete frequency interval 2–150 kHz that are known to effect smart meters and possible narrowband susceptibilities it is therefore suggested to also apply the immunity criteria of MIL-STD-461 to smart meter devices and to extend time domain testing by means of damped sinusoidal pulses towards lower frequencies down to at least 10 kHz.

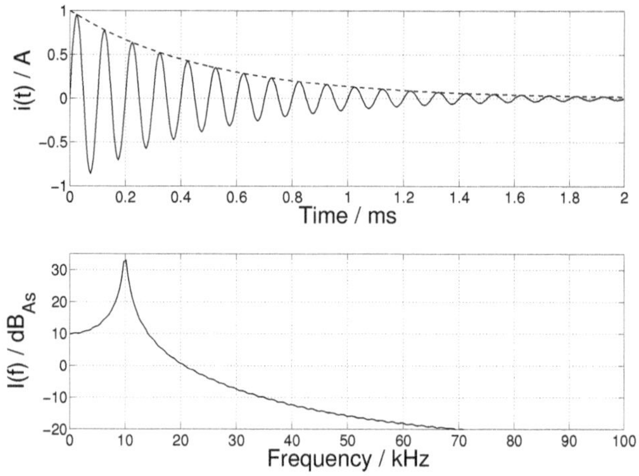

Figure 4. Damped sinusoidal pulse of frequency $f_0 = 10\,\mathrm{kHz}$ and its associated spectrum.

5 Conclusions

At present, EMC standards for smart meter and other smart grid devices still are under discussion to better take into account the increasing occurrences of conducted disturbances in the frequency range 2–150 kHz. The discussions focus on frequency domain test methods which are aimed to test within the resonance region of a system. However, smart grid devices mainly are digital systems such that physical resonance models do not necessarily apply. It has been shown by the example of an immunity test of a particular smart meter that corresponding susceptibilies can be very narrowband and can be missed during a frequency sweep if the step width is too large. Therefore it is recommended to complement immunity testing of smart grid devices by time domain tests and associated continuous amplitude spectra along the lines of MIL-STD-461.

Acknowledgements. The author thanks EWE AG Oldenburg for providing smart meter devices for investigation. Measurement support of EMC Services TUEV Nord by T. Weber also is gratefully acknowledged.

Edited by: F. Sabath

References

Baum, C. E.: The Singularity Expansion Method, in: Transient Electromagnetic Fields, edited by: Felsen, L. B., Springer, Heidelberg, 129–179, 1976.

CENELEC SC 205A: Study Report on Electromagnetic Interference between Electrical Equipment/Systems in the Frequency Range below 150 kHz, SC205A/SEC0260/R, April, 2010.

CENELEC SC 205A: Study Report on Electromagnetic Interference between Electrical Equipment/Systems in the Frequency Range below 150 kHz, Ed. 2, SC205A/SEC0339/R, April, 2013.

Chauveau, J., de Beaucoudrey, N., and Saillard, J.: Characterization of Perfectly Conducting Targets in Resonance Domain with their Quality of Resonance, Progress in Electromagnetic Research, 46, 165–188, 2007.

Gronwald, F., Stadler, T., and Kebel, R.: Compensation of Unavailable Test Frequencies during Immunity Measurements, in: Proceedings of ESA Workshop on Aerospace EMC, (ESA, SP-702), 2012.

IEC 61000-4-2: Electromagnetic compatibility (EMC) – Part 4-2: Testing and measurement techniques – Electrostatic discharge immunity test, 2008.

IEC 61000-4-3: Electromagnetic compatibility (EMC) – Part 4-3: Testing and measurement techniques – Radiated, radio-frequency, electromagnetic field immunity test, 2010.

IEC 61000-4-4: Electromagnetic compatibility (EMC) – Part 4-4: Testing and measurement techniques – Electrical fast transient/burst immunity test, 2012.

IEC 61000-4-5: Electromagnetic compatibility (EMC) – Part 4-5: Testing and measurement techniques – Surge immunity test, 2005.

IEC 61000-4-6: Electromagnetic compatibility (EMC) – Part 4-6: Testing and measurement techniques – Immunity to conducted disturbances, induced by radio-frequency fields, 2008.

IEC 61000-4-12: Electromagnetic compatibility (EMC) – Part 4-12: Testing and measurement techniques – Ring wave immunity test, 2006.

IEC 61000-4-19: Electromagnetic compatibility (EMC) – Part 4-19: Testing and measurement techniques – Test for immunity to conducted, differential mode disturbances in the frequency range from 2 kHz to 150 kHz at a.c. ports, in preparation, 2014.

IEC 62052-11: Electricity metering equipment (AC) – General requirements, tests and test conditions – Part 11: Metering equipment, 2003.

Jäkel, B.: Störungen auf industriellen Niederspannungsnetzen im Frequenzbereich bis 150 kHz, in: Elektromagnetische Verträglichkeit EMV 2004, VDE Verlag, Berlin, 743–750, 2004.

Kirchhof, J.: Grenzwertlücke – Wechselrichter stört Elektrizitätszähler, in: Elektromagnetische Verträglichkeit EMV 2010, VDE Verlag, Berlin, 97–104, 2010.

MIL-STD-461F: Requirements for the control of electromagnetic interference characteristics of subsystems and equipment, 2007.

Paul, C. R.: Introduction to Electromagnetic Compatibility, 2nd ed., Wiley, New Jersey, USA, 2006.

Smolenski, R.: Conducted Electromagnetic Interference (EMI) in Smart Grids, Springer, London, 2012.

Strzelecki, R. and Benysek, G. (Eds.): Power Electronics in Smart Electrical Energy Networks, Springer, London, United Kingdom, 2008.

Permissions

All chapters in this book were first published in ARS, by Copernicus Publications; hereby published with permission under the Creative Commons Attribution License or equivalent. Every chapter published in this book has been scrutinized by our experts. Their significance has been extensively debated. The topics covered herein carry significant findings which will fuel the growth of the discipline. They may even be implemented as practical applications or may be referred to as a beginning point for another development.

The contributors of this book come from diverse backgrounds, making this book a truly international effort. This book will bring forth new frontiers with its revolutionizing research information and detailed analysis of the nascent developments around the world.

We would like to thank all the contributing authors for lending their expertise to make the book truly unique. They have played a crucial role in the development of this book. Without their invaluable contributions this book wouldn't have been possible. They have made vital efforts to compile up to date information on the varied aspects of this subject to make this book a valuable addition to the collection of many professionals and students.

This book was conceptualized with the vision of imparting up-to-date information and advanced data in this field. To ensure the same, a matchless editorial board was set up. Every individual on the board went through rigorous rounds of assessment to prove their worth. After which they invested a large part of their time researching and compiling the most relevant data for our readers.

The editorial board has been involved in producing this book since its inception. They have spent rigorous hours researching and exploring the diverse topics which have resulted in the successful publishing of this book. They have passed on their knowledge of decades through this book. To expedite this challenging task, the publisher supported the team at every step. A small team of assistant editors was also appointed to further simplify the editing procedure and attain best results for the readers.

Apart from the editorial board, the designing team has also invested a significant amount of their time in understanding the subject and creating the most relevant covers. They scrutinized every image to scout for the most suitable representation of the subject and create an appropriate cover for the book.

The publishing team has been an ardent support to the editorial, designing and production team. Their endless efforts to recruit the best for this project, has resulted in the accomplishment of this book. They are a veteran in the field of academics and their pool of knowledge is as vast as their experience in printing. Their expertise and guidance has proved useful at every step. Their uncompromising quality standards have made this book an exceptional effort. Their encouragement from time to time has been an inspiration for everyone.

The publisher and the editorial board hope that this book will prove to be a valuable piece of knowledge for researchers, students, practitioners and scholars across the globe.

List of Contributors

C. Löcker
FGAN-Research Institute for High Frequency Physics and Radar Techniques (FHR), Neuenahrer Str. 20, 53343 Wachtberg-Werthhoven, Germany

T. Vaupel
FGAN-Research Institute for High Frequency Physics and Radar Techniques (FHR), Neuenahrer Str. 20, 53343 Wachtberg-Werthhoven, Germany

T. F. Eibert
FGAN-Research Institute for High Frequency Physics and Radar Techniques (FHR), Neuenahrer Str. 20, 53343 Wachtberg-Werthhoven, Germany

O. Klemp
University of Hannover, Department of High Frequency Technology and Radio Systems, Hannover, Germany

H. Eul
University of Hannover, Department of High Frequency Technology and Radio Systems, Hannover, Germany

M. Rahman
Fachgebiet Theoretische Elektrotechnik, Fachbereich Elektrotechnik/Informatik, Universität Kassel (UNIK), 34109 Kassel, Germany

R. Marklein
Fachgebiet Theoretische Elektrotechnik, Fachbereich Elektrotechnik/Informatik, Universität Kassel (UNIK), 34109 Kassel, Germany

M. Bresch
Fraunhofer-Institut für Mikroelektronische Schaltungen & Systeme, Finkenstraße 61, 47057 Duisburg, Germany
Infineon Technologies Austria AG, Siemensstrasse 2, A-9500 Villach, Austria

J. Shi
Fraunhofer-Institut für Mikroelektronische Schaltungen & Systeme, Finkenstraße 61, 47057 Duisburg, Germany

R. Kokozinski
Fraunhofer-Institut für Mikroelektronische Schaltungen & Systeme, Finkenstraße 61, 47057 Duisburg, Germany

R. Herschmann
Leibniz Universität Hannover, Institute of Radiofrequency and Microwave Engineering, Appelstraße 9a, 30167 Hannover, Germany
Smart Devices GmbH & Co. KG, Schönebecker Allee 2, 30823 Garbsen, Germany

O. Büchel
Fuba Automotive GmbH & Co. KG, Tec Center, 31162 Bad Salzdetfurth, Germany

M. Jelen
Fachgebiet Höchstfrequenztechnik, Technische Universität München, Germany

E. M. Biebl
Fachgebiet Höchstfrequenztechnik, Technische Universität München, Germany

S. Tejero
Institute for High Frequency Engineering, Technische Universität München, 80290 München, Germany

S. Schelkshorn
Institute for High Frequency Engineering, Technische Universität München, 80290 München, Germany

J. Detlefsen
Institute for High Frequency Engineering, Technische Universität München, 80290 München, Germany

V. C. Motrescu
Institute of General Electrical Engineering, University of Rostock, Albert-Einstein-Str. 2, 18059 Rostock, Germany

U. van Rienen
Institute of General Electrical Engineering, University of Rostock, Albert-Einstein-Str. 2, 18059 Rostock, Germany

R. Appali
Institute of General Electrical Engineering, Chair of Electromagnetic Field Theory, University of Rostock, Justus-von-Liebig-Weg 2, 18059 Rostock, Germany

S. Petersen
Institute of General Electrical Engineering, Chair of Electromagnetic Field Theory, University of Rostock, Justus-von-Liebig-Weg 2, 18059 Rostock, Germany

U. van Rienen
Institute of General Electrical Engineering, Chair of Electromagnetic Field Theory, University of Rostock, Justus-von-Liebig-Weg 2, 18059 Rostock, Germany

A. Dallinger
Technische Universität München, Lehrstuhl für Hochfrequenztechnik, Fachgebiet Hochfrequente Felder und Schaltungen, D-80333 M¨unchen, Germany

S. Schelkshorn
Technische Universität München, Lehrstuhl für Hochfrequenztechnik, Fachgebiet Hochfrequente Felder und Schaltungen, D-80333 München, Germany

J. Detlefsen
Technische Universität München, Lehrstuhl für Hochfrequenztechnik, Fachgebiet Hochfrequente Felder und Schaltungen, D-80333 München, Germany

A. Patrovsky
Technische Universität München, Fachgebiet Höchstfrequenztechnik, Arcisstraße 21, 80333 München, Germany
Ecole Polytechnique, University of Montreal, C.P. 6079 Montreal, H3C 3A7 Canada

E. M. Biebl
Ecole Polytechnique, University of Montreal, C.P. 6079 Montreal, H3C 3A7 Canada

T. Kut
Institute of Fundamentals of Electrical Engineering, Helmut Schmidt University, Hamburg, Germany

B. Chand
Institute of Fundamentals of Electrical Engineering, Helmut Schmidt University, Hamburg, Germany

A. Lücken
Institute of Electrical Power Systems, Helmut Schmidt University, Hamburg, Germany

S. Dickmann
Institute of Fundamentals of Electrical Engineering, Helmut Schmidt University, Hamburg, Germany

D. Schulz
Institute of Electrical Power Systems, Helmut Schmidt University, Hamburg, Germany

K. Körber
Christian-Albrechts-Universität zu Kiel, Computational Electromagnetics Group, Kaiserstr. 2, 24143 Kiel, Germany

L. Klinkenbusch
Christian-Albrechts-Universität zu Kiel, Computational Electromagnetics Group, Kaiserstr. 2, 24143 Kiel, Germany

B. Chand
Fundamentals of Electrical Engineering, Helmut-Schmidt-University/University of the Federal Armed Forces, Hamburg, Germany

K. Hähre
Power Electronics, Helmut-Schmidt-University/ University of the Federal Armed Forces, Hamburg, Germany

J. Keghie
Fundamentals of Electrical Engineering, Helmut-Schmidt-University/University of the Federal Armed Forces, Hamburg, Germany

S. Dickmann
Fundamentals of Electrical Engineering, Helmut-Schmidt-University/University of the Federal Armed Forces, Hamburg, Germany

C. Deselaers
Hamburg University of Technology, 21079 Hamburg, Germany

U. Bergmann
Jungheinrich AG & Co. KG, 22844 Norderstedt, Germany

F. Gronwald
Hamburg University of Technology, 21079 Hamburg, Germany

M. O. Olbrich
Technische Universität München, Fachgebiet Hchstfrequenztechnik, Arcisstr. 21, 80290 München, Germany

E. M. Biebl
Technische Universität München, Fachgebiet Höchstfrequenztechnik, Arcisstr. 21, 80290 München, Germany

R. Marklein
Fachgebiet Theoretische Elektrotechnik, Fachbereich Elektrotechnik / Informatik, Universität Kassel (UNIK), 34109 Kassel, Germany

K. J. Langenberg
Fachgebiet Theoretische Elektrotechnik, Fachbereich Elektrotechnik / Informatik, Universität Kassel (UNIK), 34109 Kassel, Germany

K. Mayer
Fachgebiet Theoretische Elektrotechnik, Fachbereich Elektrotechnik / Informatik, Universität Kassel (UNIK), 34109 Kassel, Germany

J. Miao
Fachgebiet Theoretische Elektrotechnik, Fachbereich Elektrotechnik / Informatik, Universität Kassel (UNIK), 34109 Kassel, Germany

A. Shlivinski
Fachgebiet Theoretische Elektrotechnik, Fachbereich Elektrotechnik / Informatik, Universität Kassel (UNIK), 34109 Kassel, Germany

A. Zimmer
Fachgebiet Theoretische Elektrotechnik, Fachbereich Elektrotechnik / Informatik, Universität Kassel (UNIK), 34109 Kassel, Germany

W. Müller
Fraunhofer-Institut für Zerstörungsfreie Prüfverfahren (IZFP), Universität des Saarlandes, Gebäude 37, 66123 Saarbrücken, Germany

V. Schmitz
Fraunhofer-Institut für Zerstörungsfreie Prüfverfahren (IZFP), Universität des Saarlandes, Gebäude 37, 66123 Saarbrücken, Germany

C. Kohl
Bundesanstalt für Materialforschung und Prüfung (BAM), 12200 Berlin, Germany

U. Mletzko
Materialprüfanstalt (MPA) Stuttgart, Universität Stuttgart, 70511 Stuttgart, Germany

U. Jakobus
EM Software & Systems - S.A. (Pty) Ltd, PO Box 1354, Stellenbosch 7599, South Africa

J. van Tonder
EM Software & Systems - S.A. (Pty) Ltd, PO Box 1354, Stellenbosch 7599, South Africa

M. O. Olbrich
Fachgebiet Höchstfrequenztechnik, Technische Universität München, 80290 München, Germany

L. Huang
Fachgebiet Höchstfrequenztechnik, Technische Universität München, 80290 München, Germany

E. M. Biebl
Fachgebiet Höchstfrequenztechnik, Technische Universität München, 80290 München, Germany

A. Tzoulis
FGAN-Research Institute for High-Frequency Physics and Radar Techniques (FHR), 53343 Wachtberg-Werthhoven, Neuenahrer Str. 20, Germany

T. F. Eibert
FGAN-Research Institute for High-Frequency Physics and Radar Techniques (FHR), 53343 Wachtberg-Werthhoven, Neuenahrer Str. 20, Germany

B. Schetelig
Faculty of Electrical Engineering, Helmut-Schmidt-University/University of the Federal Armed Forces Hamburg, Germany

J. Keghie
Faculty of Electrical Engineering, Helmut-Schmidt-University/University of the Federal Armed Forces Hamburg, Germany

R. Kanyou Nana
Faculty of Electrical Engineering, Helmut-Schmidt-University/University of the Federal Armed Forces Hamburg, Germany

L.-O. Fichte
Faculty of Electrical Engineering, Helmut-Schmidt-University/University of the Federal Armed Forces Hamburg, Germany

S. Potthast
Bundeswehr Research Institute for Protective Technologies and NBC Protection (WIS), Munster, Germany

S. Dickmann
Faculty of Electrical Engineering, Helmut-Schmidt-University/University of the Federal Armed Forces Hamburg, Germany

J. Hagmann
Helmut-Schmidt-Universität/Universität der Bundeswehr Hamburg, Hamburg, Germany

L.-O. Fichte
Helmut-Schmidt-Universität/Universität der Bundeswehr Hamburg, Hamburg, Germany

S. Dickmann
Helmut-Schmidt-Universität/Universität der Bundeswehr Hamburg, Hamburg, Germany

M. Schaarschmidt
Wehrwissenschaftliches Institut für Schutztechnologien – ABC-Schutz (WIS), Munster, Germany

S. Potthast
Wehrwissenschaftliches Institut für Schutztechnologien – ABC-Schutz (WIS), Munster, Germany

O. Klemp
Universität Hannover, Institut für Hochfrequenztechnik und Funksysteme, Hannover, Germany

G. Armbrecht
Universität Hannover, Institut für Hochfrequenztechnik und Funksysteme, Hannover, Germany

H. Eul
Universität Hannover, Institut für Hochfrequenztechnik und Funksysteme, Hannover, Germany

G. Armbrecht
Universität Hannover, Institut für Hochfrequenztechnik und Funksysteme, Hannover, Germany

O. Klemp
Universität Hannover, Institut für Hochfrequenztechnik und Funksysteme, Hannover, Germany

H. Eul
Universität Hannover, Institut für Hochfrequenztechnik und Funksysteme, Hannover, Germany

E. Genender
Institute of Electrical Engineering and Measurement Technology, Leibniz Universität Hannover, Hannover, Germany

M. Mleczko
Institute of Electrical Engineering and Measurement Technology, Leibniz Universität Hannover, Hannover, Germany

O. Döring
Institute of Electrical Engineering and Measurement Technology, Leibniz Universität Hannover, Hannover, Germany

H. Garbe
Institute of Electrical Engineering and Measurement Technology, Leibniz Universität Hannover, Hannover, Germany

S. Potthast
Bundeswehr Research Institute for Protective Technologies and NBC-Protection Munster, Germany

M. Gonser
Corporate Sector Research and Advance Engineering, Robert Bosch GmbH, Stuttgart, Germany

C. Keller
Corporate Sector Research and Advance Engineering, Robert Bosch GmbH, Stuttgart, Germany

J. Hansen
Corporate Sector Research and Advance Engineering, Robert Bosch GmbH, Stuttgart, Germany

R. Weigel
Institute for Electronics Engineering, University of Erlangen-Nürnberg, Germany

F. Daschner
Microwave Laboratory, Christian-Albrechts-University of Kiel, Kaiserstr. 2, 24143 Kiel, Germany

R. Knöchel
Microwave Laboratory, Christian-Albrechts-University of Kiel, Kaiserstr. 2, 24143 Kiel, Germany

E. Foca
Chair for General Materials Science, Christian-Albrechts-University of Kiel, Kaiserstr. 2, 24143 Kiel, Germany

J. Carstensen
Chair for General Materials Science, Christian-Albrechts-University of Kiel, Kaiserstr. 2, 24143 Kiel, Germany

V. V. Sergentu
Institute of Applied Physics, Academy of Sciences of Moldova, 5 Academy str., 2028 Chisinau, Moldova

H. Föll
Microwave Laboratory, Christian-Albrechts-University of Kiel, Kaiserstr. 2, 24143 Kiel, Germany

I. M. Tiginyanu
Institute of Applied Physics, Academy of Sciences of Moldova, 5 Academy str., 2028 Chisinau, Moldova

T. Kut
Institute of Fundamentals of Electrical Engineering, Helmut Schmidt University, Hamburg, Germany

A. Lücken
Institute of Electrical Power Systems, Helmut Schmidt University, Hamburg, Germany

S. Dickmann
Institute of Fundamentals of Electrical Engineering, Helmut Schmidt University, Hamburg, Germany

D. Schulz
Institute of Electrical Power Systems, Helmut Schmidt University, Hamburg, Germany

A. Jessner
Max-Planck-Institut fuer Radioastronomie, Auf dem Huegel 69, 53121 Bonn, Germany

B. Chand
Fundamentals of Electrical Engineering, Helmut Schmidt University / University of the Federal Armed Forces Hamburg, Germany

T. Kut
Fundamentals of Electrical Engineering, Helmut Schmidt University / University of the Federal Armed Forces Hamburg, Germany

S. Dickmann
Fundamentals of Electrical Engineering, Helmut Schmidt University / University of the Federal Armed Forces Hamburg, Germany

R. Rambousky
Bundeswehr Research Institute for Protective Technologies and NBC Protection (WIS), Munster, Germany

S. Tkachenko
Otto-von-Guericke University Magdeburg, Magdeburg, Germany

J. Nitsch
Otto-von-Guericke University Magdeburg, Magdeburg, Germany

C. Bommaraju
Fachgebiet Theoretische Elektrotechnik, Fachbereich Elektrotechnik/Informatik, Universität Kassel (UNIK), 34109 Kassel, Germany

R. Marklein
Fachgebiet Theoretische Elektrotechnik, Fachbereich Elektrotechnik/Informatik, Universität Kassel (UNIK), 34109 Kassel, Germany

P. K. Chinta
Fachgebiet Theoretische Elektrotechnik, Fachbereich Elektrotechnik/Informatik, Universität Kassel (UNIK), 34109 Kassel, Germany

W. Hafla
University of Stuttgart, Institute for Theory of Electrical Engineering, Pfaffenwaldring 47, 70569 Stuttgart, Germany

A. Buchau
University of Stuttgart, Institute for Theory of Electrical Engineering, Pfaffenwaldring 47, 70569 Stuttgart, Germany

F. Groh
University of Stuttgart, Institute for Theory of Electrical Engineering, Pfaffenwaldring 47, 70569 Stuttgart, Germany

W. M. Rucker
University of Stuttgart, Institute for Theory of Electrical Engineering, Pfaffenwaldring 47, 70569 Stuttgart, Germany

R. H. Rasshofer
BMW Group Research and Technology, Hanauer Str. 46, 80992 Munich, Germany

K. Gresser
BMW Group Research and Technology, Hanauer Str. 46, 80992 Munich, Germany

T. Steinmetz
Chair for Theory of Electrical Engineering and Computational Electromagnetics, Helmut-Schmidt-University, University of the Federal Armed Forces Hamburg, P.O. Box 700822 D-22008 Hamburg, Germany

G. Wimmer
Chair for Theory of Electrical Engineering and Computational Electromagnetics, Helmut-Schmidt-University, University of the Federal Armed Forces Hamburg, P.O. Box 700822 D-22008 Hamburg, Germany

M. Clemens
Chair for Theory of Electrical Engineering and Computational Electromagnetics, Helmut-Schmidt-University, University of the Federal Armed Forces Hamburg, P.O. Box 700822 D-22008 Hamburg, Germany

U. Siart
Institute for High-Frequency Engineering, Technische Universität, Arcisstr. 1, 80333 München, Germany

S. Tejero
Institute for High-Frequency Engineering, Technische Universität, Arcisstr. 1, 80333 München, Germany

J. Detlefsen
Institute for High-Frequency Engineering, Technische Universität, Arcisstr. 1, 80333 München, Germany

J. Anatzki
Institute of Electromagnetic Theory, Hamburg University of Technology, Harburger Schloss Str. 20, 21079 Hamburg, Germany

F. Gronwald
Institute of Electromagnetic Theory, Hamburg University of Technology, Harburger Schloss Str. 20, 21079 Hamburg, Germany

M. Frick
Engineering Electromagnetic Compatibility, Robert Bosch GmbH, Reutlingen, Germany

R. Eidher
Engineering Electromagnetic Compatibility, Robert Bosch GmbH, Reutlingen, Germany

R. Weigel
Institute for Electronics Engineering, Friedrich-Alexander-University of Erlangen-Nuremberg, Germany

S. Tejero
Institute for High Frequency Engineering, Technische Universität München, D-80290 München, Munich, Germany

U. Siart
Institute for High Frequency Engineering, Technische Universität München, D-80290 München, Munich, Germany

J. Detlefsen
Institute for High Frequency Engineering, Technische Universität München, D-80290 München, Munich, Germany

W. Hafla
Institute for Theory of Electrical Engineering, Stuttgart, Germany

A. Buchau
Institute for Theory of Electrical Engineering, Stuttgart, Germany

W. M. Rucker
Institute for Theory of Electrical Engineering, Stuttgart, Germany

A´ . Leibinger
Automotive Electronics, Robert Bosch GmbH, Reutlingen, Germany

A´ . Hajdu
Automotive Electronics, Robert Bosch GmbH, Reutlingen, Germany

C. Hoffmann
Technische Universität München, Lehrstuhl für Hochfrequenztechnik, Munich, Germany

P. Russer
Technische Universität München, Lehrstuhl für Nanoelektronik, Munich, Germany

T. F. Eibert
Institute of Radio Frequency Technology, University of Stuttgart, Germany

F. Gronwald
Institute of Electromagnetic Theory, Hamburg University of Technology, Harburger Schloss Str. 20, 21079 Hamburg, Germany